MARKETS, GAMES, AND STRATEGIC BEHAVIOR

SECOND EDITION

MARKETS, GAMES, AND STRATEGIC BEHAVIOR

An Introduction to Experimental Economics

SECOND EDITION

Charles A. Holt

PRINCETON UNIVERSITY PRESS

Princeton and Oxford

Copyright © 2019 by Princeton University Press

Published by Princeton University Press
41 William Street, Princeton, New Jersey 08540
6 Oxford Street, Woodstock, Oxfordshire OX20 1TR

press.princeton.edu

LCCN 2018933742
ISBN 978-0-691-17924-7

British Library Cataloging-in-Publication Data is available

Editorial: Joe Jackson and Samantha Nader
Production Editorial: Debbie Tegarden
Text Design: Carmina Alvarez
Production: Erin Suydam
Publicity: Tayler Lord
Copyeditor: Karen Verde

This book has been composed in ITC Stone Serif and Avenir

Printed on acid-free paper. ∞

Printed in the United States of America

10 9 8 7 6 5 4 3 2 1

Dedication

A friend once asked me who my intellectual hero was, and without hesitation, I mentioned Vernon Smith (then at Arizona, currently at Chapman). His research has always served as an inspiration for me. His original market experiments were done in a context that satisfied none of the unrealistic perfect-information and large-numbers assumptions that I had to memorize as a student, yet the results were quite consistent with standard supply and demand conditions. On the other hand, his experiments for durable assets (stock shares or houses) show how exuberant prices can deviate wildly from fundamental values in boom times with easy credit. His 2002 Nobel Prize in Economics (together with Danny Kahneman) is richly deserved, and the effects of his work pervade many parts of this book. Even after 60 years of running experiments, Vernon continues to develop his deep insights about how markets work, along with his philosophical convictions about liberty and personal freedom.

I have been inspired by the seminal contributions and personal enthusiasm of coauthors Al Roth and Tom Palfrey. In addition, I want to honor the memory of Jack Repcheck, formerly at Princeton University Press, who edited the Davis and Holt (1993) Experimental Economics and the Kagel and Roth (1995) Handbook of Experimental Economics. Jack's boundless optimism and energy were apparent as soon as I called him and mentioned the earlier book. His response was "you made my day," and he flew down to Charlottesville the same week with a contract and lots of encouragement. Jack continued to promote and encourage experimental economics book projects throughout his career, a tradition that has been continued by successive Princeton University Press editors, including Joe Jackson, who is editing this revision. I also wish to dedicate the book to the many Virginia students (some listed in the acknowledgments section) who have worked in the lab and helped with this project in many ways but most of all with their enthusiasm and encouragement.

Finally, I would like to express my gratitude for the mentoring and encouragement I received from my doctoral dissertation advisors Ed Prescott and the late Morris Degroot at Carnegie Mellon, from Tom Sargent and the late Leo Hurwicz at the University of Minnesota, and the late Roger Sherman at the University of Virginia.

Contents

Economics is enjoying a resurgence of interest in behavioral considerations, i.e., in the study of how people actually make decisions when rationality and foresight are limited, and when psychological and social considerations may play a role. Experimental techniques are increasingly used to study markets, games, and other strategic situations. The mounting excitement about experimental results is reflected in a string of recent Nobel Prizes. New subdisciplines are arising, e.g., behavioral game theory, behavioral law and economics, behavioral finance, behavioral political economy, and neuro-economics. Laboratory and field experiments provide key empirical guideposts for developments in these areas.

This book combines a mix of theory and behavioral insights with active classroom learning exercises. The chapters use a leadoff experiment as an organizing device to introduce the central concepts and results. The classroom games set up simple economic situations, e.g., a market or auction, which highlight several related economic ideas. Each chapter provides a relatively short (15–20 page) reading for a particular class, in a one-a-day approach. The reading can serve as a supplement to other material, or it can be assigned in conjunction with an in-class "experiment" in which students play the game with each other. Doing the experiments *before* the assigned reading enhances their teaching value. Many of the games can be run in class "by hand," with dice or playing cards. A larger class can be divided into teams of 2–4 students, which facilitates the collection and announcement of results. Team decisions are the norm in many professional masters programs, since team discussions allow students to clarify strategic insights and learn from each other.

The appendix contains sample sets of instructions for a few hand-run games that are particularly well adapted for classroom use. In addition, the broader selection of about 60 games are available free of charge at the Veconlab site that has been programmed by the author:

http://veconlab.econ.virginia.edu/admin.htm (for instructor setup),
http://veconlab.econ.virginia.edu/login.htm (for participant login).

The site for instructors can be located with a Google search for "veconlab admin," and the student login site can be found with a search for "veconlab login." Some similar experiments are available on a University of Exeter (FEELE) server and on a commercial Moblab site, which features sliders that are optimized for small screens. Wireless connections are standard on campuses, and the typical student backpack contains multiple devices (laptops, iPads, cell phones) that connect to a browser. It is certainly not necessary for class to meet in a lab anymore. Nevertheless, it is often effective to have students work in pairs, which helps with group learning and discussion and reduces "surfing." Such discussions are not as important for non-interactive individual decisions, e.g., a choice between two gambles. In this case, the web-based Veconlab programs can be accessed by students individually before class. Running experiments after hours is also easy for games like the ultimatum, battle of sexes, and guessing games that are only played once, since students can then read instructions and enter a decision before their partners have logged in.

Web-based programs can be set up and run from any standard browser that is connected to the Internet, without loading additional software. The programs have fully integrated instructions that automatically conform to the features selected by the instructor in the setup process. The instructor data displays can provide records of decisions, earnings, round-by-round data averages, and in some cases, theoretical calculations. There is an extensive menu of setup options for each game that lets the instructor select parameters, e.g., the numbers of buyers, sellers, decision rounds, fixed payments, payoffs, etc. There are many possible treatment configurations for each of the 60 online games. The author regularly teaches classes of 30–45 students who design their own experiments and run them on the other students at the end of one class, followed by a formal presentation of results at the beginning of the next class.

To reiterate, the class discussion can reach a higher level and expand to include recent research if (1) students participate in an experiment at the end of the prior class, and (2) students do the reading in advance, which can be incentivized by assigning an open-book multiple choice quiz (ten questions per chapter, available from the author on request). *The approach is based on learning by doing and teaching by doing.*

The introductory chapter provides a summary of a "pit market" experiment that can be run by hand with playing cards, along with information about the development of experimental economics as a field. Market instructions are provided in the experiment instructions appendix for chapter 1, found at the end of the book. It is effective to run a pit market on the first day of class; then the subsequent reading in chapter 1 can serve as a tie-in to key concepts. The second chapter pertains to price discovery and adjustment in several different types of market institutions, including the commonly used double auctions and call markets. The remaining chapters are grouped by category (decisions, games, social preferences, markets, and auctions). An alternative to proceeding part by part would be to cover some basics in key chapters and then pick and choose among the remaining topics. For basics, I would suggest chapters 1, 2, 20, and 24 on markets; chapters 3 and 4 on risk aversion and prospect theory; chapters 8–11 on simple games; and chapters 14–16 on bargaining, trust, and voluntary contributions. Then other chapters could be selected based on the focus of the course. The pick-and-choose process also works well with incorporating outside readings.

This book is designed to be a primary text for a course in experimental economics or behavioral game theory. Each chapter is based on a key experiment, which is presented with a measured amount of theory and related examples. Innovative field experiments are included wherever possible. The chapters are relatively self-contained, which makes it possible to choose selections tailored to serve as a supplement for a particular course. Many of the experimental designs may be of interest to non-economists, e.g., students of political science, anthropology, and psychology, as well as anyone interested in behavioral finance or behavioral law and economics. Finally, the book could serve as an organizing device for a postgraduate course with supplemental readings from current research.

I have tried to keep the text uncluttered, with no footnotes. Mathematical arguments are simple, since experiments are based on parametric cases that distinguish alternative theories. Calculus is used sparingly, with discrete examples and graphs that provide the intuition behind more general results. References to other papers are often confined to an "extensions and further reading" section at the end of each chapter. For more extensive surveys of the literature, see Kagel and Roth's (1995, 2016) *Handbook of Experimental Economics*, volumes 1–2, and the Plott and Smith (2008) *Handbook of Experimental Economics Results*, which are pitched at a level appropriate for advanced undergraduates, graduate students, and researchers in the field.

Notes on the Revised Edition

The book has been updated and reorganized, with the introductory chapters followed by a progression from simple to complex interactions: individual decisions (chapters 3–7), games (8–12), methodology (13), social preferences and public choice (14–19), markets, finance, and macroeconomics (20–25), and auctions and mechanisms (26–30). This sequence has the advantage of moving important topics like prospect theory and behavioral game theory closer to the beginning of the book, so that these ideas can be used in subsequent treatments of more complex market, macro, and auction interactions. The number of chapters is reduced, but there are new chapters on prospect theory (chapter 4), belief elicitation and ambiguity aversion (6), social dilemmas (11), tournaments (12), methodology and nonparametric testing (13), macroeconomics (25), combinatorial and two-sided auctions (29), and rank-based matching mechanisms (30). In addition to these chapters, there is new material on risk preference measures ("ink bomb" and portfolio choice measures), infinitely repeated games with random termination, graphical analysis of games with curved quantal responses, endogenous groupings and exclusion in social dilemmas, distinguishing reciprocity from altruism in trust games, principal-agent sharing contracts, unraveling in insurance markets, distinguishing regret and risk aversion in auctions, and auction design for emissions permit markets. Chapters from the previous edition have been completely revised, updated, and in some cases combined with material from deleted chapters (discrimination, prediction markets, information cascades).

There is a "Note to the Instructor(s)" at the beginning of each chapter that provides guidance on class experiments. Key insights are italicized and separated from the text for emphasis. As mentioned above, the chapter-specific multiple-choice questions on the author's website are simple enough to be done in advance to ensure that students finish the reading before class. The applications that are covered in more detail are usually selected to match available class experiments. The "extensions" sections direct the student to recent, interesting experiments that were not included due to space limitations. The end-of-chapter problems feature some non-mechanical or design-type problems, along with hints for selected problems (but not answers) in an appendix at the end of the book. The hints are detailed enough so that the students will know whether or not they are on the right track.

Acknowledgments

The revision benefited from editing and content suggestions made by a former Veconlab research assistant, Lexi Schubert. She continued to read and comment on revised chapters even while working after she graduated. Her prior experience as an Economics Teaching Fellow and student in Experimental Economics enabled her to offer insightful suggestions from both perspectives—student and instructor. She suggested the separate "notes for the instructor" paragraphs, the italicized section summaries, and many clarifications.

I would also like to thank Lisa Anderson, AJ Bostian, Sean Sullivan, and Angela Smith for many helpful suggestions on this book and on the software that it utilizes. In particular, the methodology chapter (13) is heavily influenced by joint work with Sean Sullivan (University of Iowa Law).

Much of what I know about these topics is the result of joint research projects with Jacob Goeree (UNSW) and Doug Davis (VCU), and other collaborators: Lisa Anderson (William and Mary), Simon Anderson (Virginia), Olivier Armantier (New York Federal Reserve Bank), Jordi Brandts (Autonomous University of Barcelona), Dallas Burtraw (RFF), Monica Capra (Claremont Graduate School), Irene Comeig (University of Valencia), Catherine Eckel (Texas A&M), Roland Fryer (Harvard), Rosario Gomez (Malaga), Cathleen Johnson (Arizona), Susan Laury (Georgia State), John Ledyard (Caltech), Erica Myers (Illinois), Tom Palfrey (Caltech), Karen Palmer (RFF), Charlie Plott (Caltech), Laura Razzolini

(Alabama), David Reiley (Pandora and Berkeley), Al Roth (Stanford), David Schmidtz (Arizona), Andy Schotter (NYU), Roman Sheremeta (Case Western), the late Roger Sherman, Karti Sieberg (Tampere in Finland), and Anne Villamil (Iowa). The insurance company example in section 27.3 was provided by Ann Musser, and Lee Coppock (UVA) suggested the relevance of the Xiaogang village contract described in section 18.4. Ann Talman came up with the "stripped down poker" terminology for the card game discussed at the end of chapter 10. Special thanks to Charles Noussair (Arizona) for suggesting the reordering of the chapters for the second edition. Others who have offered advice on particular chapters include: Robert Bruner (University of Virginia Darden School), Ted Burns (University of Virginia Medical School), Juan Camilo Cardenas (Universidad de los Andes), Jeff Carpenter and his undergraduate class (Middlebury), Gary Charness (Santa Barbara), James Cox (Georgia State), Nick Feltovich (Monash), Dan Fragiadakis (Texas A&M), Dan Friedman (UC Santa Cruz), Jens Großer (Florida State), Sven Grüner (Martin Luther Univeristy), Tanga Macdaniel (Appalachian State), James Murphy (University of Alaska), Regan Petrie (Texas A&M), Charlie Plott (Caltech), Andrea Robbett (Middlebury), Tim Salmon (SMU), Fernando Solis Soberon (ITAM), John Spraggon (University of Massachusetts), and Martha Stancill (FCC).

Finally, I was fortunate to have an unusually talented and enthusiastic group of current and former Virginia students who read parts of the manuscript: Andrew Barr (Texas A&M), Clement Bohr (Northwestern), Hanna Charankevich (UVA), Kari Elasson, Vadim Elenev (Johns Hopkins), Grace Finley (AEI and Council of Economic Advisors), Sherry Forbes (Stormfish Scientific), Kendall Fox Handler (IAC), Erin Golub, Kevin Hare (UVA), Mai Hassan (Michigan Political Science), Greg Herrington (various, "serial entrepreneur"), Shelley Johnson Webb (Intel), Katya Khmennitskaya (UVA), Caroline Korndorfer (UVA), Loren Langan (St. Thomas), Julie Lerner Macklowe (Vbeauté), Yunbo Liu (Duke), Sara Hoseini Makarem (UVA), Alex Mackay (Harvard), Courtney Mallow (Chemonics), Kurt Mitman (Stockholm University, who read all chapters in the first edition), Mandy Pallais (Harvard), Uliana Popova (Moody's Analytics), Anna Roram (National Women's Business Council), Stacy Roshan (Bullis School), Daniel Savelle (UVA), Mike Schreck (Analysis Group), Lexi Schubert (Virginia Retirement System and tutor for US Olympic Ski Team), Karl Schurter (Penn State), Emily Snow (Boston Consulting), Jeanna Composti Sondag (JP Morgan), Michelle Song (Stanford), Sarah Tulman (Farm Credit Administration), Alex Watkins (UVA), Katie Johnson Wick (Abilene Christian University), Maria Winchell (UVA), Sijia Yang (Yale, Fin-tech), and Laura Young (UVA). A former economics major, Greg Herrington, pointed me toward a book on web-based programming and provided a sample PHP script, which is how I started writing the Veconlab programs.

Most important, two former graduate students, Sean Sullivan and AJ Bostian, set up and maintained the Linux Veconlab web server, which has recorded more than *one million student participant logins* in the 15 years since 2003! This server and the associated software were funded with various grants from the National Science Foundation (SES 0094800 and 1459918). Finally, I would like to acknowledge support from the University of Virginia Bankard Fund, the UVA Quantitative Collaborative, and a 2017-8 4VA grant.

MARKETS, GAMES, AND STRATEGIC BEHAVIOR

SECOND EDITION

1

Introduction

Like other scientists, economists observe naturally occurring data patterns and then try to construct explanations. The resulting theories are then evaluated in terms of factors like plausibility, generality, and predictive success. As is the case in any science, it is often difficult to sort out cause and effect when many factors are changing at the same time. Thus, there may be several reasonable theories that are roughly consistent with the same observations. Without a laboratory to control for extraneous factors, economists often "test" their theories by gauging reactions of colleagues (Keynes, 1936). In such an environment, theories may gain support on the basis of mathematical elegance, persuasion, and focal events in economic history like the Great Depression. Theories may fall from fashion, but the absence of sharp empirical tests leaves an unsettling clutter of plausible alternatives. For example, economists are fond of using the word equilibrium preceded by a juicy adjective (e.g., proper, perfect, divine, or universally divine). This clutter is often not apparent in refined textbook presentations.

The development of sophisticated econometric methods has added an important discipline to the process of devising and evaluating theoretical models. Nevertheless, any statistical analysis of naturally occurring economic data is typically based on a host of auxiliary assumptions. Economics has only recently moved in the direction of becoming an experimental science in the sense that key theories and policy recommendations are suspect if they cannot provide intended results in controlled laboratory and field experiments. This book provides an introduction to the experimental study of economic behavior, organized around games and markets that can be implemented in class.

Notes for the Instructor and Students: The chapter will describe a market simulation that can be done in class with playing cards and a decision sheet from the Pit Market Instructions in appendix 2 at the end of the book. The tone will differ from that of other chapters in that it will describe how *you* (yes, you) could run a class experiment and guide the discussion afterward. This advice is particularly relevant for classes with team presentations based on class experiments and related material collected from Internet searches. If your class is not adapted for this format, just think of the teaching advice as being relevant for when you need to explain some economics concept to work colleagues, students from other majors, or in your younger brother's or sister's high school economics class.

1.1   Smith

It is always useful to start with a sense of historical perspective. Economics did not exist as an academic discipline in the eighteenth century, but the issues like tariffs and trade were as important then as they are today. Thomas Jefferson wrote that he lamented the lack of understanding of "political economy" among the colonists. He recognized and resented the effects of monopoly (as implemented by royal grants of exclusive marketing power) and even wrote to James Madison suggesting inclusion of "freedom of monopoly" in the Bill of Rights. Jefferson was aware of the Scottish philosopher, Adam Smith, and *Wealth of Nations*, which was written in the same year as the Declaration of Independence. Jefferson considered this book to be tedious and wordy, or as he put it, "prolix." This reaction reflects Smith's methodology of thinking based on detailed observations. In fact, Adam Smith wrote *Wealth of Nations* after a European trip, taken while tutoring one of his wealthy students. This trip provided Smith with the opportunity to observe and compare different economies. In France, he encountered the view that the source of wealth was vast fertile farmland, and that the wealth flowed toward Paris, the heart. He was also exposed to the Spanish view that wealth originated with gold and silver. Adam Smith considered these ideas and wondered about what could be the source of wealth in a country like England. His explanation was that England was a commercial nation of shopkeepers and merchants, with a tendency to "truck, barter, and exchange one thing for another." Smith explicitly recognized the importance of fairness in the *language* of trade, which separates humans from animals: "Nobody ever saw a dog make a fair and deliberate exchange of one bone for another with another dog." Each voluntary exchange creates wealth in the sense that both people benefit, so extensive markets with flourishing trade create considerable wealth. In *Theory of Moral Sentiments*, he extended the discussion to include social exchanges with family, friends, and neighbors, exchanges based on pro-social attitudes like altruism, reciprocity, etc.

Smith's detailed accounts of specific markets, prices, and exogenous events like a "public mourning" show a clear understanding of the forces of supply and demand that move prices temporarily or keep them down to cost levels in the long run. (Graphical representations would come later.) But Smith's deepest insights about market systems are about how traders, following their own self-interest, are led to promote the common good, even though that was not their intention. His belief in the power of the "invisible hand" was balanced by a healthy skepticism of those who "affected to trade for the common good." Smith clearly recognized the dangerous effects of the pursuit of self-interest in the political sphere. Although a philosopher, Smith lobbied Parliament for the lifting of tariffs, a development that provided a major boost to the English economy.

Adam Smith casts a long shadow. When the author was a college sophomore many decades ago, his Economics professor, John Gunn, noted that he had studied under Jacob Viner at Princeton, who had studied under . . . and so forth, back to Alfred Marshall and eventually Adam Smith! Even today, introductory books echo Smith's distinctions between land, labor and capital, specialization of labor, the extent of the market, etc. The textbooks from 50 years ago painted a picture of a highly idealized market. Students had to memorize a set of assumptions for perfect competition, which included "an infinity of buyers and sellers," "perfect information about market conditions," and the like. *At that time, Vernon Smith had begun running experiments at Purdue that did not satisfy any of the perfectness assumptions.* There were small numbers of buyers and sellers, and they had no prior knowledge of each others' values or costs, although there was good information about current bids, asks, and recent transactions prices.

1.2 A Class Pit Market

As a graduate student, Vernon Smith had participated in a market simulation in a Harvard class taught by Edward Chamberlin (1948), who argued that his class experiments highlight failures of the standard model of perfect competition. Before discussing such a market simulation, it is useful to actually run one. Experimental economists, including this author, often begin class on the first day with an experiment. The easiest way to proceed and maximize active student involvement is to distribute a numbered playing card to each person, after dividing the deck(s) into two stacks—one for seller's costs (clubs and spades) and another for buyer's values (hearts and diamonds). The pit market instructions in the appendix at the end of the book should be read out loud to ensure that everybody is on the same page and ends at the same time. Those instructions explain the process:

> Buyers and sellers will meet in the center of the room (or other designated area) and negotiate during a 5-minute trading period. When a buyer and a seller agree on a price, they will come together to the front of the room to report the price, which will be announced to all. Then the buyer and the seller will turn in their cards, return to their original seats, and wait for the trading period to end.

For example, if a buyer with a 7 of hearts makes a trade at $5 with a seller who has a 4 of clubs, then the buyer earns $7 − $5 = $2, and the seller earns $5 − $4 = $1. Traders typically hold their cards so that they cannot be seen, but

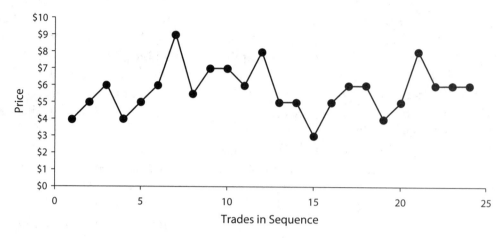

Figure 1.1. Transactions Prices for a Class Pit Market

when they come to the front to make a report, the cards are checked to be sure that the price is no lower than the cost or no higher than the value. The price should be called out, so that others who are still negotiating are aware of the "going" prices. It is useful to pre-select a student assistant or two to help with the checking and announcement process.

Figure 1.1 shows the price sequence for a market with about 60 public policy students at the University of Virginia on the first day of summer "math camp." Prices tend to converge to about $6 (the average was $5.73 last year and $5.57 this year). Sometimes it is necessary to repeat, collecting and re-shuffling buyers' and sellers' cards, and redistributing before the next trading period.

At this point, it is useful to reveal the cards used, which in this case were:

Buyers's Cards: 5, 6, 7, 8, 9, and 10 (five cards of each number)
Seller's Cards: 2, 3, 4, 5, 6, and 7 (five cards of each number)

The discussion can be focused on why the prices are near observed levels. The most important thing is to lead the discussion instead of just announcing the correct economic prediction. Consider the question: at a price of $7, would there be a larger number of willing buyers or of willing sellers? And with more willing sellers at $7, what do you think will happen to the price (they will undercut each others' prices). The parallel question is: at a price of $3, would there be a larger number of willing buyers or willing sellers? And what would tend to happen to the price? Then the question is: at what price would there be no pressures, upward or downward, on price? This is essentially a question about finding a price at which the quantity demanded equals the quantity supplied, but it is better if the students figure this out in the discussion process. The point is that the discussion of an experiment can be structured to maximize the benefits of having students discover the principles for themselves. As students

Table 1.1. Ordered Values and Costs for the Class Pit Market

Buyers' Cards	Quantity Demanded		Sellers' Cards	Quantity Supplied
$11 (0 cards)	0		$11 (0 cards)	30
$10 (5 cards)	5		$10 (0 cards)	30
$9 (5 cards)	10		$9 (0 cards)	30
$8 (5 cards)	15		$8 (0 cards)	30
$7 (5 cards)	20	excess supply	$7 (5 cards)	30
$6 (5 cards)	**25**	**equilibrium**	**$6 (5 cards)**	**25**
$5 (5 cards)	30	excess demand	$5 (5 cards)	20
$4 (0 cards)	30		$4 (5 cards)	15
$3 (0 cards)	30		$3 (5 cards)	10
$2 (0 cards)	30		$2 (5 cards)	5
$1 (0 cards)	30		$1 (0 cards)	0

of experimental economics, you will have opportunities to use experiments in class, and it helps to plan a structured de-briefing discussion.

Economics textbooks typically show demand and supply as lines, so it is useful to summarize the results of the previous discussion by organizing the values from high to low, and organizing the costs from low to high, as in table 1.1. In the top row of the left side of the table, the price is $11, and there are 0 units demanded at that price since all buyers' values are $10 or below. At a price of $10 (or slightly below) there are 5 units demanded, since there are 5 buyers with values of $10. Coming down the left side of the table, as the price falls to $5 or below, all 30 units are demanded. For sellers, the quantity supplied is 0 units at a price of $1 in the bottom row, right side, since all sellers have costs that are higher. As the price goes up, moving up the right side of the table, more and more units are supplied. Notice that there is excess demand at low prices below $6 and excess supply at prices above $6, with equality (equilibrium) at $6.

The prices and quantities trace out a supply line with a vertical intercept of $1 at 0 units (bottom row of the right side of the table), and a slope of 1/5, since the line rises by $1 in subsequent rows for each 5-unit increase in quantity, i.e., $P = 1 + 0.2Q$. Similarly, the prices and quantities for buyers trace out a demand line with a vertical intercept of $11 at 0 units (top row, left side of the table) and a negative slope of $-1/5$, i.e., $P = 11 - 0.2Q$. The intersections of these lines, solved by substitution, are at $Q = 25$ and $P = 6$, which are approximately at the observed price and quantity levels. (The familiar looking supply and demand lines mask little "steps" due to the discreteness of units or 5-unit blocks, to be clarified later.)

From Adam Smith's perspective, however, the key point is not the price prediction, but rather that the trades create wealth in the sense that both buyer and seller benefit from a voluntary trade. Each trade adds several dollars to the total benefit, and the transactions price merely determines how that "surplus" between value and cost is divided. Participants in the market focus on their own gains from trade, but the big picture is in terms of the total gains from trade from all transactions combined. In fact, the equilibrium that occurs at the intersection of supply and demand *maximizes* the gains from trade. This realization can be driven home by considering the question of which sellers are excluded if all trades are at a price of $6 (i.e., those with higher costs). The reader should think of an example of a market for a service requiring low skills (lawn care), and thinking about how the going price excluded those with high skills and high opportunity costs for their time. Parallel observations apply to the exclusion of low-value buyers. If the cards for untraded units in a pit market are kept separate, they are generally the high costs and low values, and exceptions typically involve small losses, e.g., if a seller with a $6 cost displaces one with a cost of $4, the loss is $2. This year, the remaining cards were all numbered 5, 7, and 6, so there were no efficiency losses.

The standard measure of performance in a market experiment is the actual earnings (gains from trade), which represents the wealth from all of the voluntary transactions. This total is expressed as a percentage of the maximum, which provides an efficiency measure. The maximum is determined by differences between values and costs of units that are predicted to be traded in equilibrium. These included values and costs are listed in table 1.2. First consider the column on the left side, with values of $10 and costs of $2. If all 5 of the $10 value units trade, and all 5 of the $2 cost units trade, even though those people do not necessarily trade with each other, then the surplus is the difference between value and cost (10–2) times the number of units, 5, for a total of $40 in surplus, as shown in the bottom row. Similar calculations add amounts of $30, $20, and $10, for a total of $100 in surplus. (Note that these surplus calculations do not depend on who trades with whom, as long as those with values above the equilibrium buy from those with costs below.) If some of the low-value and high-cost units in the columns on the right side of table 1.2 do not actually trade, the loss in surplus would be several dollars, which would be small relative to the total surplus. And notice that adding in excluded units with costs of $7 and values of $5 would reduce efficiency. Efficiency is often quite high in these markets. To summarize:

Table 1.2. Included Values and Costs at a Price of $6

Included Values	10	9	8	7	6
Included Costs	2	3	4	5	6
Surplus	(10−2)5=40	(9−3)5=30	(8−4)5=20	(7−5)5=10	(6−6)5=0

> **Market Trading:** *In equilibrium, the price provides a bright line boundary that excludes low-value buyer units and high-cost seller units, and thereby tends to maximize the gains from trade as measured by market efficiency.*

1.3 Early Developments in Experimental Economics

Markets

The value and cost configuration used in the previous section's pit market is symmetric. Vernon Smith (1962, 1964) considered asymmetric designs that might cause prices to start too low (if supply is relatively steep) or too high (if demand is steep). But unlike Chamberlin, he would conduct a series of market trading periods with identical values and costs. In addition, Smith used a "double auction" that collected all buyers' bid prices and sellers' asking prices into a single auction process. Double auction procedures (discussed in the next chapter) provide a price signal that is more organized than the decentralized negotiations in a pit market. A major advantage of the experimental approach is that it let Smith measure the surplus achieved, as a percentage of the theoretical maximum. Values and costs are typically not observed in naturally occurring markets, but are induced in laboratory experiments. Thus, controlled experiments permit better measurements. Smith observed efficient competitive outcomes as prices converged in a series of double auctions, even with as few as 6–10 traders. This result was significant, since the classical "large numbers" assumptions were not realistic approximations for most market settings.

An introductory class today will begin with discussions of the benefits of competitive market allocations, but the narrative switches to various imperfections like asymmetries in quality information that can cause markets to fail, as documented by George Akerlof's (1970) analysis of the market for low-quality "lemons." These observations have motivated experimental studies of market failures in insurance and other settings where the selection of those who make purchases or sales is "adverse" to those on the other side of the transactions.

Today, many of the market experiments are focused on the design and testing of new trading institutions, e.g., auctions for broadcast spectrum, water, or emissions permits to limit greenhouse gases. Auction design is an area where game theory and experimentation continue to have a major impact on public policy. In situations where price is not appropriate, e.g., allocation of slots in schools, there has also been theoretical and related work on matching mechanisms based on ranked preference lists submitted by participants. Students will recognize a typical sorority rush procedure as an application of matching mechanisms. Al Roth's 2012 Nobel Prize was given in recognition of his theoretical and experimental work on matching mechanisms. Another common application of market experiments is the study of asset markets and macroeconomics issues with interrelated markets.

Game Theory

A parallel development is based on game-theoretic models of strategic interactions. In a "matching pennies" game, for example, each player chooses heads or tails with the prior knowledge that one will win a sum of money when the coins match, and the other will win when the coins do not match. Similarly, an accountant will want to be especially well prepared when an audit occurs, but the auditor wants to catch cases when the accountant is not prepared. Each person's optimal decision in such situations depends on what the other player is expected to do. The systematic study of strategic interactions began with John von Neumann and Oscar Morgenstern's (1944) *Theory of Games and Economic Behavior*. They asserted that standard economic theory of competitive markets did not apply to the bilateral and small-group interactions that make up a significant part of economic activity. Their "solution" was incomplete, except for the case of "zero-sum" games in which one person's loss is another's gain. While the zero-sum assumption may apply to some extremely competitive situations, like sports contests or matching pennies games, it does not apply to many economic situations where all players might prefer some outcomes to others.

Economists and mathematicians at the RAND Corporation in Santa Monica, California began trying to apply game-theoretic reasoning to military tactics at the dawn of the Cold War. In many strategic scenarios, it is easy to imagine that the "winner" may be much worse off than would be the case in the absence of nuclear war. At about this time, a young graduate student at Princeton entered John von Neumann's office with a notion of equilibrium that applies to a wide class of games, including the special case of those that satisfy the zero-sum property. John Nash's notion of equilibrium (and the half-page proof that it generally exists) were recognized by the Nobel Prize committee about 50 years later. With the Nash equilibrium as its keystone, game theory has recently achieved the central role that von Neumann and Morgenstern envisioned. Indeed, with the exception of supply and demand, the "Nash equilibrium" is probably used as often today as any other construct in economics.

A *Nash equilibrium* is a set of strategies, one for each player, with the property that nobody could increase their payoff by unilaterally deviating from their planned action given the strategies being used by the other players. To illustrate this idea, consider the most famous simple game, known as a prisoner's dilemma, or more generally, a *social dilemma*. Suppose that there are two producers who sell a product that is needed by the other. The delivered product can be of high or low quality, with a high-quality delivery costing more. The value of a high-quality delivery is 3 for the recipient, the value of a low-quality delivery is only 0 for the recipient, and the producer's costs are 1 for high and 0

Table 1.3. A Prisoner's Dilemma (Row's payoff,
Column's payoff)

	Column Player	
Row Player	High	Low
High	2, 2	−1, 3
Low	3, − 1	0, 0

for low. The payoff situation can be represented as a matrix in table 1.2, where the row player's decisions, High or Low, are listed on the left, and the column player's decisions, also High or Low, are listed across the top. Therefore, if both deliver high to each other, the payoffs are 3 (from the other's high-quality delivery) minus the cost of 1 (from the person's own delivery to the other), or 2 for each person, as shown in the upper left corner of table 1.3. In the lower-right corner, the payoffs are 0 (for receiving low quality and incurring no cost of delivering it to the other). The asymmetric payoffs on the "off" diagonal are for cases in which one person receives high quality, worth 3, at no cost, and the other incurs the cost of 1 but receives low quality, as indicated by the (3, −1) outcome. For a single play of this game, a Nash equilibrium would be a pair of strategies, one for each person, for which neither would have a unilateral incentive to change. Notice that (Low, Low) is a Nash equilibrium, since if the other is going to deliver Low anyway, one's own payoff only goes down by incurring the cost of high delivery. Moreover, this is the only Nash equilibrium. For example, (High, High) is not a Nash equilibrium, since each would have an incentive to accept the other's generosity, but cut cost on one's own delivery to earn 3 instead of 2. In general, a prisoner's dilemma is a 2 × 2 game with a unique Nash equilibrium, and another non-equilibrium outcome that involves higher payoffs for both players than they receive in the equilibrium. More generally, the term *social dilemma* refers to games with two or more decisions, for which the unique equilibrium provides lower payoffs for each player than can be achieved with another outcome.

John Nash's equilibrium definition and existence proof caught the attention of researchers at the RAND Corporation headquarters in Santa Monica, who knew that the Princeton graduate student was scheduled to visit RAND that summer. Two RAND mathematicians immediately conducted a laboratory experiment designed to stress-test Nash's new theory. Nash's thesis advisor was in the same building when he noticed the payoffs for the experiment written on a blackboard. He found the game interesting and made up a story of two prisoners facing a dilemma of whether or not to make a confession. According to the story, not confessing benefits both, but the prosecutor makes promises and threats that provide each person with a unilateral incentive to confess, which is the Nash equilibrium. This story was used in a presentation to the psychology department at Stanford, and the "prisoner's dilemma" became the most commonly discussed paradigm in the new field of game theory. The actual experiment involved multiple plays (over 100) of the same game with the same two players.

It is easy to implement a prisoner's dilemma in class by writing the payoffs on the board and giving each person two playing cards: hearts or diamonds correspond to the cooperative decision (deliver high quality), and clubs or spades correspond to the uncooperative decision ("defect" with low quality). Then pairs of people can be asked to show a card simultaneously. A Nash equilibrium would have the property that neither person would have an incentive to choose a different card after seeing the other's card decision, a kind of "announcement test." The point would be to highlight the tension between the socially optimal, cooperative outcome, and the privately optimal defect outcome, and to discuss features of ongoing business relationships that would help solve this problem.

One problem with the notion of a Nash equilibrium, which arises in multi-stage games, is that it might involve a threat that is not credible. Consider an *ultimatum bargaining game* in which one player makes a take-it-or-leave-it proposal, that the other must either accept or reject. In particular, suppose that the proposer can offer a fair split (2 for each) or an unfair split (3 for the proposer, 1 for the responder). The responder observes the initial proposal and must decide whether to accept the proposal or reject, in which case both receive 0. This game has two stages, with the proposer moving first and the responder moving second after seeing the proposer's decision. A *strategy* specifies a decision for each possible contingency. In other words, a strategy is a plan of action that could be given to an assistant to play the game, with no need for the assistant to check back about what to do. The proposer has only two strategies, which are labeled fair and unfair. The responder's strategy, in contrast, must specify decisions in two contingencies, so the responder has four strategies: (accept fair, accept unfair), (accept fair, reject unfair), (reject fair, accept unfair), and (reject fair, reject unfair). If the responder's strategy is (accept fair, accept unfair), the proposer's best response is unfair, which would be accepted and is a Nash equilibrium. If the responder's strategy is (accept fair, reject unfair), the proposer's best response is to choose fair and avoid the 0 payoff from rejection. This is also a Nash equilibrium, but it requires a responder, who receives an unfair proposal (an offer of only 1), to reject it and end up with 0. The problem is that the responder might have trouble making this rejection threat credible. Another way to think about the situation is to consider the second stage as a game (a "subgame") in which only the responder has a decision, to choose between a payoff of 1 or 0, given the proposal that was made earlier. A rejection is not an equilibrium if attention is restricted to that subgame.

One of the major advances in game theory was Reinhard Selten's work on *subgame perfection*, which involves ruling out Nash equilibria that are not also equilibria in the subgames. The point here is not to fully develop the precise theoretical definitions or illustrate them with experiments, but to point out an important connection with laboratory experiments. The author always imagined that Selten, who also started doing experiments in the 1950s, was wearing his "theory hat" while working on subgame perfection, and forgetting about experiments. The realization that Selten's insight was actually motivated by experiments was communicated by him in a "witness seminar" discussion of a group of early contributors to the field, held in Amsterdam several years ago:

> I wanted to tell you a story of how the sub-game perfect equilibrium came about. What we did first, in my early experiments, I also did experiments with situations where I didn't know or where there was no theory for it. There were many oligopoly situations, which didn't involve clear theory. For example, I looked at the situation of oligopoly with demand inertia. Demand inertia means that future demand or future sales depend on prior sales. . . . It was quite complex when we explored it. My associate *Otwin Becker* and I tried to make a theory for this experiment. What should be the theoretical solution? Then, I think I simplified it completely. I simplified it to a high degree and kept demand in it. And I computed the equilibrium. [But] I suddenly found that this was not the only one. There were many other equilibria. And then I invented the idea of sub-game perfectness in order to single out the one equilibrium. (Svorenčík and Mass, 2016, p. 155)

This passage highlights an essential advantage of experimentation: the setup can be controlled enough to allow the application of relevant theory, and looking at the resulting data motivated changes in theory, making it more behaviorally relevant.

Social Preferences

After his initial exposure to the RAND experiment results in 1950, John Nash briefly considered the implications for bargaining behavior. Bargaining had been one of Nash's interests ever since taking an international trade course in college and realizing that economists did not have a good way to model it. He soon gave up on bargaining experiments, presumably because there was no well-developed theory of fairness at that time, at least among economists. Initial results of subsequent experiments with take-it-or-leave-it "ultimatum" offers were sharply at odds with the predictions of subgame perfection, as summarized in Selten's account in the Amsterdam Witness seminar:

> This is a psychologist-led experiment where the subject played against a computer, but didn't know it was a computer. The computer was

programmed to have a fixed concession rate. And they played for 20 periods. They made alternated offers and the subject was the last one to accept or reject the offer. We saw that the subject often left their $3 on the table or so. And it all resulted in a conflict. I was completely surprised about this. The psychologists were not surprised at all. They did not think anything about this, but I was surprised about it. And I discussed it with *Werner Güth* and later he made then this experiment. *Werner* simplified the whole thing to just one period to the ultimatum game, which happens at the end of this game. And of course, he got the result that very low offers are not accepted. But it was foreshadowed in these psychological experiments. It was not even remarked by these people that there was something extraordinary happening. (Svorenčík and Mass, 2016, p. 156)

Today, there is a large literature on ultimatum bargaining and social dilemma games that focus on fairness, reciprocity, altruism, and other factors that affect behavior. In particular, Elinor Ostrom's work combined laboratory and field studies of how small groups solve resource management issues. Trained as a political scientist, she was the first woman to win a Nobel Prize in economics, in 2009. Today there is a Social Dilemmas Workshop that meets regularly to discuss this line of research that Ostrom initiated.

Bounded Rationality

Reinhard Selten was inspired by earlier work on *bounded rationality*, a term that originated with Herb Simon and was the basis for his Nobel Prize. Although trained as a political scientist, Simon spent his career at Carnegie Mellon in the business school, and later in the Psychology and Computer Science departments. Simon stressed that decision makers often rely on rules of thumb or *heuristics*. He favored studies of actual behavior, with a focus on adaptive responses to events and situations. The Carnegie School in the 1950s had a behavioral focus, with some experimentation and a year-long business game used for teaching in the MBA program, which was one of the first computerized market simulations.

When the author arrived at Carnegie Mellon as a graduate student in 1970, this behavioral/experimental focus was dismissed by graduate students as being dated (relative to the exciting work on rational expectations being done by Carnegie Mellon economists Bob Lucas and Ed Prescott). In fact, Lucas once credited his own work on rational expectations to feeling exasperated with notions of adaptive learning and process, with little focus on steady states and final outcomes. Although the author wrote a theoretical thesis on auctions under Prescott with a rational-expectations assumption that "closed" the model, he also worked on behavioral economics projects with Richard Cyert (a Simon coauthor who was then president of the university) and statistician

Morris Degroot. One of those projects was motivated by Cyert's observation that investment decisions made by CEOs and boards of directors were, at the time, typically framed by the amounts of retained earnings, which seemed to be more important than classical interest rate considerations. This is an example of a behavioral bias later identified as *mental accounting*, where sources of funds are used to constrain uses. Experiments played an important role in the documentation of mental accounting and other biases discussed in Richard Thaler's (1992) book, *The Winner's Curse*, and in his "Anomalies" column in the *Journal of Economic Perspectives* (Thaler, 1988, 1989; Tversky and Thaler, 1990). The insights summarized in this book and the research it cites provide the basis for Thaler's 2017 Nobel Prize.

Much of the experimental economics research being done today builds on Simon's notions of bounded rationality and the behavioral insights of psychologists and others who studied actual business behavior. Anyone who has looked at experimental data will have noticed situations in which people respond to strong incentives, although some randomness is apparent, especially with weaker incentives. Psychologists, who would ask people to identify the brighter light or louder sound, came to model such behavioral response as being probabilistic. Today, probabilistic choice models (logit, probit, etc.) are standard in econometric work when there are discrete choices, e.g., whether or not to enroll in a treatment or rehabilitation program. This work was pioneered by Dan McFadden, who was awarded a Nobel prize in Economics for it. The games used in experiments also typically have discrete choices, and incorporating probabilistic responses to incentive differences can be done in a manner that generalizes the notion of a Nash equilibrium, which is known as a *quantal response equilibrium* (McKelvey and Palfrey, 1995), an idea that will explain seemingly anomalous deviations from Nash predictions. There are several places in this book in which sharp "best response" lines in figures will be replaced by curved "better response" lines.

There have also been important advances in understanding the heuristics that people use when playing a game just once, when past observation on others' decisions is not available. For example, think of a "level 0" person as being totally random, a "level 1" player as someone who makes a best response to a level 0 player, a level 2 player makes a best response to a level 1, etc. Even though many economic interactions are repeated, this is often not the case, especially in politics, law, or military conflict, and *level-k thinking* models have been used in the analysis of experimental data for games played once.

Game theory has been developed and applied in disciplines like law, politics, and sociology. In economics, game theory has had a major impact on public policy, especially in the design of auctions and market mechanisms. Experiments have stimulated the development of a more behaviorally relevant theory for subdisciplines like behavioral finance, behavioral law and economics, and behavioral business operations ("B-Ops"). Indeed, game theory is the closest thing there is today to a unified theory of social science.

Decisions and Risk

A game or market may involve relatively complex interactions between multiple people. Sometimes it is useful to study key aspects of individual behavior in isolation. It is straightforward to set up a simple decision experiment by giving a person a choice between gambles or "lotteries," e.g., between a sure $10 and a coin flip that yields $30 in the event of heads and $0 otherwise. The expected value of the lottery is calculated from products of payoffs and their associated probabilities, e.g., $30(1/2) + 0(1/2) = 15$. Which would you choose in this case? What if, instead, the choice were between a sure $100,000 and a coin flip that provides a 50-50 chance of $0 and $300,000? Risk aversion is indicated by a preference for a sure outcome, even though it has a lower expected payoff. The intuition for risk aversion is analogous to diminishing marginal utility, i.e., the third $100,000 is not as important as the first. Risk is a feature of many games, and von Neumann and Morgenstern (1944) developed a theory based on the expected value of a nonlinear utility function.

Not long after the formalization of expected utility, Allais (1953) presented anomalous results that show up in choices between pairs of lotteries. The *Allais paradox* subsequently generated an outpouring of experimental work and was the basis for his Nobel Prize. At about the same time, Harry Markowitz (1952) noticed that people seem to have a reference point at the current or normal wealth level, and that they treat risks above and below that point differently, with losses being more salient. He even offered a formal definition of what has come to be known as *loss aversion*. Experimental methods were not well developed at that time, and Markowitz based his conclusions on having approached colleagues and asking them questions like: "Would you rather owe me $1 or have a 1/10 chance of owing me $10?" He used both gains and losses, at various scales. Psychologists Danny Kahneman and Amos Tversky (1979) further developed these ideas and others, e.g., over- or underweighting of extreme probabilities. The result, known as *prospect theory*, was shown to explain a wide range of anomalies in experimental data, and was the basis for Kahneman's 2002 Nobel Prize in Economic Science.

1.4   Advantages of the Experimental Methods

It is important to emphasize that the overall design of an experiment should address an important policy or theoretical issue. The setting should be simple enough so that the results can be interpreted without the need to explore alternative explanations. A clear focus is often achieved by having *treatments* of primary interest that can be compared with a baseline or *control* condition. These themes will be further developed in the subsequent chapters, which include a chapter on methodology and statistical testing.

A *laboratory experiment* is done in a controlled setting, e.g., a closed room with visual separation and minimal outside distractions. Subjects are typically exposed to different treatment conditions, e.g., a sealed bid or an ascending bid auction. A *within-subjects* design has the same person (or group of people in a market or game) being exposed to two or more treatments, so that each group is its own control. A *between-subjects* design exposes each group or person to a single treatment. In this case, a typical "between" design would be to recruit 10 separate groups for one treatment and another 10 groups for the other, with no interaction across groups. Within-subjects designs are attractive when there is considerable heterogeneity between individuals, so that it is important for each person or group to serve as its own control. But between-subjects designs should be considered if there are "sequence effects" that cause outcomes for one treatment to bias those of a second treatment with the same subjects that follows.

A *field experiment* takes place in a natural setting, in which treatment conditions are implemented without the subjects being aware that they are in an experiment. For example, potential voters might be approached with either a phone call or a knock on the door, with the same message about the citizen obligation to vote. For both laboratory and field experiments, the exogenous assignment of treatments is essential to making inferences about causality. An ex post study of get-out-the-vote methods used in actual elections could be biased, for example, if political operatives target the knock-on-the-door resources to districts that are expected to be close races. Practical considerations generally dictate between-subjects designs for field experiments. Field experiments obviously provide more realistic context and subject selection, although there may be some loss of control due to extraneous events in the field environment. Moreover, many key variables, like values and costs, cannot typically be measured in the field. And the lab can be used to create a "perfect storm" that stresses the performance of proposed policies or auction procedures. Roughly speaking, lab experiments are generally better for evaluating theoretical predictions and stress tests of policy proposals, whereas field experiments may be better for evaluating policies under "normal" conditions. A mix of lab and field treatments is sometimes quite effective if it can be accomplished.

With this terminology, it is useful to highlight some of the advantages of experimental methodology:

Motivation: Monetary or other payoffs can be used to induce preferences that place people in environments that correspond to theoretical models or proposed policies.

Control: Experimental trials can be conducted under identical conditions, without the need to adjust for systematic or unexpected changes in weather or economic conditions that complicate the interpretation of results.

Replication: Laboratory trials are typically repeated with new groups of participants, which smooths the effects of random variations due to personalities, etc. Replication for field experiments is more difficult, but sometimes can be done by going to different locations or target populations.

Economy: Obviously, laboratory tests are considerably less costly than large-scale trials with otherwise untested conditions.

Measurement: In the laboratory, it is possible to "look inside the box" and measure things like efficiency using specified values and costs, which are generally more difficult to observe or estimate with data from ongoing markets. Secondary measurement is sometimes an option with field experiments, e.g., using follow-up surveys, or using data provided by charitable foundations about their donor base for a field experiment exposing different people to different donation match treatments.

Discovery: Everyone has heard the phrase that "correlation does not imply causation," e.g., if there is a third factor that is driving both the supposed "cause" and "effect." By holding other factors constant, the effects of a treatment change can be isolated.

Exploration: It is possible to "think outside the box" and use experiments to test new markets or political institutions that have not been used previously. This advantage is amplified by advances in information processing and social media technology that permit novel types of political and economic interactions, e.g., emissions permit sales that are responsive to changes in market conditions.

Stress Testing: It is important to evaluate the performance of alternative types of markets or mechanisms under adverse "perfect storm" conditions.

Demonstration: Experiments can be used to show that a procedure is feasible and to give policy makers the confidence to try it. Experiments can also be used for teaching; there is often no better way to understand a process than to experience it. Experiments that implement simple situations provide a counterbalance to more abstract economic models and graphs.

1.5   *Experimental Economics* and the Economic Science Association (ESA)

In the 1980s, Vernon Smith and his colleagues and students at Arizona established the first large experimental economics laboratory and began the process of developing computerized interfaces for experiments. A community formed around a series of conferences in Tucson. The Economic Science Association (ESA) was founded at one of those conferences in 1986, and the subsequent presidents constitute a partial list of key contributors: Vernon Smith, Charlie Plott, Ray Battalio, Elizabeth Hoffman, Charlie Holt, Bob Forsythe, Tom Palfrey, Jim Cox, Andy Schotter, Colin Camerer, Ernst Fehr, John Kagel, Jim Andreoni, Tim Cason, Al Roth, Jacob Goeree, Yan Chen, and Cathy Eckel (president-elect in 2017!).

The author approached the ESA members in the mid-1990s, with the idea of starting a journal, but the idea was initially rejected. The fear was that a journal would be too specialized and that important ideas would become marginalized and have less impact on the development of economic thinking. At this time, a series of small experimental economics conferences were held at the University of Amsterdam, which also had a lab. After a couple of attempts to start a journal with ESA, the author teamed up with one of the Amsterdam conference organizers, Arthur Schram, and began making plans to edit a journal that would be published with a Dutch publisher, Kluwer. The four advisory editors were Vernon Smith, Reinhard Selden, Al Roth, and Charlie Plott. At this point the ESA officials were asked if they wanted this to be their journal. Tom Palfrey, then president, responded that *yes*, they would be willing for it to be *one* of their journals! Moreover, in the process of gathering support from European ESA members, he negotiated agreements that confirmed regular European ESA meetings that rotated with those in the United States. In addition, the Kluwer representative, Zac Rolnik, agreed to provide the journal free of charge for several years to members of a long-standing German experimental economics association.

Experimental Economics was launched in 1998, and the second ESA journal, the *Journal of the Economic Science Association*, began in 2016. *Experimental Economics*, which received about one submission per week when it started 20 years ago, now receives about one submission per day. The journal is quite selective, the "impact factor" is high, and its size has more than doubled. Membership in the ESA includes access to these publications, as well as a listserv for queries and announcements of upcoming conferences. The ESA welcomes student members and meets in the United States and Europe each year, along with regional meetings in Asia and the Pacific.

Figure 1.2 provides a perspective on the explosive growth in the experimental economics literature. The upper gray line is based on the author's count, beginning with Chamberlin's 1948 paper. The black line begins when the *Journal of Economic Literature* classifications were revised in 1991, and it includes only those papers that list the "design of experiments" code, which excludes books and most papers in collected volumes and in other disciplines. The annual number of publications has more than tripled since the first edition of this book was published in 2007, and many of the new ideas have been incorporated in the chapters that follow.

Many exciting developments are in the works. Economics experiments are being integrated into high school and introductory economics. Theorists

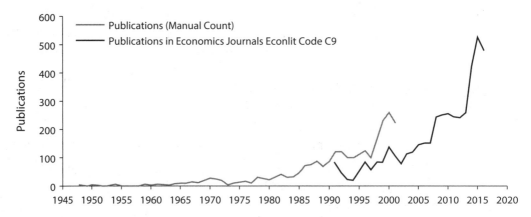

Figure 1.2. Publications in Experimental Economics

are looking at laboratory results for applications and tests of their ideas, and policy makers are increasingly willing to consider how proposed mechanisms perform in controlled tests before risking a full-scale implementation. As indicated in the final chapters of this book, experimental methods have been used to design large auctions (e.g., the FCC spectrum auctions and emissions permit auctions) and systems for matching people with jobs (e.g., medical residents and hospitals). There are also experimental economics subfields in law, business, macroeconomics, finance, and other areas. The ever-expanding list of Nobel laureates with behavioral interests is a reminder of the impact that this work has had. Economics is well on its way to becoming an experimental science!

Chapter 1 Problems

(Hints for all problems are provided in appendix 1 at the end of the book, but you should try to work the problems before turning to the appendix for help.)

1. Use the supply and demand formulas given as approximations for the setup in table 1.1 to solve for price and quantity. These formulas were: $P = 11 - 0.2Q$ for demand and $P = 1 + 0.2Q$ for supply.

2. Consider a market with 8 buyers and 8 sellers. The buyers' values are 10, 10, 10, 10, 4, 4, 4, 4. The sellers' costs are: 2, 2, 2, 2, 8, 8, 8, 8. At a price of 9, would there be excess supply or excess demand? (Explain briefly.)

3. For the market structure in problem 2, at a price of 3, would there be excess supply or excess demand? (Explain briefly.)

4. For the setup in problem 2, at a price of 6, would there be excess supply or excess demand? (Explain briefly.) At this price, how many units would trade, and what would be the total surplus (sum of value-cost differences for traded units)?

5. (non-mechanical) For the setup in problem 2, how would it be possible for 8 units to trade if prices for some trades could be higher than prices for others? Hint: think about how you might put buyers and sellers into separate groups to get more units traded.

6. For the 8 units traded at different prices, as in the answer to problem 5, what would the total surplus (sum of value-cost differences) be? What would the efficiency measure be?

2

Price Discovery and Exclusion

When buyers and sellers can communicate openly as in a trading "pit," the price and quantity outcomes can be predictable and efficient. Deviations from theoretical predictions tend to be relatively small and may be due to informational imperfections.

> **Note to the Instructor:** The discussion should be preceded by a class experiment. If the class experiment for the previous chapter was a pit market with playing cards, then a good choice for this chapter would be to use the Call Market program listed under the Markets menu on the Veconlab site. This program allows the instructor to change treatments (floor, ceiling, tax, demand shift, chat, etc.) on the fly after pressing the Stop button to end a round. The graph (accessed from the instructor "admin" interface) also provides efficiency calculations for each round.

2.1 Pit Markets

Chamberlin (1948) set up the first market experiment by letting students with buyer or seller roles negotiate trading prices. The purpose was to illustrate systematic deviations from the standard theory of perfect competition. Ironically, this experiment is most useful today in terms of what factors it suggests are needed to promote efficient, competitive market outcomes.

Each seller was given a card with a dollar amount or "cost" written on it. For example, one seller may have a cost of $2, and another may have a cost of $8. The seller earns the difference between the sale price and the cost, so the low-cost seller would be searching for a price above $2 and the high-cost seller would be searching for a price above $8. The cost is not incurred unless a sale is made, i.e., the product is "made to order." Sales below cost are typically not permitted, and such sales would result in a loss in any case. Similarly, each buyer was given a card with a dollar amount or "value" written on it. A buyer with a value of $10, for example, would earn the difference if a price below this amount could be negotiated. A buyer with a lower value, say $4, would refuse prices above that level, since a purchase above value would result in negative earnings.

The market is composed of groups of buyers and sellers who can negotiate trades with each other, either bilaterally or in larger groups. In addition to the "structural" elements of the market (numbers of buyers and sellers, and their values and costs), we must consider the nature of the market price negotiations. A *market institution* is a full specification of the rules of trade. For example, one might let sellers "post" catalogue prices and then let buyers contact sellers if they wish to purchase at a posted price, with discounts not permitted. This institution, known as a "posted-offer auction," is sometimes used in laboratory studies of retail markets. The asymmetry, with one side posting and the other responding, is common when there are many people on the responding side and few on the posting side. Posting on the "thin" side may conserve on information costs, and agents on the thin side may have the market power to impose prices on a take-it-or-leave-it basis. In contrast, Chamberlin used an institution that was symmetric and less structured; he let buyers and sellers mix together and negotiate bilaterally or in small groups, much as traders of futures contracts interact in a trading "pit." Sometimes Chamberlin announced transactions prices as they occurred, much as the market officials in organized grain exchanges watch from a "pulpit" over the trading pit and record contract prices that are posted electronically and flashed to other markets around the world. At other times, Chamberlin did not announce prices as they occurred, which may have resulted in more decentralized trading negotiations.

Suppose that the market structure is as shown in table 2.1. There are four buyers with values of $10 and four buyers with values of $4. Similarly, there are four sellers with costs of $2 and four sellers with costs of $8. Figure 2.1 shows the results of a classroom pit market experiment done with this setup. Participants were University of Virginia education school students who were interested in new approaches to teaching economics at the secondary school level. When a buyer and a seller agreed on a price, they came together to the recording desk, where the price was checked to ensure that it was no lower than the seller's cost and no higher than the buyer's value. Each dot in the figure corresponds to a

Table 2.1. A Market Example

	Values		Costs
Buyer 1	$10	Seller 9	$2
Buyer 2	$10	Seller 10	$2
Buyer 3	$10	Seller 11	$2
Buyer 4	$10	Seller 12	$2
Buyer 5	$4	Seller 13	$8
Buyer 6	$4	Seller 14	$8
Buyer 7	$4	Seller 15	$8
Buyer 8	$4	Seller 16	$8

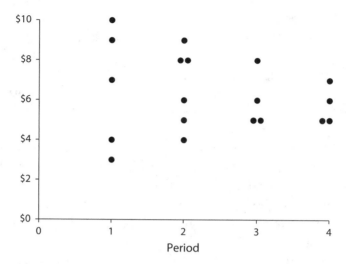

Figure 2.1. A Contract Price Sequence for the Design in
Table 2.1

trade, so there were five units traded in round 1, six units in round 2, and four units after that. The prices were variable in the first two periods and then stabilized in the $5–$7 range.

Consider the question of why the prices converged to this range, with a transactions quantity of four units per period. Notice that the quantity could have been as high as eight. For example, suppose that the four high-value ($10) buyers negotiated with high-cost ($8) sellers and agreed on prices of $9. Similarly, suppose that the negotiated prices were $3 for sales from the low-cost ($2) sellers to the low-value ($4) buyers. In this scenario, all of the sellers' units sell, the quantity is eight, and each person earns $1 for the period. Prices, however, would be quite variable, in the $8–$10 range for one group and in the $2–$4 range for the other. Figure 2.1 reveals a high price variability initially, as some of the high-cost units are sold at high prices and some of the low-value units are purchased at low prices. By the third period, prices converged to the $5–$7 range that prevents high-cost sellers from selling at a profit and low-value buyers from buying at a profit. The result is that only the four low-cost units are sold to the four high-value buyers.

The intuitive reason for the low price dispersion in later periods is fairly obvious. At a price of $9, all eight sellers would be willing (and perhaps eager) to sell, but only the four high-value buyers would be willing to buy, and perhaps not so eager given the low buyer earnings at that high price. This creates a competitive situation in which sellers may try to lower prices to get a sale, causing a price decline. Conversely, suppose that prices began in the $3 range. At these low prices, all eight buyers would be willing to buy, but only the four low-cost sellers would be willing to sell. This gives sellers the power to raise prices without losing sales.

Consider what happens with prices in the intermediate range, from $4 to $8. At a price of $6, for example, each low-cost seller earns $4 (= $6 − $2), and each high-value buyer earns $4 (= $10 − $6). Together, the eight people who make trades earn a total of $32. These earnings are much higher than $1 per person that could be earned with price dispersion, for a total of $1 times 16 people = $16. In this example, the effect of reduced price dispersion is to reduce quantity by half and to double total earnings. The low price dispersion benefits buyers and sellers as a group, since total earnings rise, but the excluded high-cost sellers and low-value buyers are worse off. Nevertheless, economic efficiency is increased by these exclusions. The sellers have high costs because the opportunity costs of the resources they use are high, i.e., the value of the resources they would employ is higher in alternative uses. And buyers with low values are not willing to pay an amount that covers the opportunity cost of the production needed to serve them.

As noted in chapter 1, one way to measure the efficiency of a market is to compare the actual earnings of all participants with the maximum possible earnings. It is a simple calculation to verify that $32 is the highest total earnings level that can be achieved by any combination of trades in this market. The efficiency was 100% for the final two periods shown in figure 2.1, since the four units that traded in each of those periods involved low costs and high values. If a fifth unit had traded, as happened in the first period, the aggregate earnings must go down, since the high cost ($8) for the fifth unit exceeds the low value ($4) for this unit. Thus the earnings total goes down by the difference, $4, reducing earnings from the maximum of $32 to $28. In this case, the outcome with price dispersion and five units traded only has an efficiency level of 28/32 = 87.5%. The trading of a sixth unit in the second period reduced efficiency even more, to 75%.

The operation of this market can be illustrated with the standard supply and demand graph. First consider a seller with a cost of $2. This seller would be unwilling to supply any units at prices below this cost, and would offer the entire capacity (1 unit) at higher prices. Thus the seller's individual supply function has a "step" at $2. The total quantity supplied by all sellers in the market is zero for prices below $2, but market supply jumps to four units at slightly higher prices as the four low-cost sellers offer their units. The supply function has another step at $8 when the four high-cost sellers offer their units at prices slightly above this high-cost step. This resulting market supply function, with steps at each of the two cost levels, is shown by the solid line in figure 2.2. The market demand function is constructed analogously. At prices above $10, no buyer is willing to purchase, but the quantity demanded jumps to four units at prices slightly below this value. The demand function has another step at $4, as shown by the dashed line in figure 2.2. Demand is vertical at prices below $4, since all buyers will wish to purchase at any lower price. The supply and demand

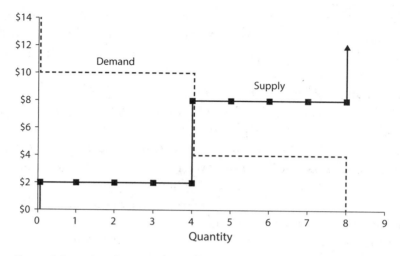

Figure 2.2. A Simple Market Design

functions overlap at a quantity of 4 in the range of prices from $4 to $8. At any price in this region, the quantity supplied equals the quantity demanded and the result is a competitive equilibrium. At lower prices, there is excess demand, which would drive prices up. At prices above the region of overlap, there is excess supply that would tend to drive prices back down.

The fact that the maximum aggregate earnings are $32 for this design can be seen directly from figure 2.2. Suppose that the price is $6 for all trades. The value of the first unit on the left is $10, and since the buyer pays $6, the surplus value is the difference, or $4. The surplus on the second, third, and fourth units is also $4, so the "consumers' surplus" is the sum of the surpluses on individual consumers' units, or $16. Notice that consumers' surplus is the area under the demand curve and above the price paid. Since sellers earn the difference between price and cost, the "producers' surplus" is the area (also $16) above the supply curve and below the price. Thus the "total surplus" is the sum of these areas, which equals the area under demand and above supply (to the left of the intersection). This total surplus area is four times the vertical distance ($10 − $2), or $32. The total surplus is actually independent of the particular prices at which the units trade. For example, a higher price would reduce consumers' surplus and increase producers' surplus, but the total area would remain fixed at $32. Adding a fifth unit would reduce surplus since the cost ($8) is greater than the value ($4).

These types of surplus calculations do not depend on the particular forms of the demand and supply functions in figure 2.2. Individual surplus amounts on each unit are the difference between the value of the unit (the height of demand) and the cost of the unit (the height of supply). Thus the area between demand and supply to the left of the intersection represents the maximum possible total surplus, even if demand and supply have more steps than the example in figure 2.2.

> **Pit Market Trading:** *When prices of transactions are announced as they occur, this public information tends to reduce dispersion and exclude high-cost sellers and low-value buyers. In this manner, adjustment to a price that equates the quantities supplied and demanded tends to increase total value created, which shows up as traders' earnings, even though maximization of total surplus is not the intention or goal of any individual trader.*

Figure 2.3 shows the results of a classroom experiment with nine buyers and nine sellers, using a design with more steps and an asymmetric structure. Notice that demand is relatively "flat" on the left side, and therefore the competitive price range (from $7 to $8) is relatively high. For prices in the competitive range, the consumers' surplus will be much smaller than the producers' surplus. The right side shows the results of two periods of pit market trading, with the prices plotted in the order of trade. Prices start at about $5, in the middle of the range between the lowest cost and the highest value. The prices seem to be converging to the competitive range from below, which could be due to buyer resistance to increasingly unequal earnings. In both periods, all seven of the higher value ($10 and $9) units were purchased, and all seven of the lower cost ($2 and $7) units were sold. As before, prices stayed in a range needed to exclude the high-cost and low-value units, and efficiency was 100% in both periods.

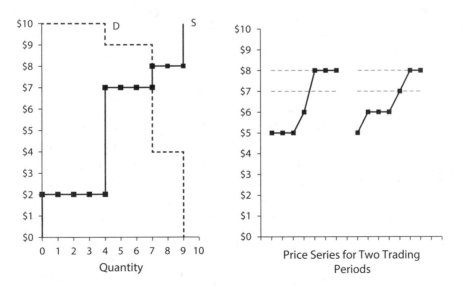

Figure 2.3. Demand (dashed line), Supply (solid line), and Transactions Prices (connected dots) for a Pit Market

The asymmetric structure in figure 2.3 was used to ensure that prices did not start in the equilibrium range, in order to illustrate some typical features of price adjustments. The first units that traded in each period were the ones with $10 values and $2 costs on the left side of the supply and demand figure. After these initial transactions in the $5–$7 range, the remaining traders were closer to the competitive "margin." These remaining traders tend to be marginal buyers with values of $9 and $4. The marginal sellers remaining tend to be those with costs of $7 and $8. Clearly these units will have to sell for prices above $7, which forces the prices closer to the competitive prediction at the end of the period. When sellers who sold early in the period see these higher prices, they may hold out for higher prices in the next period. Similarly, buyers will come to expect prices to rise later in the period, so they will scramble to buy early, which will tend to drive prices up earlier in each subsequent period. To summarize:

> Price Convergence: *The convergence in a pit market is influenced by the tendency for the highly profitable units (on the left side of demand and supply) to trade early, leaving traders "at the margin" where price negotiations tend to be near competitive levels. Price dispersion is narrowed in subsequent periods as traders come to expect the higher prices at the end of the period.*

2.2 Vernon Smith's Double Auction

The negotiations in the class experiments discussed above took place in a central area that served as the trading pit, although some people tended to break off in pairs to finalize deals. Even so, the participants could hear the public offers being made by others, a process that should reduce price dispersion. A high-value buyer, who is better off paying up to $10 instead of not trading, may not be willing to pay such high prices when some sellers are making lower offers. Similarly, a low-cost seller will be less willing to accept a low price when other buyers are observed to pay more. In this manner, good market information about the going prices will tend to reduce price dispersion. A high dispersion is needed for high-cost sellers and low-value buyers to be able to find trading partners, so less dispersion will tend to exclude these "extra-marginal" traders.

Chamberlin did report some tendency for the markets to yield "too many" trades relative to competitive predictions, which he attributed to the dispersion that can result from small group negotiations. In order to evaluate this conjecture, he took the value and cost cards from his experiment and used them to *simulate* a *decentralized* trading process. These simulations were not laboratory experiments with student traders; they were mechanical, the way one would do computer simulations today.

Table 2.2. An Eight-Trader Example

	Values		Costs
Buyer 1	$10	Seller 5	$2
Buyer 2	$10	Seller 6	$2
Buyer 3	$4	Seller 7	$8
Buyer 4	$4	Seller 8	$8

For groups of size two, Chamberlin would shuffle the cost and value cards, and then he would match one cost with one value to "make" a trade at an intermediate price if value exceeded cost. Cards for trades not made were returned to the deck to be reshuffled and rematched. It is useful to see how this random matching would work in a simple example with only four buyers and four sellers, as shown in table 2.2. The random pairing process would result in only two trades if the two low-cost units were matched with the two high-value units. It would result in four trades when both low-cost units were matched with the low-value units and high-cost units were matched with high-value units. The intermediate cases would result in three trades, for example when the value/cost combinations are: **$10/$2, $10/$8, $4/$2**, and $4/$8. The three value/cost pairs that result in a trade are shown in bold. To summarize, random matches in this example produce a quantity of trades of either two, three, or four, and some simulations should convince you that on average there will be three units traded for groups of size 2.

For groups larger than two traders, Chamberlin would shuffle and deal the cards into groups and would calculate the competitive equilibrium quantity for each group, as determined by the intersection of supply and demand for that group alone. In the four-buyer/four-seller example in table 2.2, there is only one group of size 8, i.e., all eight traders, and the equilibrium for all eight is the competitive quantity of 2 units. This illustrates Chamberlin's general finding that quantity tended to decrease with larger group size in his simulations.

Notice the relationship between the use of simulations and laboratory experiments with human participants. The experiment provided the empirical regularity (excess quantity) that motivated a theoretical model (competitive equilibrium for subgroups), and the simulation confirmed that the same regularity would be produced by this model. Computer simulations can be used to derive properties of models that are too complex to solve analytically, which is often the case for models of out-of-equilibrium behavior and dynamic adjustment. The methodological order can be reversed, with computer simulations used to derive predictions that are tested with laboratory experiments using human participants.

The simulation analysis given above suggests that market efficiency will be higher when price information is centralized, so that all traders know the

Table 2.3. A Price Negotiation Sequence

	Bid	Ask	
Buyer 2	$3.00		
		$8.00	Seller 5
		$7.50	Seller 6
Buyer 1	$4.00		
Buyer 1	$4.50		
		$6.00	Seller 7
Buyer 2	Accepts $6.00		

"going" levels of bid, ask, and transactions prices. Vernon Smith, who attended some of Chamberlin's experiments when he was a student at Harvard, used this intuition to design a trading institution that promoted efficiency. After some classroom experiments of his own, he began using a "double auction" in which buyers made bids and could outbid each others' bids, whereas sellers made offers or "asks" and could undercut each others' price offers. At any time, all traders could see the highest outstanding bid and the lowest outstanding ask. Buyers could raise the current best bid at any time, and sellers could undercut the current best ask at any time. In this manner, the bid/ask spread would typically diminish until someone closed a contract by accepting the terms from the other side of the market, i.e., until a seller accepted a buyer's bid or a buyer accepted a seller's ask. This is called a *double auction*, since it involves both buyers (bidding up, as in an auction for antiques) and sellers (bidding down, as would occur if contractors undercut each others' prices). A trade occurs when these processes meet, i.e., when a buyer accepts a seller's ask or when a seller accepts a buyer's bid.

Table 2.3 shows a typical sequence of bids and asks in a double auction. Buyer 2 bids $3.00, and seller 5 offers to sell for $8.00. Seller 6 comes in with an ask price of $7.50, and buyer 1 bids $4.00 and then $4.50. At this point seller 7 offers to sell at $6.00, which buyer 2 accepts.

In a double auction, there is public information about the bid/ask spread and about past contract prices. This information creates a large-group setting that tends to diminish price variability and increase efficiency. Smith (1962, 1964) reported that this double auction resulted in efficiency measures of more than 90%, even with relatively small numbers of traders (4–6) and with no information about others' values and costs. The double auction tends to be somewhat more centralized than a pit market, which does not close off the possibility of bilateral negotiations that are not observed by others. Double auction trading more closely resembles trading for securities on the New York Stock Exchange, where the specialist collects bids and asks, and all trades come across the "ticker tape." To summarize:

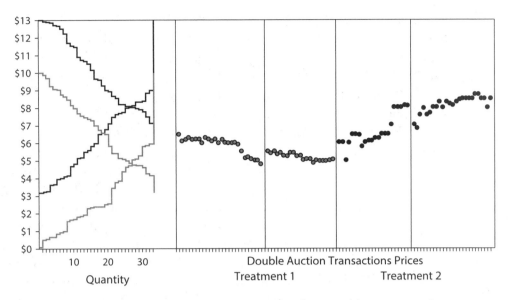

Figure 2.4. Demand, Supply, and Transactions Prices for a Double Auction with a Structural Change after the Second Period

> **Double Auction Trading:** *The centralized nature of a double auction provides a bright line "going price" that results in highly efficient allocations as high-cost sellers and low-value buyers are excluded.*

Besides introducing a centralized posting of all bids, asks, and trading prices, Smith introduced a second key feature into his early market experiments: the repetition of trading in successive market "periods" or "trading days." The effects of repetition are apparent in figure 2.4, which was done with the Veconlab web-based interface in a classroom setting. The supply and demand curves for the first two periods are shown at the bottom on the left side; the values and costs that determine the locations of the steps were randomly generated. There were four trading periods, which are delineated by the vertical lines that separate the dots that show the transactions price sequences. The prices converge to near competitive levels in the second period. A new set of random draws from higher intervals resulted in the upward demand and supply shift in period 3. This shift causes a steady rise in prices, and convergence to equilibrium is observed by the end.

2.3 Call Market Trading

Imagine an auction in which there is a fixed quantity Q being sold, and buyers make bids that are submitted during the trading period. At any point in time, the highest Q bids are provisionally accepted, and the clearing price is the highest rejected bid. For example, if Q = 3 and the bids are 10, 9, 8, 7, and 6, the three highest bidders would be listed as provisionally accepted. Bids can be changed until the market is officially closed ("called"), at which time the provisionally accepted bids become the accepted bids. If the bids listed above were still standing at the call time, the three highest bidders (with bids of 10, 9, and 8) would receive a unit of the commodity, but they would only have to pay the highest rejected bid of 7. This is known as a *uniform price auction* since the winning bidders all pay the same, market-clearing price. Note that the market-clearing price is essentially where the bid array crosses the vertical supply at Q. Uniform price auctions are widely used for the sale of fixed quantities of licenses or emissions permits, e.g., with the Regional Greenhouse Gas Initiative (RGGI) that is administered by nine northeast states in the United States. Similarly, greenhouse gas emissions in California/Quebec and in the EU are also regulated by sale in uniform price auctions, as will be discussed in a later chapter on multi-unit auctions. Such auctions were originally designed and tested by experimental economists, who compared efficiency and other performance measures across different auction formats, e.g., with "pay-as-bid" auctions in which the winning bidders pay their own bid amounts.

A *call market* is simply a two-sided uniform price auction in which bids are ranked from high to low, and provisional asks are ranked from low to high. These bid and ask rankings form "revealed" demand and supply arrays, which are crossed to determine the market-clearing price. For example, consider the screen display in table 2.4 for the buyer with ID 1, whose bids for 3 units are

Table 2.4. Call Market Bid/Ask Book

Provisional Price = $7.60

ID	Bid	—	ASK	ID
1 (you)	$10.99	—	$7.05	6
3	$8.51	—	$7.48	12
5	$8.00	—	$7.55	2
11	$7.70	—	$7.55	4
9	$7.60	—	$7.59	10
7	$7.60	—	$7.60	8
7	$7.60	—	$8.05	6
5	$7.50	—	$8.10	8
3	$6.50	—	$10.00	2
11	$6.00	—	$10.00	10
1 (you)	$5.99	—	$11.00	2
9	$5.00	—	$12.10	6
1 (you)	$4.99	—	$12.10	8
3	$4.00	—	$15.00	10

listed in the left column with the "(you)" notation. This person had values for 3 units of $11, $6, and $5, and the bids were a penny below value in each case.

Remember that ID 1 in the table would only have to pay the market-clearing price, which at this point is $7.60, so a bid of $10.99 offers some protection from a sudden price increase in the final seconds. Bids that are provisionally accepted are shown in bold in the figure (in green in the Veconlab subject display). The provisional price cannot be lower than $7.60, since the bids are below the asks in lower rows. This bidder ended up buying one unit for $7.60, and earning $11 − $7.60 = $3.40 for the period. The incentives to be included at the margin in the final seconds make this a competitive institution. Ties in price offers are broken by a random number assignment.

Figure 2.5 shows the results of the first 4 periods of trading with a call market run on the Veconlab, with 6 buyers and 6 sellers. Each buyer received 3 unit values—one high, one medium, and one low. Similarly, each seller received 3 unit costs that ranged from low to medium to high. Values and costs were re-shuffled between each round, holding aggregate demand and supply constant even though individuals received different values or costs each round. The values for all buyers were uniformly spaced, as were the sellers' costs, so as to generate approximately linear demand and supply arrays, making it possible to use standard welfare loss triangles to calculate theoretical surplus values.

For each round, the dots in the figure indicate units traded. Supply and demand cross at a quantity of 12 and a narrow price band around $12. Prices were a bit above $12 in the first round, with only 10 units traded, which reduced

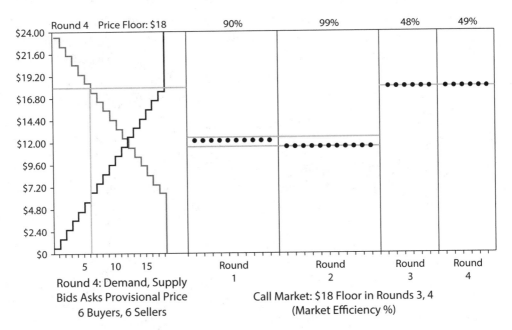

Figure 2.5. Veconlab Results for a Call Market

efficiency to 90%, as shown by the percentage at the top of the vertical slot for round 1. The transactions quantity rises to 11 in the second round, with an efficiency of 99%.

The call market program lets the administrator make treatment changes on the fly, and after observing convergence in this case, the administrator imposed a binding price floor of $18 in the third and fourth rounds. The prediction for this floor is that prices will be exactly at $18, with only 6 units traded, for a quantity reduction of 50%. The 6 buyer units with highest values will be purchased, so that would tend to mitigate the efficiency loss associated with the floor. In fact, the efficiency loss was quite severe, with efficiencies of 48% and 49% in those rounds.

During debriefing after a similar outcome, the class was asked why efficiency was so low. Answers included things like "bounded rationality," "overconfidence," and various other biases. The next question to the class was: "at a price floor of $18, which sellers wish to sell?" The answer is that all of them do, whereas only the 6 high-value buyers wish to buy at this high price. The next question was which 6 seller units actually get sold if they are all offered for sale. The answer is that sales of units with price ties were decided at random, using a random number assigned to each trader. In particular, the sellers with low costs and high profit margins are eager to cut price, but the binding floor prevented price from doing its job in terms of seller selection. The result is that some high-cost seller units end up being sold, and efficiency suffered as a result. In this symmetric design, if only the 6 units with lowest costs had been traded, the efficiency would have been 75% (problem 4), a calculation that would match standard textbook considerations of the welfare loss triangle. This "standard" efficiency prediction is much higher than the approximate 50% efficiency that was caused by imperfect seller selection at the floor. As will be seen in subsequent chapters, a monopoly also results in reductions in market efficiency due to output reduction, but those losses are mitigated by the incentives a monopolist has to organize production efficiently by using the lowest cost capacity units. To summarize:

Price Controls: *A binding price floor prevents price from operating to exclude high-cost sellers, with severe efficiency consequences. These efficiency losses are much larger than those caused by comparable output reductions due to monopoly, since production is organized efficiently in a monopoly.*

There are various political and economic reasons for imposing price controls in some cases, but experiments highlight the efficiency consequences of those controls. In this case, the experiment provides the chance to look "inside the box" and determine which seller units are sold, a process that may be much more difficult in naturally occurring markets with costs that are not directly observed.

2.4   Extensions and Further Reading

Finley, Holt, and Snow (2018) provide an analysis of efficiency losses observed in experiments with price controls, based on insights developed by Colander, Gaastra, and Rothschild (2010). The idea is that the usual marginal cost curve, with low-cost units used first, will shift up with random rationing at a binding price floor. In other words, imagine a "random-rationed marginal cost" line that lies above the array of individual unit costs. The resulting higher costs are driving the large welfare losses observed in experiments.

The simple experiments presented in this chapter can be varied in numerous directions. An upward shift in demand, accomplished by increasing buyers' values, should raise prices. An increase in the number of sellers would tend to shift supply outward, which has the effect of lowering prices. The imposition of a $1 per-unit tax on buyers would (in theory) shift demand down by $1. To see this, note that if one's value is $10, but a tax of $1 must be paid upon purchase, then the net value is only $9. All demand steps would shift down by $1 in this manner. Alternatively, a $1 tax per unit imposed on sellers would shift supply up by $1, since the tax is analogous to a $1 increase in cost. Some extensions will be considered in the chapter on market power in the market experiments section of this book.

Of the various market institutions, Chamberlin's *pit market* corresponds most closely to trading of futures contracts in a trading pit, whereas Smith's *double auction* is more like the trading of securities on the New York Stock Exchange. A *call market* (with a price determined by the intersection of bid and offer arrays) is sometimes used in electronic trading, or to provide opening prices on traditional stock exchanges. A *posted-offer auction* (where sellers set prices on a take-it-or-leave-it basis) is more like a retail market with many buyers and few sellers. Davis and Holt (1993, chapters 3 and 4) provide a detailed survey of seminal experiments comparing double auction and posted-offer market trading.

These different trading institutions have different properties and applications, and outcomes need not match competitive predictions. Nevertheless, it is useful to consider outcomes in terms of market efficiency, measured as the percentage of maximum earnings achieved by the trades that are made. In evaluating alternative designs for ways to auction off some licenses, for example, one might want to consider the efficiency of the allocations along with the amounts of revenue generated for the seller. For example, see Holt et al. (2007) for an example of a consulting report with experiments used to design the structure of greenhouse gas emissions auctions for the RGGI. Those uniform price auctions have been run quarterly since the program started in 2008.

The theoretical predictions discussed in this chapter are derived from an analysis of supply and demand. In the competitive model, all buyers and sellers take

price as given. This model may not provide good predictions when some traders perceive themselves as being price makers, with "power" to push prices in
their favor. A single-seller monopolist will typically be able to raise prices above
competitive levels, and a small group of sellers with enough of a concentration
of market capacity may be able to raise prices as well. Experiments indicate that
the exercising of market power is more difficult than it first seems, because when
a seller holds back on sales to push prices up, it may try to sell those units later,
which drives the price back down. The exercise of this type of market power is
discussed in a subsequent chapter on market institutions and power.

Taking advantage of market power is a strategic decision, and the relevant
theoretical models are those of *game theory*, which is the analysis of interrelated
strategic decisions. The word "interrelated" is critical here, since the amount
that one firm may wish to raise price depends on how much others are expected
to raise prices. Moreover, a unilateral increase in a seller's price can be risky if
the other sellers do not do the same, so risk attitudes may have a role to play in
markets and auctions. We will return to game theory experiments in chapter 8,
after first reviewing results from individual decision experiments involving risk
and information in chapters 3–7 that follow.

Chapter 2 Problems

1. Suppose that all of the numbered diamonds and spades from a deck of cards
    (excluding ace, king, queen, and jack) are used to set up a market. The diamonds
    determine demand, e.g., a 10 represents a buyer with a redemption value of $10.
    Similarly, the spades determine supply. There are 9 buyers and 9 sellers, each
    with a single card.

    (a) Graph supply and demand and derive the competitive price and quantity
        predictions.
    (b) What is the predicted effect of a supply shift that results from removing
        the 3, 4, and 5 of spades?

2. (non-mechanical) When people are asked to come up with alternative theories
    about why prices in a pit market stabilize at a particular level observed in class,
    they sometimes suggest taking the average of all buyers' values and sellers' costs.

    (a) Devise and graph two experimental designs (list all buyers' values and all
        sellers' costs) that have the same competitive price prediction but the com-
        bined average of the buyer values and seller costs is different.
    (b) Devise and graph two experimental designs for which the average of buyer
        values equals the average of seller costs in each treatment, but for which
        the competitive price predictions differ between the two treatments.

3. (non-mechanical) Next consider the following explanation, which was suggested in a recent experimental economics class: "Rank all buyer values from high to low and find the median (middle) value. Then rank all seller costs from low to high and find the median cost. The price should be the average of the median value and the median cost." By giving buyers more units than sellers, it is possible to create a design where the median cost and the median value are both equal to the lowest of all values and costs. Use this idea to set up a supply and demand array where the competitive price prediction is considerably higher than the prediction based on medians.

4. (high school geometry) Suppose that the price floor of $18 in figure 2.5 had been imposed by a single seller who controls all seller units. Use linear approximations of supply and demand to show that the resulting welfare loss (triangle areas) would be only about 25%, instead of the 50% loss associated with the price floor in the experiment.

PART I

Individual Decisions

Risk Aversion, Prospect Theory, and Learning

The next several chapters cover experiments involving the decisions of single individuals under conditions of risk, ambiguity, or situations where learning is required. This includes methods of obtaining measures of risk aversion or subjective beliefs.

Chapter 3 considers individual decisions in situations with random elements that affect money payoffs. This permits a discussion of expected money value for someone who is "neutral" toward risk. Non-neutral attitudes (risk aversion or risk seeking) are the main topics, and different methods of measuring risk preferences are discussed. Most individuals are risk averse or risk neutral, although some are risk seeking.

Chapter 4 takes a more behavioral approach, with the various components of prospect theory like reference points (from which gains and losses are measured) and heightened sensitivity to losses and small probabilities.

Probability assessments are often made in the context of combining prior beliefs with new information. These separate sources of information can be controlled in experiments, and the results can be compared with theoretical predictions derived from statistical models via Bayes' rule, as discussed in chapter 5. Different methods of eliciting a person's subjective beliefs are considered in chapter 6, where the focus is on methods that do not require auxiliary assumptions or measures of risk aversion. Chapter 7 pertains to individuals making decisions in a sequence of periods, in which learning and adjustment are possible. When people can observe others' prior decisions, learning takes on a social dimension, which can result in herding and "information cascades."

3

Risk and Decision Making

Risky decisions are characterized by a set of consequences or "prizes" and the associated probabilities. When the prizes are monetary, it is straightforward to calculate the expected money value of each decision. For example, an investment option that will pay either $1,000 or $2,000 with equal probability (1/2) would have an expected payoff of: $(1/2)(\$1,000) + (1/2)(\$2,000) = \$1,500$. A person who is neutral to risk would prefer this gamble to a sure amount of $1,499. A risk-neutral person will select the decision with the highest expected payoff, whereas a risk-averse person is willing to accept a lower expected payoff in order to reduce risk. This chapter discusses the procedures and results for several of the most commonly used ways of measuring risk preferences, beginning with a structured "price list" menu of choices that will be used to illustrate the concepts of expected value maximization and risk aversion. Other procedures to be discussed include investment tasks (dividing money between safe and risky assets) and a "bomb" method that highlights a risk versus return tradeoff.

Note to the Instructor: Variations on the choice menu experiment can be conducted prior to class discussions, using the Veconlab software (select the Lottery-Choice experiment listed under the Decisions menu). This web-based game permits individuals to complete the task on their own "after hours," thereby conserving class time. The program provides graphical displays that the instructor can access from a Graph button, which can be projected during subsequent discussion. Alternative investment task procedures are provided by the Veconlab Investment Game (continuous allocation) or the "investment portfolio" option for the Value Elicitation game (with six discrete portfolio options). Instructions for a hand-run version of the choice menu done with 10-sided dice are provided in the Class Experiment chapter 3 instructions appendix at the end of the book. This appendix also contains instructions for a risk aversion measure that is motivated by an "ink bomb" story that will also be covered in the reading.

3.1 Who Wants to Be a Millionaire?

Imagine that you are a contestant on the game show *Who Wants to Be a Millionaire?* You are at the $500,000 point and the question is on a topic that you know nothing about. Fortunately, you have saved the "fifty-fifty" option that rules out two answers, leaving two that turn out to be unfamiliar. At this point, you figure you only have a one-half chance of guessing correctly, which would make you into a millionaire. If you guess incorrectly, you receive the safety level of $32,000. Or you can fold and take the sure $500,000. In thinking about whether to take the $500,000 and fold, you decide to calculate the expected money value of the guess option. With probability 0.5 you earn $32,000, and with probability 0.5 you earn $1 million, so the expected value of a guess is:

$$0.5(\$32,000) + 0.5(\$1,000,000) = \$16,000 + \$500,000 = \$516,000.$$

This expected payoff is greater than the sure $500,000 from stopping, but the trouble with guessing is that you *either* get 32K *or* $1 million, not the average. The issue is whether the $16,000 increase in average payoff is worth the risk. If you love risk, then there is no problem; take the guess. If you are neutral to risk, the extra 16K in the average payoff should cause you to take the risk. Alternatively, you might reason that the possible low payoff, 32K, would be gone in 6 months, and only a prize of 500K or greater would be large enough to bring about a change in your lifestyle (new all-terrain vehicle, tropical spring vacation, etc.), which may lead you to fold. The person who folds would be classified as being *risk averse*, because the risk for this person is sufficiently bad that it is not worth the extra 16K in average payoff associated with the guess.

Now consider a more extreme case. You have the chance to secure a sure $1 million. The alternative is to take a coin flip, which provides a prize of $3 million for heads, and nothing for tails. If you believe that the coin is fair, then you are choosing between two gambles:

> *Safe Gamble:* $1 million for sure.
> *Risky Gamble:* $3 million with probability 0.5, 0 with probability 0.5.

As before, the expected value of the risky lottery can be calculated by multiplying the probabilities with associated payoffs, which yields an expected payoff of $1.5 million. When asked about this choice, most people will select the safe million. Notice that for these people, the extra $500,000 in expected payoff is not worth the risk of ending up with nothing. An economist would call this risk aversion, but to a layman the reason is intuitive: the first million provides a major change in lifestyle. A second million is an equally large money amount, but it is hard for most of us to imagine how much *additional* benefit this would provide. Roughly speaking, the additional (marginal) utility of the second million dollars is much lower than the utility of the first million. The marginal

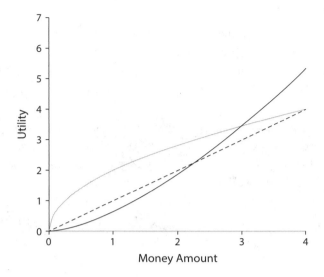

Figure 3.1. Utility Functions for Risk Neutrality (dashed straight), Risk Aversion (concave gray curved line), and Risk Seeking (convex dark curved line)

utility of the third million is likely to be even lower, although you would probably have a lot more people wanting to be your friend. A utility function with a diminishing marginal utility is one with a curved "uphill" shape, as shown by the gray line in figure 3.1 that is bowed above the dashed 45-degree line, which is indicative of the diminishing marginal utility. This feature is also apparent from the fact that the slope of the gray line utility function near the origin is high, and it diminishes as we move to the right. The more curved the utility function, the faster the utility of an additional million diminishes.

The diminishing-marginal-utility hypothesis was first suggested by Daniel Bernoulli (1738). There are many functions with this property. One is the class of power functions: $U(x) = x^{1-r}/(1-r)$ where x is money income and r is a measure of risk aversion that is less than 1. The $1-r$ term in the denominator is just a scaling convention; the curvature is determined by the exponent. Notice that when $r = 0$, the exponent in the utility function is 1, so we have the linear function: $U(x) = x^1 = x$. This function has no curvature, and hence no diminishing marginal utility for income. For such a person, the second million is just as good as the first, etc., so the person is neutral to risk. The dashed line in figure 3.1 shows the utility for a risk-neutral person. Alternatively, if the risk aversion measure r is increased from 0 to 0.5, we have $U(x) = 2x^{0.5}$, which is the square root function in the figure (after being scaled by a factor $1/(1-r)$ which is 2 in this case). Further increases in r result in more curvature, and in this sense r is a measure of the extent to which marginal utility of additional money income decreases. This measure is often called the coefficient of *relative risk aversion*.

When r is negative, the function exhibits increasing marginal utility of utility increments, which corresponds to risk-seeking preferences. Functions with decreasing marginal utility, like the gray line in figure 3.1, are referred to as *concave* functions. Conversely, functions with increasing marginal utility, like the dark line, are referred to as *convex* functions. To summarize:

> Risk Preference and Utility: *The straight dashed line in figure 3.1 is for the case where the utility of money equals the money amount, and $r = 0$. This linear utility represents the case of risk neutrality, i.e., a person who only cares about the expected money value of a gamble. The gray line with a diminishing slope to the right is for the case of $r > 0$, which indicates risk aversion. The dark line with an upward curvature, i.e., marginal utility increasing with movements to the right, is for the case where $r < 0$, which indicates risk seeking.*

In all of the examples considered up to this point, you have been asked to think about what you would do if you had to choose between gambles involving millions of dollars. The payoffs were (unfortunately) hypothetical. This raises the issue of whether what you say is what you would actually do if you faced the real situation, e.g., if you were really in the final stage of the series of questions on *Who Wants to Be a Millionaire*? It would be fortunate if we did not really have to pay large sums of money to find out how people would behave in high-payoff situations. The possibility of a *hypothetical bias*, i.e., the proposition that behavior might be dramatically different when high hypothetical stakes become real, is echoed in the 1993 film *An Indecent Proposal* (Paramount Pictures, 1993):

> John (a.k.a. Robert) Suppose I were to offer you one million dollars for one
>   night with your wife.
> David: (a.k.a. Woody) I'd assume you were kidding.
> John: Let's pretend I'm not. What would you say?
> Diana (a.k.a. Demi) He'd tell you to go to hell.
> John: I didn't hear him.
> David: I'd tell you to go to hell.
> John: That's just a reflex answer because you view it as hypothetical. But
>   let's say there were real money behind it. I'm not kidding. A million
>   dollars. Now, the night would come and go, but the money could last
>   a lifetime. Think of it—a million dollars. A lifetime of security for one
>   night. And don't answer right away. But consider it—seriously.

In the film, John's proposal was ultimately accepted, which is the Hollywood answer to the incentives question. On a more scientific note, the effects of scaled-up incentives can be investigated with experimental techniques. Before returning to this issue, it is useful to consider the results of a lottery-choice experiment designed to evaluate risk attitudes, which is the topic of the next section.

3.2 A Simple Lottery-Choice Experiment

In all of the cases discussed above, the safe lottery is a sure amount of money, with no risk at all. A sure option may have a special attraction, which is sometimes referred to as a "certainty bias." One way to avoid this bias is to consider a case where both lotteries have random outcomes, but one is riskier than the other. In particular, suppose that option A pays either $40 or $32, each with probability one-half, and that option B pays either $77 or $2, each with probability one-half. First, we calculate the expected values:

Option A: 0.5($40) + 0.5($32) = $20 + $16 = $36.00
Option B: 0.5($77) + 0.5($2) = $38.50 + $1 = $39.50.

In this case, option B has a higher expected value, by $3.50, but a lot more risk since the payoff spread from $77 to $2 is almost ten times as large as the spread from $32 to $40.

This choice was on a menu of choices used in an experiment conducted by Holt and Laury (2002). There were about 200 participants from several universities, including undergraduates, MBA students, business school faculty, and a dean. Even though option B had a higher expected payoff when each prize is equally likely, 84% of the participants selected the safe option, which indicates some risk aversion.

The payoff probabilities were implemented by the throw of a ten-sided die. This allowed the researchers to alter the probability of the high payoff in one-tenth increments. A part of the menu of choices is shown in table 3.1, where the probability of the high payoff ($40 or $77) is one-tenth in decision 1, four-tenths in decision 4, etc. Notice that decision 10 is essentially a rationality check, where the probability of the high payoff is 1, so it is a choice between $40 for sure and $77 for sure. The subjects indicated a preference for all ten decisions, and then one decision was selected at random, ex post, to determine earnings. In particular, after all decisions were made, a ten-sided die was thrown to determine the relevant decision, and then a second throw of the ten-sided die determined the subject's earnings for the selected decision. This procedure has the advantage of providing data on all ten decisions without any "wealth effects." Such wealth effects could come into play, for example, if a person wins $77 on one decision and this makes them more willing to take a risk on the subsequent decision.

The expected payoffs associated with each possible choice are shown under "Risk Neutrality" in the second and third columns of table 3.2. First, look in

Table 3.1. A Menu of Lottery Choices Used to Evaluate Risk Preferences

	Option A	Option B	Your Choice A or B
Decision 1	$40.00 if throw of die is 1 $32.00 if throw of die is 2–10	$77.00 if throw of die is 1 $2.00 if throw of die is 2–10	_____
Decision 2	$40.00 if throw of die is 1–2 $32.00 if throw of die is 3–10	$77.00 if throw of die is 1–2 $2.00 if throw of die is 3–10	_____
….			
Decision 4	$40.00 if throw of die is 1–4 $32.00 if throw of die is 5–10	$77.00 if throw of die is 1–4 $2.00 if throw of die is 5–10	_____
Decision 5	$40.00 if throw of die is 1–5 $32.00 if throw of die is 6–10	$77.00 if throw of die is 1–5 $2.00 if throw of die is 6–10	_____
Decision 6	$40.00 if throw of die is 1–6 $32.00 if throw of die is 7–10	$77.00 if throw of die is 1–6 $2.00 if throw of die is 7–10	_____
….			
Decision 10	$40.00 if throw of die is 1–10	$77.00 if throw of die is 1–10	_____

the fifth row where the probabilities are 0.5. This is the choice between options that have expected values of $36 and $39.50, as calculated above. The other expected values are determined in the same manner, by multiplying probabilities by the associated payoffs, and adding up these products.

The best decision when the probability of the high payoff is only 0.1 (in the top row of table 3.2) is obvious, since the safe decision also has a higher expected value, or $32.80, versus $9.50 for the risky lottery. In fact, 98% of the subjects selected the safe lottery (option A) in this choice. Similarly, the last choice is between sure amounts of money, and all subjects chose option B in this case. The expected payoffs are higher for the top four choices, as shown by the boldface numbers in columns for risk neutrality. A risk-neutral person, who by definition only cares about expected values regardless of risk, would make four safe choices in this menu. In fact, the average number of safe choices was six, not four, which indicates some risk aversion. As can be seen from the 0.6 row of the table, the typical safe choice for this option involves giving up about $10 in expected value in order to reduce the risk. About two-thirds of the people made the safe choice for this decision, and 40% chose safe in decision 7.

Table 3.2. Optimal Decisions for Risk Neutrality and Risk Aversion

Probability of the High Payoff	Risk Neutrality Expected Payoffs for $U(x) = x$		Risk Aversion ($r = 0.5$) Expected Utilities for $U(x) = x^{1/2}$	
	Safe $40 or $32	Risky $77 or $2	Safe $40 or $32	Risky $77 or $2
0.1	**$32.80**	$9.50	5.72	2.15
0.2	**$33.60**	$17.00	5.79	2.89
0.3	**$34.40**	$24.50	5.86	3.62
0.4	**$35.20**	$32.00	5.92	4.36
0.5	$36.00	**$39.50**	5.99	5.09
0.6	$36.80	**$47.00**	6.06	5.83
0.7	$37.60	**$54.50**	6.12	6.57
0.8	$38.40	**$62.00**	6.19	7.30
0.9	$39.20	**$69.50**	6.26	8.04
1.0	$40.00	**$77.00**	6.32	8.77

The intuitive effect of risk aversion is to diminish the utility associated with higher earnings levels, as can be seen from the curvature for the "square root" utility function in figure 3.1. With nonlinear utility, the calculation of a person's *expected utility* is analogous to the calculation of expected money value. For example, the safe option A for decision 5 is one-half of $40 and a one-half chance of $32. Recall that the *expected value of the payoff* is found by adding up the money prize amounts after they have been multiplied by the associated probabilities:

Expected payoff (safe option) = 0.5($40) + 0.5($32) = $20 + $16 = 36.

The *expected utility* of this option is obtained by replacing the money amounts, $40 and $32, with the utilities of these amounts, which we will denote by $U(40)$ and $U(32)$. If the utility function is the square root function, then $U(40) = (40)^{1/2}$ = 6.32 and $U(32) = (32)^{1/2} = 5.66$. (For simplicity, the normalization of dividing by $1 - r$ has been omitted, as it will not matter in the comparisons with all terms scaled in the same manner.) Since each prize is equally likely, we take the average:

Expected utility (safe option) = 0.5 $U(40)$ + 0.5 $U(32)$
= 0.5(6.32) + 0.5(5.66) = 5.99.

Thus, 5.99 is the expected utility for the safe option when the probabilities are 0.5, as shown in the 0.5 row of table 3.2 in the column under "Risk Aversion" for the Safe Option. Similarly, the expected utility for the risky option, with payoffs of $77 and $2, is 5.09. The other expected utilities for all ten decisions

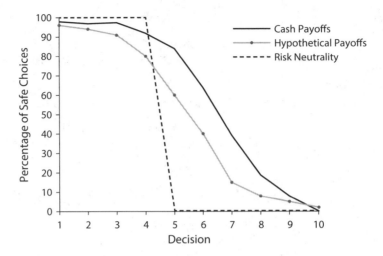

Figure 3.2. Percentages of Safe Choices with Real Incentives (20×) and Hypothetical Incentives. *Source*: Holt and Laury (2002).

are listed in the right-side columns of table 3.2. The safe option has the higher expected utility for the top six decisions. Hence, the theoretical prediction for someone with this utility function is to choose six safe options before crossing over to the risky option. Recall that the analogous prediction for risk neutrality is four safe choices, and that the data for this treatment exhibit six safe choices on average. In this sense, the square root utility function provides a better fit to the data than the linear utility function that corresponds to risk neutrality.

The thick solid line in figure 3.2 tracks the percentages of safe choices, where the decision number is listed on the horizontal axis. The behavior predicted for a risk-neutral person is represented by the dashed line, which stays at the top (100% safe choices) for the first four decisions, and then shifts to the bottom (no safe choices) for the last six decisions. The thick dark line representing actual behavior is generally above this dashed line, which indicates a tendency to make more safe choices. There is some randomness in actual choices, however, so that choice percentages do not quite reach 100% on the left side of the figure.

This real-choice experiment was preceded by a *hypothetical* choice task, in which subjects made the same ten choices with the understanding that they would not be paid their earnings for that part. The data averages for the hypothetical choices are plotted as the gray line in the figure. This gray line lies below the dark line, indicating less risk aversion when choices have no real impact. The average number of safe choices was about six with real payments, and about five with hypothetical payments. Even without payments, subjects were a little risk averse as compared with the risk-neutral prediction of four safe choices, but they could not imagine how they really would behave when they had to face real consequences. Second, with hypothetical incentives, there may be a tendency for people to think less carefully, which may produce "noise" in the data. In particular, for decision 10, the gray line is slightly higher, corresponding to the fact that 2% of the people chose the sure $40 over the sure (but hypothetical) $77.

The question of whether or not to pay subjects is one of the issues that divides research in experimental economics from work on similar issues by psychologists (see Hertwig and Ortmann, 2001, for a provocative survey of practices in psychology, with about 30 comments and an authors' reply). One justification for using high hypothetical payoffs is realism. Two prominent psychologists provided a thoughtful justification for using hypothetical incentives:

> Experimental studies typically involve contrived gambles for small stakes, and a large number of repetitions of very similar problems. These features of laboratory gambling complicate the interpretation of the results and restrict their generality. By default, the method of hypothetical choices emerges as the simplest procedure by which a large number of theoretical questions can be investigated. The use of the method relies on the assumption that people often know how they would behave in actual situations of choice, and on the further assumption that the subjects have no special reason to disguise their true preferences. (Kahneman and Tversky, 1979, p. 265)

Of course, there are many documented cases where hypothetical and real-incentive choices coincide, but it is dangerous to assume that real incentives are not needed in choices involving risk. In addition, it is quite reasonable to question the validity of using low-stakes, repetitive tasks to study important high-stakes decisions.

3.3 Payoff Scale, Order, and Demographics Effects

A key aspect of the Holt and Laury design is that it permitted an examination of large changes in payoff scale. Since individuals have widely differing attitudes toward risk, each person was asked to make decisions for several different scales. After a trainer exercise to acquaint them with the dice and random selection procedures, all participants began by making ten decisions for a low real-payoff choice menu, where all money amounts were 1/20 of the level shown in table 3.1. Thus, the payoffs for the risky option were $3.85 and $0.10, and the possible payoffs for the safer option were $2.00 and $1.60. These low payoffs will be referred to as the "1×" treatment, and the payoffs in table 3.1 will be called the 20× payoffs. Other treatments are designated similarly as multiples of the low payoff level. The initial low-payoff choice was followed by a choice menu with high *hypothetical* payoffs (20×, 40×, or 90×), followed by the same menu with high *real* payoffs (20×, 40×, or 90×), and ending with a second 1× real choice.

Table 3.3. Average Numbers of Safe Choices: Order and Incentive Effects

Experiment	Incentives	1×	20×	50×	90×
Holt and Laury (2002)	Real	5.2[a]	6.0[c]	6.8[c]	7.2[c]
208 subjects		5.3[d]			
	Hypothetical		4.9[b]	5.1[b]	5.3[b]
Holt and Laury (2005)	Real	5.7[a]	6.7[a]		
168 subjects	Hypothetical	5.6[a]	5.7[a]		

Key: Superscripts indicate Order ([a] = 1st, [b] = 2nd, [c] = 3rd, [d] = 4th)

The average numbers of safe choices, shown in the top two rows of table 3.3, indicate that the number of safe choices increases steadily as real payoffs are scaled up, but this incentive effect is not observed as hypothetical payoffs are scaled up. Finally, note that the letter superscripts in the table indicate the order in which the decision was made (*a* first, *b* second, etc.). The 1× treatment done in the fourth position in the order (*d* superscript) yielded an average of 5.3 safe choices, as compared with 5.2 safe choices when this treatment was done first. This "return to baseline" suggests that risk aversion may not be affected by the order in which the decision was made, although this inference will be reconsidered below.

The dramatic effects of payoff scale are shown in table 3.4, which presents the choice proportions for one of the high payoff (90×) paired lotteries made by 17 subjects. Note that 32% of the subjects selected the safe lottery, even though its expected value was more than $100 lower! These people were not willing to take *any* risk at these high payoffs. In all treatments, the experimenters had to go to each person's desk to throw the dice, and in the process it became apparent that there was a lot of stress and excitement with the high-stakes decisions. For some, this was revealed by changes in skin color around their necks.

Harrison et al. (2005) correctly point out that the Holt and Laury *within-subjects* design combines order and payoff scale effects. As noted in chapter 1, within-subjects designs can be problematic if there are sequence effects. A comparison of the 1× decisions (done first and last) with those of higher scales is complicated by the change in order, as are comparisons of high hypothetical and high real payoffs (done in orders 2 and 3 respectively). The presence of order effects is supported by the Harrison et al. results. They found moderately higher numbers of safe choices for a 10× scale treatment that follows the 1× treatment (6.4) than when the 10× treatment is done first (6.0). Such order effects call into question the real versus hypothetical comparisons, although it is still possible to make inferences about the scale effects from 20× to 40× to 90× in the Holt and Laury experiment, since these were made in the same order.

Table 3.4. Choice Percentages for a High Payoff Choice

Lottery A	Lottery B
$200.00 if throw of die is 1–9 $160.00 if throw of die is 10	$336.50 if throw of die is 1–9 $9.00 if throw of die is 10
Chosen by 38% **(expected payoff = $196)**	**Chosen by 62%** **(expected payoff = $303.75)**

In response to these issues, Holt and Laury (2005) ran a follow-up study with a 2 × 2 *between-subjects* design: real or hypothetical payoffs and 1× or 20× payoffs, as shown in the bottom part of table 3.3. Each participant began with an unpaid lottery-choice trainer (done with choices between $3 and gambles involving $1 or $6 as before). Then each person completed a single menu of choices from one of the four treatment cells. Therefore, all decisions were made in the *same* order. Again, the scaling up of real payoffs from 1× to 20× caused a sharp increase in the average number of safe choices (from 5.7 to 6.7), whereas this incentive effect was not observed with a scaling up of hypothetical choices. To summarize:

Risk Aversion Effects: *Subjects are generally risk averse when presented with the menu of safe-risky choices, although a small fraction seem to be risk neutral or risk seeking. Risk aversion increases sharply as real payoff amounts are scaled up. In contrast, changes in the scale of hypothetical payoffs have little effect.*

These conclusions are apparent from the graph of the distributions of the number of safe choices for each of the ten decisions, as shown in figure 3.3 where the gray lines are hypothetical and the black lines are for actual money payoffs. In this context, it does not seem to matter much whether or not one used real money when the scale of payoffs is low (thin lines), but observing no difference and then inferring that money payoffs do not matter with high stakes (thick lines) would be incorrect.

Harrison, Lau, and Rutstrom (2007) used the Holt and Laury procedure to estimate risk preferences for a *representative sample* of 253 adults in Denmark, using high payoffs comparable to those in table 3.1. (The payoffs were about eight times higher, but only one person in ten was actually paid.) One advantage of an experiment with careful sampling is that it may be possible to make inferences about the effects of social policies for the whole country. The conclusion of this study is that the Danes are risk averse, with a relative risk aversion measure of about 0.67. This number is similar to the value of 0.5 used to construct table 3.2. Most demographic variables had no significant effect, with the

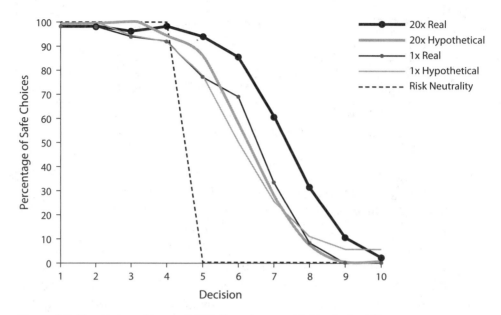

Figure 3.3. The Holt and Laury (2005) Experiment with No Order Effects

exception that middle-aged and educated people tended to be less risk averse. There was no gender effect, as is typically the case for this choice menu. Many studies of risk aversion that use different measurement methods, however, do find that women tend to be more risk averse. But there can be issues in terms of "confirmation bias" (of preconceived views) as argued by Nelson (2016). There can also be "selection bias" in terms of what is mentioned in published reports. Moreover, different types of risk aversion tasks may implement environments that interact with gender differences, e.g., whether or not losses are perceived to be greater for the risky choice (more on this below). Gender effects are important because of the implications for the types of investment advice given to women. Gender effects do appear in some survey questionnaires of a willingness to take risk. For example, Dohmen et al. (2011) conducted a large survey of Germans and found less risk aversion for men and for people who are taller, more educated, and younger.

3.4 Investment Task Measures

From a financial decision-making perspective, it is natural to infer risk aversion from an investment allocation between safe and risky assets. A sequence of papers, beginning with Gneezy and Potters (1997), present subjects with an explicit investment decision, where the rate of return for the risky asset is *independent* of the amount invested. For example, a subject might be given an amount, $10, that can be allocated between a safe and risky asset. The safe asset returns the amount invested, so it would be possible to keep the $10. The risky asset triples the investment with probability one-half, returning $0 otherwise. This presentation is intuitive and easy for subjects to understand. With an amount x invested in the risky asset, the return is $(1/2)(0) + (1/2)(3x) = 1.5x$. Hence, each dollar invested in the risky asset provides an expected net gain, so a risk-neutral or risk-seeking person would invest everything in the risky asset. Gneezy and Potters used the average amount invested in the safe asset as a measure of risk aversion.

The investment portfolio approach was also used by Eckel and Grossman (2002) in their seminal study of gender effects on risk and loss aversion. Recall that subjects in the Gneezy and Potters experiment could divide the initial endowment between safe and risky assets in an essentially continuous manner. In contrast, Eckel and Grossman presented subjects with five possible gambles, which correspond to investing various proportions in the risky asset. The "all safe" portfolio in the top row of table 3.5 corresponds to leaving an initial stake of $16 in a safe asset. The option in the second row is analogous to taking $4 out of the safe account (leaving only $12 in safety) in order to obtain a 0.5 chance of a tripled return ($4 × 3 = $12), zero otherwise. The payoff outcomes from both safe and risky assets combined would be $12 (remaining stake) with probability 0.5 and $24 (stake plus gain) with probability 0.5, as shown in the equally likely payoffs column. Moving down the table, higher amounts are taken out of the safe asset and put into an asset that either triples the investment or fails completely. Portfolio 3 involves an equal split, with a return of $8 from that safe asset and a possible tripling of the $8 put into the risky asset, as can be seen from the third column. The bottom row corresponds to investing the entire $16 in a risky asset, which yields either $0 or 3 × $16 = $48 with equal probability. Since each outcome in the second column is equally likely, the expected payoff for each gamble is the average of the two numbers in the Payoffs column. Obviously, a person who is risk neutral or risk seeking would invest all in the risky asset (bottom row), which has the highest expected payoff.

Table 3.5. Eckel and Grossman Gambles with a Portfolio Choice Interpretation

Portfolio	Equally Likely Payoffs	Investment Allocation and Gamble Structure	CRRA Range	Male Choice Rate	Female Choice Rate
1	$16 or $16	$16	> 2	0.02	0.08
2	$12 or $24	$12 + 0.5 chance of 3×$4	0.67 – 2	0.16	0.19
3	$8 or $32	$8 + 0.5 chance of 3×$8	0.38 – 0.67	0.24	0.42
4	**$4** or $40	$4 + 0.5 chance of 3×$12	0.2 – 0.38	0.23	0.18
5	**$0** or $48	0.5 chance of 3×$16	< 0.2	0.35	0.14

The Eckel-Grossman procedure is simple to administer, since what the subjects are shown is the list of alternative investments in the second column of table 3.5, from which they must pick a single row option. Instead of using rows in a table, the investment options are sometimes conveyed as pie charts in a circular array. In either case, subjects are not provided with explicit investment terminology, which may (or may not) make the task less transparent, but which has the distinct advantage of avoiding the focal "equal split" terminology seen in the middle row. For purposes of comparison with other tasks, the CRRA column shows the range of values for a coefficient of constant relative risk aversion, r, that would correspond to each choice. This task (without additional modifications discussed below) does not distinguish risk neutrality and risk seeking, since all people in these categories would select the gamble in the bottom row with the highest expected value.

The loss treatment involved shifting all payoffs in the table down by $6. This shift produces possible losses for gambles with pre-transformation payoffs of $4 and $0 in the two bottom rows, which are shown in boldface in the second column. The idea here is that subjects might be more sensitive to actual losses, which show up as negative payoffs in the menu, and that aversion to those losses might cause the options with negative payoffs to be selected less often. Earnings were equalized by providing an additional $6 fixed payment in the loss treatment. With $0 as a point of reference, loss aversion (to negative payoffs) would result in more choices of portfolios that do not offer the possibility of a loss even after the downward shift in payoffs, i.e., portfolios in the top three rows.

The choice proportions made in the loss treatment, however, were not significantly different from the choice proportions for the no-loss treatment (data averages not shown). Therefore, the table only shows combined data for both treatments, separated by gender in the final two columns. The modal decision for the 104 male subjects was the all-risky portfolio in the bottom row, whereas the modal decision for the 96 females was the equal split. Eckel and Grossman concluded that the observed gender differences are statistically and economically significant in this environment. Finally, the absence of a treatment difference is not necessarily evidence against loss aversion, since the fixed $6 payment in the loss treatment essentially negated losses incurred from the selected gambles. Moreover, the risky portfolios in the no-loss treatment might be perceived as having losses *relative to the safe payoff*, even though none of the earnings are actually negative.

Both of the investment task designs, the continuous Gneezy-Potters task and the discrete choice Eckel-Grossman task, have the desirable property that the procedures are quite simple and involve only a single decision. One benefit of this simplicity is that the investment tasks are easy to administer in the field, especially in cases where speed is important. On the other hand, a possible downside feature of this simplicity is that it is easy to determine the expected payoff maximizing choice. In the second column of table 3.5, for example, it is only necessary to add up the two payoff amounts, a sum that increases from $32 to $36 in the second row, and to $48 in the bottom row. This suggests that more mathematically oriented subjects might tend to choose the bottom decision that happens to correspond to risk neutrality or risk seeking. The implication is that studies of gender effects might also consider some measure of "numeracy."

Gender differences are generally observed with investment tasks, discrete or continuous (Charness and Gneezy, 2012), but not with choice menus as in table 3.1. Filippin and Crosetto (2016) surveyed 54 studies that used the Holt-Laury menu (more than 7,000 subjects). Gender effects only showed up in about 10% of those studies, and there is no significant gender effect in the pooled data. They conclude that gender effects are more likely to be observed in tasks with a safe return and fixed probabilities (like the investment tasks). The presence of a safe return may cause low positive payoffs from a risky option to be viewed as losses in a relative sense, which might have differential effects on choices made by men and women.

Binswanger (1980) first used a discrete set of six investment options that can be constructed from dividing a cash amount between a safe and risky asset. Instead of using a fixed return, e.g., tripling, the return on the risky asset was very high for small amounts invested, and the return was smaller as with larger investments, as noted by Holt and Laury (2014). This had the effect of keeping choices from piling up at the all-safe or all-risky boundaries. Binswanger dealt with the possibility of risk seeking by adding a couple of dominated alternatives, with lower expected values and higher risk. Similarly, Eckel and Grossman (2008c) expanded the menu of choices to add an investment option that is dominated in the sense that it has more risk and a lower expected value than the all-risky option at the bottom of table 3.5. Such a dominated option would only be selected by a risk seeker.

Cohn et al. (2015) contains an interesting application of the Gneezy Potters investment task in a laboratory experiment done in the field. The subjects (financial professionals attending a trade fair) were primed by showing them documents and price trends for a particular security that were moving upward sharply in the "boom" treatment. A downward price trajectory was used in the "bust" treatment. The investment task involved allocating 300 Swiss francs, between a safe and a risky asset (note that the investment task terminology is close to the financial market terminology used in the treatment manipulation). It was important to use a simple and quick procedure, so that effects of priming, if any, would be less likely to wear off. The result was that those who were primed with the downward price trend allocated more francs to the safe asset (55% safe) than those who were exposed to the bust treatment (42% safe).

3.5 The Bomb Task

One dimension of risk aversion measures is the degree of "hotness." A balloon method involves a subject watching a balloon expand, with a burst more likely as it gets larger. The subject earns $0 if the balloon pops, and if the process is stopped in time, earnings are proportional to the size of the balloon at the point it was stopped. This is clearly an emotionally "hot" procedure, although observation of extreme risk seeking can be truncated by the pop. Crosetto and Filippin (2013) developed an equivalent numerical method, known as the bomb risk elicitation task, BRET. This method has 100 boxes, in which the subject selects a subset of boxes, knowing that a bomb is located in one of the boxes and earnings will be zero if the bomb is in a box that is selected. Otherwise, earnings are proportional to the number of boxes selected. To the extent that the subject thinks about which boxes might avoid the bomb, the procedure retains some of the excitement of the balloon method, and there are no truncation issues. The procedure can be conducted in a static setting by letting subjects mark boxes, or in a dynamic setting by having boxes vanish at a rate of 1 per second, until the subject presses a stop button.

Consider a 12-box, "ink bomb" version that has been integrated into several Veconlab programs. The instructions, found in the chapter 3 instructions appendix, are extremely short and easy to grasp. The key sentences are:

> After you have made your choices, a 12-sided die with sides marked 1 through 12, will be thrown for you at your desk. Each integer: 1, 2, . . . 12, is equally likely. If the throw of the die matches one of the boxes that you have checked, then your earnings will be $0 for this task. If the throw of the die corresponds to one of the boxes that you did NOT check, then you earn an integer number of dollars that equals the number of checked boxes.

□1   □2   □3   □4   □5   □6   □7   □8   □9   □10   □11   □12

> Think of it this way: each box contains a dollar bill, but one of the boxes also contains an ink bomb that destroys all of the dollars in all boxes if the box with the bomb is opened.

It can be shown that a risk-neutral person would mark half of the boxes. To see the intuition, let p be the proportion of boxes marked, and the associated probability of getting the payoff is $1 - p$ which is the probability that the ink bomb is in one of the boxes not marked. Since the possible payoff is proportional to the number of marked boxes, the expected payoff is proportional to the product of these terms: $p(1 - p)$. It is obvious from the symmetry that this expected payoff is maximized at $p = \frac{1}{2}$, i.e., marking half of the 12 boxes. (Alternatively, note that the expected payoff of $p - p^2$ has a derivative of $1 - 2p$, which is zero when $p = \frac{1}{2}$.) A risk-averse person would mark fewer than six boxes and take less

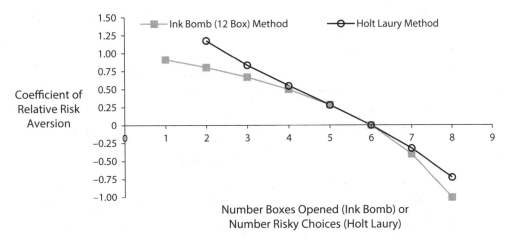

Figure 3.4. Implied Relative Risk Aversion Midpoints for Ink Bomb and Holt-Laury Menu

risk. A risk-loving person would mark more than six boxes and take more risk in order to have a chance at a higher possible payoff.

Figure 3.4 shows a comparison of the implied levels of relative risk aversion for the ink bomb and Holt-Laury menu. The horizontal axis plots the riskiness of the choices: the number of boxes checked for the bomb task or the number of risky option B choices selected in the choice menu. Both measures cross 0 (risk neutrality) at 6 boxes (ink bomb) or 6 risky and 4 safe choices (choice menu). Both lines are closely aligned, especially in the most relevant ranges of risk aversion between 0 (risk neutrality) and 0.5 (square root utility). The bomb task is simpler and no more prone to mathematical shortcuts, and it carries more emotional charge.

Using the 100-box version, Crosetto and Filippin report that most subjects are risk averse or risk neutral, and that risk aversion is increasing in payoff scale but unaffected by gender differences. Their explanation for the lack of a gender effect is based on the idea that standard investment tasks offer the option of a sure money payoff (the safe asset) so that the 0 payoff from the risky asset is coded as a loss, which might trigger gender differences if women tend to be more loss averse. They note that there is no sure safe option in the bomb task or in the Holt-Laury choice menu, which may be why gender effects are not observed. Loss aversion will be considered in the next chapter. To summarize:

Risk Aversion Measures: *Investment task measures, discrete or continuous, are simple, although mathematical calculations are somewhat focal. The bomb measure is also simple to administer and comprehend, but it is less mathematically transparent, and it preserves some of the excitement from balloon tasks that motivated it. Gender effects are generally observed with investment task measures, but not with the Holt-Laury choice menu and ink bomb measures.*

The benefit of the numerical connection between the boxes marked and underlying payoff probabilities is that it is possible to determine an implied coefficient of relative risk aversion for each possible number of boxes marked. Suppose that the utility function exhibits constant relative risk aversion, i.e., $u(w) = (w)^{1-r}/(1-r)$ for money income w. The derivation of the connection between risk aversion r and the number of boxes checked, x, is sketched in problem 9 (and associated hints). If the bomb task has N boxes, then

(3.1) $r = \frac{N-2x}{N-x}$ (implied relative risk aversion r with x boxes marked).

Note that if half of the boxes are marked, $x = N/2$, then the numerator in (3.1) is 0, i.e., a risk-neutral person would mark half of the boxes. This formula was used to determine the vertical points on the 12-box ink bomb line in figure 3.4.

3.6 Extensions, Concerns, and a Word of Caution

The precision of the mathematical utility formulas used in this chapter is misleading. Risk aversion is driven by a mix of emotions, with elements of fear, regret, caution, etc., which cannot be fully captured in a simple formula. Risk aversion differs from person to person, and may change in response to economic or psychological events. Economists tend to use simple functional forms like the power function for constant relative risk aversion, even though the r parameter in that function tends to be higher for high payoff scales. Therefore, constant relative risk aversion measures may be useful for comparisons involving the same payoff scale, but comparisons can be misleading if payoff scales for the behaviors being compared are quite different. In any case, it is important to keep the diminishing-marginal-utility intuition for risk aversion in mind, since this factor plays such a prominent role in investments, insurance, and other types of economic decisions.

There is a large literature on risk preferences, and many issues have been passed over here, e.g., whether utility should be a function of final wealth or of income (gains or losses from current wealth). In this chapter, we treat the utility as a function of income (gains and losses) instead of final wealth. There is experimental and theoretical evidence for this (Rabin, 2000; Cox and Vjollca, 2001). Appropriate reference points that separate gains from losses will be discussed in the next chapter.

In a provocative book entitled *Risky Curves: On the Empirical Failure of Expected Utility Theory*, Friedman et al. (2014) argue that different measures of risk aversion are not strongly correlated and are not very predictive of behavior in risky situations. The reader may also wish to peruse Eckel's (2016) thoughtful and nuanced review of this book. Friedman et al. note that risk of low or negative payoffs is considered to be much more serious than risk associated with high payoffs, a topic that will be revisited in the context of "upside" and "downside" risk in the next chapter. Deck et al. (2013) did find a significant ($p = 0.01$) within-subjects correlation between Holt-Laury (tabular) and Eckel-Grossman (investment task) measures of risk aversion, with a correlation of 0.27, although there is no correlation between either of those measures and two more dynamic ("balloon" and "deal-or-no-deal") tasks. One takeaway is that risk preferences may be multidimensional in a way that existing measurement tasks do not pick up.

Sometimes one gets the impression that other emotions can dominate risk preferences, e.g., the urge to speculate during an asset price surge may diminish natural caution. For example, there is no significant correlation between measured risk preferences and laboratory asset shares held at the peak of a price bubble, as will be noted in chapter 24. On the other hand, presorting by risk aversion can produce groupings that have reinforcing effects. For example, more risk-averse groups tend to bid higher in auctions and tournament competitions (chapters 12 and 26).

The discussion of risk aversion measurement has been within the confines of expected utility theory, which assumes that probabilities are perceived correctly. Any misperception of probabilities in the evaluation of the *safe* options on the left side of table 3.1 is unlikely to have much effect, since the payoffs for the safe options are closer together. For the *risky* options on the right side, in contrast, the possible overweighting of low probabilities could cause people to pick safer options, especially if the low probabilities of the very low $2 payoff on the risky side are overperceived in the bottom rows. This probability weighting effect on crossover points is discussed in an extensive Holt and Laury (2014) survey written for the *Handbook of Risk and Uncertainty*. Drichoutis and Lusk (2016) have offered a suggestion for adding another choice menu to differentiate probability and utility curvature parameters, a task that is also undertaken in Comeig, Holt, and Jaramillo (2016) to be discussed in the next chapter. Another issue is that truncating a choice menu at one end, top or bottom, will change the location of typical crossover rows (see Holt and Laury, 2014 for references). The point is that specific numerical risk aversion estimates should not be taken too seriously. Nevertheless, relative comparisons, in which any bias applies to both things being compared, can be quite informative. In addition, structured "price list" menus analogous to table 3.1 will be used in subsequent chapters to measure ambiguity aversion, time preference, subjective beliefs, etc.

A key aspect of menu-based risk elicitation tasks is the assumption that choices between a pair of gambles will be *independent* of the probability that actual payoffs will be determined by a different choice pair. Holt (1986) notes that a violation of this "independence" assumption of expected utility theory might bias or invalidate such comparisons. This is a behavioral issue that motivated Brown and Healy (2018) to compare:

(1) *Random Selection:* multiple paired gambles in rows of an ordered table, with one of the 20 rows randomly selected ex post for payment

(2) *Single Decision:* a choice between two options to be paid for sure, even
    though the option is embedded in the same table list, but with the
    surrounding rows being unavailable

A significantly higher proportion of subjects selected the risky option for the
single decision that would be paid for sure, as compared with the proportion
who selected the risky option for that same choice pair when it was embedded
in the table. This difference in choice proportions for the designated row, however, was not observed when the table presentation in (1) was replaced by a presentation of the individual gamble pairs, one at a time, in random order, with
one selected ex post at random for payment. The authors therefore recommend
random presentation of individual gamble pairs, one at a time.

The actual table of choice pairs was similar to table 3.1 in the sense that
subjects should exhibit a *single switch* from safe to risky at some point as one
moves to lower rows in the table (except in cases of indifference). It is worth
noting, however, that presenting pairs of gambles in a random order results in
significantly more multiple switches ("reversals"). In fact, about 30% of the
subjects exhibited reversals in the sequential presentation, as compared with
no observed reversals at all for the table presentation. Moreover, the sequential presentation resulted in 4 individuals out of 60 selecting a lower payoff for
the pair that involves no uncertainty at all, whereas no subjects made this irrational choice when it was presented in the bottom row of the table. Eckel et
al. (2005) also tried presenting the individual gamble choice pairs in sequence
and rejected the idea due to the increased proportions of multiple crossings. A
final issue (not directly addressed in the Brown and Healy paper) is whether the
distribution of subjects' risk aversion classifications would be different between
the tabular and random-sequence presentations.

Li (2017) recently developed an appealing alternative approach that bypasses
the need for any random selection. In the first stage, two alternative gambles
are presented, and the subject's choice is then "kept" and compared with a new
option. Thus, the choice in each stage is between the "winner" from the previous stage and a new option. The option selected in the final stage determines
the subject's earnings, so there is no random selection at the end. The clever
insight is that options can be structured to take the subject through all choice
pairs in a table, such as the Holt and Laury list in table 3.1. The sequential process begins with a choice between options A and B in the top row. If the subject
selects option A, then the second choice is between that first row option A and
the better version of option A (with a higher probability of the high payoff) in
the second row, which the subject would presumably select. At this point, the
subject's current "winning" choice is option A in row 2, and the next choice
is between options A and B in row 2. Conversely, if the subject selects option
B, the next choice is between that selection and the better version of option B

in the next lower row in the table. In this manner, the experimenter can guide the subject through a sequence of choices that includes all ten paired choices in the rows of the Holt-Laury table. Each choice is payoff relevant since it can determine the final payoff if it is designated as preferred to options encountered subsequently. This sequential tournament-like approach is known as the *accumulative best choice* (*ABC*) method. Li (2017) also provides an experimental test of the ABC method, which is promising in that it outperforms the use of a table with random ex post selection of a row for payoffs, as in the Holt-Laury test. In particular, there is a higher correspondence between individual choices in the ABC sequence and comparable choices for a single pair of gambles than is observed for a comparison between choices in the table and the choice for a single pair of gambles to be paid for sure.

Holt and Laury (2014) discuss how the non-integer payoffs and multiple comparisons in their table were intended to make it more difficult to instantly "see" the expected-payoff-maximizing crossover point, so that intuition and emotion might have more effect. As noted above, the "bomb" task is even less transparent mathematically, but the various investment portfolio tasks with 0.5 probabilities are more transparent (just add the two possible payoffs for each alternative portfolio). One possible concern with providing subjects with a small number of simple paired choices, presented one at a time, is that subjects may be more likely to rely on expected-value heuristics. This reliance on expected values could be diminished if subjects encounter a large number of such paired choices in sequence (too much effort to do all calculations). Conversely, the effects of mechanical calculations could be enhanced for those with more quantitative majors, e.g., men. For example, Fehr-Duda et al. (2011) report a probability perception experiment in which about 40% of male subjects mention using expected value as a decision criterion, whereas only a "negligible number" of females mention this behavior. A risk aversion task that simplifies expected payoff comparisons might reduce measured risk aversion, or it might generate gender differences that are merely artifacts of the measurement method. This concern could turn out to be unfounded, but it is important to bring it to the attention of those who are using and evaluating risk elicitation procedures.

Chapter 3 Problems

1. For the square root utility function, find the expected utility of the risky lottery for decision 6 in table 3.1, and check your answer with the appropriate entry in table 3.2.

2. Consider the quadratic utility function $U(x) = x^2$. Sketch the shape of this function in a figure analogous to figure 3.1. Does this function exhibit

increasing or decreasing marginal utility? Is this shape indicative of risk aversion?

3.   Suppose that a deck with all face cards removed is used to determine a money payoff, e.g., a draw of 2 of clubs would pay $2, etc. Write down the nine possible money payoffs and the probability associated with each. What is the expected value of a single draw from the deck, assuming that it has been well shuffled?

4.   Calculate the expected payoffs for the pair of lotteries in table 3.4. Which lottery would be chosen by a risk-neutral person? Which would be chosen by someone with "square root" utility for income from the choice?

5.   Consider the choice menu shown below, in which the choice is between the sure payoff in the left column or the equal probability lottery in the middle column. The choices at the top and bottom are obvious; please explain.

Payoff	Lottery	Your Choice	
$1	1/2 of $1, 1/2 of $6	☐ Payoff	☐ Lottery
$2	1/2 of $1, 1/2 of $6	☐ Payoff	☐ Lottery
$3	1/2 of $1, 1/2 of $6	☐ Payoff	☐ Lottery
$4	1/2 of $1, 1/2 of $6	☐ Payoff	☐ Lottery
$5	1/2 of $1, 1/2 of $6	☐ Payoff	☐ Lottery
$6	1/2 of $1, 1/2 of $6	☐ Payoff	☐ Lottery

6.   The menu in problem 5 is sometimes called a "price list." Please explain the intuition behind this terminology.

7.   (non-mechanical) How can a price list menu determine the money value of an economic good, like access to a proposed nature trail? If you ask people what they are willing to pay for a good, they tend to provide lower amounts than if you phrase the question in terms of the least amount one would be willing to accept for giving up the good. This is termed a "willingness-to-pay/ willingness-to-accept bias." Do you think this bias might affect choices made in the price list shown in the table for problem 5?

8.   For the choice menu in problem 5, a risk-neutral person will start with lottery in the top row and switch to the sure payoff in which row? What about a person with square root utility?

9.   (calculus derivation of ink bomb risk aversion inference) With N boxes in total, if x of them are marked, then the probability of avoiding the bomb is

$\frac{N-x}{N}$ and the associated utility is $\frac{x^{1-r}}{1-r}$. Since the utility of the 0 payoff is 0, the expected utility can be written as a product of the payoff probability and the utility: $\frac{(N-x)x^{1-r}}{N(1-r)}$. The subject knows N and risk aversion r, and must choose x. Maximizing this expected utility can be done by finding the value of x for which the derivative (slope) of expected utility is zero, since a zero slope occurs at a flat spot as would be the case at the top of a hill. Find an equation that characterizes the optimal number of boxes to check, x. At this point, you will have solved for x as a function of r. From the experimenter's perspective, however, the subject's maximizing choice of x is observed, and the risk aversion r must be inferred. Thus the next step is to invert the solution to express r as a function of x, in order to verify equation (3.1) in the text.

4

Prospect Theory and Anomalies

Choices between gambles with money payoffs may produce anxiety and other emotional reactions, especially if these choices result in significant gains or losses. Utility theory implies that such choices, even the difficult and stressful ones, can be modeled as the maximization of a mathematical function. Actual decisions sometimes deviate from precise mathematical predictions, and this chapter considers some common anomalies and biases, e.g., higher sensitivity to losses or the misperception of extreme probabilities.

The predominant approach to the study of decision making in risky situations is *expected utility theory*. Expected utility calculations are sums of products of probabilities and associated utilities for various prize amounts. With a utility function $u(x)$ and two possible payoffs, x_1 and x_2, the expected utility would be $p_1 u(x_1) + (1 - p_1)u(x_2)$, where p_1 is the probability of receiving x_1. The probabilities enter linearly, which precludes overweighting of low probabilities. With this approach, nonlinearities in utility permit an explanation of risk aversion. This model received a formal foundation in von Neumann and Morgenstern's (1944) book on game theory, which specifies a set of assumptions ("axioms") that imply behavior consistent with the maximization of the expected utility.

Almost from the beginning, economists were concerned that observed behavior seemed to contradict some of the model's predictions. The most famous contradiction, the Allais paradox, will be discussed in this chapter. Such anomalies have stimulated work on alternative approaches. The main focus here is on the most commonly mentioned alternative, *prospect theory*. A few examples have been selected to help the reader interpret experiment results and gain an understanding of biases associated with the key features of prospect theory: loss aversion, misperception of probabilities, and context-dependent risk preferences.

Note to the Instructor: The list of pairwise lottery choices in the "pen and paper" instructions appendix for this chapter is set up to evaluate the Allais paradox and comparisons of "upside risk" and "downside risk." Alternatively, the Veconlab Pairwise Lottery-Choice program (on the Decisions menu) can be used to record risky decisions between paired alternatives—no need for paper instructions or dice.

4.1 Harry Markowitz and the Utility of Wealth

The earliest progress with modeling risk aversion may be Daniel Bernoulli's (1738) paper, which predates the development of economics and psychology as distinct sciences. Bernoulli argued that the maximization of the *expected value of monetary payoffs* is not a reasonable description of actual behavior because it takes no account of risks associated with small probabilities and low payoffs. He considered a lottery in which a fair coin is flipped until "heads" is first obtained. If the first flip is heads, the process stops with a payoff of $2, which happens with probability 1/2. But if the first heads is encountered on the second flip, which happens with probability 1/4, the payoff is doubled, to 4, and it is doubled again after each time that tails is observed. The expected value of this lottery is $(1/2)2 + (1/4)4 \ldots + (1/2^n)2^n + \ldots$, which is an infinite sum of ones. The Bernoulli paradox is that people would not be willing to pay infinite or even very large amounts in exchange for a random lottery with an infinite expected value. His resolution was to suggest that the utility function is nonlinear, with a concave shape that underweights high payoffs. (Concavity and convexity were discussed in the previous chapter in the context of figure 3.1.)

Milton Friedman, who had worked on decision theory during World War II, returned to this issue in collaboration with Leonard Savage, one of the founders of decision theory. Their 1948 paper argued that the utility function should contain both convex and concave segments to explain why some people simultaneously purchase *both* insurance against losses *and* lottery tickets for large gains.

Harry Markowitz, who studied under both Friedman and Savage at Chicago, had always been interested in risk, and his 1952 *Journal of Finance* paper formalized portfolio theory based on a tradeoff between expected return and risk (measured by variance). Although this paper was the primary basis for a Nobel Prize 40 years later, there was another, more relevant paper that Markowitz (1952) published that same year in the *Journal of Political Economy*. He extended the Friedman-Savage model to cover losses as well as gains from a current level of wealth. The earliest stirrings of experimental economics were "in the air" in those days, and Markowitz began by asking friends and colleagues to choose between simple gambles involving gains or losses. He asked people to choose between:

(1) 10 cents with certainty or one chance in ten of getting $1
(2) $1 with certainty or one chance in ten of getting $10
(3) $10 with certainty or one chance in ten of getting $100
(4) $100 with certainty or one chance in ten of getting $1,000
(5) $1,000 with certainty or one chance in ten of getting $10,000
(6) $1,000,000 with certainty or one chance in ten of getting $10,000,000

(7) owe 10 cents or one chance in ten of owing $1

(8) owe $1 or one chance in ten of owing $10

...

(12) owe $1,000,000 or one chance in ten of owing $10,000,000

The typical response was to select the risky option for small gains, in questions 1–3, and then switch to the safe outcome at some point as stakes increased, with everyone choosing the safe million dollars in question 6. Markowitz concluded that people are risk seeking for small- to moderate gains relative to a reference point. He labeled the reference point "customary wealth," defined to be current wealth, unless the person had experienced recent windfall gains or losses. Besides finding risk-seeking behavior for small gains and risk aversion for large gains, he found the opposite pattern with losses, using an analogously structured set of hypothetical questions. For example, even though people were risk seeking with small gains, they tended to be risk averse with small losses, e.g., preferring a sure loss of $1 to a one in ten chance of owing $10. Similarly, even though they were risk averse for large gains, they were risk seeking for large losses. In other words, Markowitz identified four different risk preference regions, with some risk seeking and risk aversion for gains, and with some risk aversion and risk seeking for losses. He went on to make a number of conjectures about the shape of utility functions in terms of curvature and inflection points.

Markowitz also realized that losses loom larger than gains: "Generally people avoid symmetric bets. This suggests that the curve falls faster to the left of the origin than it rises to the right of the origin" (1952, p. 154). At a later point, this logic is stated in terms of absolute value notation in what today would be considered a formal definition of loss aversion: "We may also assume that $|U(-X)| > U(X)$, $X > 0$ (where $X = 0$ is customary wealth)" (p. 155). His utility graphs were explicitly constructed to be monotonic but bounded, in order to avoid the Bernoulli paradox. This tradition of considering a reference point that is affected by customary wealth levels has been followed by most psychologists and experimental economists ever since. To summarize:

Reference Points, Risk Preferences, and Loss Aversion: *Markowitz (1952) identified a major deviation from the standard Friedman-Savage utility-of-wealth model: risk preferences are different in the gain and loss domains, with the reference point division being "customary wealth." He also identified loss aversion, based on a tendency for a loss from current wealth to have more impact than an equal gain above it. Finally, he identified an alternating fourfold pattern of risk aversion or -seeking for both gains and losses.*

4.2 Probability Weighting in Prospect Theory

The next key development, about 30 years later, was the Kahneman and Tversky (1979) paper that introduced *prospect theory*. This theory was motivated by experimental evidence for choices involving gains and losses, in the Markowitz tradition. In addition to the notions of a reference point, loss aversion, and gain/loss-dependent risk preferences, Kahneman and Tversky incorporated a notion of probability weighting, which had been considered earlier by others, e.g., Edwards (1962). Probability weighting is based on the idea that low probabilities tend to be overweighted, e.g., a probability of 0.1 might be treated as if it were actually higher, say 0.2. To summarize:

Probability Weighting: If p is the actual probability of an event and the weighted probability is denoted w(p), then overweighting of low probabilities would be indicated by a weighting function for which w(p) > p for small values of p. Similarly, underweighting of high probabilities would be indicated by w(p) < p for relatively high values of p.

In figure 4.1, with the actual probability on the horizontal axis, the dashed 45-degree line would correspond to the case in which probabilities are not misperceived at all. Prospect theory predicts a probability weighting function that starts at the origin and rises above the 45-degree dashed line for low

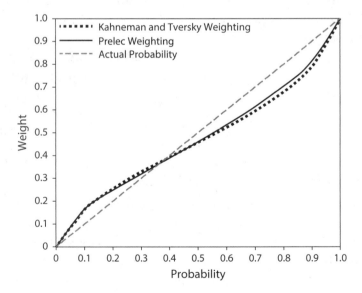

Figure 4.1. Alternative Probability Weighting Functions with a Typical Weight (ω = 0.7)

probabilities, falls below it for high probabilities, and ends up on the 45-degree line in the upper-right corner. For example, this would imply that one overestimates the small chance of winning a lottery for $1 million, and conversely, underestimates the chance of losing. The weighting function should be increasing in p, i.e., higher probabilities are perceived to be higher. The notion that the probability weighting function should return to the 45-degree line at the upper-right corner is based on the intuition that it is difficult to misperceive a probability of 1. The resulting "inverse S" pattern is exhibited by the curved lines in figure 4.1, which represent the two most commonly used parameterizations of probability weighting functions.

The curved dotted line in figure 4.1 shows the shape of a parametric form of the weighting function introduced by Kahneman and Tversky (1979), whereas the solid curved line shows a similar shape for the most commonly used alternative that is due to Prelec (1998). The formulas for these functions are shown in equation (4.1), where the weighting parameter ω determines the degree of curvature.

$$(4.1) \qquad w(p) = \frac{p^\omega}{[p^\omega + (1-p)^\omega]^{1/\omega}} \quad \text{(Kahneman and Tversky)}$$

$$w(p) = \exp(-(-\ln p)^\omega) \quad \text{(Prelec)}$$

If $\omega = 1$, the curvature vanishes for both functions in (4.1), with $w(p) = p$. As ω decreases below 1, the curves in the figure would exhibit more pronounced curvature. The value of the weighting parameter is often found to be close to 0.7, which produces the "inverse S" shape in the figure. For this particular parameter value, the two formulas are almost exactly equivalent; both imply overweighting of low probabilities and underweighting of high probabilities.

The usefulness of probability weighting can be seen by using it to develop an alternative explanation for Markowitz's observation that people prefer a one in ten chance of $10 to a sure $1. If the 1/10 is overweighted, then even a risk-neutral or slightly risk-averse person might choose the risky option.

4.3 Loss Aversion in Prospect Theory

Another key component of prospect theory is the notion of "loss aversion," which was modeled with a parameter $\lambda > 1$ that adds weight to losses. For example, consider an increasing utility function for a payoff x that is represented by $U(x)$ defined for both positive and negative values of x. This allows the utility curvature, concavity or convexity (marginal utility that is decreasing or increasing) to be different for gains and losses. If the reference point is $x = 0$, then utility can be normalized by letting $U(0) = 0$ so that gains have positive utilities and losses have negative utilities. Then loss aversion can be modeled by

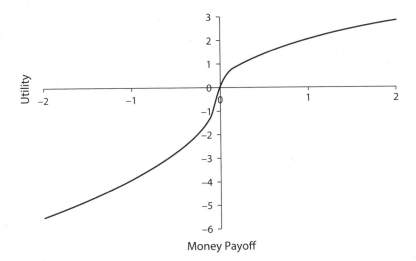

Figure 4.2. A Typical Prospect Theory Utility Function with Loss Aversion ($\lambda > 1$)

using $U(x)$ when x is positive, and $\lambda U(x)$ when x is negative. The multiplication of the negative utility for a loss by $\lambda > 1$ causes the negative utility for a loss to go down. In this manner, the utility function is steeper to the left than to the right, as indicated in figure 4.2 and which corresponds to Markowitz's original conjecture. The consensus of people who have evaluated the effects of loss aversion in experiments is that λ is approximately 2, so that losses are about twice as significant as gains (de Palma et al., 2008).

Figure 4.2 shows a typical prospect theory utility function. The reference point is at 0, and utility declines more sharply to the left than it rises to the right. The diminishing marginal utility on the right indicates a tendency to be risk averse for gains, and the increasing marginal utility to the left indicates a tendency to be risk preferring for losses. The "kink" at 0 is due to the presence of loss aversion that pushes negative utilities down by a factor of $\lambda > 1$. The difference in curvature between the left and right sides of figure 4.2 suggests a "reflection effect," with risk aversion for gains and risk seeking for losses, but intervening effects of probability weighting will introduce other factors that influence risk preference, especially in the case of relatively high or low probabilities. With intermediate probabilities in the 0.4 to 0.6 range, probability weighting has less impact, so utility curvature will tend to dominate in these cases.

There are some additional details of prospect theory that will be discussed below, but an admittedly broad-brush summary is:

Prospect Theory: *Kahneman and Tversky incorporated earlier insights about reference points, context-specific risk preferences depending on gains or losses, loss aversion, and probability weighting. The main features are:*

> *(1) losses have more impact than gains from a point of reference: $\lambda > 1$;*
> *(2) low probabilities are overweighted, high probabilities are underweighted;*
> *(3) utility exhibits risk aversion for gains and risk seeking for losses.*

Kahneman and Tversky's (1979) prospect theory paper is one of the most widely cited papers in the economics literature, and it was a primary impetus for Danny Kahneman's 2003 Nobel Prize. Even though the major components of prospect theory had been proposed by others, the genius of prospect theory is the combination of these elements and qualitative features into a unified theoretical perspective that can explain a wide range of behavioral anomalies. The most famous of these anomalies is the Allais paradox, to be considered next.

4.4 The Allais Paradox

Consider a choice between a sure 3,000 and a 0.8 chance of winning 4,000. This can be thought of as a choice between two "lotteries" that yield random earnings, represented as the choice between S1 and R1 in the top row of table 4.1.

When given a choice, people may prefer some lotteries to others, and economists assume that these preferences can be represented by the expected value of a utility function, i.e., that the expected utility is higher for the lottery selected than for the lottery not selected. This is not an assumption that people actually think about utility calculations, but rather, that choices can be represented by (are consistent with) rankings provided by the utility function.

Consider the decision a risk-neutral person would make by comparing the expected values for gambles S1 and R1 in the top row of table 4.1. An expected payoff comparison would favor risky option R1, since 0.8(4,000) = 3,200, which is higher than the 3,000 for the safe option S1 in that row. In this situation, Kahneman and Tversky reported that 80% of the subjects chose the safe option, which indicates some risk aversion. (Payoffs, in Israeli pounds, were

Table 4.1. Allais Paradox with Hypothetical Payoffs

	Lottery Labels (Choice Percentages)		
3,000 with probability 1.0	**S1** (80%)	**R1** (20%)	4,000 with probability 0.8 0 with probability 0.2
3,000 with probability 0.25 0 with probability 0.75	**S2** (35%)	**R2** (65%)	4,000 with probability 0.2 0 with probability 0.8

Source: Kahneman and Tversky (1979).

hypothetical.) A person who is not neutral to risk would have preferences represented by a utility function with some curvature. The decision of an expected utility maximizer who prefers the safe option could be represented:

$$(4.2) \qquad U(3,000) > 0.8U(4,000) + 0.2U(0).$$

Suppose that there is a three-fourths chance that all gains from either lottery will be confiscated, i.e., that is a 0.75 chance of earning $0 and only a 0.25 chance of obtaining the payoffs for the two lotteries in table 4.1. With probabilistic confiscation introduced in this manner, a majority (65%) chose the riskier option that could provide the 4,000 payoff, even though 80% of the subjects chose the sure 3,000 payoff in the initial pairwise choice with no confiscation.

This switch in preferences is an example of the Allais paradox. To see that this switch cannot be explained by expected utility theory, multiply the probabilities on both sides of (4.2) by 0.25:

$$(4.3) \qquad 0.25U(3,000) > 0.2U(4,000) + 0.05U(0).$$

In order to make the probabilities on each side sum to 1, add the $0.75U(0)$ that corresponds to confiscation to both sides of (4.3) to obtain:

$$(4.4) \qquad 0.25U(3,000) + 0.75U(0) > 0.2U(4,000) + 0.8U(0).$$

The direction of the inequality remains unchanged by this equal addition to each side. This inequality implies that the same person (who initially preferred lottery S1 to lottery R1) would prefer a one-fourth chance of 3,000 to a one-fifth chance of 4,000. These latter two lotteries are labeled S2 and R2 in the second row of table 4.1. Any reversal of this preference pattern, e.g., preferring S1 to R1 in the top row and R2 to S2 in the second row, would violate expected utility theory. A risk-neutral person, for example, would prefer the lottery with the possibility of winning 4,000 (R1 and R2) in both cases.

The intuition underlying these predictions can be seen by reexamining equation (4.4). The left side is the expected utility of a one-fourth chance of 3,000 and a three-fourths chance of 0. Equivalently, we can think of the left side as a one-fourth chance of lottery S1 (which gives 3,000) and a three-fourths chance of 0. Although it is not so transparent, the right side of (4.3) can be expressed analogously as a one-fourth chance of lottery R1 and a three-fourths chance of 0. Thus, the implication of the inequality in (4.4) is that a one-fourth chance of lottery S1 is preferred to a one-fourth chance of lottery R1. The mathematics of expected utility implied that if you prefer lottery S1 to lottery R1 as in (4.2), then you prefer a one-fourth chance of lottery S1 to a one-fourth chance of lottery R1 as in (4.4). The intuition for this prediction is that an "extra" three-fourths chance of 0 was added to *both* sides of the equation in going from (4.2) to (4.4). This extra probability of 0 dilutes the chances of winning in both S1 and R1, but this added probability of 0 is a common, and hence "irrelevant," addition. One of the basic axioms used to motivate expected utility is the assumption of *independence of irrelevant alternatives.*

As intuitive as the phrase "independence of irrelevant alternatives" may sound, a significant fraction of the Kahneman and Tversky subjects violated the predictions of this axiom. As indicated above, 80% chose S1 over R1 in the top row, but 65% chose R2 (the diluted version of R1) over S2 (the diluted version of S1) in the second row. This Allais paradox behavior is inconsistent with expected utility theory, as originally noted by the French economist, Maurice Allais (1953), who first proposed these types of paired lottery-choice situations. Anomalous behavior in Allais paradox situations has also been reported for experiments in which the money prizes were paid in cash (e.g., Starmer and Sugden, 1989, 1991, and de Palma et al., 2008). Battalio, Kagel, and MacDonald (1985) even observed similar choice patterns with rats that could choose between levers that provided food pellets on a random basis.

The de Palma et al. (2008) Allais paradox data are summarized in table 4.2, which has the same format as the previous table. Each subject in the experiment was presented with three pairs of choices, two of which are shown in the two rows of the table (the third choice pair, not shown, had an option that dominated S1 to check for violations of dominance). Subjects, who had completed an auction experiment, were told that one of the three choices would be selected ex post for payment. Notice that the dilution of the $3 and $4 payoffs in the bottom row has the effect of reducing incentives, which was expected to generate more randomness in choice.

The experiment results are listed in the table as percentages in parentheses. The main result is a tendency for people to choose S1 in the top row and R2 in the bottom. Out of 72 subjects, there were 35 who violated expected utility theory by reversing the choice direction in the two rows. Of these, 30 exhibited the standard Allais paradox pattern, with only 5 who exhibited the reverse violation (R1 and S2).

One of the many coauthors on this paper was Dan McFadden, who won a Nobel Prize in Economics for his application of probabilistic choice to economic

Table 4.2. Allais Paradox Experiment with Cash Payoffs, 72 Subjects

	Lottery Labels Choice Percentages		
$3.00 with probability 1.0	S1 (56%)	R1 (44%)	$4.00* with probability 0.8 $0 with probability 0.2
$3.00 with probability 0.25 $0 with probability 0.75	S2 (11%)	R2 (89%)	$4.00* with probability 0.2 $0 with probability 0.8

Source: de Palma et al. (2008).
Key: * half of the subjects made the choice with $4.20 instead of $4.00.

decision like traffic routes. The ex ante expectation of at least one of the coauthors was that subjects would choose the safe option S1 in the top row of table 4.2 as in previous studies, but that with real incentives, the same directional preference of S2 over R2 would be observed. Moreover, the expectation was that with weaker incentives, the S2/R2 choice percentages would be pulled away from S1 in the direction of a 50-50 split. The data dramatically overshot this expectation, with only 11% S2 choices, which showed that the Allais paradox could not be attributed to low incentives and random effects. To summarize:

Allais Paradox: *In experiments with financial incentives, a sizeable percentage of subjects exhibit a pattern of reversing a preference for a safe payoff with lower expected value to a risky payoff when both lotteries are altered in a manner that does not change the expected utility prediction. An overwhelming proportion of those who do switch their choices exhibit a switch in the direction predicted by the Allais paradox, a result that cannot be due to added randomness in decision making when incentives are diluted.*

There are several ways in which prospect theory might explain the Allais paradox pattern. Recall that the theory is based on a reference point, from which gains and losses are evaluated, and gains are treated differently from losses. The relevant reference point is less precise and more context-specific than Markowitz's notion of "customary wealth." For the choice between S1 and R1 in table 4.1, for example, the presence of a sure payoff of 3,000 in S1 might cause the 0 payoff in R1 to be "coded" as a loss. Utility is assumed to decline more sharply with losses from the reference point than it rises with gains. In this sense, the 0 payoff in lottery R1, if coded as a loss, would shift decisions away from R1. But in row 2, where there is no longer a safe option on either side, the 0 payoff might not be coded as a loss, and comparisons might be switched in favor of the higher expected value provided by R2 over S2.

An alternative explanation of the Allais paradox pattern could be based on probability weighting, with low probabilities being overweighted as indicated by the left side of figure 4.1. Note that lottery S1 on the left side of table 4.1 is a sure thing, so no misperception of probability is possible. The 0.8 chance of a 4,000 payoff for R1 on the right, however, may be affected. Suppose that the 0.8 probability of 4,000 is treated as if it were lower, say 0.7. This misperception would enhance the attractiveness of lottery S1, so that even a risk-neutral person might prefer S1 if the probability of getting 4,000 was misperceived in this manner. Next consider the choice that results when both lotteries are diluted by a three-fourths chance of a 0 payoff. Note that the 0.25 probability of the 3,000 gain on the left side of equation (4.3) is about the same as the 0.2 probability of the 4,000 on the right. In other words, 0.25 and 0.2 are located close to each other, so that a smooth probability weighting function will overweight them both by more or less the same amount. Here a probability weighting function will have little effect; a risk-neutral person will prefer the diluted version of lottery R, even though the same person might prefer the undiluted version of lottery S when the high probability of the 4,000 payoff for lottery R is underweighted. To summarize:

> **Prospect Theory and the Allais Paradox:** *The choice pattern in the standard Allais paradox is inconsistent with classical expected utility theory that stipulates linear (unweighted) probabilities and no role for loss aversion. But if the 0.8 probability of the high 4,000 payoff in gamble R1 is underweighted, then the attractiveness of that gamble is diminished, which could explain the tendency for subjects to choose the sure 3,000 option, for which probability weighting is not an issue. Similarly, the 0 payoff for the R1 lottery might be coded as a loss relative to the sure 3,000 payoff in gamble S1, which would also diminish the attractiveness of the risky R1 gamble. The gambles in the other choice pair, R2 and S2, are relatively unaffected by probability weighting since the "diluted" probabilities are close together. Similarly, those choices may not be sensitive to loss aversion, since there is no sure payoff from which to measure losses. These observations suggest how the Allais paradox choice pattern might be explained by elements of Prospect Theory.*

4.5 The Fourfold Pattern

Prospect theory has several components, and it is sometimes hard to identify the separate effects. For example, a utility function with diminishing marginal utility will create a tendency for risk aversion. But nonlinear probability weighting can also affect risk preferences. Consider, for example, Markowitz's finding that people tend to be risk preferring for small to moderate gambles of the type: $1 for sure or a one in ten chance of $10. His explanation was based on utility curvature. An alternative prospect theory explanation could be based on overweighting the small 1/10 probability of the high gain, which could cause a risk neutral or slightly risk averse person to gamble. Conversely, Markowitz observed that people tended to avoid the risk for small losses in choices between a sure loss of $1 or a one in ten chance of losing $10. Overweighting the 1/10 probability of the large loss would drive people away from the risk. With prospect theory, utility curvature also matters in the sense that diminishing marginal utility dominates when the gains are very large. In a comparison of $1,000,000 for sure and a 1/10 chance of $10,000,000, the 1/10 chance may be overweighted a little, but the extra $9 million is not worth that much more if one already is a millionaire. To summarize:

Fourfold Pattern: *The interactions between probability weighting and utility curvature result in:*

> *(1) risk seeking for low probabilities of moderate gains,*
> *(2) risk aversion for low probabilities of moderate losses,*
> *(3) risk aversion when gains have moderate probabilities or are very large,*
> *(4) risk seeking when losses have moderate probabilities or are very large.*

4.6 ## Upside versus Downside Risk

These observations suggest that risk preference is a multidimensional concept. As a consequence, it is useful to distinguish probability perception effects from utility curvature effects. For example, the act of purchasing insurance against a low payoff removes the effect of the low payoff, but at a cost. So the safe option here involves just paying the premium and enjoying a money payoff that is known in advance. The risky option involves not paying the premium, but receiving a low payoff, e.g., a low harvest value) if the adverse event occurs. If the probability of the adverse event is 1/10 and is overweighted, then even a risk-neutral person (linear utility) would purchase the insurance. In fact, risk aversion is typically observed in laboratory experiments with insurance purchase decisions, especially with women (Bediou et al., 2013).

But if overweighting of low probabilities is significant, then the same people who purchase insurance against low payoffs or losses might be willing to take on upside risks. More specifically, if women are observed to be more likely to purchase insurance than men, as in some experiments, then those women might take upside risks more often than men, unless the payoffs are so large as to cause utility curvature to dominate probability weighting effects.

These predictions about upside and downside risk were tested in a recent experiment conducted with the Veconlab Pairwise Lottery-Choice program (Comeig et al., 2016). Subjects were given a list of decisions (pairwise choices) in each of two treatments, with one of the decisions in each treatment to be randomly selected to determine payoffs for that treatment at the end. The decisions were presented in a random order that differed from subject to subject. Each decision involved a choice between a safe option, where the two payoffs were close together, and a risky option with extreme payoffs—high and low. For the downside-risk pair shown in the top part of table 4.3, the extreme payoff is 25 cents, as shown in bold. For the upside-risk pair in the bottom part of the table, the extreme payoff is 2500 cents ($25). Notice that the payoffs for the safe option B on the right side are closer together than the payoffs for the risky option A.

Table 4.3. Downside versus Upside Risk

	Option A:	Option B:
Downside Risk	9 in 10 chances of 664	9 in 10 chances of 547
	1 in 10 chances of **25**	1 in 10 chances of 275
	Option A:	Option B:
Upside Risk	9 in 10 chances of 389	9 in 10 chances of 511
	1 in 10 chances of **2,500**	1 in 10 chances of 600

In all choice pairs, the payoffs are structured so that the expected money value of the risky option (labeled as option A on the left side of the table) is 80 cents higher than for the safe option. The probabilities associated with the extreme payoff were 1/10 in some cases and 1/3 in others. The numbers in all paired choices were penny amounts selected to make expected value comparisons hard to discern. Finally, one of the treatments involved high stakes in the sense that all payoffs were increased by a factor of 5. With the 5× payoff scale, for example, the upside risk option A would have a high payoff of $125 instead of $25 shown in table 4.3. In half of the sessions, the 5× treatment was first, and in the other half, the 5× payoff followed the 1× payoff scale. There were equal numbers of men and women, balanced across treatment orders.

The main conclusion is that subjects who tend to select the safer option when faced with the downside risk of a low payoff also tend to select the riskier option when faced with the upside risk of a high payoff. This is despite the fact that the expected payoff differences between the options are the same in each case (80 cents). This conclusion is indicated by the proportions of risky choices shown in figure 4.3, where the bars for the downside risk questions (first and third clusters) are uniformly lower than for comparable upside risk questions. There is a large gender difference for pairs of dark and gray bars on the left (downside risk, 1× scale) as indicated by a comparison of the proportions of risky choices for males (dark bars) and females (gray bars). This gender

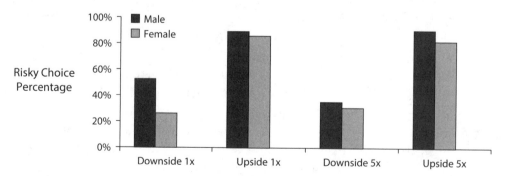

Figure 4.3. Male versus Female Percentage of Risky Choices by Risk Type (upside or downside) Payoff Scale (1× scale or 5× scale)

difference turns out to be statistically significant, which is not the case for any of the small gender differences for the other three treatment categories (upside risk with low stakes and both upside and downside with high stakes).

Upside and Downside Risk Experiment Results: *The proportions of risky choices are higher for low-probability upside risk than is the case for low-probability downside risk. With downside risk, male subjects tend to select the riskier option more often than females with low stakes, but this difference goes away with high stakes. With upside risk, there is no significant difference between male and female proportions of riskier choices, regardless of payoff scale.*

This study provides another example of how gender differences in risk aversion can depend on context, especially if different factors (utility curvature and probability weighting) are involved. The authors estimated risk aversion and probability weighting parameters for the Kahneman and Tversky weighting function in equation (4.1). The weighting parameter estimates for both men and women were very similar (with weights of about 0.7), but the utility function estimates indicated more risk aversion for women.

4.7 Extensions and Further Reading

As mentioned above, the dominant method of modeling choice under risk in economics and finance involves expected utility, applied either to gains and losses or to final wealth. The final wealth approach involves a stronger type of rationality in the sense that people can see past gains and losses and focus on the variable that determines consumption opportunities (final wealth). The Camerer (1989) and Battalio et al. (1990) experiments provide strong evidence that decisions are framed in terms of gains and losses, and that people do not "integrate" gains and losses into a final asset position. Indeed, there is little if any experimental evidence for such "asset integration." Rabin (2000) and Rabin and Thaler (2001) also provide a theoretical argument against using expected utility as a function of final wealth, the argument being that the risk aversion needed to explain choices involving small payoffs implies absurd levels of risk aversion for choices involving large amounts of money. Most analyses of risk aversion in laboratory experiments have, in fact, always been done in terms of gains and losses (Binswanger, 1980; Kachelmeier and Shehata, 1992; Goeree, Holt, and Palfrey, 2002, 2003).

Even if expected utility is modeled in terms of gains and losses from a reference point, there is the issue of whether to incorporate other elements like nonlinear probability weighting and loss aversion. The intuition behind loss aversion is quite appealing, but there are many cases in which it is not apparent. Part of the problem in implementation is that the appropriate reference point is not always obvious. For example, the presence of a safe option in the Eckel and Grossman investment task might mean that payoffs below this level are coded as losses, even if none of the actual payoffs in the no-loss treatment were negative.

Some, like Camerer (1995), have urged economists to give up on expected utility theory in favor of prospect theory and other alternatives. Rabin and Thaler (2001) expressed the hope that they had written the final paper that discusses the expected utility hypothesis, referring to it as the "ex-hypothesis" with the same tone that is sometimes used in talking about an ex-spouse. Other economists like Hey (1995) maintain that the expected utility model outperforms the alternatives, especially when decision errors are explicitly modeled in the process of estimation. Such errors would permit decisions to go in either direction due to randomness, but a preponderance of decisions should be in the direction of higher expected utility. It should be noted, however, that the incentivized Allais paradox experiment summarized in table 4.2 cannot be explained with decision errors. In spite of all of the controversy, expected utility continues to be widely used, either implicitly by assuming risk neutrality or explicitly by modeling risk aversion.

Some may find the mixed evidence on these issues to be worrisome, but to an experimentalist it provides an exciting area for new research, especially for important high-stakes decisions. One way to run such experiments is to go to countries where using high incentives would not be so expensive. For example, Binswanger (1980) studied the choices of farmers in Bangladesh when the prize amounts sometimes involved more than a month's salary. Similarly, Kachelmeier and Shehata (1992) performed lottery-choice experiments in rural China with high payoffs, and subsequently in the United States and Canada. They found that the method of asking the question has a large impact on the way people value lotteries. When people were asked for a selling price (the least amount of money they would accept to sell the lottery), people tended to give a higher answer, on average. Such answers indicate a high value for the risky lottery, and hence a preference for risk. In contrast, when the *same people* were asked to specify the most they would be willing to pay for a risky lottery, they tended to give a much lower number, which would seem to indicate risk aversion. The incentive structure was such that the optimal decision was to provide a "true" money value in both treatments (similar elicitation tasks will be explained in the next chapter on belief elicitation). Despite the truthful elicitation incentives, people seem to go into a bargaining mode when presented with a pricing task, demanding high selling prices and offering low buying prices. The nature of this *willingness-to-pay/willingness-to-accept bias* (WTP-WTA) is not well understood, at least not beyond the simple bargaining mode intuition provided here (Coursey, Hovis, and Schulze, 1987). Nevertheless, it is important for policy makers to be aware of WTP/WTA biases. Studies of non-market goods (like air and water quality) may have estimates of environmental benefits that could vary by 100% depending on how the question is asked. Given the strong nature of this WTP/WTA bias, it is usually advisable to avoid using market pricing terminology when eliciting valuations.

An alternative explanation of the WTP/WTA bias is based on an *endowment effect*, i.e., being "endowed" with a commodity tends to increase a person's value for the commodity. The presence of an endowment effect seems more credible when the commodity is a physical object, like a coffee mug. A number of additional biases have been documented in the psychology literature on judgment and decision making, including loss aversion that plays such a major role in prospect theory. For example, there may be a tendency for people to be overconfident about their judgments. A related notion is the idea of *confirmation bias*, which is a tendency to seek out and recall information that conforms to previous beliefs. Bayes' rule, discussed in the next chapter, provides an unbiased statistical procedure for combining prior beliefs and new information. In contrast, a confirmation bias would involve paying too much attention to new information that confirms prior beliefs, and a neglect of contrary evidence. Some of the systematic types of judgmental errors will be discussed at length in later chapters, such as the "winner's curse" in auctions for prizes of unknown value. For further discussion of anomalies, see Camerer (1995).

Finally, it should be noted that the original version of prospect theory is a behavioral or *descriptive* theory, not a *prescriptive* theory about how decisions should be made. In contrast, expected utility is a prescriptive theory for making optimal decisions under some specific conditions that satisfy the axioms that were used to derive it. A behavioral theory that involves biases (like overweighting low probabilities) may yield implausible predictions in some circumstances, and this is the case for prospect theory. It is widely known, for example, that for any utility theory involving nonlinear probability weighting, it is possible to specify two lotteries for which one dominates the other and yet the prediction (sum of products of weighted probabilities and associated utilities) is that the dominated lottery is selected. (Roughly speaking, a lottery is dominated by another if it yields payoffs that are no higher and sometimes lower than the dominating lottery.) Kahneman and Tversky were well aware of this issue, and they suggested an "editing" phase in which dominated prospects are first removed before the others are compared. While this editing process may seem ad hoc, to the author it seems like a reasonable compromise to make with a behavioral theory. There is a technical fix for the dominance issue that was developed in the economics literature. This alternative was incorporated into prospect theory by (Tversky and Kahneman, 1992) in their paper on *cumulative prospect theory*. The main features of this theory are described in the appendix that follows, which is more technical.

Appendix: A Note on Cumulative Prospect Theory

The cumulative prospect theory approach uses the weighting function to weight *cumulative* probabilities instead of weighting individual probabilities. In other words, the fix essentially treats the weighting function $w(p)$ like a cumulative distribution function that rises from 0 to 1, so that the vertical increments sum to 1, and these increments are used to weight the associated utilities. (Then the weights are calculated as differences in $w(p)$, which is analogous to calculating probabilities as differences in a cumulative probability distribution function.) This alternative is widely used (de Palma et al., 2008), especially by behavioral economists who estimate parameters of prospect theory models (as with Comeig et al., 2016).

The cumulative prospect theory fix is implemented by first ranking the possible payoffs for a gamble from high to low, where the subscript indicates the rank, so the largest is x_1. Then the probability weight is assigned to $U(x_1)$ in the normal manner, as $w(p_1)$, e.g., using one of the weighting functions in equation (4.1). To ensure that the weights sum to 1, the other weights must sum to $1 - w(p_1)$. So if there are two possible payoffs for a gamble, x_1 and x_2, with probabilities p_1 and $1 - p_1$, the weights would be $w(p_1)$ and $1 - w(p_1)$. Hence the expected value of the weighted utility of the gamble would be: $w(p_1)U(x_1) + [\mathbf{1 - w(p_1)}]U(x_2)$ with cumulative weighting, instead of $w(p_1)U(x_1) + \mathbf{w(p_2)}U(x_2)$ in the original version of prospect theory, where the difference is shown in bold. This seems reasonable, if the probability of the largest payoff is low, it will be overweighted, and the high probability of the other payoff will be underweighted. In this case, it probably does not matter which version of prospect theory is used to generate theoretical predictions, although weighting parameter estimates will differ.

For the case of more than two possible payoffs, the Cumulative Prospect Theory adjustments are also done with differences (increments) in the original $w(p)$ function to ensure that the weights sum to 1. With three outcomes, the weight for the utility of the highest payoff $U(x_1)$ would be $w(p_1)$ as before. The weight for the second ranked payoff, $U(x_2)$ would be the difference $w(p_1 + p_2) - w(p_1)$, and weight for $U(x_3)$ would be the residual $1 - w(p_1 + p_2)$. Note that these weights sum 1.

To the author, the weighting procedure used in *cumulative prospect theory* seems arbitrary when there are more than two outcomes. For example, suppose that the three payoffs for a lottery are $10.01, $10, and $0, with probabilities of 1/3 each. For the standard "inverse S"-shaped weighting function that underweights 2/3, the adjusted weights could be quite different for each outcome, *even though all three probabilities are equal to 1/3*. For example, with a weighting parameter of 0.7, the Prelec weighting function in figure 4.1 would only slightly overweight the probability of 1/3, since $w(0.33) = 0.34$ in this case. As a result, the cumulative prospect theory weight for the high payoff of $10.01 would be $w(p_1) = 0.34$. The *cumulative* prospect theory weighted probability for the second highest payoff is calculated as the increment in the simple weighting function: $w(p_1 + p_2) - w(p_1)$ = $w(0.67) - w(0.33) = 0.25$. Finally, the weight associated with the lowest payoff of $0 is calculated as the final increment in the simple weighting function: $w(p_1 + p_2 + p_3) - w(p_1 + p_2) = 1 - w(p_1 + p_2) = 1 - w(0.67) = 0.41$. In this case, the three cumulative prospect theory weights (0.34, 0.25, and 0.41) are substantially different, even though the actual probabilities of the three payoffs are all equal to 1/3. It is not clear whether this difference in cumulative weights will cause problems in predicting actual behavior (problem 5), but anomalous outcomes seem to be possible. In the author's opinion, prospect theory should be thought of as a behavioral, descriptive theory and not as a prescriptive theory, so the technical "correction" is not needed. And in some cases, the technical correction might even be misleading in terms of describing actual behavior.

Chapter 4 Problems

1. Show that a risk-neutral person would prefer a 0.8 chance of winning 4,000 to a sure payment of 3,000, and that the same person would prefer a 0.2 chance of winning 4,000 to a 0.25 chance of winning 3,000. In each case, the alternative payoff is 0.

2. If the probability of the 4,000 payoff for lottery R1 in table 4.1 is replaced by 0.7, show that a risk-neutral person would prefer lottery S1.

3. (non-mechanical) In "directed search" models, the workers see the wages posted by employers and have to decide simultaneously which employer to approach for employment. If an employer has more applicants than positions, then the limited positions are allocated at random among the applicants. Suppose that there are two employers, each with one position posted, and one of the wage postings is five times higher than the other. Would you expect to see more workers apply to the high-wage position or to the low-wage position? Speculate on the possible effect of probability weighting on the nature of the directed search.

4. Verify that the two probability weighting functions in (4.1) reduce to linearity: $w(p) = p$ when $\omega = 1$.

5. (non-mechanical, open-ended) Use the observations made in the final paragraph of this chapter (or similar arguments) to design a set of gamble choices that you think might produce results that are inconsistent with cumulative prospect theory predictions, and explain the intuition behind your proposed test.

5

Bayes' Rule

Learning is an important aspect of adjustments in markets and games. The simplest setting is one in which a person starts with some initial belief about an unknown condition, e.g., whether a company will declare bankruptcy, and then observes new information, e.g., a sales report. The initial belief and the new information are somehow combined in the process of forming a new belief. This chapter pertains to the basic theory of how belief probabilities are updated after the arrival of new information, which is known as Bayesian learning. The discussion is based on a simple frequency-based or "counting" heuristic to explain *Bayes' rule*, which is a mathematical formula for updating beliefs.

> **Note to the Instructor:** The belief elicitation experiment used in this chapter to illustrate Bayes' rule biases can be run with the Veconlab "Bayes Rule" program on the Decisions menu, by selecting the BDM option. (The other options, QSR and Lottery Choice, will be discussed in chapter 6.) The web-based Bayes' rule program is quick and easy to run, and it provides automatic calculations and a graph of average elicited probabilities as a function of the Bayes' prediction.

5.1 Introduction

Suppose you have just received a test result indicating that you have a rare disease. The *base rate* or incidence of the disease for those in your socioeconomic group is a tenth of a percent (0.001). Unfortunately, the disease is life-threatening, but you have some hope because the test is capable of producing a *false positive* reading. Your doctor tells you that if you do have the disease, the test will come back positive 100% of the time, but if you do not have the disease, there is a 1% chance of a false positive. The issue is to use this information to determine your chances of having the rare disease, given a positive test result. PLEASE write down a guess on a scrap of paper NOW, so that you do not forget:

My guess for the chances in 100 of having the disease: ____

Confronted with this problem, most people will conclude that it is more likely than not that the person actually has the disease, but such a guess would be seriously incorrect. The 1% false positive rate means that testing 1,000 randomly selected people will generate about 10 positive results (1%), but on average only one person out of 1,000 actually has the disease and receives a true positive result. With 10 false positives and 1 true positive, the chances of having the disease after seeing a positive test result are only about 1/11, *even after you have tested positive with a test that is correct 99 times out of 100*. This example illustrates the dramatic effect of prior information about the "base rate" of some attribute in the population. This example also indicates how one might set up a simple frequency-based counting rule that will provide approximately correct probability calculations:

  (i)   Select a hypothetical sample of people (say 1,000).
 (ii)   Use the base rate to determine how many of those, on average, would have the disease by multiplying the base rate and the sample size (e.g., 0.001 times 1,000 = 1).
(iii)   Next, calculate the expected number of positive test results that would come from someone who has the disease, i.e., true positives. (In the example, on average one person in 1,000 has the disease and the test would pick this up, so think of this as 1 "true positive" from an infected person.
 (iv)   Subtract the expected number of infected people (ii) from the sample size to estimate the number that are not infected (1,000 − 1 = 999).
  (v)   Estimate the number of false positive test results that would come from people who are not infected. (The false positive rate is 1/100 in the example, so the number of "false positives" from people who are not infected is 999/100, which is 9.99, or approximately 10.)
 (vi)   Calculate the chances of having the disease given the positive test result by taking the ratio of the number of true positives (from ii) to the total numbers of positives, i.e., the sum of true positives (from ii) and false positives (from v). For the example, this ratio is 1/(1 + 9.99), which is about 1/11, or 9%.

Calculations such as those given above are an example of the use of *Bayes' rule*. This chapter introduces this rule, which is an optimal procedure for using prior information, like a population base rate, together with new information, e.g., a test result. As the incorrect answers to the disease question suggest, decisions and inferences in such situations may be seriously biased. A related issue is the extent to which people are able to correct for potential biases in market situations where the incentives are high and one is able to learn from past experience. And in some (but not all) situations, those who do not correct for biases lose business to those who do.

When acquiring new information, it is useful to distinguish three factors: the initial beliefs, the information obtained, and the new beliefs after seeing the information. If the initial *prior* beliefs are strongly held, then the new *posterior* beliefs are not likely to change very much, unless the new information is very good. Hence, learning involves considering both the prior belief and the new information, in terms of the reliability of each. For example, passing a lie detector test will not eliminate suspicion if the investigator is almost positive that the suspect is guilty. On the other hand, very good information may overwhelm prior beliefs, as would happen with the discovery of DNA evidence that clears a person who has already been convicted. Bayes' rule provides a mathematical way of handling diverse sources of information. The Bayesian perspective is useful because it dictates how to evaluate different sources of information, based on their reliability. This perspective may even be valuable when it is undesirable or inappropriate to use certain types of prior information, e.g., how crime rates differ by race or other demographics in a jury trial for a serious crime. In such cases, knowing how prior information would be used (by a Bayesian) may make it easier to guard against a bias derived from such information. Finally, Bayesian calculations provide a benchmark from which biases can be measured.

5.2   A Simple Example and a Counting Heuristic

The simplest informational problem is deciding which of two possible situations or "states of nature" is relevant, e.g., guilty or innocent, infected or not, defective or not, etc. To be specific, suppose that there are two cups or "urns" that contain equally sized amber (*a*) and blue (*b*) marbles, as shown in figure 5.1. Cup A has two ambers and one blue, and cup B has one amber and two blues. A fair coin is flipped to choose one of these cups. The cup selection is hidden, so the prior information is that each cup is equally likely. Then the subject is shown several draws of marbles from the selected cup. Each time a draw is made, it is returned to the cup, so draws are *with replacement* from a cup with contents that do not change.

Suppose the first draw is amber and you are asked to report the probability that cup A is being used, which will be represented as $\Pr(A|a)$. An answer of 1/2 is sometimes encountered, since it can be justified by the argument that each urn was equally likely to be selected. But if each cup was equally likely beforehand, what was learned from the draw?

Another commonly reported probability for cup A following an amber draw is 1/3, since this cup has a one-half chance of being used, and if used, there is a 2/3 chance of drawing an amber. Then we multiply 1/2 and 2/3 to get 1/3. *A little math can be dangerous!* This 1/3 probability is clearly wrong, since the cups were

Cup A   Cup B

a, a, b   *a, b, b*   Figure 5.1. A Two-Cup Example

equally likely ex ante, and the amber draw is more likely when cup A is used, so the probability of cup A should be greater than 1/2 after seeing an amber draw. Another problem with this answer is that analogous reasoning requires the probability of cup B to be 1/2 times 1/3, or 1/6. This yields an inconsistency, since if the probability of cup A is 1/3 and the probability of cup B is 1/6, where does the rest of the probability go? These probabilities (1/3 and 1/6) sum to 1/2, so we should double them (to 2/3 and 1/3), which are the correct probabilities for cups A and B after the draw of one amber marble. This scaling up of probabilities will be seen later as a part of the mathematical formula for Bayes' rule (the "pesky denominator" that you will soon encounter).

A close look at where the amber marbles are in figure 5.1 makes it clear why the probability of cup A being used is 2/3 after the draw of an amber marble.

There are two marbles labeled *a* on the left and one on the right. All six marbles are equally likely to be drawn before the die is thrown to select one of the cups. No amber marble is more likely to be chosen than any other, and two of the three amber marbles are in cup A. It follows that the posterior probability of cup A given an amber draw is 2/3. In other words, there are two *true positive a* marbles (in the A cup) and one *false positive a* marble (in the B cup), so the probability of the A cup is the ratio of the number of possible true positives to the number of all positives, whether true or not, or $Pr(A|a) = 2/3$.

The calculations in the previous paragraphs are a special case of Bayes' rule with equal prior probabilities for each cup. Now consider what can be done when the probabilities are not equal. In particular, suppose that the first marble drawn (amber) is returned to the cup, and the decision maker is told that a second draw is to be made from the *same* cup (A or B) originally selected by the throw of the die. Having already seen an amber, the person's beliefs *before* the second draw are that the probability of cup A is 2/3 and the probability of cup B is 1/3, so the person thinks that it is twice as likely that cup A is being used after observing one *a* draw.

The next step is to figure out how the previous paragraph's method of counting marbles (when the two cups were initially equally likely) can be modified for the new situation where one cup is twice as likely as the other one. In order to create a situation that corresponds to the new beliefs, we want to somehow get twice as many possible draws coming from the cup that is twice as likely. Even though the physical number of marbles in each cup has not changed, we can represent these beliefs by thinking of cup A as having twice as many marbles as cup B, *with each marble in either cup having the same chance of being drawn*. These posterior beliefs are represented in figure 5.2, where the proportions of amber and blue marbles are the same as they were in cups A and B respectively. Even though the physical number of marbles is unchanged at six, the prior corresponds to a case in which the marbles in figure 5.2 (real and imagined) are numbered from one to nine, with one of the nine marbles chosen randomly.

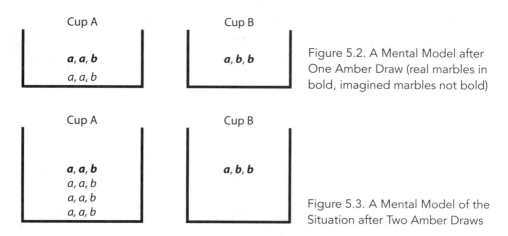

Figure 5.2. A Mental Model after One Amber Draw (real marbles in bold, imagined marbles not bold)

Figure 5.3. A Mental Model of the Situation after Two Amber Draws

When the posterior beliefs after an amber draw are represented in figure 5.2, it is clear that a blue on the second draw is equally likely to have come from either cup, since each cup contains two marbles with "b" labels. Thus, the posterior probability for cup A after a blue on the second draw is 1/2. This result is consistent with intuition based on symmetry, since the prior probabilities for each cup were initially 1/2, and the draws of an amber (first) and a blue (second) are balanced. A mixed sample in the opposite order (blue first, then amber) would, of course, have the same effect.

Suppose instead that the two draws were amber, with the two cups being equally likely to be used ex ante. As before, the posterior belief after the first amber draw can be represented by the cups in figure 5.2. Since four of the five (real or imagined) amber marbles are in cup A, the posterior probability of cup A after seeing a second amber draw is 4/5. After two amber draws, cup A is, therefore, four times as likely as cup B, since 4/5 is four times as large as 1/5. To represent these posterior beliefs in terms of colored marbles that are equally likely to be drawn, we need to have four times as many rows on the cup A side of table 5.2 as there are on the cup B side. Thus, we would need to add two more imagined rows of three marbles under cup A in figure 5.2, holding the proportions of amber and blue marbles fixed. Doing this would provide the representation in figure 5.3.

At this point, you might consider the probability of cup A after seeing two ambers and a blue. To answer this question, you should consider how many "b" labels, real or imagined, are in cup A in figure 5.3, and how many are in cup B.

5.3 Relating the Counting Heuristic to Bayes' Rule

Up to this point, the analysis has been intuitive, but it is now time to be a little more analytical. Some more notation will help make the connection between the counting heuristic and a mathematical formula for Bayes' rule. Suppose there are N marbles in each cup. The marbles will be a or b as before, and we will use the letter *s* to represent a specific color, so the "sample" *s* can be either *a* or *b*, i.e., amber or blue. What we want to know is the probability of cup A given the draw of a marble of color *s*. When *s* is amber and the contents are as shown in figure 5.1, we already know the answer (2/3), but our goal here is to find a general formula for the probability of cup A given a draw of color *s*. This probability is denoted by Pr(A|*s*), which reads "the probability of A given *s*." This formula should be general enough to allow for different proportions of colored marbles, and for differences in the prior probability of each cup.

Consider Pr(*s*|A), the probability of *s* given A, which reverses the order of the A and the *s*. Thus Pr(*s*|A) denotes the fraction of marbles in cup A that are of color *s*, where *s* is either amber or blue. Similarly, P(*s*|B) is the fraction of marbles in cup B that are of color *s*. For example, if there are ten marbles in cup A and if Pr(*s*|A) = 0.6, then there must be six marbles of color *s* in the cup (calculated as 0.6 times 10). In general, there are a total of Pr(*s*|A)N marbles of color *s* in cup A, and there are Pr(*s*|B)N marbles of color *s* in cup B. If each cup is equally likely to be selected, then each of the 2N marbles in the two cups is equally likely to be drawn ex ante (before the cup is selected). Suppose the marble drawn is of color *s*. In other words, *s* is the "sample" information. The posterior probability that the cup is A given that the draw is *s*, denoted Pr(A|*s*), is just the ratio of the number of color *s* marbles in cup A to the total number of marbles of this color in both cups:

(5.1) $$\Pr(A|s) = \frac{\text{Number of Color } s \text{ Marbles in Cup A}}{\text{Number of Color } s \text{ Marbles in Both Cups}},$$

which can be expressed:

(5.2) $$\Pr(A|s) = \frac{\Pr(s|A)N}{\Pr(s|A)N + \Pr(s|B)N}.$$

You can think of the numerator of equation (5.2) as the number of true positives, and the sum in the denominator as being the total number of positives, true and false. It is worth emphasizing that this formula is only valid for the case of equal prior probabilities and equal numbers of marbles in each urn. Nothing is changed if we divide both numerator and denominator of the right side on (5.2) by 2N, which is the total number of marbles in both cups, which yields a formula for calculating the posterior when the priors are 1/2:

(5.3) $$\Pr(A|s) = \frac{\Pr(s|A)\left(\frac{1}{2}\right)}{\Pr(s|A)\left(\frac{1}{2}\right) + \Pr(s|B)\left(\frac{1}{2}\right)} \quad \text{(for priors of ½)}.$$

A person who has seen one or more draws may not have prior probabilities of 1/2, so this formula must be generalized. This involves replacing the (1/2) terms on the right side of the equation with the new prior probabilities, denoted Pr(A) and Pr(B). This is Bayes' rule:

(5.4) $Pr(A|s) = \dfrac{Pr(s|A)\,Pr(A)}{Pr(s|A)\,Pr(A) + Pr(s|B)\,Pr(B)}$ (Bayes' rule).

For the previous example with equal priors, $Pr(A) = 1/2$, $Pr(\boldsymbol{a}|A) = 2/3$, and $Pr(\boldsymbol{a}|B) = 1/3$, so equation (5.4) implies that the posterior probability following an amber draw is:

$$Pr(A|a) = \frac{\frac{2}{3}\frac{1}{2}}{\frac{2}{3}\frac{1}{2} + \frac{1}{3}\frac{1}{2}} = \frac{1/3}{1/2} = \frac{2}{3}$$

Similarly, the probability of cup B is calculated: $Pr(B|\boldsymbol{a}) = (1/6)/(2/6 + 1/6) = 1/3$. Notice that the denominators in both of the previous calculations are 1/2, so dividing by 1/2 scales up the numerator by a factor of 2, which makes the probabilities add up to one. This is the rationale for the denominator that is often forgotten.

To summarize, if there is a prior probability of 1/2 that each cup is used, and if the cups contain equal numbers of colored marbles, then the posterior probabilities can be calculated as ratios of numbers of marbles of the color drawn, as in equation (5.1). If the marble drawn is of color s, then the posterior that the draw was from cup A is the number of color s marbles in cup A divided by the total number of color s marbles in both cups. When the prior probabilities or numbers of marbles in the cups are unequal, then the 1/2 terms in (5.3) are replaced by the prior probabilities, as in Bayes' rule (5.4). In general, the sample s could consist of multiple draws with replacement, and the formula in (5.4) still applies. To summarize:

Bayes' Rule: *This probability of an event A after seeing the sample information s in (5.4) is the ratio of the chances of seeing sample information s when A is true to the chances of seeing s whether or not A is true.*

For example, suppose that the sample consists of two $\boldsymbol{a}$ draws and one $\boldsymbol{b}$ draw, with replacement, from the cups in figure 5.1. Thus the observed sample s is either $\boldsymbol{aab}$, $\boldsymbol{aba}$, or $\boldsymbol{baa}$ depending on the order of the outcomes. Each of these outcomes has a probability that is computed as a product of the probabilities for each independent draw conditional on cup A. For example, $Pr(\boldsymbol{aab}|A) = (2/3)(2/3)(1/3) = 4/27$. Since there are three outcomes listed above with only a single $\boldsymbol{b}$ draw, $Pr(s|A) = 3(2/3)(2/3)(1/3) = 4/9$. A similar consideration of the three outcome orders can be used to show that $Pr(s|B) = 3(1/3)(1/3)(2/3) = 2/9$. Substituting these into Bayes' rule (5.4) and using the probabilities of 1/2 yields the correct answer of $Pr(A|s) = 2/3$ when the sample is two $\boldsymbol{a}$ draws and one $\boldsymbol{b}$ draw. (The intuition is: since each cup is equally likely and the contents are symmetric, one of the $\boldsymbol{a}$ draws "cancels" one of the $\boldsymbol{b}$ draws, so the posterior probability of cup A after seeing two *a* draws and one *b* is the same as it would be for a *single* draw that turned to be an *a*.) More practice is provided with problems 1 and 2 (with suggestions listed in appendix 1).

5.4   Experimental Results

Nobody would expect that something so noisy as the formation of beliefs would adhere strictly to a mathematical formula, and experiments have been directed toward finding the nature of systematic biases. The disease example mentioned earlier suggests that, in some contexts, people may underweight prior information based on population base rates.

This *base rate bias* was the motivation behind some experiments reported by Kahneman and Tversky (1973), who gave subjects lists of brief descriptions of people who were either lawyers or engineers. The subjects were told that the descriptions had been selected at random from a sample that contained 70% lawyers and 30% engineers. Subjects were asked to report the chances out of 100 that the description pertained to a lawyer. A second group was given some of the same descriptions, but with the information that the descriptions had been selected from a sample that contained 30% lawyers and 70% engineers. Respondents had no trouble with descriptions that obviously described one occupation or another. Some of the descriptions were intentionally neutral, with phrases like "he is highly motivated" or "will be successful in his career." Even though the original sample was stated to have contained 70% engineers, the modal response for such neutral descriptions involved probabilities of near a half. This observed behavior is insensitive to the prior information about the proportions of each occupation, a type of base rate bias. Notice that the same base rate bias is present in answers to the question at the beginning of the chapter in which the base rate information, 1 in 1,000 who have the disease, typically is not fully accounted for in assessing the chance of having the disease after a positive test result.

Grether (1980) pointed out several potential procedural problems with the Kahneman and Tversky experiment. There was *deception* to the extent that the descriptions had been made up. Moreover there is an *incentive* issue; even if people "bought into" the context, they would have no external motivation to think about the problem carefully. The information conveyed in the descriptions is hard to evaluate in terms of factors that comprise Bayes' rule formula. In other words, it is difficult to determine an appropriate guess about the probability of a particular description conditional on the occupation. Grether's experiments were done using bingo cages and two types of objects. One of the biases that he considered is known as *representativeness bias*. In the two-cup example discussed earlier, a sample of three draws that yields two ambers and one blue has the same proportions as cup A, and in this sense the sample looks representative of cup A. We saw that the probability of cup A after such a sample would be 2/3, and a person who reports a higher probability, say 80%, may be doing so due to representativeness bias.

Notice that a sample of two ambers and one blue makes cup A more likely. If you only ask someone which cup is *more likely*, an answer of A cannot distinguish between Bayesian behavior and a strong representativeness bias, which also favors cup A. Grether cleverly got around this problem by introducing some asymmetries in prior beliefs that make it possible for representativeness to indicate a cup that is less likely given Bayes' rule. In this manner, a sample of two ambers and one blue would look like the contents of cup A, but if the prior probability of A is small enough, the Bayesian probability of cup A would be less than one-half. Therefore, representativeness and Bayes' rule would have differing predictions when a person is asked which cup is more likely.

A binary choice question about which cup is more likely makes it easy to provide incentives: simply offer a cash prize if the cup actually used turns out to be the one the person said is more likely. This is the procedure that Grether used, with a $15 prize for a correct prediction and a $5 prize otherwise. When representativeness and Bayesian calculations indicated the same answer, subjects tended to give the correct answer about 80% of the time (with some variation depending on the specific sample). This percentage fell to about 60% when representativeness and Bayes' rule suggested different answers.

If base rates are low and a person is prone to probability weighting as discussed in the previous chapter, a low base rate will be overweighted, i.e., perceived as being higher than the actual probability determined by Bayes' rule. Of course, nonlinear probability weighting might also affect other probability perceptions that are used in Bayes' rule calculations, e.g., probabilities associated with a given sample for a particular event. To summarize:

> **Informational Biases:** *A base rate bias refers to the tendency to ignore or underweight prior information about population averages or "base rates." A low base rate might also be overweighted due to nonlinear probability weighting that is determined by the inverse-S shaped weighting function. Representativeness bias refers to the tendency to assign probabilities that are too high (relative to Bayes' rule) when the sample looks like (or is representative of) the population for one of the events, even though the prior probability of that event may be small.*

5.5 Bayes' Rule with Elicited Probabilities

Sometimes it is useful to ask subjects to report a probability, instead of just saying which event is more likely. This can be phrased as a question about the "chances out of 100 that the cup used is A." The issue here is how to provide incentives for people to think carefully. A useful approach, pioneered by Becker, DeGroot, and Marschak (1964), is a way of eliciting utility that was later adapted for eliciting a probability. Subjects often find this "BDM" method to be confusing, but it is based on a simple idea:

> Suppose that you send your friends to a fruit stand, and they ask you whether you prefer apples or oranges in case both are available. You would have no incentive to lie about your preference, since telling the truth allows your friend to make the best decision on your behalf.

The BDM method asks a subject to report the "chances in 100" for some event that will later either occur or not. This procedure makes a choice for the subject on the basis of the submitted probability, and is structured to ensure that the subject should report truthfully so that the best choice (from the subject's perspective) will be made. There are two payoff methods. Using the "event lottery," the subject will be paid an amount, say $10, if the event occurs. Alternatively, the subject will be paid using a "dice lottery" that pays $10 with a probability N out of 100. The event lottery is better for the subject if the chances in 100 for the event are greater than N, and the event lottery is worse if those chances are lower than N. The experimenter is essentially in the position of the friend going to the fruit stand, which can be explained:

> You should truthfully report your belief about the chances in 100 that the event will occur, so that the procedure can select the option that turns out to be best for you, the event lottery or the dice lottery. The particular dice lottery is not known in advance; the actual value of N will be determined (after you report) by two throws of a 10-sided die, first for the tens digit and second for the ones digit.

To convince yourself that the subject is motivated to tell the truth in this situation, consider what might happen otherwise. Suppose that the subject thinks the chances for the particular event, e.g., the cup used is A, are 50 out of 100 and incorrectly reports that the chances are 75 out of 100. Thus, the cup A lottery actually provides a one-half chance of $10. If the experimenter then throws a 7 and a 0, then $N = 70$, and the dice lottery would yield a 70% chance of $10, which is a much better prospect than the 50-50 chance based on the subject's actual beliefs. But since the subject incorrectly reported the chances for cup A to be 75 out of 100, the experimenter would reject the dice lottery and base the subject's earnings on the cup A lottery, which gives a 20% lower chance of winning. A symmetric argument can be made for why it would be a mistake to report that the chances of cup A are less than the chances that correspond to the subject's beliefs (problem 3).

Since the BDM method only requires comparisons of probabilities of a money payoff of $10, and since a higher probability is always preferred, the method is not based on any assumptions about risk aversion or preference. The avoidance of risk aversion issues is, of course, a key advantage of the BDM method, but the complexity associated with dice and event lotteries is a concern. Probability elicitation is useful when the researcher wants a numerical measure of probability, e.g., to assess a Bayesian information processing, as opposed to the qualitative data obtained by asking which event is more likely. The disadvantage is that the elicitation process itself is not perfect in the sense that the measurements may contain more "noise" due to confusion than would be the case with simple binary comparisons about which event is more likely.

Table 5.1 shows results for two subjects in a research experiment that used the BDM method, with cash prize amounts of $1.00 instead of $10.00. The two cups, A and B, each contained three marbles with the contents as shown in figure 5.1. The experiment consisted of three parts. The first part was done largely to acquaint people with the procedures, which are admittedly complicated. Then there were ten rounds with asymmetric probabilities (a two-thirds chance of using cup A) and ten rounds with symmetric probabilities (a one-half chance of using cup A). The order of the symmetric and asymmetric treatments was reversed with different groups of people.

The information and decisions for subjects 1 and 2 in this experiment are shown for rounds 21–27, where the prior probability was 1/2 for each cup. First consider subject 1 on the left. In round 21, there were no draws, and the None (0.50) in the Draw column indicates that the correct Bayesian probability of cup A is 0.50. The subject's response in the Elicited Probability column was 0.49, as is also the case for this person in round 26 where the draws, **ab**, cancelled each other out. The other predictions can be derived using the counting heuristic or with Bayes' rule, as described above. In round 22, for example, there was only one draw, **a**, and the Bayesian probability of cup A is 0.67, since two of the three **a** marbles from both cups are located in cup A. Subject 1 reports a probability 0.65, which is quite accurate. This person was unusually accurate, with the largest deviation from the theoretical prediction being in round 24, where the reported probability of 0.25 for cup A with a sample of **bab** is lower than the Bayes' rule prediction of 1/3. Which bias is exhibited here?

Subject 2 is less predictable. The **ab** and **ba** samples, which leave each cup being equally likely, resulted in answers of 0.60 and 0.30 in rounds 23 and 26. The average of these answers is not far off, but the dispersion is atypically large for such an easy inference task, as compared with others in the sample. This person was fairly accurate with single draws of **b** (round 22) and of **a** (round 25), but the three-draw samples show behavior that is consistent with "representativeness bias." In round 27, for example, the **aab** sample matches cup A, and

Table 5.1. Elicited Probabilities for Two Subjects in a Bayes' Rule Experiment

Cup A: *a, a, b*			Cup B: *a, b, b*		
Subject 1			Subject 2		
Round	Draw (Bayes)	Elicited Probability	Round	Draw	Elicited Probability
21	None (0.50)	0.49	21	None (0.50)	0.50
22	*a* (0.67)	0.65	22	*b* (0.33)	0.30
23	*bb* (0.20)	0.18	23	*ba* (0.50)	0.60
24	*bab* (0.33)	**0.25**	24	*aba* (0.67)	0.70
25	*a* (0.67)	0.65	25	*a* (0.67)	0.65
26	*ab* (0.50)	0.49	26	*ab* (0.50)	0.30
27	*bba* (0.33)	0.33	27	*aab* (0.67)	**0.80**

the elicited probability of 0.80 for cup A for subject 2 is higher than the actual probability of 0.67.

Figure 5.4 shows some aggregate results for all 22 subjects in the symmetric and asymmetric treatments combined. The horizontal axis plots the Bayesian probability of cup A, which varies from 0.05 for a sample of **bbbb** in the symmetric treatment (with equal priors) to 0.97 for a sample of ***aaaa*** in the asymmetric treatment (where the prior probability of cup A was 2/3). The elicited probabilities are shown in the vertical dimension. The 45-degree dotted line shows the Bayes' prediction, the solid line connects the average of the elicited probabilities, and the dashed line connects the medians.

In the aggregate, Bayes' rule does quite well. A slight upward bias on the left side of the figure and a slight downward bias on the right side are roughly consistent with nonlinear probability weighting, as discussed in the previous chapter. Alternatively, those biases might be due to the fact that there is more "room" for random error in the upward direction on the left and in the downward direction on the right. This error-based conjecture is motivated by the observation that the medians (thick dashed line) are generally closer to Bayesian predictions. For example, one person got confused and reported a probability of

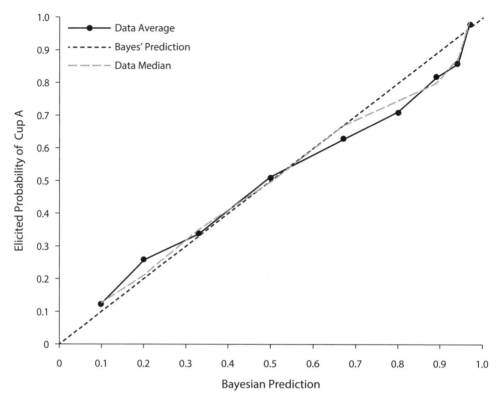

Figure 5.4. Elicited Probabilities versus Bayes' Predictions. *Source*: Holt and Smith (2009).

0.01 for cup A after observing draws of **aa** in the symmetric treatment, which should lead to a posterior of 0.8. On the right side of the figure, there is more "room" for extreme errors in the downward direction. To see this, imagine a vertical line that could be drawn up from the point 0.8 on the horizontal axis of figure 5.4. If people with no clue who put marks more or less equally spaced on this vertical line, more of the marks will be below the 45-degree dashed line, since there is more room below. Random errors of this type will tend to pull average reported probabilities down below the 45-degee line on the right side of the figure, and the reverse effect (upward bias) would occur on the left side. Averages are much more sensitive to extreme errors than are medians, which may explain why the averages show more of a deviation from the 45-degree dashed line.

5.6   A Follow-up Experiment with a Rare Event

The data from the previous section involved prior probabilities in the 0.33 to 0.66 range, and in the aggregate, Bayes' rule provides reasonable predictions when subjects have a chance to make decisions and learn. This raises the issue of how subjects do when the prior probability of the event is low, as was the case for the rare disease example of base rate bias at the beginning of the chapter. For this test, a prior rate of 0.04 for the cup A was used, and this "amber cup" only contained amber (**a**) marbles. On average, in 100 trials there should be four times that cup A used, and the draw would be **a** each time, since it only contains amber marbles. Therefore, in 100 trials, there would be four true positives. Thus cup B would be used 96 times on average in 100 trials. In order to obtain a posterior of only 0.1 (as with the initial disease example), there would have to be a 9-to-1 ratio of false positives to the four true positives, so there would have to be 36 false positives from the 96 times that cup B is used. Therefore, the false positive rate was set at $36/96 = 3/8$. In other words, the probability of a draw of **a** from cup B was set to 3/8 in the experiment to be discussed.

The payoffs and procedures were similar to those used in the hand-run experiments reported in the previous section, except that the computer permitted running more rounds, and the prior probability was maintained at the same level in all three sets of 20 decisions, to avoid sequence effects. Also, prize amounts were doubled to $2. As before, subjects saw either no draws, 1 draw, 2 draws, 3 draws, or 4 draws (with replacement) prior to reporting the chances of 100 that cup A is being used. The BDM elicitation procedure was implemented with the Veconlab Bayes' rule program.

The posterior probability of cup A varies from 0 (if even a single **b** draw is ever observed) to 0.1 (a single draw that is **a**), up to 0.68 (four **a** draws in a row and no **b** draws). Figure 5.5 shows the Veconlab graph display of means and medians of all decisions made by the 24 subjects in this extreme prior treatment (60 decisions each). There is an upward bias, especially for the mean (dark line with connected dots), but the overall picture is surprisingly accurate. For example, consider the case of one **a** draw, which induces a posterior of 0.1 for red (as with the disease example at the start of the chapter, where the disease was rarer and the test was more accurate than the treatment used here). Looking vertically up from the 0.1 point on the horizontal axis of the graph, we see that the mean report was below 0.3 and the median was just slightly above the Bayes' prediction of 0.1.

One reaction to observed biases is to begin with Bayesian calculations and then model observed patterns. Grether (1992) and Goeree et al. (2007) use a generalization of Bayes' rule that raises the conditional signal probabilities to a power, say β. For the case of two events, A and B, the probabilities associated with a sample s would be raised to a power β, so the formula for Bayes' rule in equation (5.4) would have the terms, $(\Pr(s|A))^{\beta}$ and $(\Pr(s|B))^{\beta}$ in the numerator and denominator on the right side. This generalization reduces to Bayes' rule when $\beta = 1$. When $\beta < 1$, this formula would put "too much" weight on

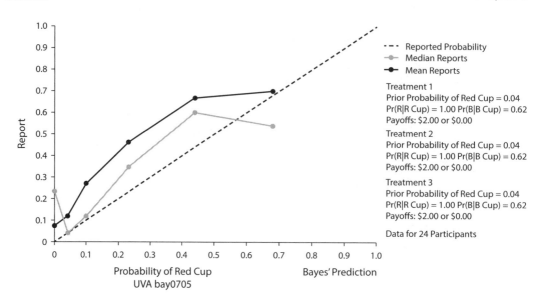

Figure 5.5. Elicited Probabilities and Bayes' Predictions for a Low Probability Event. *Key*: Mean (dark line with dots), median (gray line). *Source*: Holt and Smith (2009).

the sample relative to the prior information. For example, if $\beta = 1/2$, then the formula would require taking the square root of the signal probabilities, and the square root of a fraction is greater than the fraction itself. (To illustrate, note that $(1/2)(1/2) = 1/4$, so the square root of $1/4$ is $1/2$, which is greater than $1/4$.) Using data from a social learning ("information cascade") experiment, Goeree et al. (2007) estimate values for β that are significantly less than 1.

Holt and Smith (2009) used data from all treatments of the experiment described in this section to estimate a probability weighting function with the typical shape (overweighting for low probabilities, underweighting for high probabilities). The probability weighting function was specified to be the form suggested by Kahneman and Tversky (1979), shown in the top line of equation (4.1) in the previous chapter. The parameter estimate for the Bayes' rule experiment turned out to be $\omega = 0.713$ with a standard error of (0.024). This is essentially the same as the 0.7 parameter value that was used to construct the dotted probability weighting curve in figure 4.1. The authors also estimated a base rate bias parameter, $\beta = 1.027$, which is not significantly different from 1. As noted in the previous paragraph, this parameter value indicates the absence of a base rate bias. To summarize:

Bayes' Rule Experiment Results: *With symmetric prior probabilities of 1/2, probabilities elicited using financial incentives (BDM method) tend to track Bayes' rule predictions relatively well. When one event has a very low prior probability,* *there is an upward bias. An econometric analysis suggests that the bias is due to "normal" (inverse S) probability weighting, which overweights low probabilities.*

5.7 Extensions

It is well known that elicited beliefs may deviate dramatically from Bayes' rule predictions, especially in extreme cases like the disease example discussed in the first section. Instruction in the mathematics of conditional probability calculations may not help much, and such skills are quickly forgotten. On the other hand, the presentation of the problem in terms of frequencies and the use of counting heuristics helps people make good probability assessments in new situations (Gigerenzer and Hoffrage, 1995, 1998; Anderson and Holt, 1996a).

Several alternatives to the BDM procedure will be discussed in the next chapter, where the focus is on possible sources of bias in the elicitation procedure. The most commonly used alternative, the "quadratic scoring rule," requires an assumption of risk neutrality or some other adjustment.

One source of bias that might be hard to model is the effect of emotion, or "affect," as it is referred to in academic literature. This *affect effect* identified by Charness and Levin (2005) is a "win-stay, lose-switch" heuristic that can cause the optimal choice under Bayes' rule to "feel wrong." Suppose, for example, there are two possible salespeople who can be sent out on a job, one of whom is unavailable the current week. The other one is sent and the sales performance comes back good. The decision for the second week is whether to send the first person back out ("win-stay") or to switch. The reason that the win-stay approach is not necessarily best is that the good performance of the first salesperson could be due to favorable market conditions, and it may be the case that the other salesperson has the experience to garner high sales under such favorable conditions. In this situation, the optimal decision could be to switch even though the first person was successful. The experiment design generated cases where the best decision was to switch after a good outcome and to stay with the initial decision after a bad outcome (see problem 7). Roughly speaking, about half of the subjects' decisions were inconsistent with this optimal decision, with deviations leaning in the direction of what is implied by the "win-stay, lose-switch" heuristic. For other documented violations of Bayes' rule, see Zizzo et al. (2000) and Ouwersloot, Nijkam, and Rietveld (1998).

Although there are systematic deviations from the predictions of Bayes' rule, there is no widely accepted alternative model of how information is actually processed in a wide array of situations. At present, economists tend to use Bayes' rule or parameterized generalizations to derive predictions, although there is some renewed interest in non-Bayesian models like that of reinforcement learning, to be discussed in chapter 7.

Problems for Chapter 5

1. Consider the setup in figure 5.1 for which the two cups are equally likely to be used, but with the contents altered so that cup A has three ambers and a blue (**aaab**), and cup B has two of each (**bbaa**). What is the probability for cup A after seeing a **b** draw?

2. Consider the setup in problem 1, but with two observed draws with replacement, which turn out to be balanced (one **a** and one **b**). What is the posterior probability for cup A?

3. Suppose we are using the BDM elicitation scheme described in section 5.5, and that the subject's beliefs after seeing the observed draws are that cups A and B are equally likely. Show why it would be bad for the person to report that the chances for cup A are 25 out of 100.

4. Twenty-one Virginia students participated in a classroom Bayes' rule experiment, but with only seven rounds, with cash payments ($3–$4 per person). Each cup was equally likely to be used. Participants claimed that they did have time to do mathematical calculations. The draw sequences and the average elicited probabilities are shown in the table.

    Mean and Median Elicited Probabilities for Cup A

		Cup A = {a, a, b}		Cup B = {a, b, b}			
Draw:	No Draws	a	b	bb	ab	abb	bbb
Mean	0.5	0.68	0.33	0.24	0.47	0.28	0.13
Median	0.5	0.67	0.33	0.2	0.50	0.30	0.11

    (a) Calculate the Bayes' posterior for each of the seven sample outcomes shown in the seven columns.
    (b) How would you summarize the deviations from Bayesian predictions? Is there evidence for the representativeness bias?

5. Suppose that a cook produces three pancakes, one that is burnt on one side, another that is burnt on both sides, and a third that is not burnt on either side. The cook chooses the pancake at random, with each one having equal

probability of being the one to be served. Then the cook flips the pancake high so that each side is equally likely to be the one that shows. All you know (besides the way the pancake was selected) is that the one on your plate is showing a burnt side on top. What is the probability that you have the pancake with only one burnt side? Explain.

6.   A woman of age 40 undergoes a screening for breast cancer, and the test comes back positive. The rate of previously undiagnosed breast cancer for a woman in this category is 10 per 1,000. The test is fairly accurate in the sense that if she has cancer, it will produce a positive result 80% of the time. For a woman who does not have cancer, the test will produce a positive reading in only 10% of the cases. What is the probability that the patient actually has cancer if the test is positive? (In one study, more than nine out of ten physicians in Germany gave incorrect answers to this question, and the typical answer was off the mark by a factor of 10.)

7.   Consider a decision maker who has two decisions, one Moderate and one Extreme. The best decision depends on the unknown "state" of the world, which is equally likely to be "Good" or "Bad." Think of the particular combination of state and decision as a cup with six marbles, marked H for high payoff or L for low payoff, as shown in the accompanying table. For any given cup, one marble is drawn at random to determine the payoff. For example, if the Moderate decision is taken and the state is Good, then the chances are 4 out of 6 of getting a high payoff. Suppose that you are forced to choose Moderate and the outcome is H in round 1, and there is a second (and final) round *in which the state will stay the same as in round 1*. What is the probability that the state is Good? Is it best to stay with Moderate or switch after getting the high payoff, H, initially? Explain using Bayes' rule.

	Good State (1/2)	Bad State (/2)
Extreme Decision	H H H H H H	L L L L L L
Moderate Decision	H H H H L L	H H L L L L

8.   Did subject 1 tend to exhibit representative bias in round 24 of table 5.1?

6

Belief Elicitation and Ambiguity Aversion

Subjective probabilities or "beliefs" are key components of any theory of decision making under risk. For example, a decision to attend a small committee meeting to vote on an issue might depend on one's belief about the chances that the vote will make a difference. Elicitation of beliefs is important for experimental economists, and several alternative elicitation methods will be discussed, including the commonly used "quadratic scoring rule" (QSR) that was originally devised for extracting information about weather forecasts. The quadratic scoring rule involves a choice between a list of gambles over monetary payoffs, and this procedure is most effective when payoffs are small or some other device is used to induce risk neutrality. Alternatively, a simple "Lottery-Choice" (LC) menu for belief elicitation that is patterned after the risk aversion menus from chapter 3 will be described. The choice menu has variations in probabilities, with only two possible money payoffs, which avoids the need to correct or adjust for risk preferences.

The second part pertains to some applications of value and belief elicitation. For example, the money value of an object can be elicited using a price list menu of money values on one side of a table, which can be compared with the option of owning the commodity on the right. The object could be an economic commodity or a risky prospect. Value elicitation for risky lotteries would identify risk aversion if the elicited value turns out to be less than the expected value, as calculated from money payoffs and associated probabilities. When the probabilities are not known, value elicitation can be used to identify ambiguity aversion, which would be indicated if the elicited value of the ambiguous lottery is even lower than the elicited value of the lottery with known probabilities. An even more direct method of assessing ambiguity aversion is to elicit the belief associated with the event from the crossover point of a price list with probabilities.

Note to the Instructor: The Veconlab Bayes' rule program, on the Veconlab Decisions menu, has options for the quadratic scoring rule (QSR) and the Lottery-Choice (LC) menu-based belief elicitations, in addition to the BDM method.

6.1   Belief Elicitation

One way to elicit a person's beliefs is to simply ask an un-incentivized question, e.g., what are the chances in 100 that a Republican candidate will win in an up-coming election? The "chances in 100" terminology is easy to understand and does not trigger mathematical stress the way a word like "probability" might. In any case, serious financial incentives to provide an accurate answer might help reduce noise in responses due to statements of beliefs that might be "expressive" of one's political preferences. For example, a higher payoff could be promised to the person making the forecast if the reported chances are high and the candidate wins, or if the reported chances are low and the candidate loses.

The use of incentivized elicitation procedures is the norm in research experiments, but there are some problems. Making a forecast in the shadow of possible payments is itself a risky decision, so the forecast may depend on the person's risk aversion. Moreover, risk aversion is not measured with enough precision to have confidence in a procedure that requires adjusting a person's forecast for some estimate of that person's risk aversion. One solution is to use small payments to avoid triggering high-risk aversion that is associated with high stakes, and to just assume risk neutrality. Another solution is to use a procedure for which there are only two possible payments, e.g., $0 and $10, as was the case with the BDM method described in the previous chapter to elicit beliefs in a Bayes' rule experiment. With only two payoffs, there are no issues with utility curvature, since it takes 3 points to generate a curve. This chapter will discuss the implementation and application of several alternative belief elicitation methods.

6.2   Quadratic Scoring Rule Assuming Risk Neutrality

The most common approach to eliciting a probability with incentives is the quadratic scoring rule (QSR). The forecaster reports a "chances in 100" assessment, which is divided by 100 to obtain a probability denoted by R. The forecaster is paid a fixed amount of money, say $1, minus a penalty for placing high probability on an event that does *not* occur. If the reported event probability is R, the *penalty* is $(1 - R)^2$ if the event occurs and R^2 if the event does not occur. So if the person is sure that the event will occur, then the best response is to report 100 ($R = 1$) to avoid any penalty. Conversely, the best response is to report 0 if the person is sure the event will not occur. To summarize the QSR incentives:

(6.1)   Quadratic Scoring Rule Payoffs:
$$1 - (1 - R)^2 \text{ if the event occurs,}$$
$$1 - R^2 \text{ if the event does not occur.}$$

If the person's subjective probability of the event is denoted by p, then a person's expected utility would be calculated by multiplying the utilities of the payoffs in (6.1) by the associated probabilities of p and $1 - p$. The expected utility is:

(6.2) $pU[1 - (1 - R)^2] + (1 - p)U[1 - R^2]$ (expected utility with QSR)

An extremely risk-averse person would tend to choose reports close to 0.5, which would equalize the QSR payoffs determined by (6.1) at 0.75, so the forecaster would receive this amount regardless of whether or not the event occurs. In other words, extreme risk aversion would cause the forecaster to remove all risk by submitting equal forecasts at 1/2, even if the person thought one of the events was more likely. This uninformative report is obviously the worst-case outcome for the person seeking the information.

Next consider the best case, in which the forecaster is risk neutral and the U notation can be dropped from (6.2). Since the fixed payoff is 1 regardless of which event occurs, the expected payoff is 1 minus the expected penalties:

(6.3) $1 - p(1 - R)^2 - (1 - p)R^2$ (expected money payoff with QSR)

The way to maximize the expected payoff is to take the derivative of (6.3) with respect to R (the choice variable) and set this derivative equal to 0. Intuitively, the derivative is the slope, and the highest point on a hill is where the hill gets flat, i.e., at the maximum. The derivative formula for a quadratic function, $\frac{d}{dx}(x^2) = 2x$ can be used to obtain the equation in (6.4):

(6.4) $2p(1 - R) - 2(1 - p)R = 0$

The terms involving $2Rp$ cancel each other out, and the other terms can be rearranged and simplified to show that $R = p$, i.e., the expected-payoff-maximizing forecast is the person's true subjective probability.

As the previous discussion of best and worst cases suggests, the effect of risk aversion would be to pull the forecast toward 0.5, the safety point for the forecaster. Some researchers dodge this issue by instructing subjects that reporting their actual beliefs is best for them in the sense of maximizing their "expected payoff," an instruction that is technically true but obviously a little misleading since subjects will not notice the difference between utility and payoff. Researchers have even offered to show subjects a derivation of the expected payoff result afterward if they wish to see it. Most experimental economists, however, consider such instructions to be marginally deceptive.

One alternative to the quadratic scoring rule is the logarithmic scoring rule, in which a natural log function is used to construct the penalties for reporting a low probability for an event that occurs afterward. The natural log of a very low probability can be quite negative, so some way to deal with large losses is required, e.g., by putting a lower bound on the reported probability. The natural log of 0.01 is about $-\$4.61$, so a lower bound of 1 chance in 100 would work.

Table 6.1. A Simplified Quadratic Scoring Rule Menu with a $1 Fixed Payoff

Chances in 100 for Event A	Payoff for Event A	Payoff for Event B
0	0	1
10	0.19	0.99
20	0.36	0.96
30	0.51	0.91
40	0.64	0.84
50	0.75	0.75
60	0.84	0.64
70	0.91	0.51
80	0.96	0.36
90	0.99	0.19
100	1	0

If different scoring rules are being compared, it is important to apply the same cutoff to all procedures (Palfrey and Wang, 2009). Like the QSR, this scoring rule also induces truthful belief revelation if the person is risk neutral (problem 5), but not otherwise.

The quadratic (or logarithmic) scoring rule can be implemented by providing a table with 101 rows, one for each possible "chances in 100" report: 0 in 100, 1 in 100, ... 50 in 100, ...100 in 100. Each row lists the payoff if the event occurs and if it does not. Table 6.1 shows payoffs for a simplified QSR table with just 11 possible rows, and a fixed payment of $1. The payoffs were computed using the QSR formula in expression (6.1). Notice payoffs are 0.75 for each event in the 50 chances row, as indicated by the earlier discussion. This table would be extended to have more rows, e.g., 101 rows in four columns in the Veconlab Bayes' rule program when the QSR setup option is used.

To get a feel for how the QSR works, note that if one thinks event A will never occur, then the right column is relevant and the best decision is to choose 0 chances in the top row, to earn $1 with no penalty. If the subjective probability of A is 0.2, then the expected payoff is calculated by 0.2 times the event A payoff and 0.8 times the event B payoff. For the top row, this would be 0.2(0) + 0.8(1) = 0.8. The same calculation could be done for all rows to find the best response for a belief of 0.2. This best decision would be the 20 chances response, which yields 0.2(0.36) + (0.8)(0.96) = 0.84. These calculations reveal an issue with the QSR: since the payoffs are quadratic, the incentives are fairly flat near the maximum. When one's belief is 0.2, there is only a 1 cent expected payoff gain from reporting truthfully (0.2) as compared with reporting 0.3 or 0.1 (problem 2). In this case, the safe response of 50 chances removes all risk and only reduces the payoff from an expected value of 84 cents (with risk) to a sure

75 cents! Scaling up the payoffs will increase incentives, but with the drawback of enhancing the risk and possible effects of risk aversion. To summarize:

> **Quadratic Scoring Rule:** *The payoff is reduced by a penalty, which is the square of the reported probability for the event that does NOT occur. For a risk-neutral person, the optimal decision is to report one's belief truthfully. The incentives are flat near the maximum, so deviations are not very costly. Risk aversion will bias the report toward equal probabilities (50 chances in 100) that result in no risk.*

Despite the possible effects of payoff flatness and risk aversion, the quadratic scoring rule is widely used and has provided valuable insights in the laboratory and the field. Duffy and Tavits (2008), for example, used it to elicit beliefs about the probability that one's vote would make or break a tie in a voting experiment. They observed that subjects were more likely to pay a cost of voting if they placed a higher probability that their vote would be pivotal. The elicited probabilities were closer to 0.5 than observed proportions of pivotal votes, which could have been due to risk aversion or other factors. As mentioned earlier, one way to mitigate extreme effects of risk aversion is to use low payoffs, as in the Duffy and Tavits procedure, where the maximum a subject could earn by making a correct guess was 10 cents.

The risk aversion issue is considered directly by Armantier and Treich (2013) in a prize-winning paper, with payoffs scaled from low (1×) to high (10×), to high hypothetical (10×, not paid). Even the low-incentive payoffs were quite substantial, obtained by multiplying the quadratic scoring rule payoffs in (6.1) by about $9 US given the exchange rate reported in the paper. The high-incentive payoffs were ten times that level. The quadratic scoring rule was used to elicit event probabilities determined by outcomes of throws of a pair of ten-sided dice, so the actual probabilities of various events like "the two throws sum to 4" were objectively known, subject to thinking errors. This setup has the advantage of not forcing subjects into unfamiliar Bayes' rule calculations. Figure 6.1 is constructed with the objective probability on the horizontal axis and elicited probability averages on the vertical axis, so correct reports would fall along the dashed 45-degree line. The thin dark data line showing average responses with 1× payoffs is clearly too flat in the sense that it starts above the dashed line for low probabilities and ends below for high probabilities. The "pull-to-center" tendency, however, is much stronger with high (10×) payoffs, as shown by the line with dark dots. This bias is eliminated by using hypothetical payoffs in a quadratic scoring rule, where the forecast averages (gray line) are close to the 45-degree line. The responses in the hypothetical payoff treatment, however, exhibit considerably more noise, with systematically higher numbers of extreme responses (close to 0 or 1), responses on the wrong side of a half, and responses that reverse underlying objective probabilities.

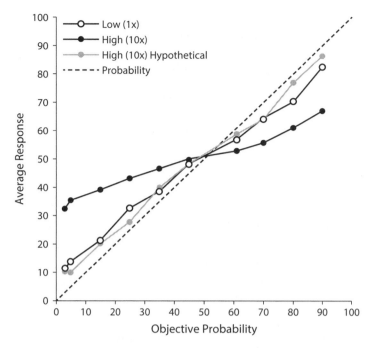

Figure 6.1. Beliefs Elicited with a Quadratic Scoring Rule for Hypothetical Payoffs, Low Payoffs (1× scale), and High Payoffs (10× scale). *Source*: Armantier and Treich (2013).

From a procedural perspective, it is important to note that the effects of payoff scale differences in this paper were accentuated by the procedure of selecting only one of 30 elicitation tasks ex post to determine a subject's earnings. When a single decision is used, there is a lot of risk in terms of final payment. Moreover, the authors had to raise the payoff scale enough to allow for the reasonable earnings levels, on average, which tends to enhance risk, and hence, strengthen any pull-to-center bias associated with a combination of risk aversion and high stakes. When the nonsensical responses for the hypothetical payoffs are considered, the best results for the Armantier and Treich data in figure 6.1 are for the low incentive treatment, which has less of a risk aversion bias than with high incentives, along with less noise than with the hypothetical incentives. The low incentives option is the approach typically followed by researchers who have used the QSR, e.g., Duffy and Tavits. To summarize:

Payoff Scale Experiment: *There is a considerable pull-to-center bias associated with using the quadratic scoring rule, especially with high payoffs that trigger risk aversion, as indicated by* figure 6.1. *This bias is not easily circumvented by using hypothetical payoffs, which introduces different biases based on noise and apparent decision error.*

The authors included other treatments that indicate additional sources of bias. For example, suppose a person has a "stake" in the outcome of an event, i.e., they have an outside payment in addition to whatever incentive is provided by the elicitation mechanism. In this case, a risk-averse person may wish to adjust their probability reports in a manner that smooths their payoffs for the two events to some extent. This type of behavior was observed in the experiment as well, especially for a treatment with a high outside payment ("high stakes").

6.3 Elicitation Methods That Are Impervious to Risk Preference

As noted in the introduction, the alternative to assuming or ignoring risk aversion is to use an elicitation task that only offers two possible payoffs, e.g., $0 and $1, which dodges the effects of utility curvature. This is the effect of the Becker, DeGroot, and Marschak (BDM) procedure discussed in the previous chapter's Bayes' rule experiment. The BDM intuition was explained previously, so the focus here is on what subjects actually see in instructions, which are necessarily devoid of context and suggestive examples. The Veconlab Bayes' rule program first describes the contents of the red cup (2 red marbles and a blue in the default settings) and the blue cup (2 blue marbles and a red), and the way draws are made with replacement. It also mentions the two possible payoffs, a high payoff of $2.00 and a lower payoff of $0. Then the N lottery is explained:

> **N Lottery Payoff Method:** The alternative method will be to use a lottery that has N chances out of 100 of providing the high payoff, where the exact number N is a randomly generated number between 0 and 100. Each value of N in this range is equally likely, so the N lottery will be undesirable if N is low and it will be desirable if N is high. The way the N lottery is played is to obtain a second random number, say B, that is equally likely to be any one of 100 integers: 0, 1, . . . 99. If the second number is less than N, then the N lottery yields the high payoff. So if $N = 1$, there is a 1 in 100 chance of getting the high payoff, if $N = 2$ there is a 2 in 100 chance, etc.

Then the main reason for truthfully reporting the "chances in 100" that the red cup is being used is explained briefly on one page and then summarized:

> **Payoffs:** The two possible payoffs in each round are: $2.00 and $0.00.
> **Payoff Method:** There is an N lottery that has N chances in 100 of providing the higher payoff. There is a Red Cup Lottery that provides the higher payoff if the red cup is used. The computer will select the best payoff method based on your assessment (P) of the chances in 100 that the red cup is being used this round. In particular, the N lottery will only be relevant if it provides a greater chance of the high payoff.

Helpful Hint: Your decision should be your best guess about the chances out of 100 that the red cup is being used, so that the computer can select the payoff option, *N* lottery or red cup lottery, that is best for you.

Grether's (1992) seminal Bayes' rule experiments were run both with and without BDM incentives, using bingo cages and transparent randomization procedures. The cages had different proportions of colored bingo balls. After seeing a draw, but not which cage was used, the subject would be asked which cage was most likely to have been used. Subjects in some sessions were also asked to provide a probability assessment of the chances that one cage is being used, with the BDM incentives to induce truthful revelation. Behavior was responsive to relevant sample information as predicted, but there were three times as many "nonsense or incoherent responses" without monetary incentives, e.g., putting a higher probability on the cage previously designated as being *less* likely.

Other types of bias have also been observed with BDM. One possibility is that responses could be biased upward if the process is perceived to be analogous to the selection of a selling price, i.e., a "willingness to accept" bias, but this is avoided by using neutral terminology that abstracts away from buying or selling. There was some noise and heterogeneity in observed behavior, and Grether's (1992) regressions yielded somewhat low correlations between actual probabilities and those elicited with the BDM procedure.

Despite the great effort taken in the preceding paragraphs to explain why the BDM method induces truthful reporting, the reader who struggled while reading it will not be surprised to hear that subjects often tense up trying to figure out what to do. In experiments with children, the BDM method has even been reported to cause some tears (based on unofficial reports by authors). One researcher even went so far as to say that "BDM is a dirty word in experimental economics." To summarize:

> **BDM Belief Elicitation:** *This procedure requires subjects to name a probability cutoff for determining whether to be paid if an event occurs (the Event Lottery) or whether to be paid with Random Lottery opportunity that arises after the belief is reported. The BDM method induces truthful belief reporting, without any need to rely on assumptions about risk neutrality. This property also helps the researcher guide subjects by telling them that truthful reporting is in their best interests. The downside is that BDM procedures may be difficult for subjects to comprehend.*

The abstraction of the BDM procedure can be made more transparent by presenting subjects with a structured set of pairwise choices, in the spirit of the Holt and Laury risk aversion menu. In particular, the choice menu is structured to implement the intuition from Savage (1971) that the subjective probability

Table 6.2. Lottery-Choice Menu: Initial Coarse Grid Table

Random Lottery			Red Cup Lottery
0 chances in 100 of $2.00	☐	☑	$2.00 if the Red Cup was used
10 chances in 100 of $2.00	☐	☐	$2.00 if the Red Cup was used
20 chances in 100 of $2.00	☐	☐	$2.00 if the Red Cup was used
30 chances in 100 of $2.00	☐	☐	$2.00 if the Red Cup was used
40 chances in 100 of $2.00	☐	☐	$2.00 if the Red Cup was used
50 chances in 100 of $2.00	☐	☐	$2.00 if the Red Cup was used
60 chances in 100 of $2.00	☐	☐	$2.00 if the Red Cup was used
70 chances in 100 of $2.00	☐	☐	$2.00 if the Red Cup was used
80 chances in 100 of $2.00	☐	☐	$2.00 if the Red Cup was used
90 chances in 100 of $2.00	☐	☐	$2.00 if the Red Cup was used
100 chances in 100 of $2.00	☑	☐	$2.00 if the Red Cup was used

of an event is defined to be the number p for which a person is indifferent between getting paid $1 if the event occurs and getting $1 with probability p. Holt and Smith (2016) used this approach by presenting subjects with a menu of choices between being paid $2 if the event occurs or being paid $2 with probability p, where p was incremented in each subsequent row of the list shown in table 6.2. In their symmetric design, the event was the cup, red or blue, that was used, with two red marbles and one blue in the red cup and two blues and one red in the blue cup.

The Red Cup Lottery is listed in all rows on the right. When the Random Lottery on the left offers 0 chance of payment (top row), the subject would typically choose the Red Cup Lottery on the right, which is pre-marked. When the Random Lottery on the left offers 100 chances in 100 of the $2 payoff (bottom row), the subject would prefer that sure $2 payment to the Event Lottery (as marked).

A subject should come down the right side of table 6.2 and switch to the left when the Random Lottery offers a higher chance of payoff than their beliefs for the Event Lottery. Subjects made decisions separately for all rows, to encourage careful thinking, but if the decisions did not implement a single crossover, they were sent back to re-choose. A complicated menu with too many rows may tend to yield extreme or noisy responses, and therefore, the authors used a two-stage approach, with a coarse grid followed by a fine grid in the range of the initial crossover point. If the subject crossed over when the Random Lottery offers 60 chances in 100, then the fine grid lottery would be analogous to table 6.2, but with the 11 rows having Random Lotteries offering 50, 52, 53, . . . 60 chances of the $2 payoff. Some others have used coarse grids (Andreoni and Sanchez, 2014, and Trautmann and van de Kuilen, 2015a), but the fine grid is especially useful for belief elicitation when the subject may have very precise beliefs based on probability calculations.

Notice that choice menus only have two money payoffs in each row (the "$0 otherwise" payoff is spelled out in the instructions). In this manner, the elicitation of a *probability equivalent* crossover point can be used to estimate a person's subjective belief in a manner that is understandable and yet does not require a separate procedure for estimating and removing effects of nonlinear risk preferences. Instead of choosing a single chances-in-100 number, the subject makes a choice between the event and random lotteries for each row, with the relevant row selected ex post to determine payoffs. If the relevant row turns out to be the crossover row for the coarse grid, a randomly selected row for the fine grid is used.

> **Lottery-Choice Menu Elicitation:** *These menus help subjects make decisions, line by line, about whether they prefer to be paid if an event occurs or with a Random Lottery. The subject has an incentive to cross over from one side of the menu to the other in a row that corresponds to their subjective belief. This method is simpler and more transparent than a BDM procedure with the same incentives, and like BDM, it does not require an assumption of risk neutrality.*

6.4 A Comparison of Elicitation Methods

The Lottery-Choice menu was used in an experiment with a symmetric setup, in which the red cup had two red marbles and one blue, and the blue cup had two blues and one red. Each cup had a 1/2 chance of being used, so the priors are also symmetric. Subjects made 30 decisions, with up to three draws with replacement from the cup being used for that round. In this between-subjects design, each person used the same elicitation procedure in all rounds, with equal numbers of subjects using the Lottery-Choice menu, BDM (report a number), and QSR. Some researchers have recently used a quadratic scoring rule without "chances in 100" terminology, and this "QSR Numbers" treatment was added as a fourth treatment at the suggestion of a reviewer. The justification for the QSR Numbers treatment is that the use of "chances in 100" may be misleading in that case, since the optimal report under QSR need not be one's true belief about the chances that the red cup is used. This logic does not apply to the Lottery-Choice and BDM methods, which were done with chances terminology, as was the "QSR Chances" treatment.

The predictions for all four methods were clustered along the 45% line in graphs with the correct Bayesian decision on the horizontal axis. The left

Table 6.3. Pairwise Comparisons of Performance Measures

	Correct Decisions Within ±1	Average Absolute Deviation	Boundary Decisions (0 or 100)	Wrong Side of One-Half	Significance Tests for Paired Comparisons
Lottery-Choice	36%**	7%	3%**	5%	** (0.05) for Test
BDM	20%	10%	10%	8%	of Lottery Choice vs. BDM
QSR Chances	17%	9%*	15%	3%	* (0.1) for test
QSR Numbers	11%	12%	9%	6%	of Chances vs. Numbers

column of table 6.3 shows that the Lottery-Choice Method had the highest percentage of correct predictions (within 1%). Individuals in all treatments faced the same sequence of random draws and cups used, so the average absolute deviation from Bayes' prediction is also a useful summary measure. The average absolute deviation shown in the second column was lowest for the Lottery-Choice menu, although these differences are not statistically significant, except that the Lottery-Choice and QSR Chances treatments are significantly lower than the QSR Numbers. Another dimension in which the Lottery-Choice menu performed well is in terms of the incidence of boundary decisions, which are never optimal. The proportion of boundary decisions (i.e., 0 or 100), shown in the third column, was lowest for Lottery Choice and highest for QSR Chances, although the QSR Chances had the lowest percentage of reports that were on the wrong side of one half (when the Bayes prediction was not one-half). The asterisks in table 6.3 pertain to pairwise tests, between Lottery Choice and BDM in the top two rows, and between QSR Chances and Numbers in the bottom two. The overall message is that in all four dimensions, the Lottery-Choice menu (with choices in all rows, one row selected at random ex post), does better than the BDM (with a single chances-in-100 report). Moreover, the QSR Chances generally does better (in three of the four categories) than the QSR Numbers with no "chances-in-100" terminology.

Belief Elicitation Methods: *In a between-subjects comparison of elicitation methods, all four methods (Lottery-Choice menu, BDM, and QSR with and without chances-in-100 terminology) were unbiased in terms of how they tracked Bayes' rule predictions with symmetric priors. The Lottery-Choice menu did outperform the single decision BDM in terms of percentages of correct decisions, low average absolute deviation, low incidence of boundary predictions, and low incidence of decisions on the wrong side of one half. The QSR with chances-in-100 terminology generally did better than the QSR with only numbers terminology.*

6.5   BDM Value Elicitation: How To and How Not To

All four belief elicitation methods discussed in the previous section were basically unbiased, with elicited beliefs tracking Bayes' predictions, and without systematic upward or downward biases in this symmetric-prior, low-payoff setting. One reason that BDM, in particular, is unbiased is that it is not framed in terms of a buying or selling price. Such framing effects are clearly illustrated in a classic paper by Kachelmeier and Shehata (1992) that involved eliciting a money value "certainty equivalent" for lotteries. They ran some of their treatments in China for substantial stakes (analogous to $10 and $100 for the two payoff levels used). Subjects were asked to provide a *minimum selling price* for lotteries that offered various probabilities of winning one of those amounts. The surprising finding was that the selling prices were greater than the expected values of the lotteries, which indicated risk-seeking behavior (although there was less risk seeking with high stakes). The risk-seeking behavior that they observed was at odds with previous findings that subjects are generally risk averse.

Kachelmeier and Shehata followed up with a between-subjects treatment designed to explain the unexpected risk seeking encountered in China. The follow-up treatment was done in the United States, and it involved a choice between a 0.5 chance of $20, nothing otherwise. Subjects were asked to write down a *minimum selling price* for the lottery, with the understanding that a bid between $0 and $20 would be randomly generated, and that the subject would sell the lottery *for the bid amount* if it exceeded the selling price, and would keep the lottery, which would provide a 50% chance of $20 otherwise. Before the outcome was revealed, the same subjects were asked to write a *maximum purchase price* for an equivalent lottery (that also pays $20 with probability 1/2). Then a random offer to sell would be generated in the relevant range, with a purchase made *at the offer price* if it was below the subject's stated maximum purchase price. The usual BDM incentive is for the subject's decision to reveal the true value of the lottery, since that way the experimenter can make the decision of whether to buy or sell that is best for the subject (problem 6). This "certainty equivalent" value of the lottery is the same, in theory, regardless of whether the buying or selling version of BDM is used. A risk-neutral person would value the lottery at its expected money value, or 0.5 times $20 = $10. A risk-averse person would reveal a certainty equivalent value below $10, and a risk-preferring person would reveal a certainty equivalent value above $10. The trouble was that the same people were both risk averse and risk preferring! The average selling price was about $11 (risk preferring), whereas the average purchase price was only about half as high, at $5.50, indicating substantial risk aversion in the buy frame. This result was highly significant. The point is that the elicitation method can bias the results dramatically.

This type of buy-sell difference has been documented in the environmental economics literature in which BDM methods have been used to elicit values for environmental goods. There it is known as the WTA-WTP, or "willingness-to-accept/willingness-to-pay" bias. This bias could be due to an "endowment effect" if ownership of the lottery tends to enhance its perceived value, but it is doubtful that just a statement of ownership in the instructions is strong enough to have this effect. There is the possibility that subjects in a market setting misinterpret the BDM incentives and slip into a market mentality of trying to buy low or sell high. Regardless of what the source of the bias is, it is essential to avoid market terminology in elicitation tasks, despite the fact that a market frame makes it easier to explain the random number in terms of a bid or sell offer.

As was the case with belief elicitation, a BDM procedure for value elicitation can be explained without any market terminology, by having the subject report a dollar value of a gamble that will be compared with a randomly generated amount. The subject will be paid with the gamble outcome or the randomly generated amount, whichever is better for the subject based on the reported value. Subjects are probably more likely to be confused by a procedure without any market overlay, which can be solved by providing a choice menu that lets the subject make a choice in each row between the gamble, which stays the same, and dollar amounts that increase coming down the menu, as shown in table 6.4. The crossover point would determine the certainty equivalent value of the gamble. A table with more rows, e.g., 20, would provide a more precise measure. A person who crosses over before the $10 row puts a low value on the risky gamble on the right side, and hence is risk averse. A person who crosses at

Table 6.4. A "Price List" Choice Menu for Eliciting a Certainty Equivalent Value

Money Amount			Gamble
$0	☐	☑	50 chances in 100 of $20
$2	☐	☐	50 chances in 100 of $20
$4	☐	☐	50 chances in 100 of $20
$6	☐	☐	50 chances in 100 of $20
$8	☐	☐	50 chances in 100 of $20
$10	☐	☐	50 chances in 100 of $20
$12	☐	☐	50 chances in 100 of $20
$14	☐	☐	50 chances in 100 of $20
$16	☐	☐	50 chances in 100 of $20
$18	☐	☐	50 chances in 100 of $20
$20	☑	☐	50 chances in 100 of $20

$10 is risk neutral, since that is the expected value of the gamble on the right. As was the case with the probability equivalent table 6.2 used in belief elicitation, it would be best to avoid multiple crossings by requiring a single crossing, which is what the Veconlab software does in the Value Elicitation program.

It is important to remember that biased results obtained with market terminology for whatever reason (endowment effect, loss aversion, buy-sell negotiation impulses) are not an indication that the notion of risk aversion or risk preference is somehow flawed, since this bias can be avoided. To summarize:

Proper BDM Implementation: *When the BDM method is framed as a minimum selling price, the commonly observed responses in an experiment are above expected value, which indicates risk seeking. Conversely, and when the method is framed as a maximum purchase price, the commonly observed responses below expected value indicate risk aversion. The market framing bias is predictable and easy to avoid, using neutral terminology and/or a structured choice menu as was done previously with belief elicitation.*

6.6   Ambiguity Aversion

The gambles or lotteries considered up to this point have all involved known probabilities. Daniel Ellsberg, who is best known for leaking the Pentagon Papers, wrote a 1961 paper suggesting that people are averse to making choices in situations of unknown probability. This paper has been very influential, as indicated by the large subsequent literature on ambiguity aversion (7,000 plus citations on Google Scholar). Ambiguity aversion is indicated if a person is willing to sacrifice some payoff advantage in order to avoid an ambiguous bet with unknown probabilities. Conversely, some people might be neutral to ambiguity, treating all probabilities the same, irrespective of whether they are known or estimated. And others might be ambiguity loving. Investments in common stocks can be considered to be ambiguous bets, which suggests a close relationship with related notions of risk and loss aversion. The consensus from decades of research is that ambiguity aversion is prevalent for a majority of subjects, especially for gains. This conclusion, however, has been challenged by a number of recent papers by experimental economists. A summary of several of these papers will help the reader think about interesting procedural issues related to elicitation and beliefs.

The setup for a standard ambiguity aversion experiment is unusually difficult for a first-time reader to grasp, especially in the way it is summarized in introductions to papers written for specialists. So it is good to see the text as subjects see it the first time. The following paragraph is taken from a particularly clearly written set of instructions used by Charness, Karni, and Levin (2013):

Consider six containers that have 36 chips in each. Each container will have a different (but known) number of Red chips, with the remaining chips in the container being either Blue or Green; you will not be told how many chips are Blue and how many chips are Green.

You'll be faced with a table with six rows, with each row representing one container. The first row will ask you to make a choice when there are 9 Red chips in the container (container 9), and so 27 Blue or Green chips; the second row to consider has 10 Red chips in the container (container 10), and so 26 Blue or Green chips; and so on. . . .

For each row, your task is to choose one of the three colors to bet on. After people make their choices, we will randomly draw a number from 9–14 to determine which line in the table is to be implemented (played). Once a line is selected it determines which container will be used. We will draw one chip from that container and pay $10.00 to each person who picked that color.

Experiment Decision Sheet

Red Numbers	Blue or Green Numbers	Bet on Red [R], Bet on Green [G], or Bet on Blue [B]		
9	27	R	G	B
10	26	R	G	B
11	25	R	G	B
12	24	R	G	B
13	23	R	G	B
14	22	R	G	B

Notice that the decision table with six rows is essentially a choice menu, with one row chosen ex post to be used. Since there are 36 chips in each envelope, betting on red is rational whenever the number of red chips is greater than 12, since this would leave an average of fewer than 12 chips for each of the other two colors. A person who always bets on red when there are more than 12 chips and always bets on blue or green when there are fewer than 12 red chips is classified as *ambiguity neutral*. Consider a person who starts betting on red for rows with fewer than 12 red chips and continues betting on red as the number of reds increases. This person is giving up something in terms of expected payoffs by choosing red when there are fewer than one-third red chips, and hence such a person is classified as *ambiguity averse*. A person who does not start betting on red until there are more than 13 red chips is *ambiguity seeking*. In this experiment a majority, about 60%, of subjects were classified as ambiguity neutral, with only 8% ambiguity averse, 12% ambiguity seeking, and the rest classified as incoherent due to multiple switches. The authors defend the incoherence classification:

Simply because a subject chooses red when the number of red slips of paper [chips] in the envelope is 10 doesn't necessarily imply that the subject is ambiguity averse. For example, the same subject may choose green when the number of red slips in the envelope is 11; these choices are inconsistent with respect to ambiguity preferences.

The authors attributed the low rate of ambiguity aversion to letting the subjects choose between blue or green ambiguous bets, which may reduce fear of unfair manipulation by the experimenter (some previous researchers have done this, others have not). There was more ambiguity aversion when this subject choice was removed in another treatment. Also, they stressed transparency in the procedures, ensuring that the experiment administrator did not know the proportions of blue and green being used in various envelopes, and offering to show subjects the contents of the envelopes afterward. Even the use of physical objects like cards could add to credibility as compared with computerized implementations.

To summarize, the Charness et al. procedure was to observe where a subject would cross over from an ambiguous bet on a color (choice of blue or green) to an unambiguous bet on red as the number of red chips is increased. The same procedure can be used if there are only two possible colors, e.g., blue and white. Then the point of probability indifference could be determined in a manner analogous to the belief elicitation menu in table 6.2, by substituting an ambiguous lottery for the Event Lottery, as shown in table 6.5. The ambiguous lottery on the right side pays off if the chip drawn is blue, but the proportion of blue

Table 6.5. Coarse Lottery-Choice Menu for Risk versus Ambiguity

Random Lottery, Known Chances		Lottery with ? Chances in 100 of Blue	
0 chances in 100 of $2.00	☐	☑	$2.00 if draw is blue
10 chances in 100 of $2.00	☐	☐	$2.00 if draw is blue
20 chances in 100 of $2.00	☐	☐	$2.00 if draw is blue
30 chances in 100 of $2.00	☐	☐	$2.00 if draw is blue
40 chances in 100 of $2.00	☐	☐	$2.00 if draw is blue
50 chances in 100 of $2.00	☐	☐	$2.00 if draw is blue
60 chances in 100 of $2.00	☐	☐	$2.00 if draw is blue
70 chances in 100 of $2.00	☐	☐	$2.00 if draw is blue
80 chances in 100 of $2.00	☐	☐	$2.00 if draw is blue
90 chances in 100 of $2.00	☐	☐	$2.00 if draw is blue
100 chances in 100 of $2.00	☑	☐	$2.00 if draw is blue

chips is not specified. This coarse-grid menu would be followed by a fine-grid menu with chances on the left that increment by 1 instead of 10. In this "two-color" context, ambiguity aversion would be indicated by a crossover to the Random Lottery when it has less than 50 chances in 100 of paying off. This is the setup used in the Veconlab Value Elicitation program (Ambiguity Aversion setup option).

Several papers have measured ambiguity aversion as a probability equivalent. Binmore, Stewart, and Voorhoeve (2012, fn. 3) cite some of these but point out that those papers just asked subjects for a probability that made the subject indifferent, instead of observing the subject's actual decisions to infer indifference, as is done with a choice menu. Instead of using a choice menu like table 6.5, Binmore et al. used an iterative "titration" method that is common in psychology research. The way titration works is that the subject begins with a choice between a 50% payoff gamble and the ambiguous gamble on the right side. If the subject chooses the 50% payoff gamble, as would be the case with ambiguity aversion, then the next choice might be between the ambiguous gamble and known probability bet with a lower payoff probability, say 30%. If the ambiguous gamble is selected at that stage, the payoff percentage for the known probability bet would be raised, etc. It typically does not take many pairwise comparisons for this procedure to converge if it is going to converge. Binmore et al. note that "there is a risk that subjects might not always answer truthfully" because they prefer to be paid on bets that can only be reached by lying. For example, choosing the ambiguous bet in several comparisons would raise the payoff percentage for the bet with known probabilities above 50%. However, the authors report that "we found no evidence of learning in comparing subjects' behavior in later stages of the experiment." A choice menu with all rows showing (as in table 6.5) would avoid these strategic manipulation issues that could arise with titration.

Titration issues aside, Binmore et al. put subjects into two probability indifference tasks and classified them as ambiguity averse if both tasks indicated ambiguity aversion. They found much less ambiguity aversion than the 60% rate that is standard in the literature. They attribute this to their two-task design, which reduces the chances that someone exhibiting noisy behavior will be classified as ambiguity averse. They also stress the importance of procedures that clarify the setting and reduce perceptions of possible deception. For example, they used physical devices, i.e., a stack of cards constructed with a mix of red and black or white cards. These cards were shuffled by a machine in the presence of the subjects, with each subject making a draw from the top to determine the payoff.

Another study that used titration to elicit a probability equivalent is the one by Dimmock, Kouwenberg, and Wakker (2016). Their setup was fully computerized (no physical objects), which might trigger distrust, but which enabled them to use a larger and more representative subject pool with online connections.

They find significant amounts of ambiguity aversion. A demographic questionnaire allowed them to determine whether measured ambiguity aversion was correlated with participation in a stock market (it was not).

Even though the literature does not provide unambiguous evidence about the prevalence of ambiguity aversion, even after five decades of research, it does highlight some interesting procedural issues. In addition to procedural differences, some differences in results may be due to the sensitivity of ambiguity to context, e.g., gains versus losses, which options are being compared, etc. And, it is hoped that students will have experienced an ambiguity aversion experiment in class, where trust issues are not present, to form their own opinions. To summarize:

Ambiguity Aversion Experiments: *The established result that a majority of subjects are averse to ambiguity (for gains) has been challenged by recent papers that elicit probability equivalents. The use of probability indifference measures avoids the need to estimate or correct for the related concept of risk aversion. Some of these authors have argued that ambiguity aversion is diminished by using physical objects (e.g., chips or cards) that make procedures more transparent and less sensitive to trust issues. In any case, it seems advisable to base conclusions on actual decisions in paired comparisons and not to simply ask for a probability that results in indifference between risky and ambiguous bets.*

6.7 Extensions

Another way to get around the issue of risk aversion when eliciting beliefs with the QSR is to make payments in terms of lottery tickets that offer a chance of winning a fixed prize, say $10, where the chances of winning are proportional to one's lottery ticket holdings. In this "binary lottery" case, there are only two possible payoffs, $10 and $0, so utility curvature does not matter. Another way to see this is to notice that expected payoff is proportional to the probability of winning, so it is linear in the number of tickets acquired, and linearity corresponds to risk neutrality. Hossain and Okui (2013) and Harrison, Martinez-Correa, and Swarthout (2014) report promising results with this approach. The evidence is mixed, as indicated by the provocative title of the Selten, Sadrieh, and Abbink (1999) paper: "Money Does Not Induce Risk Neutral Behavior, But Binary Lotteries Do Even Worse." One clear advantage of this binary lottery method is that it is possible to state in the instructions that truthful reporting is optimal, although the underlying reason, maximizing the expected number of lottery tickets for the final payoff, may be difficult to explain.

A different approach to dealing with risk aversion is to use the standard quadratic scoring rule, elicit risk preferences with a different task, and then attempt to de-bias the elicited beliefs for the estimated degree of risk aversion. This approach has the disadvantage of introducing a second source of bias (to an already difficult task) due to possible randomness in risk aversion estimates at the individual level.

Two-stage lottery choice menus were successfully implemented in two doctoral dissertations. In a laboratory study of voting games with random voting costs, Tulman (2013) found that voter turnout decisions were correlated with elicited beliefs about whether one's own vote would be pivotal and swing the vote in favor on one's preferred outcome (or create a tie). In a threshold voluntary contributions experiment (a setup to be discussed in chapter 16), Schreck (2013) found that contributions to a charity were strongly correlated with elicited beliefs about whether the person's own contribution would trigger a matching contribution that would result if total contributions reached a target level. This dissertation provides a clean set of instructions and lottery-choice menus for hand-run elicitation (pages 82–86).

There is a large and growing literature on belief elicitation, which is surveyed in Schotter and Trevino (2014). The literature on ambiguity is surveyed in Trautmann and van de Kuilen (2015a). One interesting finding is the tendency for ambiguity to diminish for decisions made by groups of subjects, e.g., Charness et al. (2013) and Keck, Diecidue, and Budescu (2014).

Chapter 6 Problems

1. Use the QSR payoff formulas in equation (6.1) to check the payoff numbers 0.36 and 0.96 shown in the "20 chances" row of table 6.1.

2. When the subjective probability for event A is 20 chances in 100, use the numbers in table 6.1 to calculate the subject's expected payoff for reporting incorrect beliefs of either 10 or 30, and verify that the subject's expected payoff would be even higher for reporting truthfully (20 chances in 100).

3. Would risk seeking tend to bias responses in a quadratic scoring rule, and if so, conjecture on the direction of the bias.

4. (non-mechanical, calculus) How could a quadratic scoring rule be used to elicit a probability distribution over three events, assuming risk neutrality?

5. (calculus) An alternative to the quadratic scoring rule is a logarithmic scoring rule. Since a reported probability R is less than 1, the natural log of R, $\ln(R)$ will be negative, with no lower bound if R is very close to zero. Earnings for a subject who reports R would be:

$F + \ln(R)$ if the event in question is observed, and
$F + \ln(1 - R)$ if the event is not observed

The $\ln(\cdot)$ terms are negative, so there are penalties that severely penalize the subject if the outcome that actually occurs had been reported to have a low probability. Show that this scoring rule provides an incentive for truthful reporting of the belief probability for the event, which you can denote by p.

6. (non-mechanical) Suppose that a subject is endowed with a lottery that pays $20 with probability 1/2, and a BDM procedure is used to elicit the minimum selling price. In particular, a bid will be randomly selected from the interval from $0 to $20, and the lottery will be sold (at the bid price) if the bid exceeds the reported minimum selling price; otherwise the lottery is kept and the subject's earning is determined by a coin flip. If the subject is indifferent between the lottery and a money amount V, explain why the subject should report a minimum selling price equal to V.

7. (non-mechanical) First, explain what crossover pattern in table 6.4 would indicate risk aversion. Second, how could table 6.4 be modified to test for ambiguity aversion, while keeping the money amount price list on the left intact? Would it be necessary to control for risk aversion, and if so, how might this be done?

7

Individual and Social Learning

Perhaps the simplest prediction problem involves guessing which of two random events will occur. The probabilities of the two events are fixed but not known, so a person can learn about these probabilities by observing the relative frequencies of the two events. One of the earliest biases recorded in the psychology literature was the tendency for individuals to predict each event with a frequency that approximately matches the fraction of times that the event has occurred up to that point. This bias, known as "probability matching," has been widely accepted as being evidence of irrationality, despite Siegel's experiments in the 1960s, which tell a different story. These experiments provide an important methodological lesson for how experiments should be conducted. The results are also used to begin a discussion of learning models that may explain paths of adjustment to some "steady state" where systematic changes in behavior have ceased.

The second part of the chapter pertains to social learning, i.e., situations in which people can make inferences from what others have done in the past. Suppose that individuals may receive different information about some unknown event, like whether or not a newly patented drug will be effective. This information is then used in making a decision, like whether or not to invest in the company that developed the new drug. If decisions are made in sequence, then the second and subsequent decision makers can observe and learn from earlier decisions. The *informational dilemma* occurs when one's private information suggests a decision that is different from what others have done before. An "information cascade" forms when people follow the consensus decision regardless of their own private information.

Notes for the Instructor: Binary prediction tasks should be run with the Veconlab Probability Matching program (listed on the Decisions menu), which permits larger numbers of rounds and which displays averages graphically. The information cascade could be implemented with draws from cups and throws of dice, as described in Anderson and Holt (1996b). A better option for class is to use the Veconlab Information Cascade game, which maintains anonymity and handles all of the random signals and events automatically based on probabilities specified by the defaults.

7.1   Being Treated Like a Rat

Before the days of computers, the procedures for binary prediction tasks in psychology experiments sometimes seemed like a setup for rats or pigeons that had been scaled up for humans. For example, Siegel and Goldstein (1959) had subjects come individually to a lab to be seated at a desk with a plywood screen that separated the working area from the experimenter on the other side. The screen contained two lightbulbs, one on the left and one on the right, and a third, smaller light in the center that was used to signal that the next decision must be made. When the signal light went on, the subject recorded a prediction by pressing one of two levers (left or right). Then one of the lights was illuminated, and the subject would receive reinforcement (if any) based on whether or not the prediction was correct. When the next trial was ready, the signal light would come on, and the process would be repeated, perhaps hundreds of times. No information was provided about the relative likelihood of the two events (left and right), although sometimes people were told how the events were generated, e.g., by using a printed list. In fact, one of the events would be set to occur more often, e.g., 75% of the time. The process generating these events was not always random, e.g., sometimes the events were rigged so that in each block of 20 trials, exactly 15 would result in the more likely event (as in problem 5).

By the time of the Siegel and Goldstein experiments in the late 1950s, psychologists had already been studying similar problems for more than two decades. The results indicated a curious pattern: the proportion of times that subjects predicted each event roughly matched the frequency with which the events occurred. For example, if left occurred three-fourths of the time, then subjects would come to learn this by experience and then would tend to predict left three-fourths of the time.

The astute reader may have already figured out the best thing to do in such a situation, but a formal analysis will help ensure that the conclusion does not rely on hidden assumptions like risk neutrality. Let us assume that the events really are independent random realizations, not done in blocks of 20. Let U_C denote the utility of the reward for a correct prediction, and let U_I denote the utility of the reward for an incorrect prediction. The rewards could be "external" (money payments, food), "internal" (psychological self-reinforcement), or some combination. The only assumption is that there is some preference for making a correct prediction: $U_C > U_I$. These utilities may even be changing over time, depending on rewards already received; the only assumption is that an additional correct prediction is preferred.

Once a number of trials have passed, the person will have figured out which event is more likely, so let p denote the subjective probability that represents those beliefs, with $p > 1/2$. There are two decisions: predict the more likely event and predict the less likely event. Each decision yields a lottery:

Predict more likely event: U_C with probability p
                                                U_I with probability $1 - p$

Predict less likely event: U_I with probability p
                                                U_C with probability $1 - p$

Thus the expected utility for predicting the more likely event is higher if

$$pU_C + (1 - p)U_I > pU_I + (1 - p)U_C$$

or equivalently,

$$(2p - 1)U_C > (2p - 1)U_I$$

which is always the case since $p > 1/2$ and $U_C > U_I$. Although animals may become satiated with food pellets and other physical rewards, economists have no trouble with a non-satiation assumption for money rewards. Note that this argument does not depend on any assumption about risk attitudes, since the two possible payoffs are the same in each case, i.e., there is no more "spread" in one case than in the other. With only two possible outcomes, the only role of probability is to determine whether the better outcome has a higher probability, so risk aversion does not matter. Another way to think about this is that if there are only two relevant points, then they essentially constitute a straight line. Since the higher reward is for a correct prediction, the more likely event should always be predicted, i.e., with probability 1. Probability matching is irrational as long as there is no satiation, $U_C > U_I$, and no outside utility associated with predicting an unlikely event correctly.

Despite the implication that a subject should predict the more likely event 100% of the time after it becomes clear which event that is, this behavior may not be observed in the laboratory. In a survey of the probability matching literature, the psychologist Fantino (1998, pp. 360–361) concludes: "human subjects do not behave optimally. Instead they match the proportion of their choices to the probability of reinforcement. . . . This behavior is perplexing given that non-humans are quite adept at optimal behavior in this situation." As evidence for the higher degree of rational behavior in animals, Fantino cites a 1996 study that reported choice frequencies for the more likely event to be well above the probability matching predictions in most treatments conducted with chicks and rats. Before concluding that the animals are more rational than humans, it will be instructive to review a particularly well-designed probability matching experiment.

7.2. Siegel and Goldstein's Experiments

Sidney Siegel is the psychologist who has had the largest direct and indirect impact on experimental methods in economics. His early work provides a high standard of careful reporting and procedures, appropriate statistical techniques, and the use of financial incentives where appropriate. His 1956 book on non-parametric statistics for experimental data will be featured in chapter 13 on methodology. His experiments on probability matching are a good example of careful procedures. In one experiment, 36 male Penn State students were allowed to make predictions for 100 trials, and then 12 of these were brought back on a later day to make predictions in 200 more trials. The proportions of predictions for the more likely event are shown in figure 7.1, with each point being the average over 20 trials.

The 12 subjects in the "no-pay" treatment were simply told to "do your best" to predict which lightbulb would be illuminated. These averages are plotted as the heavy dashed line, which begins at about 0.5, as would be expected in early trials with no information about which event is more likely. The proportion of predictions for the more likely event converges to about 0.75 as predicted by probability matching, with a leveling off at about trial 100.

In the "pay-loss" treatment, 12 participants received 5 cents for each correct prediction, and they lost 5 cents for each incorrect decision. The 20-trial averages are plotted as the thick solid line in the figure. The line converges to a level of about 0.9. A third "pay" treatment offered a 5-cent reward but no loss for an incorrect prediction, and the results (not shown) are in between the other two treatments, and above 0.75. Clearly, incentives matter, and approximate probability matching is not observed with incentives in this context.

It would be misleading to conclude that incentives always matter, or that probability matching will be observed in no-pay treatments using different

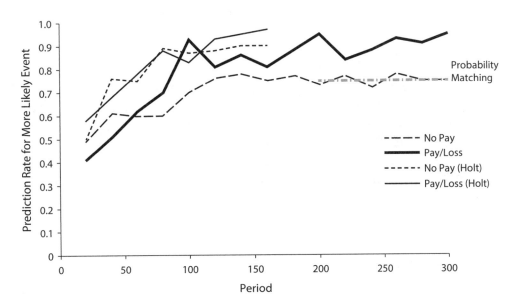

Figure 7.1. Prediction Proportions for the Event with Frequency 0.75. *Source*: Siegel, Siegel, and Andrews (1964) for dark lines, and Holt (1992) for light lines.

procedures. The two thin lines in the upper left part of figure 7.1 show 20-trial averages for an experiment (Holt, 1992) in which six students in no-pay and pay/loss treatments made decisions using computers, where the events were determined by a random number generator. The instructions matched those reported in Siegel, Siegel, and Andrews (1964), except that colored boxes on the screen were used instead of lightbulbs. Probability matching was not observed in either the pay/loss treatment (with a reward of 10 cents and a penalty of 10 cents, shown by the thin solid line) or the no-pay treatment (thin dashed line).

The reason for the absence of probability matching in the no-pay treatment is unclear. One conjecture is that the computer interface makes the situation more anonymous and less like a matching pennies game. In the Siegel setup with the experimenter on one side of the screen, the subject might incorrectly perceive the situation as having some aspects of a game against the experimenter in which the best strategy for the subject would be to randomize the predictions, which would move choice proportions toward 1/2 (see the subsequent discussion of a matching pennies game in chapter 10).

Finally, note that Siegel's findings suggest a resolution to the paradoxical finding that rats are smarter than humans in binary prediction tasks. Since you cannot just tell a rat to "do your best," animal experiments are always run with food or drink incentives on hungry or thirsty animals. As a result, the choice proportions are closer to those of financially motivated humans. In a survey of more than 50 years of probability matching experiments, Vulkan (2000) concluded that probability matching is generally not observed in humans with real payoffs, although humans can be surprisingly slow learners in this simple setting. To summarize:

> **Probability Matching:** *This term refers to a tendency for humans to make predictions for a sequence of binary events in a proportion that approximately matches the empirical frequency of those events. This matching behavior is not rational if correct predictions are rewarded. Probability matching is typically not observed with animals that are motivated by food or liquids, and it is generally not observed with human subjects who are financially motivated. In both cases, choice probabilities tend to converge slowly toward the extreme of predicting the more likely event each time.*

7.3   A Simple Model of Belief Learning

Probability matching experiments provide a useful data set for the study of learning behavior. Given the symmetry, a person's initial "prior" beliefs ought to be that each event is equally likely, but the first observation should raise the probability of the event that was just observed. One way to model this learning is to let initial beliefs for the probability of events L and R be calculated as:

$$(7.1) \qquad \Pr(L) = \frac{\alpha}{\alpha + \alpha} \quad \text{and} \quad \Pr(R) = \frac{\alpha}{\alpha + \alpha} \quad \text{(prior beliefs)},$$

where α is a positive parameter to be explained below. Of course, α has no role yet, since both of the above probabilities are equal to 1/2. If L is observed, then $\Pr(L)$ should increase, so add 1 to the numerator for $\Pr(L)$. To make the two probabilities sum to 1, add 1 to the denominators for each probability expression:

$$(7.2) \qquad \Pr(L) = \frac{\alpha + 1}{\alpha + 1 + \alpha} \quad \text{and} \quad \Pr(R) = \frac{\alpha}{\alpha + 1 + \alpha} \quad \text{(after observing L)}.$$

Note that α determines how quickly the probabilities respond to the new information; a large value of α will keep these probabilities close to 1/2. Continuing to add 1 to the numerator of the probability for the event just observed, and to add 1 to the denominators, we have a formula for the probabilities after N_L observations of event L and N_R observations of event R. Let N be the total number of observations to date. Then the resulting probabilities are:

$$(7.3) \qquad \Pr(L) = \frac{\alpha + N_L}{2\alpha + N} \quad \text{and} \quad \Pr(R) = \frac{\alpha + N_R}{2\alpha + N} \quad \text{(after N observations)},$$

where $N = N_L + N_R$. (Although it is beyond the scope of this book, there is a particular combination of prior beliefs about $\Pr(L)$ on the interval from 0 to 1, and signal information that is binomial with probability $\Pr(L)$, which determines a Bayes' rule posterior distribution with a mean characterized by (7.3).)

In the early periods, the totals, N_L and N_R, might switch in magnitude, but the more likely event should soon dominate, and therefore $\Pr(L)$ will be greater than 1/2. The implication is that all people will eventually start to predict the more likely event, although "unexpected" deviations might result from randomness, e.g., probabilistic choice functions that were developed in the mathematical psychology literature, which will be discussed later, in chapters 9 and 10.

An alternative approach would be to consider "recency effects" which make probability assessments more sensitive to recently observed data. For example, recall that the sums of event observations in the belief-learning formula (7.3) weigh each observation equally. It may be reasonable to allow for "forgetting" in some contexts, so that the observation of an event like L in the most recent trial may carry more weight than something observed a long time ago. Putting more weight on recent events would be optimal if the underlying probabilities were changing over time, which is not the case with probability matching. But people might be adapted to situations where probabilities do evolve, e.g., as food sources move.

Regardless of the source of recency effects, one approach is to replace sums with weighted sums. For example, if event L were observed three times, N_L in (7.2) would be 3, which can be thought of as $1 + 1 + 1$. If the most recent observation (listed on the right in this sum) is twice as prominent as the one before it, then the prior event would get a weight of one-half, the one before that would get a weight of one-fourth, and so on. A more flexible approach is to weight the most recent event by 1, the next most recent event by ρ, the observation before that by ρ^2, etc. Then if ρ is estimated to be less than one, this would indicate a *recency effect*. A common estimate of this parameter for data from strategic games is about 3/4, which is the weight assigned to the second most recent observation. The third most recent observation only gets a weight of 9/16, and all other more distant observations have weights below a half. Roughly speaking, these weights are used to compute the person's belief as a weighted average of past observations.

7.4   Reinforcement Learning

Rewards and punishments are typically referred to as "reinforcements" in psychology experiments. One prominent theory of learning associates changes in behavior with the reinforcements actually received. For example, suppose that the person earns a reinforcement of x for each correct prediction, and nothing otherwise. If one predicts event L and is correct, then the probability of choosing L should increase, and the extent of the behavioral change may depend on the size of the reinforcement, x. One way to model this is to let the choice probability be:

$$(7.4) \qquad \Pr(\text{choose L}) = \frac{\alpha + x}{\alpha + x + \alpha} \quad \text{and} \quad \Pr(\text{choose R}) = \frac{\alpha}{\alpha + x + \alpha}.$$

As before, α is a parameter that affects the speed of adjustment. Despite the similarity with (7.2), there are some key differences. The left side of (7.4) is a choice probability, not a probability representing beliefs. With reinforcement learning, beliefs are not explicitly modeled, as is the case for the "belief learning" models considered in the previous section. The second difference is that the x in (7.4) represents reinforcement, not an integer count as in (7.2).

Of course, reinforcement is a broad term, which can include both physical entities like food pellets given to rats, as well as psychological feelings associated with success or failure. One way to implement this idea for experiments with money payments is to make the simplifying assumption that reinforcement is measured by earnings. Suppose that event L has been predicted N_L times and that the predictions have sometimes been correct and sometimes not. Then the total earnings for predicting L, denoted e_L, would be less than $x N_L$. Similarly, let e_R be the total earnings from correct R predictions. The choice probabilities would be:

$$(7.5) \qquad \Pr(\text{choose L}) = \frac{\alpha + e_L}{2\alpha + e_L + e_R} \quad \text{and} \quad \Pr(\text{choose R}) = \frac{\alpha + e_R}{2\alpha + e_L + e_R}.$$

This reinforcement model might also explain some aspects of behavior in probability matching experiments with financial incentives. The choice probabilities would be equal initially, but a prediction of the more likely event will be correct 75% of the time, and the resulting asymmetries in reinforcement would tend to raise prediction probabilities for that event, and the total earnings for this event would tend to be larger than for the other event. The combination of these two factors, higher earnings and higher choice probabilities, would cause e_L to grow faster, so that e_R / e_L would tend to get smaller as e_L gets larger. Thus the probability of choosing L in equation (7.5) would tend to converge to 1. This convergence may be quite slow, as evidenced by some computer simulations reported by Goeree and Holt (2003a). (See problem 4 for details on how such simulations might be done.) The simulations normalized the payoff x to be 1, and used a value of 5 for α. Figure 7.2 shows the 20-period averages for simulations run for 1,000 people, each making a series of 300 predictions. The simulated averages start somewhat above the observed averages for Siegel's

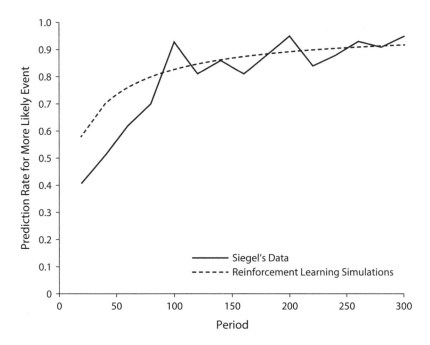

Figure 7.2. Simulated and Observed Prediction Proportions for the More Likely Event. *Source*: Goeree and Holt (2003) for the simulated data.

subjects (more like the Holt data in figure 7.1), but similar qualitative patterns are observed.

7.5 "To do exactly as your neighbors do is the only sensible rule."

Conformity is a common occurrence in social situations, as the section title implies (from Emily Post, 1927, ch. 33). People may follow others' decisions because they value conformity and fear social sanctions. For example, an economic forecaster may prefer the risk of being wrong along with others to the risk of having a deviant forecast that turns out to be inaccurate. Similarly, Keynes (1936) remarked: "Worldly wisdom teaches that it is better for reputation to fail conventionally than to succeed unconventionally."

In some cases, conformity can be rational. People may be tempted to follow others' decisions because of the belief that there is some wisdom or experience implicit in an established pattern. The possible effects of collective wisdom may be amplified in a large group. For example, someone may prefer to buy a Honda or Toyota sedan thinking that the large market shares for those models signal a lot of customer satisfaction. This raises the possibility that a choice pattern started by a few may set a precedent that triggers a string of incorrect decisions. For example, there are certain classes of stocks that are considered good investments, and herd effects can be amplified by efforts of some to anticipate where the next fads will lead. Anyone who purchased "tech" stocks in the late 1990s can attest to the dangers of following the bulls.

A particularly interesting type of bandwagon effect can develop when individuals make observable decisions in a sequence. Suppose that a person is applying for a job in an industry with a few employers who know each other. If the applicant makes a bad impression in the first two interviews and is not hired, then the third employer approached may hesitate even if the applicant makes a good impression. This employer may reason: "anyone may have an off day, but I wonder what the other two employers saw that I missed." If the joint information implied by the two previous decisions is deemed to be more informative than one's own information, it may be rational to follow the pattern set by others' decisions, in spite of contradictory evidence. A chain reaction started in this manner may take on a life of its own as subsequent employers hesitate even more. This is why first impressions can be important in the workplace. The effect of information inferred from a sequential pattern of conforming decisions is referred to as an *information cascade* (Bikhchandani, Hirschleifer, and Welch, 1992).

Since first impressions can be wrong, the interviews or tests that determine initial decisions may trigger an "incorrect" cascade that implies false information to those who follow. As John Dryden quipped: "Nor is the people's judgment always true, the most may err as grossly as the few."

7.6 A Model of Rational Learning from Others' Decisions

The discussion of cascades will be based on a very stylized model in which the two events are referred to as cup A with contents *a*, *a*, *b*, and cup B with contents *a*, *b*, *b*. Think of these cups as containing amber or blue marbles, with the proportions being correlated with the cup label. Each cup is equally likely to be selected, with its contents then being emptied into an opaque container from which draws are made privately. Each person sees one randomly drawn marble from the selected cup, and then must guess which cup is being used. Decisions are made in a pre-specified sequence, so that the first person has nothing to go on but the color of the marble drawn. Draws are made with replacement, so the second person sees a draw from the unknown cup, and must make a decision based on two things: the first decision and the second (own) draw. Like the employers who cannot sit in on others' interviews, each person can see prior decisions but not the private information that may have affected those decisions. There is no external incentive to conform in the sense that one's payoff depends only on guessing the cup being used; there is no benefit in conforming to others' (prior or subsequent) decisions.

The first person has only the observation, *a* or *b*, and the prior information that the cups are ex ante equally likely. Since two of the three *a* marbles are in cup A, the person should assess the probability of cup A to be 2/3 if an *a* is observed, and similarly for cup B. Thus, the first decision should reveal that person's information. In the experiments discussed below, about 95% of the people made the decision that corresponded to their information when they decided first.

The second person then sees a private draw, *a* or *b*, and faces an easy choice if the draw matches the previous choice. A conflict is more difficult to analyze. For example, if the first person predicts cup A, the second person who sees a *b* draw may reason: "the cups are now equally likely, since that was the initial situation and the *a* that I think the first person observed cancels out the *b* that I just saw." In such cases, the second person should be indifferent, and might choose each cup with equal probability. On the other hand, the second person may be a little cautious, being more sure about what they just saw than about what they are inferring about what the other person saw. Even a slight chance that the other person made a mistake (deliberate or not) would cause the second person to "go with their own information" in the event of a conflict. About 95% of the second decision makers behaved in this manner in the experiment described below, and we will base subsequent discussion on the assumption that this is the case.

Now the third person will have observed two decisions and one draw. There is no loss of generality in letting the first decision be labeled A, so the various

Table 7.1. Possible Inferences Made by the Third Decision Maker

Prior Decisions	Own Draw	Inferred Pr(cup A)	Decision
A, A	*a*	0.89	A (no dilemma)
A, A	*b*	0.67	A (start cascade)
A, B	*a*	0.67	A
A, B	*b*	0.33	B

possibilities are listed in table 7.1. (An analogous table could be constructed when the first decision is B.) In the top row of the table, all three pieces of information line up, and the standard Bayesian calculations would indicate that the probability of cup A is 0.89, making A the best choice (see problem 6). The case in the second row is more interesting, since the person's *own* draw is at odds with the information inferred from others' decisions. When there are two others, who each receive independent draw "signals" that are just as informative as the person's own draw, it is rational to go with the decision implied by the others' decisions. The decision in the final two rows is less difficult because the preponderance of information favors the decision that corresponds to one's own information.

The pattern of following others' behavior (seen in the top two rows of table 7.1) may have a domino effect: the next person would see three A decisions and would be tempted to follow the crowd regardless of their own draw. This logic applies to all subsequent people, so an initial pair of matching decisions (AA or BB) can start an information cascade that will be followed by all others, regardless of their private information. If the first two decisions cancel each other out (AB or BA) and the next two form an imbalance (AA or BB), then this imbalance causes the fifth person to decide to follow the majority. An example of such a situation would be: ABAA, in which case the fifth person should choose A even if the draw observed is *b*. Again, the intuition is that the first two decisions cancel out, and that the next two matching decisions are more informative than the person's own draw.

Notice that there is nothing in this discussion that requires a cascade to provide a correct prediction. There is a 1/3 chance that the first person will see the odd marble drawn from the selected cup, and will guess incorrectly. There is a 1/3 chance that the same thing will happen to the second person, so in theory, there is a (1/3)(1/3) = 1/9 chance that both of the first two people will guess incorrectly and spark an incorrect cascade. Of course, cascades may initially fail to form and then may later form when the imbalance of draws is 2 or more in one direction or the other. In either case, the aggregate information inferred from others' decisions is greater than the information inherent in any single person's draw.

7.7   Experimental Evidence

Anderson and Holt (1997) used this setup in a laboratory experiment in which people earned $2 for a correct guess, nothing otherwise. Subjects were in isolated booths, so that they could not see others' draws. Decisions (but not subject ID numbers) were announced by a third person to avoid having confidence or doubts communicated by the decision maker's tone of voice. The third person would return to the front of the room to make the announcement. Even though the third person was not visible to the subjects, this return trip helped prevent subjects from learning from slight delays that might signal indecision on the part of someone with contrary information. The marbles were light or dark, with two lights and a dark in cup A, and with two darks and a light in cup B. The cup to be selected was determined by the throw of a six-sided die, with a 1, 2, or 3 determining cup A. The die was thrown by a "monitor" selected at random from among the participants at the start of the session. There were six other subjects, so each prediction sequence consisted of six private draws and six public predictions. For each group of participants, there were 15 prediction sequences. The monitor used a random device to determine the order in which individuals saw their draws and made predictions.

When the private draw was shown to the subject in the privacy of their own booth, the experimenter would not return the marble to its cup until the subject had recorded the draw, since draws were occasionally recorded incorrectly, and if the marble was visible, this could be corrected.

At the end of the sequence, the monitor announced the cup that had actually been used, and all participants who had guessed correctly added $2 to their cumulative earnings. The monitor received a fixed payment, and all others were paid their earnings in cash. This "ball and urn" setup was designed to reduce or eliminate preferences for conformity that were not based on informational considerations.

It is possible that an imbalance of signals does not develop, e.g., alternating *a* and *b* draws, making a cascade unlikely. An imbalance did occur in about half of the prediction sequences, and cascades formed about 70% of the time in such cases. A typical cascade sequence is shown in table 7.2.

Sometimes people do deviate from the behavior that would be implied by a Bayesian analysis. For example, consider the sequence in table 7.3. The decision

Table 7.2. A Cascade

Subject	58	57	59	55	56	60
Draw	*b*	*b*	*a*	*b*	*a*	*a*
Prediction	B	B	B	B	B	B

Table 7.3. A Typical "Error"

Subject	8	9	12	10	11	7
Draw	*a*	*a*	*b*	*a*	*b*	*a*
Prediction	A	A	B	A	A	A

of subject 12 in the third position is the most commonly observed type of "error," i.e., basing a decision on one's own information even when it conflicts with the information implied by others' prior decisions. This type of error occurred in about a fourth of the cases in which it was possible. Notice that the cascade starts later, with subject 11 in the fifth position.

Once a cascade begins, the subsequent decisions convey no information about their signals (in theory), since these people are just following the crowd. In this sense, patterns of conformity are based on a few draws, and consequently, cascades may be very fragile. In particular, one person who breaks the pattern by revealing their own information may alter the decisions of those who follow.

Deviations almost always involved relying on one's own information, because a deviation in that direction is less costly than is a deviation away from both a cascade and from one's own signal information. One interesting question is whether followers in a cascade realize that many of those ahead of them may also be followers who blindly ignore their own information. To address this issue, Kübler and Weizsäcker (2004) introduced a cost of information into the standard cascade model, so that each person could decide whether to purchase a signal after seeing prior decisions in the sequence, but not after seeing whether the earlier people had purchased signals. The cost of the information was set high enough so that only the first person in the sequence should purchase information. In theory, the others should follow the initial decision, without any more information purchases. In the experiment, the first people in the sequence did not always purchase a signal, and those that followed tended to purchase too many signals, which could be justified if people do not trust those earlier in the sequence to make careful decisions. In a second treatment in which people could see whether those ahead of them had purchased signals (but not the actual signals of others), people tended to purchase even more signals than in the baseline case where prior signal purchase decisions were not observed. It seems that people became nervous when they could actually see that those ahead of them did not purchase information and were just following the crowd. This raised signal purchase rates even higher above the theoretical predictions.

Since the signals received are noisy, it is possible that a cascade may form on the wrong event. Table 7.4 shows one such "incorrect cascade." Cup B was actually being used, but the first two individuals were unfortunate and received

Table 7.4. An Incorrect Cascade (Cup B Was Used)

Subject	11	12	8	9	7	10
Draw	*a*	*a*	*b*	*b*	*b*	*b*
Prediction	A	A	A	A	A	A

misleading *a* signals, which could happen with probability $(1/3)(1/3) = 1/9$. The matching predictions of these two people caused the others to follow with a string of incorrect A predictions, which would have been frustrating given that these individuals had seen private draws that indicated the correct cup.

If the sequence in table 7.4 had been longer, one wonders if the cascade would have been corrected. After all, two-thirds of those who follow will be seeing *b* signals that go against the others' predictions. Even a correct cascade can break and reform. This type of breaking and reforming is apparent in the Goeree et al. (2007) data for very long decision sequences with groups of 20 and 40 people. Cascades invariably formed, but they were almost always broken at some point. Moreover, incorrect cascades tended to be corrected more often than not, and cascades later in the sequence were more likely to be correct than those that formed early in the sequence. Thus the main point of the paper is conveyed well by the title: "Self-correcting Cascades." To summarize:

Social Learning Experiments: *In a social learning context with sequences of observed decisions by individuals with private information, an initial pattern can trigger others to follow, irrespective of their own information. Cascades are observed in experiments, although incorrect cascades triggered by initial decisions based on noisy incorrect information are often reversed. Deviations from a pattern in the direction of one's own private information are less costly in an expected value sense than deviations in the other direction.*

7.8   Extensions

Both of the individual learning models discussed in the first part of this chapter are somewhat simple, which is part of their appeal. The reinforcement model incorporates some randomness in behavior and has the appealing feature that incentives matter. But it has less of a cognitive element; there is no reinforcement for decisions not made. For example, suppose that a person chooses L three times in a row (by chance) and is wrong each time. Since no reinforcement is received, the choice probabilities stay at 0.5, which seems like an unreasonable prediction. Obviously, people learn something in the absence of previously received reinforcement, since they realize that making a good decision may result in higher earnings in the next round. Camerer and Ho (1999) have developed a generalization of reinforcement learning that contains some elements of belief learning. Roughly speaking, observed outcomes receive partial reinforcement even if nothing is earned. See Capra et al. (1999, 2002) for estimation results of belief learning models with recency effects, using data from game theory experiments.

The information cascade story is an important paradigm for social learning, and it has many applications. The model discussed in this chapter is based on Bikhchandani et al. (1992), which contains a rich array of examples. Hung and Plott (2001) discuss cascades in voting situations, e.g., when the payoff depends on whether the majority decision is correct.

The noisy behavior model used by Anderson and Holt (1997) to explain cascade patterns is an example of a quantal response equilibrium, which will be discussed in more detail in chapter 10 on randomization. Goeree et al. (2007) also use this approach to explain the dynamic patterns of a cascade experiment in much longer choice sequences. They considered a generalized version of Bayes' rule, which permits perfect Bayesian learning as a special case. As indicated in chapter 5, their estimation results suggest that subjects tend to deviate from Bayes' rule in the direction of putting too much weight on their own signals and not enough on prior probabilities that could be constructed from prior decisions and the structure of the random draws. Celen and Kariv (2004) reach a similar conclusion from a cascade experiment with a direct elicitation of subjects' beliefs. Also, see Hück and Oechssler (2000) for a discussion of violations of Bayesian inference in a cascade context.

Some of the more interesting applications are in the area of finance. Keynes (1936) compared investment decisions with people in a guessing game who must predict which beauty contestant will receive the most votes. Each player in this game, therefore, must try to guess who is viewed as being attractive to the others, and on a deeper level, who the others will think that others will find more attractive. Similarly, investment decisions in the stock market may involve both an analysis of fundamentals and an attempt to guess what stocks will attract attention from other investors, and the result may be "herd effects" that may cause surges in prices and later corrections in the other direction. Some of these swings in investment may be due to psychological considerations, which Keynes compared with "animal spirits," but herd effects may also result from attempts to infer information from others' decisions. In such situations, it may not be irrational to follow others' decisions during upswings and downswings in prices (Christie and Huang, 1995). Bannerjee's (1992) model of herd behavior is motivated by an investment example. This model has been evaluated in the context of a laboratory experiment reported by Alsopp and Hey (2000). Other applications to finance are discussed in Devenow and Welch (1996) and Welch (1992).

A major difference between typical market settings and the cascade models is that information in markets is often conveyed by prices, which are continuous, as opposed to binary predictions made in cascade experiments. In some recent laboratory experiments, the prices are computed by a computerized "market maker" on the basis of past predictions (A or B) made by individuals who saw private signals and made predictions in sequence. A prediction in this case is a purchase of one of the two assets, A or B. The market maker's price adjustments were designed to incorporate the information content of *all* past purchase decisions, so each asset is equally attractive given the price that must be paid. It follows that the best asset to purchase for the next person in a sequence is the asset that is indicated by that person's own signal (which is not yet reflected in the prices). Thus, the prediction is that herding should not occur (assuming that prior decision makers are rational and that the prices incorporate the information in their signals). Consider the standard two-event cascade design discussed in the previous section, and suppose that the first person sees a signal and predicts A. The market maker sets the prices of the A and B assets to 2/3 and 1/3, which are the probabilities calculated with Bayes' rule, assuming that the first person saw an *a* signal. If the next person in the sequence sees a *b* signal, that person would have a posterior of 1/2 for each event, so buying the B asset would make sense because it costs less. Thus each person should, in theory, "follow their own signal." In accordance with this prediction, cascades are not typically observed in these experiments with endogenous prices (Drehmann, Oechssler, and Roider, 2005; Cipriani and Guarino, 2005). There are, however, deviations from predicted behavior in other respects. For example, people sometimes ignore their signals to buy an asset that has a very low price. This trading against the market is called "contrarian behavior," and is consistent with the "favorite/longshot bias" observed in betting on horse races.

Chapter 7 Problems

1.　The initial beliefs implied by (7.1) are that each event is equally likely. How might this equation be altered in a situation where a person has some reason to believe that one event is more likely than another, even before any draws are observed?

2.　A recent class experiment used the probability matching Veconlab software with payoffs of $0.20 for each correct prediction, $0 otherwise. Earnings averaged several dollars. There were six teams of 1–2 students, who made predictions for 20 trials only. The more likely event was not revealed, but was programmed to occur with probability 0.75. Calculate the expected earnings per trial for a team that follows perfect probability matching. How much more would a team earn per trial by being perfectly rational after learning which event is more likely?

3. Answer the two parts of problem 2 for the case where a correct answer results in a gain of $0.10 and an incorrect answer results in a loss of $0.10.

4. (uses a spreadsheet program, e.g., Excel) The discussion of long-run tendencies for the learning models was a little loose, since there are random elements in these models. One way to proceed is to simulate the learning processes implied by these models. Consider the reinforcement learning model, with initial choice probabilities of one-half each. You can simulate the initial choice by using the random number generator in Excel with the command "=rand()" which returns a random number between 0 and 1. The simulation could proceed by letting the person predict L if the random number is less than 0.5. The determination of the random event, L or R, could be done similarly, i.e., if a random number is below 0.75. Given the prediction, the event, and the reward, say 10 cents, you can use the reinforcement learning formula in (7.4) to determine the choice probabilities for the next round. This whole process can be repeated for a number of rounds by copying commands down the page to cells below for 20 rounds. Then you can do this again for a new simulation, which can be thought of as a simulation of the decision pattern of a second person. Use several simulations to determine how the choice probabilities evolve for ten rounds.

5. (non-mechanical) Consider a probability matching experiment in which each block of 20 trials is balanced to ensure that one of the events occurs in exactly 15 of the trials. Speculate on how this might change incentives for optimal prediction as the end of a block is approached. Herbert Simon, who later won a Nobel Prize for his work on bounded rationality, once remarked that data from experiments with these block designs suggest that the subjects may be smarter than the experimenters themselves. What do you think he had in mind?

6. Use the ball counting heuristic from chapter 5 to verify the Bayesian probabilities in the "inferred Pr" column of table 7.1 under the assumption that the first two decisions correctly reveal the draws seen.

7. (advanced) The discussion in this chapter was based on the assumption that the second person in a sequence would make a decision that reveals their own information, even when this information differs from the draw inferred from the first decision. The result is that the third person will always follow the pattern set by two matching decisions, because the information implied by the first two decisions is greater than the informational content of the third

person's draw. In this question, consider what happens if we alter the assumption that the second person always makes a decision that reveals the second draw. Suppose instead that the second person chooses randomly (with probabilities of 1/2) when the second draw does not match the draw inferred from the first decision. Use Bayes' rule to show that two matching decisions (AA or BB) should start a cascade even if the third draw does not match.

PART II

Behavioral Game Theory

One important role of game theory is to provide general paradigms that characterize the incentive structures of wide ranges of strategic social and economic interactions. The most standard paradigm is the prisoner's dilemma (encountered previously), which has a single Nash equilibrium with relatively low payoffs. Although there exists a better outcome for both players, the possibility of unilateral deviations (improving the individual player's welfare) prevent this preferred outcome from being realized. In contrast, a coordination game has both an equilibrium that is preferred by both players and a second equilibrium that is worse for both players. The better equilibrium may not be realized if there is more risk or if players somehow get stuck in the low-payoff equilibrium and expect it to continue. These simple dilemma and coordination games are discussed in chapter 8, along with a guessing game that provides a paradigm for thinking about what other players may do, what they think other players think they may do, and so on.

Many games involve decisions by different players made in sequence, e.g., one person makes an offer that the other must either accept or counter. In these situations, players who make initial decisions must predict the actions of subsequent decision makers. These sequential games, which are represented by an extensive form or "decision tree" structure, will be discussed in chapter 9.

The equilibria considered in chapters 8 and 9 all pertain to the use of deterministic, non-random strategies. In many situations, e.g., poker, war, and athletic or competitive business contests, it may be advantageous for a player to be unpredictable. Equilibria with randomized, "mixed" strategies are discussed in chapter 10. The final part of that chapter also considers the dual effects of explicit randomization and of behavioral "noise" that smooths out the sharp best-response functions that arise with perfect rationality.

Chapter 11 pertains to social dilemmas that are more complex than simple one-period prisoner's dilemmas, e.g., interactions with random termination. With a low probability of stopping, the resulting long horizons magnify the possible future costs of self-serving actions (defections) in the current period. In addition, the chapter considers games with more than two players, more than two decisions, and even the ability to choose partners. A simple example is a modification of the prisoner's dilemma so that a person can cooperate, defect, or break off the relationship. Another option is to scale up the interaction by essentially doing more business with a cooperative partner and less business with others.

Chapter 12 considers a special class of competitive games known as tournaments, in which players compete to be the winner. Tournaments are used to study attitudes toward competition and how such attitudes may be related to gender and other factors. A special type of tournament is a "rent-seeking" contest, in which the costs of competitive efforts (e.g., lobbying) may dissipate the value of the prize.

As one might expect, human behavior will not always conform tightly to simple mathematical models, especially in interactive situations. These models are important in terms of establishing benchmark predictions, from which deviations may be systematic and predictable. The chapters in this part of the book present a number of games in which behavior is influenced by intuitive economic forces in ways that are *not* captured by the simplest versions of game theory. Some examples are taken from Goeree and Holt (2001), "Ten Little Treasures of Game Theory and Ten Intuitive Contradictions." The "treasures" are treatments where data conform to theory, and the contradictions are produced by payoff changes with strong behavioral effects, even though these changes do not affect the Nash predictions. Since game theory is so widely used in the study of strategic situations like auctions, mergers, and legal disputes, it is fortunate that progress is being made toward understanding these anomalies. The resulting models often relax the strong game-theoretic assumptions of perfect selfishness, perfect rationality, and perfect predictions of others' decisions.

The final chapter in this part, chapter 13, contains a discussion of experiment design and nonparametric statistical test procedures. The basic methods of analyzing games introduced in these chapters will be applied in subsequent parts of the book in the context of bargaining, public goods, management of common pool resources, voting, and electoral competition. The final parts of the book provide applications to even more complex and policy-relevant situations: auctions, markets, and macroeconomic settings.

8

Some Simple Games: Competition, Coordination, and Guessing

A game with two players and two decisions can be represented by a 2 × 2 pay-off table or "matrix." Such games often highlight the conflict between incentives to compete or cooperate. This chapter introduces classic matrix games, the prisoner's dilemma and the coordination game, with the purpose of developing the idea of a Nash equilibrium. This concept is a somewhat abstract and mathematical, and the guessing game, discussed last, serves to highlight the extent to which ordinary people might come to such an equilibrium, even though initial play or play in a single round is typically not close to the Nash prediction.

Note to the Instructor: Both the Matrix Games and the Guessing Game can be run from programs with those names on the Veconlab Games menu. For the matrix game, it is possible to do a prisoner's dilemma in one treatment and a coordination game in another. Setups with random matchings are useful for matrix games, but this requires all decisions to be submitted before going to another "round," so it is best to specify a small number of rounds (even smaller for larger classes). For the Guessing Game, it is fine to put everyone into the same group or in several large groups and use the "go at your own pace" option. The guessing game is an ideal game to do only once, at least in the first treatment, as the focus is on how players learn from introspection about what others might do, not from experience with actual decisions.

8.1 Game Theory and a Prisoner's Dilemma Experiment

The Great Depression, which was the defining economic event of the twentieth century, caused a major rethinking of existing economic theories that represented the economy as a system of self-correcting markets that needed little in the way of active economic policy interventions. On the macroeconomic side, John Maynard Keynes's *The General Theory of Employment, Interest, and Money* focused on psychological elements ("animal spirits") that could cause a whole economy to become mired in a low-employment equilibrium, with no tendency for self-correction. An equilibrium is a state where there are no net forces causing further change, and Keynes's message implied that such a state may not necessarily be good. In such cases, an active policy intervention like bank deposit insurance might provide the confidence and impetus that enables people to escape from a bad equilibrium situation.

On the microeconomics side, Edward Chamberlin argued that markets may not yield efficient, competitive outcomes either. At about the same time, von Neumann and Morgenstern (1944) published *Theory of Games and Economic Behavior*. The book was motivated by the observation that a major part of economic activity involves bilateral and small-group interactions. In those circumstances, the classical assumption of non-strategic, price-taking behavior is not realistic. In light of the protracted Depression, these new theories generated considerable interest. Chamberlin's models of "monopolistic competition" were quickly incorporated into textbooks, and von Neumann and Morgenstern's book on game theory received front-page coverage in the *New York Times*.

A *game* is a mathematical model of a strategic situation in which players' payoffs depend on their own and others' decisions. A game is characterized by the players, their sets of feasible decisions, the information available at each decision point, and the payoffs (as functions of all decisions and random events). A *strategy* for a game is a complete plan of action that covers all contingencies. For example, a strategy in an auction could be an amount to bid for each possible estimate of the value of the prize. Since a strategy covers all contingencies, even those that are unlikely to occur, it is like a map that could be given to a hired employee to be "played" on behalf of the player in the game. An equilibrium is a set of strategies that is stable in some sense, i.e., with no inherent tendency for change because players have no incentive to deviate. The notion of equilibrium is most relevant in settings like repeated markets in which behavior has had a chance to "settle down" as participants have learned what to expect.

As noted in chapter 1, a Princeton graduate student, John Nash, extended the von Neumann and Morgenstern analysis by developing a notion of equilibrium for non-zero-sum games. Nash defined an equilibrium formally and showed that it exists under general conditions. This proof caught the attention of researchers at the RAND Corporation headquarters in Santa Monica, one of whom devised the story of the prisoner's dilemma. In the orginial prisoner's dilemma game story, the two suspects are separated and offered a set of threats and rewards that make it best for each to confess and essentially "rat" on the other person, whether or not the other person confesses. This story suggests that two prisoners might be bullied into confessing to a crime that they did not commit, which is a scenario from *Murder at the Margin*, written by two mysterious economists under the pseudonym Marshall Jevons (Breit and Elzinga, 1978).

Table 8.1. A Prisoner's Dilemma (Row's Payoff, Column's Payoff)

Row Player:	Column Player:	
	Not Confess	Confess
Not Confess	8, 8	0, 10
Confess	10, 0	3, 3

A game with a prisoner's dilemma structure is shown in table 8.1. The row player's Bottom decision corresponds to confession, as does the column player's Right decision. The Bottom-Right outcome, which yields payoffs of 3 for each, is worse than the Top-Left outcome, where both receive payoffs of 8. (A high payoff here corresponds to a light penalty.) The dilemma is that the "bad" confession outcome is an equilibrium in the following sense: if either person expects the other to talk, then their own best response to this belief is to confess as well. For example, consider the Row player, whose payoffs are listed on the left side of each outcome cell in the table. If Column is going to Confess, then Row either gets 0 for not confessing or 3 for choosing to Confess, so Confess is the best response to Confess. Similarly, Column's Confess decision is the best response to a belief that Row will also Confess. There is no other cell in the table with this stability property. For example, if Row thinks Column will not confess, then Row would want to confess, so the top/left (not confess) outcome is not stable. It is straightforward to show that the diagonal elements in which only one player confesses are also unstable.

The payoff numbers for the prisoner's dilemma in table 8.1 can be derived in the context of a simple example in which both players must choose between a low effort (0) and a higher effort (1). The cost of exerting the higher effort is $10, and the benefits to each person depend on the total effort for the two individuals combined. Since each person can choose an effort of 0 or 1, the total effort must be 0, 1, or 2, and the benefit per person is given in table 8.2.

To see the connection between this "production" function and the prisoner's dilemma payoff matrix, let the Confess decisions correspond to zero effort. Thus the mutual confession (bottom-right) outcome results in 0 total effort (for a per-person benefit of $3) and no cost. Therefore, the payoffs are 3 for each person in the bottom-right box of table 8.1. The top/left cell in the matrix is relevant when both choose efforts of 1 (at a cost of 10 each). The total effort is 2, so

Table 8.2. A Production Example that Results in a Prisoner's Dilemma (Effort Cost = 10)

Total Effort	0	1	2
Benefit per Person	$3	$10	$18

each earns 18 − 10 = 8, as shown in the top-left cell of the matrix. The top-right and bottom-left parts of the payoff table pertain to the case where one person receives the benefit of 10 at no cost, the other receives the benefit of 10 at a cost of 10, for a payoff of 0. Notice that each person has an incentive to "free ride" on the other's effort, since one's own effort is more costly ($10) than the marginal benefit of effort, which is 7 (= 10 − 3) for the first unit of effort and 8 (= 18 − 10) for the second unit.

Cooper, DeJong, Forsythe, and Ross (1996) conducted an experiment using the prisoner's dilemma payoffs in table 8.1, with the only change being that the payoffs were (3.5, 3.5) at the Nash equilibrium in the bottom-right box. Their matching protocol prevented individuals from being matched with the same person twice, or from being matched with anyone who had been matched with them or one of their prior partners. This *no-contagion* or "turnpike" protocol is analogous to going down a receiving line at a wedding and telling everyone the same bit of gossip about the bride and groom. Even if this story is so curious that everyone you meet in line repeats it to everyone they meet (who are behind you in the line), you will never encounter anyone ahead of you who has heard the story. In an experiment, the elimination of any type of repeated match-ing, either direct or indirect, means that nobody is able to send a "message" to future partners or to punish or reward others for cooperating. Even so, coopera-tive decisions (Top or Left) were fairly common in early rounds (43%), and the incidence of cooperation declined to about 20% in rounds 15–20.

A recent classroom experiment using the Veconlab program with the table 8.1 payoffs and random matching produced cooperation rates of about 33%, with no downward trend. A second experiment with a different class produced coop-eration rates starting at about 33% and declining to zero by the fourth period. This latter group behaved quite differently when they were matched with the same partner for five periods; cooperation rates stayed level in the 33–50% range until the final period. The "end-game effect" is not surprising since cooperation in earlier periods may consist of an effort to signal good intentions and stimulate reciprocity. These forward-looking strategies are not available in the final round.

The results of these prisoner's dilemma experiments are representative. There is typically a mixture of cooperation and defection, with the mix being somewhat sensitive to payoffs and procedural factors, and with some varia-tion across groups. In general, cooperation rates are higher when individuals are matched with the same person in a series of repeated rounds. In fact, the first prisoner's dilemma experiment run at the RAND Corporation more than 60 years ago lasted for 100 periods, and cooperative phases were interpreted as evidence against the Nash equilibrium.

A strict game-theoretic analysis with a fixed, known number of rounds would lead people to realize that there is no reason to cooperate in the final round, and knowing this, nobody would try to cooperate in the next-to-last round with the hope of stimulating final-round cooperation. Thus, there should be no cooperation in the next-to-last round, and hence there is no reason to try to stimulate such cooperation. Reasoning "backwards" from the end in this manner, one might expect no cooperation even in the very first round, at least when the total number of rounds is finite and known. Nash responded to the RAND mathematicians with a letter maintaining that it is unreasonable to expect people to engage in this many levels of iterative reasoning in an experiment with many rounds. There is an extensive literature on related topics, e.g., the effects of punishments, rewards, adaptive behavior, and various "tit-for-tat" strategies in repeated prisoner's dilemma games with no pre-announced final round, which will be discussed in later chapters.

Finally, it should be noted that there are many ways to present the payoffs for an experiment, even one as simple as the prisoner's dilemma. Both the Cooper et al. and the Veconlab experiment used a payoff matrix presentation. Alternatively, the instructions could be presented in terms of the cost of effort and the table showing the benefit per person for each possible level of total effort. This presentation is perhaps less neutral, but the economic context makes it less abstract and artificial than a matrix presentation. It is also possible to set up a hand-run prisoner's dilemma game as one where each person chooses which of two cards to play. See Capra and Holt (2000) for instructions. For example, suppose that each person has playing cards numbered 8 and 6. Playing the 6 "pulls" $6 (from the experimenter's reserve) into one's own earnings, and playing the 8 "pushes" $8 (from the experimenter's reserve) to the other person's earnings. If they both pull the $6, earnings are $6 each. Both would be better off if they played 8, yielding $8 for each. Pulling $6, however, is better from a selfish perspective, regardless of what the other person does. The best outcome from a selfish point of view is to pull $6 when the other person pushes $8, which yields a total of $14. A consideration of the resulting payoff matrix (problem 2) indicates that this is clearly a prisoner's dilemma. The card presentation is quick and easy to implement in class, where students hold the cards played against their chests. A Nash equilibrium survives an "announcement test" in that neither would wish they had played a different card given the card played by the other. In this case, the unique Nash equilibrium is for each to play 6.

Prisoner's Dilemma Experiments: *Subjects in a prisoner's dilemma with random matching may try some cooperation initially, but defection tends to dominate after several rounds, especially as a pre-announced final round approaches. Cooperation is higher when subjects are rematched with the same partner.*

The prisoner's dilemma is an important paradigm in social sciences, since many business, political, and social relationships have the property that joint cooperation is good for all, but individuals have an incentive to defect, e.g., to save on effort costs, and then "free ride" off of others' cooperative efforts. These generalizations, referred to as *social dilemma games*, may involve multiple players and more complex strategies, but the same tension between joint cooperation and individual defection pervades. It is a serious error, however, to view every interaction as a prisoner's dilemma, as the game considered next indicates.

8.2 A Coordination Game

Many production processes have the property that one person's effort increases the productivity of another's effort. For example, an Internet commerce company must take orders, produce the goods, and ship them. Each sale requires all three services, so if one process is slow, the efforts of those in other activities are to some extent wasted. In terms of the two-person production function, recall that 1 unit of effort produced a benefit of $10 per person and 2 units produced a benefit of $18. Suppose that the second unit of effort makes the first one more productive in the sense that the benefit is more than doubled when a second unit of effort is added. In table 8.3, per-person benefit is $30 for 2 units of effort.

If the cost of effort remains at $10 per unit, then each person earns $30 − $10 = $20 in the case where both supply a unit of effort, as shown in the revised payoff matrix in table 8.4, where decisions have been relabeled as High and Low effort. As before, the Low effort outcome in the bottom-right cell is a Nash equilibrium, since neither person would want to change from their decision unilaterally. For example, if Row knows that Column will choose Low, then Row can get $0 from High and $3 from Low, so Row would want to choose Low, as indicated by the downward pointing arrow. Similarly, Low is a best response for Column to Row's decision to use Low, as indicated by the right arrow. In each

Table 8.3. A Production Example that Results in a Coordination Game

Total Effort	0	1	2
Benefit per Person	$3	$10	$30

Table 8.4. A Coordination Game with Low-Effort Cost
(Row's Payoff, Column's Payoff)

Row Player:	Column Player: High	Low
High	20, 20	⇓ 0, 10 ⇐
Low	⇑ 10, 0 ⇒	3, 3

case, the arrow next to a payoff indicates the directional movement that will raise that player's payoff.

There is a second Nash equilibrium in the top-left box of this modified game, which is reached when both choose High effort. To verify this, think of a situation in which we "start" in that box, and consider whether either person would want to deviate, so there are two things to check.

Step 1: If Row is thought to choose High, then Column would want to choose High in order to earn 20 instead of 10, as indicated by the left arrow.

Step 2: If Row expects Column to choose High, then Row would want to choose High, as indicated by the up arrow.

Thus, the High/High outcome in the top-left box survives an "announcement test" and would be stable. This second equilibrium yields payoffs of 20 for each, far better than the payoffs of 3 for each in the Low/Low outcome. In this case, it is the strong synergy in production that results in the high-payoff equilibrium in which each person benefits from the other's effort. This is called a *coordination game*, since the presence of multiple equilibria raises the issue of how players will guess which one will be relevant.

It used to be common for economists to simply *assume* that rational players would coordinate on the best equilibrium, if all could agree which is the best equilibrium. While it is apparent that the High/High outcome is likely to be the most frequent outcome, one example does not justify a general assumption that players will always coordinate on an equilibrium that is better for all. This assumption can be tested with laboratory experiments shown to be incorrect. In fact, players sometimes end up at the equilibrium that is the *worst* for all concerned (Van Huyck, Battalio, and Beil, 1990, 1991).

Consider the intuitive effect of increasing the cost of effort from \$10 to \$17. With this increase in the cost of effort, the payoffs in the high-effort top-left outcome are reduced to \$30 − \$17 = \$13. Moreover, the person who exerts high effort alone only receives a \$10 benefit and incurs a \$17 cost, so the payoffs for this person are −\$7, as shown in table 8.5.

Notice that High/High in the top-left box is still a Nash equilibrium: if Column is expected to play High, then Row's best response is High and vice versa, as indicated by the up and left directional arrows. But High effort is a risky decision that might result in payoffs of \$13 and −\$7, as compared with the \$10 and \$3 payoffs associated with the Bottom and Right strategies that yield the other Nash equilibrium. While the absolute payoff is higher for each person in the High/High outcome, each person has to be very sure that the other one will not deviate. In fact, this is a game where behavior is likely to converge to the Nash equilibrium that is worst for all. These payoffs were used in a classroom experiment that lasted five periods with random matching of

Table 8.5. A Coordination Game with High-Effort Cost
(Row's Payoff, Column's Payoff)

Row Player:	Column Player: High	Low
High	13, 13	⇓ –7, 10 ⇐
Low	⇑ 10, –7 ⇒	3, 3

12 participants. Participants were individuals or pairs of students at the same computer. By the fifth period, only a quarter of the decisions involved High effort. The same people then played the less risky coordination game shown in table 8.4, again with random matching. The results were quite different; all but one of the pairs ended up in the good (High, High) equilibrium outcome in all periods. To summarize:

Coordination Game Experiments: *A coordination game has multiple Nash equilibria, including one that is better for all players. Nevertheless, it is not appropriate to simply assume that behavior will somehow converge to the best outcome for all, nor is it correct to expect that behavior will end up at the equilibrium that is worst for all. Decisions in coordination game experiments sometimes converge to the high-effort outcome that yields high payoffs. But in games where the high-payoff outcome is risky, behavior might converge to the low-payoff equilibrium.*

Another type of synergy that generates a coordination game is a situation in which the joint production depends on the *minimum* of players' efforts. For example, if production requires two components, e.g., wheels and frame, then the number of units that can be shipped depends on the minimum of the quantities of the two components. These kinds of interconnections can create severe incentive problems. For example, the marketing department may not exert extra effort if they believe that production will fall below projections, and vice versa. Students who work on team projects are well aware of the incentive issues. One student quipped: "Minimum effort game? Now that's one I can play!"

For simplicity, consider a game with two players and two possible effort levels, 1 and 2. The cost of effort is 1, and the value produced is 3 times the minimum effort. So if both choose efforts of 2, the minimum is 2 and their payoffs are 2(3) minus the cost of 2, which is 4, as shown in the top-left cell of table 8.6. But choosing an effort of 2 when the other chooses 1 only yields $3 - 2 = 1$ because the minimum is 1. Finally, choosing an effort of 1 drives the minimum to 1, so the payoff is $1(3) - 1 = 2$. Notice that there are two Nash equilibria: one with high efforts and one with low efforts.

Table 8.6. A Minimum Effort Coordination Game (Row's Payoff, Column's Payoff)

Row Player:	Column Player:	
	Effort = 2	Effort = 1
Effort = 2	4, 4	$\Downarrow$ 1, 2 $\Leftarrow$
Effort = 1	$\Uparrow$ 2, 1 $\Rightarrow$	2, 2

Table 8.6 can also be used for an *N*-person game, in which case the "column efforts" represent the minimum of the *N*-1 other players' efforts, although other player's payoffs would depend on their effort levels. In this case, intuition suggests that incentive problems would be magnified, since any uncertainty about others' decisions makes it more likely that at least one of the others slips into low effort.

For these coordination games with multiple equilibria, the one with the most "attraction" may depend on payoff features (e.g., the effort cost or the number of players) that determine which equilibrium is riskier. These issues will be revisited later in this chapter.

8.3 A Guessing Game

A key aspect of most games is the need for players to guess how other players will behave in order to determine their own best decision. This aspect is somewhat obscured in a 2 × 2 game since a wide range of beliefs may lead to the same decision, and therefore, it is not possible to say much about the beliefs that stand behind any particular observed decision. In this section, we turn to a game with decisions in the range from 0 to 100, and any number of players, *N*. Each player selects a number in this interval, with the advance knowledge that the person whose decision is closest to 1/2 of the average of all *N* decisions will win a money prize. (The prize is divided equally in the event of a tie.) Thus, each person's task is to make a guess about the average, and then submit a decision that is half of that amount.

There is no way to learn from past decisions in the first round or if the game is only played once. In this case, each person must "learn" by thinking *introspectively* about what others might do, what they think others think they might do, etc. The issue is how many levels of iterated reasoning are involved. The most naïve person would not think at all and might choose randomly between 0 and 100, with an average of 50; think of this low-rationality person as a "level-0" type (Stahl and Wilson, 1995). A slightly more forward-thinking person might reason that since the decisions must be between 0 and 100, without further knowledge it might be OK to guess that the average of others' decisions would be at the midpoint, 50. This person would submit a decision of 25 in order to be at about half of the anticipated average. Think of this person as a "level 1" type, since they engaged in one level of iterated reasoning. A "level 2" type would carry this one step further and guess half of 25, or 12.5. Even higher levels of iterated thinking will lead a person to choose a decision that is closer and closer to 0. So the best decision to make depends on one's beliefs about the extent to which others are thinking iteratively in this manner, i.e., on beliefs about the others' rationality. Since the best response to any common decision (above 0) is to submit a number half of that, it follows that no common decision above 0 can be a Nash equilibrium. Clearly a common decision of 0 is a Nash equilibrium: each person gets a positive share (1/N) of the prize and there is no incentive to deviate (e.g., choosing a higher decision would result in a zero payoff).

Figure 8.1 shows data for a classroom guessing game run with the Veconlab software. The five participants made first-round decisions of 55, 29, 22, 21, and 6.25, with an average of about 27. Notice that the lowest person is one who engaged in three levels of iterated reasoning about what the others might do, and this person was closest to the target level of about 27/2 = 13.5. This person submitted the same decision in round 2 and won again; the other decisions were 20, 15, 11.5, and 10. The other people did not seem to be engaging in any precise iterated reasoning, but it is clear the person with the lowest decision is being rewarded, which will tend to pull decisions down in subsequent rounds. These decisions converge to near-Nash levels by the fifth round, although nobody selected 0 exactly.

The second treatment changes the target from 1/2 of the average to 20 + 1/2 of the average, and the prize goes to the person whose guess is closest to that

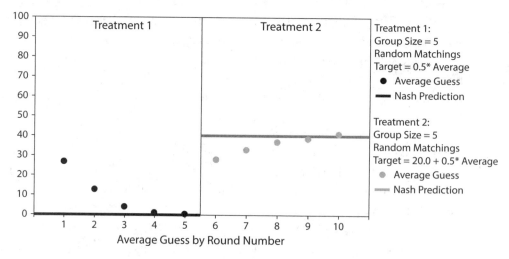

Figure 8.1. Data from a Classroom Guessing Game. (The target is 1/2 of the average in rounds 1–5, and 20 + 1/2 of the average in rounds 6–10.)

target. The instructor (a.k.a. author) who ran this in class had selected the parameters quickly, with the conjecture that adding 20 to the target would raise the Nash equilibrium from 0 to 20. When the average came in at 28, the instructor thought that it would then decline toward 20, just as it had declined toward 0 in the first treatment. But the average rose to 33 in the next round, and continued to rise to a level that is near 40. Notice that if all choose 40, then the average is 40, half of the average is 20, and 20 + half of the average is 20 + 20 = 40, which is the Nash equilibrium for this treatment (problem 6).

The guessing game was used in the Nagel (1995, 1999) experiments. She reports a widespread failure of behavior to converge to the Nash equilibrium in a one-round experiment, regardless of subject pool. Nagel discusses the degree to which some of the people go through iterations of thinking that may lead them to choose lower and lower decisions (e.g., 50, 25, 12.5, etc.). The name of the "guessing game" was motivated by one of Keynes's remarks that investors are typically in a position of trying to guess what stock others will find attractive (first iteration), or what stock that others think other investors will find attractive (second iteration), etc.

> Guessing Game Experiments: *Some people think iteratively about what others will do in a guessing game played once, but this is not uniform, and the effect of such introspection is not enough to move decisions to near-Nash levels in a single round of play. When people are able to learn from experience, behavior typically does converge to the Nash equilibrium, even in some cases when the equilibrium is not apparent to the person who designed the experiment.*

8.4   Three Components: Introspection, Learning, and Equilibrium

The prisoner's dilemma has a Nash equilibrium that is worse than the outcome that results when individuals cooperate and ignore their private incentives to defect. The dilemma is that the equilibrium is the bad outcome and the good outcome is not an equilibrium. In contrast, the coordination game has a good outcome that is also an equilibrium. The previous discussion suggests the usefulness of modeling the process that leads to equilibrium, which is especially important when there are multiple Nash equilibria and multiple rounds of play, perhaps with random matching. Therefore, it is useful to distinguish three approaches:

1. *Equilibrium*: Analysis usually starts (and sometimes ends) with locating the Nash equilibrium, which involves thinking about starting from equilibrium and considering whether any player could increase earnings from a unilateral defection. Most of the effort in

a game theory class is focused on finding and characterizing Nash equilibria.

2. *Introspection*: For a game that is only played once, or in the first round of repeated play, a player has no experience on which to form beliefs about others' behavior. The most widely used model of introspection is the "level-k" approach that allows for people to have different levels of strategic thinking, ranging from level 0 (no thinking, random play) to level 1 (thinks others are random and best responds to level 0), level 2 (thinks others are level 1 and best responds), etc. Analysis of laboratory experiments classifies most people in levels 1–3.

3. *Learning*: After the first round, subjects are able to use information they glean from observing other players' decisions to update their beliefs. Dynamics could be based on belief learning models that were discussed in the previous chapter, or alternatively, on reinforcement learning models that specify probabilistic responses that depend on cumulative earnings ratios for alternative decisions. Learning models can be used to study the direction of convergence of decisions to Nash equilibria, a process that may also explain why behavior does not converge to certain equilibria.

These three factors, equilibrium, introspection, and learning, will be considered in the context of the coordination games with low- and high-effort costs, which are reproduced in table 8.7 for easy reference and comparison.

Considering what a player would do if the other is expected to choose each of the efforts with probability 1/2 (this could be thought of as a level 1 response), this approach yields an interesting comparison between the two games in table 8.7. In the top (low-effort cost) game, the High effort has a higher expected payoff (average of 20 and 0) relative to the Low effort (average of 10 and 3). The opposite is the case for the high-effort cost game at the bottom of table 8.7, in which the additional effort cost has caused Low effort

Table 8.7. Coordination Games with Different Effort Costs

Low Effort Cost Game	High	Low
High	20, 20	0, 10
Low	10, 0	3, 3

High Effort Cost Game	High	Low
High	13, 13	–7, 10
Low	10, –7	3, 3

to have a higher expected payoff for somebody who thinks the other decision is random. When one holds the belief that the other player will choose high effort with probability 1/2, the expected payoff difference favoring High effort switches from positive in the top game to negative in the bottom game. The remainder of this section takes this observation and presents it in a graphical framework that will be useful for thinking about learning and adjustment to equilibrium.

Equilibrium and Best Responses

Let p denote the probability that the *other* player chooses high effort. Suppose that high effort is expected, $p = 1$, and consider the incentive that a player has to do the same. In the top payoff matrix of table 8.7, the payoff increase for the row player matching the other's high effort is determined from the left column as $20 - 10 = 10$. Conversely, suppose that the column player is expected to choose low effort (so $p = 0$). In this case, the payoff "increase" is $0 - 3 = -3$. In other words, if the other player is going to choose low effort, then choosing high effort results in a payoff loss, -3. These points, a payoff difference of 10 when $p = 1$ and a payoff difference of -3 when $p = 0$, are represented by diamonds on the top and bottom borders of figure 8.2, where p is on the vertical axis and the payoff difference is on the horizontal axis.

The payoff difference refers to the difference between the payoff for high effort and that for low effort. The payoff difference depends on the probability that the other chooses high effort, so the difference can be expressed: $\pi_H(p) - \pi_L(p)$, where the H and L subscripts indicate High and Low effort. To summarize:

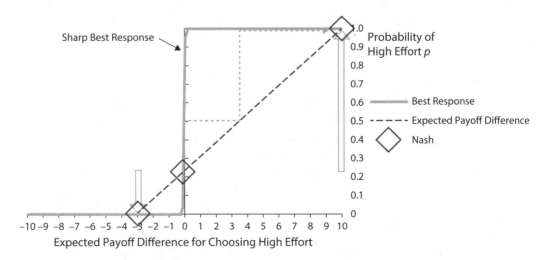

Figure 8.2. Expected Payoff Difference (dashed) Line for Low-Effort Cost Coordination Game

Payoff Differences and Best Responses: *When the expected payoff difference for high effort over low effort is positive, the best response is to choose high effort, as indicated by the gray best response line at the top on the right side of figure 8.2. Conversely, when the expected payoff difference is negative, the best response is to choose low effort, as indicated by the best response line at the bottom on the left. At an expected payoff difference of 0 in the middle, the best response crosses from bottom to top, and for this reason, this is labeled as a "sharp best response" line in the figure.*

Recall that Nash equilibria are determined by considering payoff differences for unilateral changes. When $p = 0$ at the bottom, the expected payoff difference from exerting high effort is negative, and at the top, when $p = 1$, the expected payoff difference is positive. The two diamonds at the top and bottom of the figure correspond to the two Nash equilibria: The diamond on the upper right side is for the case in which the other person is expected to choose high effort, and there is a positive +10 incentive to stay with the same high-effort decision. The lower left diamond is for the case where the other person is expected to choose low effort ($p = 0$), and there is a negative incentive, −3, to switch to high effort. (Please ignore the middle diamond for now, it is at a point where the expected payoff difference is 0, so players would be willing to randomize, and randomized strategies will be discussed in a later chapter.) To summarize the equilibrium implications:

Nash Equilibrium and Best Responses: *In a Nash equilibrium, each player is making a best response to belief probabilities about others' behavior. With only two decisions and a belief p, the Nash equilibrium is at intersections between the sharp best response line and the expected payoff difference line, as indicated by the diamonds in figure 8.2.*

Introspection

Next, consider the effects of introspection. If a player's beliefs are that the other is equally likely to choose high effort, then $p = 0.5$. In this case, choosing high effort for the top game in table 8.7 is equally likely to earn 20 or 0, for an average of 10. This expected payoff from high effort is above what can be expected by choosing low effort, which would be an average of 10 and 3 when $p = 0.5$. Thus, at $p = 0.5$, the payoff difference for high effort over low is positive, as shown in figure 8.2 by the dotted line that starts in the middle and then goes to the right to the dashed payoff difference line. Since this difference is positive, the dotted line continues up to the top, reflecting the fact that the *best response* to a positive payoff difference is to choose high effort. Hence a level 1 player, best responding to $p = 0.5$ randomness, would choose high effort. And a level 2 player would also choose "High" in response to what the level 1 player does, etc.

Learning and Basins of Attraction

In practice, learning may be a slow and irregular process, but notice that there is a wide range of beliefs, anywhere above about 0.23, for which the payoff difference is positive, which would tend to increase the proportion of players in a mix that would choose high effort. This upward tendency is represented by the up arrow on the right side of figure 8.2. However, if the belief is that the other player is unlikely to choose high effort ($p < 0.23$), then the expected payoff difference is negative, with a tendency to choose low effort indicated by the down arrow on the left. The *basin of attraction* for a specific equilibrium is the set of beliefs for which forces are pushing players toward that equilibrium. In this game, the basin of attraction for the high effort equilibrium is larger.

Consider the high-effort-cost coordination game at the bottom of table 8.7. The increased effort cost (now 17 instead of 10) has shifted the expected payoff difference line to the left by 7 in figure 8.3 from where it was in figure 8.2. There are still two equilibria for the high-effort-cost game, indicated by the diamonds at the top and bottom, but the basin of attraction for the high-effort equilibrium is now smaller. Moreover, a belief probability of $p = 0.5$ now results in a negative expected payoff difference, which would cause a level 1 player to choose low effort.

Learning and Introspection in Coordination Games: *An increase in effort cost in moving from the top to bottom games in table 8.7 has reduced the basin of attraction for the high-effort equilibrium. Moreover, the best response of a level 0 (random) player is to choose low effort in the high-cost game, the reverse of choosing high effort in the low-cost game.*

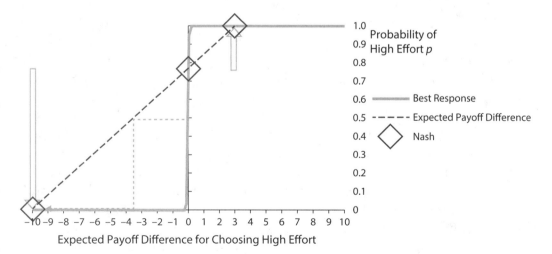

Figure 8.3. Expected Payoff Difference (dashed) Line for High-Cost Coordination Game

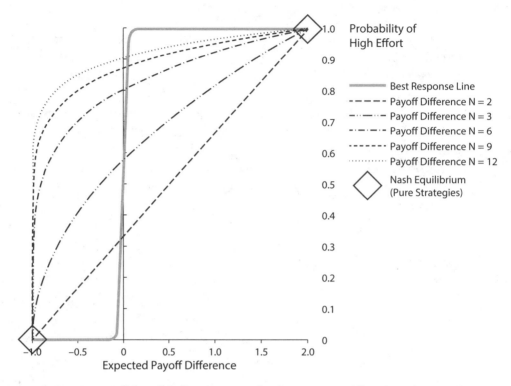

Figure 8.4. Expected Payoff Difference Lines for the Minimum Effort Coordination Game

Finally, consider the effect of increases in the numbers of players on the basin of attraction for the two Nash equilibria for the minimum effort coordination game in table 8.6. Figure 8.4 shows the expected payoff difference lines for various group sizes in the minimum effort coordination game in table 8.6. Recall that each unit increase in minimum effort raises the joint product value by 3 and it raises the cost by 1. Assuming that the other player is choosing a high effort, an an increase from 1 to 2 effort units will raise the minimum effort and add $3 - 1 = 2$ to a player's payoff. In the figure, the vertical axis is the probability that the other players choose high effort, denoted by p. So when $p = 1$ at the top of the figure, the payoff difference is 2, and all of the difference lines for various group sizes meet at the top of the figure at a payoff difference of 2. But when $p = 0$ at the bottom of the figure, the others are not choosing high effort and minimum effort will be 1. In this case, the payoff difference from increasing effort is just the ("wasted") cost incurred of -1, so all of the payoff difference lines for various group sizes start from a common point in the lower left corner of the figure, where the diamond indicates a Nash equilibrium with low effort. Similarly, the high-effort equilibrium is indicated by the diamond at the upper right corner,

With two players, the expected payoff difference is linear with respect to the probability p that the other player chooses high effort, so the dashed payoff difference line for $N = 2$ increases linearly from the lower left to the top right. The vertical line in the center of the graph is at an expected payoff difference of 0, so there is an incentive to increase effort to the right and reduce effort to the left, as indicated by the best response line that starts on the left bottom and crosses to the top on the right side of the figure. The basins of attraction for the two-person game are divided at $p = 0.33$, where the payoff difference line crosses over from negative to positive.

With three players, the situation is much more risky, since the minimum effort will not be 2 unless both of the others choose high effort, which happens with probability p^2. This quadratic effect reduces the expected payoff difference for an effort increase at any given level of p. The payoff difference lines for larger group sizes, 6, 9, and 12, shift even farther to the left, resulting in even lower basins of attraction for the high-effort equilibrium. Therefore, one would not expect to see high-effort outcomes in games with large groups of players, as was the case with the classic minimum effort experiments done by Van Huyck et al. (1990), who used large groups of 14–16 players. To summarize:

Group Size Effects in Minimum Effort Coordination Games: *An increase in the number of players needed to coordinate on a minimum effort tends to increase the risk associated with high effort. The result is that the basin of attraction for the high-effort Nash equilibrium shrinks considerably, making the low-effort outcomes more likely for large groups.*

The coordination failures for large groups captured the attention of macroeconomists, who had long speculated about the possibility that whole economies could become mired in low-productivity states, where people do not engage in high levels of market activity because no one else does. The macroeconomic implications of coordination games are discussed in Bryant (1983), Cooper and John (1998), and Romer (1996), for example. Chapter 25, the macroeconomics chapter, begins with a bank run experiment that is essentially a coordination game.

8.5 Extensions: Other Perspectives on Cooperation and Coordination

The prisoner's dilemma game can be given an alternative interpretation, i.e., that of a "public good." This is because each person's benefit depends on the total effort, including that of the other person. In effect, neither person can appropriate more than half of the benefit of their own effort, and in this sense the benefit is publicly available to both, just as national defense or police protection are freely available to all. We will consider the public goods provision problem in more detail in chapter 16 of this book. The setup allows a large number of effort levels, and the effects of changes in costs and other incentives will be considered. In this case, one person's contribution to a common payoff pool can make others better off, which creates possibilities for reciprocity and other-regarding behavior, which add incentives other than those induced by payoff differences in the game itself.

There have been hundreds of prisoner's dilemma experiments over the decades, and some interesting findings have emerged. As might be expected, cooperation is enhanced when people can choose whom to interact with and whom to exclude. Moreover, if the final period is not known, but is determined randomly, it may be possible to sustain cooperation with implicit threats of punishment for defection if the stop probability is sufficiently low. Some of these topics are covered in the chapters that follow. For a brief, non-technical discussion of John Nash's original (1950) *Proceedings of the National Academy of Sciences* paper, its subsequent impact, and relevant policy applications, see Holt and Roth (2004).

It is important to note that behavior in coordination games with multiple equilibria can be also influenced by factors other than payoff differences. For example, the maximum joint payoff has some drawing power, and the presence of a loss for the high-effort choice in one of the games would push in the other direction (problem 8). This fluidity of behavior in coordination games has made them an ideal basis for studying the effects of incentives (Brandts and Cooper, 2006) and communication (Capra et al., 2009; Brandts and Cooper, 2007). The latter paper shows that messages from a manager urging high effort and the mutual benefits often have more of an effect than effort-boosting incentives. The most effective communication was a simple message encouraging high effort and stressing mutual benefits. Bonus incentives paid for by a manager did not raise production enough to cover the manager's cost. The title tells the main story: "It's What You Say, Not What You Pay: An Experimental Study of Manager-Employee Relationships in Overcoming Coordination Failure."

Finally, there is a simple theoretical calculation, proposed by two Nobel Laureates, John Harsanyi and Reinhard Selten (1988), which involves the losses of deviating and selecting the equilibrium with the lowest product of deviation losses. Consider the low-effort-cost game at the top of table 8.7. The deviation loss (what is lost by deviating) from the high-effort equilibrium in the upper-left box is $20 - 10 = 10$, so the product of deviation losses is 100. For the low-effort equilibrium in the bottom-left box, the deviation losses are $3 - 0 = 3$ for each player, so the product is only 9. These magnitudes of deviation loss products are reversed for the high-effort-cost game at the bottom of table 8.7. The equilibrium with the highest deviation loss product is said to be *risk dominant*, which is the equilibrium "selected" by this theoretical criterion (Straub, 1995). These observations are consistent with the experimental evidence

Chapter 8 Problems

1. Suppose that the cost of a unit of effort is raised from $10 to $25 for the example based on the table shown below. Is the resulting game a coordination game or a prisoner's dilemma? Explain, and find the Nash equilibrium (or equilibria) for the new game.

Total Effort	0	1	2
Benefit per Person	$3	$10	$30

2. What is the payoff table for the prisoner's dilemma game described in section 8.1 of the text based on "pulling" $6 or "pushing" $8?

3. Each player is given a 6 of hearts and an 8 of spades. The two players select one of their cards to play. If the suit matches (both hearts or both spades), then they each are paid $6 for the case of matching hearts and $8 for the case of matching spades. Earnings are zero in the event of a mismatch. Is this a coordination game or a prisoner's dilemma? Show the payoff table and find all Nash equilibria (that do not involve random play).

4. Suppose that the payoffs for the original prisoner's dilemma in table 8.2 are altered as follows. The cost of effort remains at $10, but the per-person benefits are given in the table below. Recalculate the payoff matrix, find all Nash equilibria (that do not involve random play), and explain whether the game is a prisoner's dilemma or a coordination game.

Total Effort	0	1	2
Benefit per Person	3	5	18

5. Suppose that two people are playing a guessing game, with a prize going to the person closest to 1/2 of the average. Guesses are required to be between 0 and 100. Show that none of the following are Nash equilibria: (a) Both choose 1. (Hint, consider a deviation to 0 by one person, so that the average is 1/2, and half of the average is 1/4.) (b) One person chooses 1 and the other chooses 0. (c) One chooses x and the other chooses y, where $x > y > 0$. (Hint: If they choose these decisions, what is the average, what is the target, and is the target less than the midpoint of the range of guesses between x and y? Then use these

calculations to figure out which person would win, and whether the other person would have an incentive to deviate.)

6. Consider a guessing game with N people, and the person closest to 20 plus 1/2 of the average is awarded the prize, which is split in the event of a tie. The range of guesses is from 0 to 100. Show that 40 is a Nash equilibrium. (Hint: Suppose that N-1 people choose 40 and one person deviates to $x < 40$. Calculate the target as a function of the deviant's decision, x. The deviant will lose if the target is greater than the midpoint between x and 40, i.e., if the target is greater than $(40 + x)/2$. Check to see if this is the case.)

7. For the guessing game in problem 6, a "level 0" person would choose randomly on the interval from 0 to 100. What decision would be made by a "level 1" person who simply makes a best response to a level 0 person? Do the decisions for those who engage in successively higher levels of reasoning converge?

8. (non-mechanical) An alternative explanation for the failure to coordinate on high effort for the game in table 8.5 could be based on loss aversion, since the only negative payoff, -7, is a possible payoff if one chooses high effort. How could one modify the payoffs for the games in tables 8.4 and 8.5 to distinguish between loss aversion and sizes of basins of attraction.

9. (non-mechanical) Review the deviation loss calculations and the discussion of best responses to prior beliefs of $p = 0.5$ in figures 8.2 and 8.3. How do the structures of those figures clarify the relationship between level 1 responses and risk dominance?

10. (minimum effort game, non-mechanical) Consider a two-player game in which each player can choose an effort of 1 or 2. The cost of effort is 1 for low effort and 2 for high effort. The benefit for each player is the product of a positive value parameter, V, and the minimum of the two efforts. So if the players' efforts are both 2, they both earn $2V$. But if one player chooses an effort of 1 and the other chooses an effort of 2, the minimum is only 1, and they both earn only V. For what values of V would this setup result in a coordination game with two equilibria, one with low efforts and one with high efforts?

9

Multi-Stage Games, Noisy Behavior

The games considered up to this point have involved simultaneous decisions. Many interesting games, however, are sequential in nature, so that the first mover must try to anticipate how the subsequent decision maker(s) will react. For example, the first person may make a take-it-or-leave-it proposal for how to split a sum of money, and the second person must either accept the proposed split or reject. Another example is a labor-market interaction, where the employer first chooses a wage and the worker, seeing the wage, selects an effort level. Several common principles are used in the analysis of such games, and these will be covered in the current chapter. Somewhat paradoxically, the first person's decision is sometimes more difficult, since the optimal decision may depend on a forecast of the second person's response, whereas the person who makes the final decision does not need to forecast the other's action if it has already been observed. In this case, we begin by considering the final decision maker's choice, and then we typically work backward to consider the first person's decision, in a process that is known as *backward induction*. Of course, the extent to which this method of backward induction yields good predictions is a behavioral question, which will be evaluated in the context of laboratory experiments.

For the players in a multi-stage game, the process of forecasting others' subsequent decisions is complicated by the possibility that others make mistakes, especially when the costs of mistakes are small. The chapter ends with a consideration of probabilistic choice methods, imported from psychology, that are used to model the sensitivity of error rates to the relevant costs.

Note to the Instructor: Several of the games discussed are the default settings for the Veconlab Two-Stage Game program. There is also a Centipede Game program, with setup options that allow any number of stages. A hand-run version of the centipede game can be done easily with a roll of quarters and a tray (problem 6).

9.1   Extensive Forms and Strategies

This section will present a simple two-stage bargaining game, in which the first move is a proposal of how to divide $4, which must either be $3 for the proposer and $1 for the other, or $2 for each. The other player (responder) sees this proposal and either accepts, which implements the split, or rejects, which results in $0 for each. As defined previously, a *strategy* for such a game is a complete plan of action that covers every possible contingency. In other words, a strategy tells the player what decision to make at each stage. In fact, the instructions of a strategy are so detailed that it could be carried out by an employee or agent, who would never need to ask questions about what to do.

In the two-stage bargaining game, a strategy for the first person is which proposal to offer, i.e., ($3, $1) or ($2, $2), where the proposer's payoff is listed on the left in each payoff pair. A responder's strategy must specify a reaction to each of these proposals. Thus the responder has a number of options: (a) accept both, (b) reject both, (c) accept ($3, $1) and reject ($2, $2), or (d) reject ($3, $1) and accept ($2, $2). These strategies will be referred to as: AA, RR, AR, and RA respectively. (The first letter indicates the response, Accept or Reject, to the unequal proposal.) Similarly, the proposer's strategies will be easier to remember if they are labeled as Equal (equal split) or Unequal (unequal, favoring the proposer). A Nash equilibrium is a pair of strategies, one for each player, with the property that neither can increase their own payoff by deviating to a different strategy, assuming the other player stays with the equilibrium strategy.

One Nash equilibrium for this game is (Equal, RA), where the proposer offers an equal split, and the responder rejects the unequal proposal and accepts the equal proposal. Deviations are not profitable, since the responder is already receiving the proposal that offers the higher of the two possible payoffs, and the proposer would not want to switch to the unequal proposal given that the responder's strategy requires that it be rejected. This Nash equilibrium, however, involves a (non-credible) threat by the responder to reject an unequal proposal. This would cause both players to earn $0 in the scenario that the first player chose unequal. Such a rejection would reduce the responder's payoff from $1 to $0, which violates a notion of sequential rationality that requires play to be rational in all parts of the game. The reader might wonder how this kind of irrationality can be part of a Nash equilibrium, and the answer is that the simple notion of a Nash equilibrium only requires rationality in terms of considering unilateral deviations from the equilibrium by one player, under the assumption that the other player's strategy is unchanged. In the equilibrium being considered, (Equal, RA), the unequal proposal is not made, so the responder never has to carry out the threat to reject this proposal. Selten (1965) proposed that the rationality requirement be expanded to cover all parts of the game, which he defined as "subgames" in a somewhat more formal manner than is necessary here. He called the resulting equilibrium a *subgame perfect* Nash equilibrium. Obviously, not all Nash equilibria are subgame perfect, as the equilibrium (Equal, RA) demonstrates.

In order to find the (subgame perfect) equilibrium, consider starting in the final stage, where the responder is considering a specific proposal. Each of the possible proposals involves strictly positive amounts of money for the responder, so a rational responder who only cares about their own payoff would accept either proposal. Hence, the only responder strategy that satisfies rationality in all subgames is AA. Next consider the first stage. If the proposer predicts that the responder is rational and will accept either offer, then the proposer will demand the larger share, so the equilibrium will be (Unequal, AA). This is a Nash equilibrium, and from the way it was constructed, we know that it is subgame perfect.

The reasoning process used in the previous paragraph is an example of backward induction, which simply means analyzing a sequence of decisions by starting at the end and working backward. The concepts of sequential rationality and subgame perfection can all be defined more precisely, but the goal here is for the reader to obtain an intuitive understanding of the principles involved, based on specific examples. This understanding is useful in constructing predictions for outcomes of simple multi-stage games, which can serve as benchmarks for evaluating observed behavior in experiments. As we shall see, these benchmarks do not always yield very good predictions, for a variety of reasons to be discussed.

It is instructive to show how the two-stage bargaining game can be represented as a decision tree, which is known as the "extensive form" of a game. In figure 9.1, the game begins at the top, at the node that is labeled Proposer. This player chooses between an Equal proposal ($2, $2), or an Unequal proposal ($3, $1). The Responder has a decision node following each of these proposals. On the left, the Responder may either choose A (accept) or R (reject) in response to the Equal proposal. The same options are also available on the right. Notice that each decision node is labeled with the name of the player who makes a decision at that node, and each branch emanating from a node is labeled with the name of one of the feasible decisions. The branches end at the bottom, which shows the payoffs of the Proposer (on the left) and Responder (on the right). Those nodes that do not contain any further branches are called terminal nodes. It will be shown in chapter 14 that equal split proposals are common in ultimatum bargaining games.

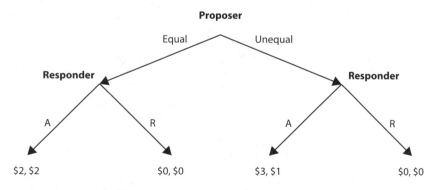

Figure 9.1. An Extensive Form Representation for the Bargaining Game

9.2 Two-Stage Trust Games

The first two experiments to be considered are based on the games shown in figure 9.2. Consider the top part, where the first player must begin by making a safe choice (*S*) or a risky choice (*R*). Decision *S* is safe in the sense that the payoffs are deterministic: 80 for the first player and 50 for the second (the first player's payoff will be listed on the left in each case). The payoffs that may result from choosing *R* depend on the second player's response, *P* or *N*. (Think of *P* as "punish.") For the game shown in the top part of figure 9.2, solving for the theoretical equilibrium is easy. We know the second player would earn 10 from choosing *P* and 70 from choosing *N*. Hence, we could expect that a rational player would choose *N*. Consequently, the first player should make the risky choice *R* as long as he trusts the second player to be rational.

In the lab experiment conducted, this game was played only once, without repetition, by pairs of subjects. Payoffs were in pennies. As indicated by the outcome percentages in the top part of the figure, 84% of the first movers were confident enough to choose *R*, and all of these received the anticipated *N* response. Note that the second movers would have reduced their earnings by 60 cents if any of them had selected *P* in response to *R*.

The game shown in the bottom part of figure 9.2 is almost identical, except that the second mover only loses 2 cents by making a *P* response, which

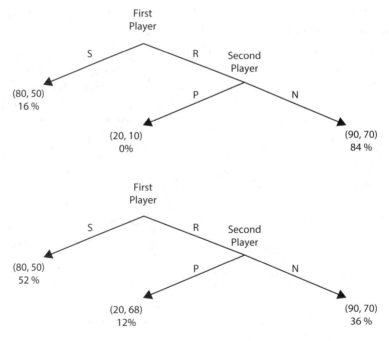

Figure 9.2. A Two-Stage Game Where Mistakes Matter, with Outcome. Percentages *Source*: Goeree and Holt (2001).

generates a payoff of 68 for the second mover instead of 70. In this case, more than half of the first movers were sufficiently concerned about the possibility of a P response that they chose the safe decision, S. This fear was well founded, since a fourth of the first movers who chose R encountered the P response in the second stage. In fact, the first players who chose S in the bottom game in figure 9.2 earned more on average than those who chose R (problem 1).

The standard game-theoretic analysis of the two games in figure 9.2 would yield the same prediction for each one. By assuming that both players are sequentially rational, the analysis is reduced to players only maximizing their own payoffs. This assumption implies that the second player will choose N, regardless of whether it increases earnings by 60 cents, as in the top game, or by only 2 cents, as in the bottom game. In each of these games, we begin by analyzing the second player's best choice (choose N), and knowing this, we can calculate the best choice for the first player (choose R). The pair of choices, R and N, is a Nash equilibrium, since neither player can increase earnings by deviating.

The Nash equilibrium (R and N) that was identified for the games in figure 9.2 is focal in the sense that it maximizes the payoffs for each player. There is another Nash outcome, (S and P). As with any Nash equilibrium (NE), this can be verified by showing that neither person has an incentive to deviate unilaterally (problem 2). In particular, the second player earns 50 in this equilibrium, and a unilateral deviation from P to N would not change the outcome, since the outcome is fully determined by the first player's S decision. In theoretical analysis, this second equilibrium is typically "ruled out" because it implies a type of behavior for the second player that is not rational in the final stage, i.e., it is not subgame perfect. Despite these arguments, S is the most commonly observed outcome in the laboratory for the bottom game in figure 9.2.

There is more tension for the games in figure 9.3. As before, there are two Nash equilibria: (S and P) and (R and N). First consider the (R and N) outcome, which yields 90 for the first player and 50 for the second. If the second player were to deviate to P, then the outcome would be (R and P) with lower payoffs for the second player, since we are considering a unilateral deviation, i.e., a deviation by one player given that the other continues to use the equilibrium decision. Similarly, given that the second player is using N, the first player cannot deviate and increase the payoff above 90, which is the maximum for the first player in any case. This equilibrium (R and N) is the one preferred by the first player. There is another Nash equilibrium (S and P); see problem 3. Here, you can think of P as indicating "punishment," since a P decision in the second stage reduces the first player's payoff from 90 to 60. The reason that such a punishment might be enacted is that the second player actually prefers the outcome on the left side of the figure for each of the two games in figure 9.3. A punishment offers a chance for the second player to obtain revenge. The difference between the two games

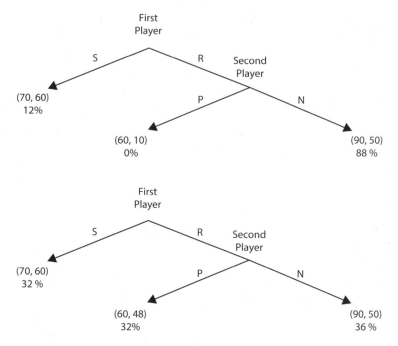

Figure 9.3. A Two-Stage Game with a Threat That Is "Not Credible."
Source: Goeree and Holt (2001).

in the figure is that the cost of punishment is 40 cents in the top game and only 2 cents in the bottom game.

The notion of subgame perfection rules out the possibility that the second mover will make a mistake in either of the games in figure 9.2, or that the second mover will carry out a costly (and hence non-credible) threat for either of the games in figure 9.3. Thus the predicted outcome would be the same (*R* and *N*) in all four games. *This prediction works well for only one of the games in each figure.*

The problem with the assumption of perfect sequential rationality is that it does not allow room for random deviations or "mistakes," which are more likely if the cost of the mistake is small (2 cents for each of the bottom games in the figures). Such deviations may be due to actual calculation and recording errors, or to un-modeled payoff variations caused by emotions, concerns about fairness, relative fairness, etc. A more systematic analysis of the costs of deviations and their effects on behavior will be undertaken in a later section. To summarize:

Two-Stage Game Experiments: *In simple games in which one player makes a decision in the first stage and the other makes a decision in the second stage, there can be multiple Nash equilibria. For example, a "punish" decision in the second stage can cause the first decision maker to pick a safe option that prevents the* *second player from moving at all. Backward induction rules out Nash equilibria that involve punishments in the second stage as they are not rational for the second mover at that point. Such punishment strategies, however, are frequently employed in experiments if the cost of punishment is low to the second mover. Often, the first mover seems to recognize this possibility in advance and choose the safe option in such games to prevent the possibility of a negative outcome.*

9.3 The Centipede Game

In the two-stage games, the first mover has to consider how the second mover will react. This may involve the first mover thinking introspectively: "What would I do in the second stage, having just seen this particular initial decision?" This process of seeing a future situation through the eyes of someone else may not be easy or natural, and the difficulty increases if an initial decision maker has to think about reactions several stages later. In such cases, it is still often straightforward, but tedious, to begin with the last decision maker, consider what is rational for that person, and work backward to the second-to-last decision maker, and then to the third-to-last decision maker, and so on. The "noise" observed in some of the two-stage games considered earlier in this chapter makes it plausible that players in longer multi-stage games may not be very predictable, at least in terms of their ability to reason via backward induction. Rosenthal (1982) proposed a sequential game with 100 stages, which serves as an extreme "stress test" of the backward induction reasoning process. Each of the 100 stages has a terminal node in the extensive form that hangs down, so the extensive form looks like an insect with 100 legs, which is why this game is called the "centipede game."

Figure 9.4 shows a truncated version of this game, with only four legs. At each node, the relevant player (Red or Blue) can either move down, which stops the game, or continue to the right, which passes the move to the other player, except in the final stage. The play begins on the left side, where the Red player must decide whether to stop the game (the downward arrow) or continue (the right arrow). A stop/down move in this first stage results in payoffs of 40 cents for Red and 10 cents for the other player, Blue. Note that the payoffs for this first terminal node are listed as (40, 10), with payoffs for the Red player shown on the left. If Red decides to continue to the right, then the next move is made by Blue, who can either continue or stop, yielding payoffs of (20, 80), where the left payoff is for Red, as before. Blue might reason that stopping gives 80 for sure, whereas a decision to continue will either yield payoffs of 40 (if Red stops in the next stage), 320 (if Blue stops by moving down in the final stage), or 160 (if Blue moves right in the final stage). Obviously, it is better for Blue to move down in the final stage, getting 320 instead of 160. Having figured out the best

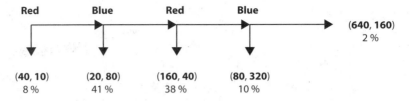

Figure 9.4. A Centipede Game. *Source:* McKelvey and Palfrey (1992).

decision in the final stage, we can begin the process of backward induction. If Blue is expected to go down in the final stage, then Red will expect a payoff of 80 if play reaches the final stage, instead of 640. Thus it would be better for Red to take the 160 and stop the game in the third stage, rather than pass to Blue and end up with only 80 in the final stage.

To summarize, we have concluded that Blue will stop in the final stage, and that Red, anticipating this, will stop in the third stage. Similar reasoning can be used to show that Blue will want to stop in the second stage, if play reaches this point, and hence that Red will want to stop in the first stage, yielding payoffs of 40 for Red and only 10 for Blue. Thus the process of backward induction yields a very specific prediction, i.e., that the game will stop in the initial stage, with relatively low payoffs for both.

McKelvey and Palfrey (1992) used this setup in an experiment, which involved groups of 20 participants divided into Red and Blue player roles. Each participant was matched with all ten people in the other role, in a series of ten games, with the payoffs shown in figure 9.4. The aggregate percentages for each outcome are shown below the payoffs at each node. Only 8% of the Red players stopped in the first stage, so the experiment provides a sharp rejection of predictions based on backward induction. Rates of continuation were somewhat lower toward the end of the session, but the incidence of stopping in the initial stage of each game remained low throughout the session. Most of the games ended in the second and third stages, but 2% of the games went on until the final stage. To summarize:

Centipede Games: *The centipede game has a sharp Nash prediction, based on backward induction, that the first player will stop the game immediately, even though payoffs may become quite high if the game proceeds to later stages. This stark prediction is not observed in laboratory experiments, where the play does not stop in the first stage very often.*

The failure of backward induction predictions in this game is probably due to the fact that even a small amount of unpredictability of others' decisions in later stages may make it optimal to continue in early stages, since the high payoffs are concentrated at the terminal nodes on the right side of the figure. It could be the case that a small proportion of people were concerned with others' payoffs ("altruists"), and that even if these types are not predominant, it would be optimal for selfish individuals to continue in early stages and have a chance of obtaining the very high payoffs in later stages. In fact, any type of noise or unpredictability, regardless of its source, would have similar effects. Notice that about one in five Blue players decides to continue in the final stage, which cuts their payoff in half but raises the Red player's payoff by a factor of 8, from 80 to 640. It is not clear whether this is due to randomness, miscalculation, altruism, or reciprocity (returning the favor). For an analysis of the effects of randomness in centipede games, see McKelvey and Palfrey (1998) and Zauner (1999). The next section presents a method of modeling random decision making that is commonly used in game theory and in econometric studies of individual decisions.

9.4 Probabilistic Choice Functions

Anyone who has taken an introductory psychology course will remember the stimulus-response diagrams. Stochastic response models were developed after researchers noticed that responses could not always be predicted with certainty. Psychologists would ask subjects to identify which of two lights is brighter or which sound is louder. When the signals (lights or sounds) were not close in intensity, almost everybody would indicate the correct answer, with any errors being caused by mistakes in recording decisions. As the two signals became closer in intensity, some people would guess incorrectly, e.g., because of bad hearing, random variations in ambient noise, or distraction and boredom. As the intensity of the signals approached equality, the proportions of guesses for each signal would approach 1/2 in the absence of measurement bias. In other words, there was not a sharp break where the stronger signal was selected with certainty, but rather a "smooth" tendency to guess the stronger signal as its intensity increased.

The work of a mathematical psychologist, Duncan Luce (1959), suggested a way to model noisy choices, i.e., by assuming that response probabilities are increasing in the stimulus intensity, i.e., the probability associated with picking one sound as being louder should be an increasing function of the decibel level of that sound. Since probabilities must sum to 1, the choice probability associated with one sound should also be a *decreasing* function of the intensity of the *other* sound.

In economics, the stimulus intensity for a given response (decision) can be thought of as the expected payoff of that decision. The high-payoff decision is almost sure to be chosen if the payoff difference is large, but if the payoff difference is small, errors in perception, calculation, or inattention might produce

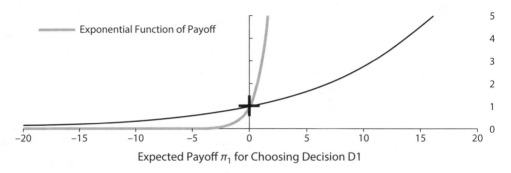

Figure 9.5. Exponential Functions with Different Precisions

a low-payoff decision. Suppose that there are two decisions, D_1 and D_2, with expected payoffs that we will represent by π_1 and π_2. In the centipede game, for example, the Blue player's payoffs could be $\pi_1 = 160$ for Pass and $\pi_2 = 320$ for Take. The next step is to find an increasing function. The commonly used function is exponential, $f(\pi_1) = \exp(\pi_1)$, or equivalently, $f(\pi_1) = e^{\pi_1}$. This function has a curved shape, like a hill that keeps getting steeper, as in figure 9.5. The gray line is an exponential function of payoff, π, and the dark line is for the case of payoffs that are scaled down by a factor of 10: $f(\pi_1) = \exp(\pi_1/10)$. Notice that the exponential function is positive, even if the payoff is negative, and that $\exp(0) = 1$.

A simple approach could be to assume that the probability of the decision is determined by the function itself: $\Pr(D_1) = f(\pi_1)$ and $\Pr(D_2) = f(\pi_2)$. The problem is that the two probabilities may not sum to 1. This is easily fixed by a simple trick with a fancy name: "normalization," i.e., division by the sum of the functions:

$$(9.1) \qquad \Pr(D_1) = \frac{f(\pi_1)}{f(\pi_1) + f(\pi_2)} \quad \text{and} \quad \Pr(D_2) = \frac{f(\pi_2)}{f(\pi_1) + f(\pi_2)}$$

If we let $f(\pi_1) = \pi_1$, and similarly for decision 2, then the choice functions would be ratios of payoffs. Such ratios are sensitive to payoff changes. But payoff ratios will not work if payoffs can be negative, so we need a function that is always positive valued. The function that is typically used in applied work on probabilistic choices is the exponential function:

$$(9.2) \qquad \Pr(D_1) = \frac{\exp(\pi_1)}{\exp(\pi_1) + \exp(\pi_2)} \quad \text{and} \quad \Pr(D_2) = \frac{\exp(\pi_2)}{\exp(\pi_1) + \exp(\pi_2)}$$

which sum to 1. The choice probabilities for each decision in this equation are increasing in the payoff for that decision and decreasing in the other payoff, since the other decision payoff only appears in the denominators. The probabilities in (9.2) will also be equal to 1/2 when the expected payoffs for each decision are equal. Otherwise, the decision with the higher expected payoff will have a higher choice probability. The probabilistic choice model that is based on exponential functions is known as the *logit model*. The logit functions implement a type of *bounded rationality* since the decision with the highest expected payoff is chosen more often, but is not chosen with probability 1. Thus, better decisions are more likely to be selected, but errors are possible, so rationality is imperfect or "bounded."

Just as we could make it harder to distinguish between the widths of two pins by making them each half as thick, we can add more noise or randomness into the choice probabilities in equation (9.2) by reducing all expected payoffs by half or more. Intuitively speaking, dividing all expected payoffs by 100 may inject more randomness, since dollars become pennies, and non-monetary factors (boredom, indifference, playfulness) may have more influence. In general, the degree of observed randomness might depend on the complexity of the game, so we introduce a parameter λ that would be small, e.g., 1/10, when there is a lot of randomness, and large when there is more precision. In general, this parameter determines the degree of rationality and would be estimated from observed decisions. This precision parameter is multiplied by all of the payoffs in (9.2) to obtain:

$$(9.3) \quad \Pr(D_1) = \frac{\exp(\lambda\pi_1)}{\exp(\lambda\pi_1) + \exp(\lambda\pi_2)} \text{ and } \Pr(D_2) = \frac{\exp(\lambda\pi_2)}{\exp(\lambda\pi_1) + \exp(\lambda\pi_2)} \quad \text{(logit)}$$

These logit choice functions have the property that as precision goes to 0, all of the payoff terms are 0. Since $\exp(0) = 1$, the choice probability ratios are equal to 1/2 in this case of complete randomness. In other words, a person who is completely random will have decisions that are totally unaffected by payoffs, and instead, appear to result from coin flipping.

If both expected payoffs are equal, the choice probabilities in equation (9.3) are 1/2, regardless of precision. This property is illustrated in figure 9.6, which shows probabilistic choices for a low precision, flatter line, and for a high precision, sharper line. Both lines pass through the midpoint with choice probabilities of 1/2 when the expected payoff difference is 0.

Logit Model: *Probabilistic Choice Functions specify choice probabilities to be increasing exponential functions of the expected payoffs for each choice, with probabilities normalized to sum to 1, as in equation (9.3). The sharpness of logit probabilistic responses to payoffs is controlled by a precision parameter, λ, with total randomness in the limit as λ goes to 0 and choice probabilities are equal for all decisions, irrespective of payoffs. As λ becomes large, choice probabilities are higher for decisions with higher payoffs. Perfect rationality is the limiting case as precision goes to infinity and the best decision is selected with probability 1.*

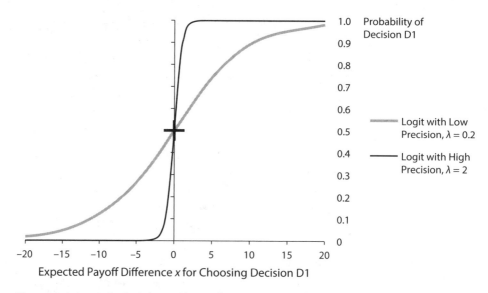

Figure 9.6. Logit Probabilistic Choice for Decision D1 for High and Low Precision

These choice functions can be used to examine the decisions made for the two-stage game shown in figure 9.3. Consider the second player in the top panel, who must decide whether to punish and earn 10 or not punish and earn 50. This large payoff difference should override the emotional response to the fact that the first player already chose a decision that reduced the second player's payoff below the highest possible value of 60. If $\lambda = 0.1$, the logit response for the second player is:

$$(9.4) \quad \Pr(P) = \frac{\exp(10\lambda)}{\exp(50\lambda) + \exp(10\lambda)} = \frac{\exp(1)}{\exp(5) + \exp(1)} = 0.02 \text{ (costly punishments)}.$$

This choice probability is close to the 100% no-punishment rate observed, conditional on getting to the second stage in the top panel of figure 9.3. The rate of punishments is much higher for the game with less costly punishments in the bottom panel of the figure, where 32% of the outcomes are P and 36% are N, so the observed frequency of punishments is $0.32/(0.36 + 0.32) = 0.47$. Using the same precision parameter value, the logit choice probability in this case is:

$$(9.5) \Pr(P) = \frac{\exp(48\lambda)}{\exp(50\lambda) + \exp(48\lambda)} = \frac{\exp(4.8)}{\exp(5) + \exp(4.8)} = 0.45 \text{ (cheap punishments)}.$$

which is close to the observed punishment rate in this case. Finally, recall that the Goeree and Holt (20001) experiment involved games that were played only once, so there is no reason to expect that the first player would have correct expectations about the second player's choice tendencies, although the data in the bottom panels suggest that the first player seemed to expect more punishments when punishment was less costly for the second player.

The closeness of observed and predicted punishment rates is remarkable and even a bit unusual, and some words of caution are needed. First, there are many other factors besides money payoffs that affect decisions: social preferences and emotions can also be very important, and can be manipulated in experiments (problem 7). On the other hand, a model with errors can be used to estimate these non-payoff effects. Second, the particular value of precision needs to be estimated, and estimation with only a few things to be explained can result in "overfitting." Even more problematic is the possibility that the precision may depend on the complexity of the decision. These issues will be discussed in later chapters, but it is worth noting that the curved response lines shown in the next chapter will also be drawn for a precision value of 0.1.

9.5 Extensions

One possible explanation of the data pattern in the centipede game experiment is that the failure of backward induction may be due to low incentives, and that behavior would be more "rational" in high-stakes experiments. Recent experiments with potential stakes of thousands of dollars indicate that this is not the case. High payoffs will cause quicker exits from the centipede game, but exits in the first stage are still not the norm (Parco, Rapoport, and Stein, 2002). Other recent work has considered how subjects in experiments learn and adjust in centipede games, see Nagel and Tang (1998) and Ponti (2002). Bornstein, Kugler, and Ziegelmeyer (2004) compare behavior of individuals and groups in centipede games. When each of the two players corresponds to a group of three subjects, the result is more competitive. In particular, three-person groups exited the centipede game sooner than individuals (groups of size 1), and in this sense, groups were more "rational." Fey, McKelvey, and Palfrey (1996) report results of a variation of the centipede game, where the payoffs sum to a constant, which generates a more competitive environment.

Chapter 9 Problems

1. Use the choice percentages shown below each outcome in the *bottom game* in figure 9.2 to show that the first players who chose S earned more on average than those who chose R.

2. Show that (S for the first player, P for the second player) is a Nash equilibrium for the top game in figure 9.2.

3. Show that (S for the first player, P for the second player) is a Nash equilibrium for each of the games in figure 9.3; i.e., check the profitability of unilateral deviations for each player.

4. Consider a game in which there is $4 to be divided, and the first mover is only permitted to make one of three proposals: (a) $3 for the first mover and $1 for the second mover, (b) $2 for each, or (c) $1 for the first mover and $3 for the second mover. The second mover is shown the proposal and can either accept, in which case it is implemented, or reject and cause each to earn $0. Show this game in extensive form. Be sure to show the payoffs for each person, with the first mover listed on the left, for each of the six terminal nodes.

5. Show that proposal (c) in problem 4 is a part of a Nash equilibrium for the game in problem 4. To finish specifying the decisions for this equilibrium, you have to say what the second mover's response is to each of the three proposals. Does behavior in this equilibrium satisfy sequential rationality?

6. Consider a game in which the instructor divides the class into two groups and takes out $8 in quarters (32 quarters). The instructor then puts a quarter into a collection tray and allows those on one side of the room to take the quarter (and somehow divide or allocate it among themselves) or to pass. A pass results in doubling the money (to 2 quarters) and giving those on the other side of the room the option to take or pass. This process of doubling the number of quarters in the tray continues until one side takes the quarters, or until one side's pass forces the instructor to put all remaining coins into the plate, which are then given to the side of the room whose decision would be the next one. Draw a figure that shows the extensive form for this game. Label the players A and B, and show payoffs for each terminal node as an ordered pair: ($ for A, $ for B). What is the predicted outcome of the game, on the basis of backward induction, perfect rationality, and selfish behavior?

7. Find the punishment rate, conditional on getting to the second stage, for each panel in figure 9.2, and compare the conditional punishment rates with the logit predictions for a precision of 0.1. Is the second player more or less prone to punish in the second stage in figure 9.2 as compared with figure 9.3? What is it about the second player's payoffs that might generate emotions that affect a proclivity to punish?

10

Randomized Strategies

Sometimes there is a strategic advantage associated with being unpredictable, much as a tennis player does not always lob in response to an opponent's charge to the net. It is easy to imagine economic situations where people would not like to be predictable. For example, a lazy manager only wants to prepare for an audit if such an audit is likely. The auditor, who is rewarded for discovering problems, would only want to focus audits on managers who may be unprepared. This chapter discusses randomized strategies in the context of simple matrix games (Matching Pennies and Battle of Sexes) and more complex Attacker-Defender and Poker-like games.

> **Notes for the Instructor:** The Veconlab 2 × 2 Matrix Game program can implement simple two-decision games like Matching Pennies. Fixed matchings are fine for games with randomized strategies in which players wish to be unpredictable. The game considered in the Misaligned Profiling section is the Veconlab Multi-Site Attacker/Defender Game (use defaults, and select the desired numbers of players and rounds). The "stripped-down" poker game discussed in the extensions section can be played with the Veconlab Signaling Game on the information menu (select poker defaults).

10.1   Symmetric Matching Pennies Games

Consider a game in which each person places a penny on a table, covering it so that the other person cannot see which side is up. By prior agreement, one person takes both pennies if the pennies match, with two "heads" or two "tails," and the other takes the pennies if they do not match. This is analogous to a soccer penalty kick, where the goalie must dive to one side or another before it is clear which way the kick will go, and the kicker cannot see which way the goalie will dive at the time of the kick. In this case, the goalie wants a match and the kicker wants a mismatch. Table 10.1 shows a game where the Row player prefers to match, with heads (Top Left) or tails (Bottom Right). Conversely, Column, whose payoff is listed second, can earn 72 with a mismatch (Top Right) or (Bottom Left).

In each of the cells of table 10.1, there is one person who would gain by altering the placement of their penny unilaterally, as indicated by the arrows.

Table 10.1. A Modified Matching Pennies Game (Row's Payoff, Column's Payoff)

	Column Player:	
Row Player:	Left (heads)	Right (tails)
Top (heads)	72, 36 ⇒	⇓ 36, 72
Bottom (tails)	⇑ 36, 72	72, 36 ⇐

For example, if they were both going to choose heads, the column player would prefer to switch to tails, as indicated by the arrow pointing to the right in the upper-left box of the payoff table. The arrows in each box indicate the direction of a unilateral payoff-increasing move, so there is no equilibrium in non-random strategies. (Non-random strategies are commonly called "pure strategies" because they are not probability-weighted mixtures of other strategies.) Notice that the arrows go in a counterclockwise circle, with no stable stopping point.

The game in table 10.1 is *balanced* in the sense that each possible decision for each player has the same set of possible payoffs, i.e., 36 and 72. In matching pennies games with this type of balance, the commonly observed result is for subjects to play each decision with an approximately equal probability (Ochs, 1994; Goeree and Holt, 2001). This tendency to choose each decision with probability 1/2 is not surprising given the simple intuition that one must be unpredictable in this game. Nevertheless, it is useful to consider the representation of the Nash equilibrium as an intersection of players' best-response functions.

In figure 10.1, the dark solid line shows the best response for the row player. Think of the horizontal axis as what Row expects Column to do. These beliefs can be thought of as a probability of Right, going from 0 on the left to 1 on the right. If Column is expected to choose Left, then Row's best response is to choose Top, so the best-response line starts at the top-left part of the figure, as shown by the dark line. If Column is expected to choose Right, then Row should play bottom, so the best-response line ends up on the bottom-right side of the graph. The crossover point is where Column's probability is exactly 0.5, since Row does better by playing Top whenever Column is more likely to choose Left. The crossover at 0.5 is due to the symmetry of payoffs, since Row earns 72 or 36 from Top, and 36 or 72 from Bottom, and the expected payoffs are equal when the probability that Column chooses Right is 0.5. (The curved gray line will be discussed in a later section.)

The column player's best-response line is derived analogously as shown by the thick dashed line on the right side panel of figure 10.1. It starts in the lower-left corner, rises to 0.5, and then crosses over horizontally to the right side, because Column would want to play Right whenever the probability of Top is above 0.5.

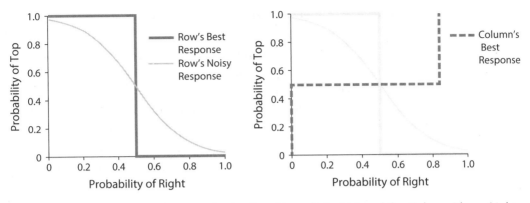

Figure 10.1. Best-Response Function for the Row Player (left side) and the Column Player (right side) in a Symmetric Matching Pennies Game

The intersection of these two lines is in the center of the graph, at the point where each probability is 0.5, which is the Nash equilibrium in mixed strategies. In other words, if each person is randomizing with probability 1/2, each person's decisions have the same expected payoffs, so they would be willing to randomize. (If one of the expected payoffs was higher, then it would be a best response to choose that decision every time.) To summarize:

Equilibrium in Matching Pennies Game: *Randomization requires indifference or equal expected payoffs. As a consequence, the row player's choice probabilities must keep the column player indifferent, and vice versa. In a symmetric matching pennies game, the only way one person will be indifferent is if the other is using equal probabilities of heads and tails, which is the equilibrium outcome.*

Think about the Nash requirement that unilateral deviations cannot increase a player's expected payoff. If one person plays heads half of the time, then playing tails will win half of the time; playing heads will win half of the time, and playing a "50–50" mix of heads and tails will win half of the time. In other words, when one player is playing randomly with probabilities of one-half, the other person cannot do any better than using the same probabilities. If each person were to announce that they would use a coin flip to decide which side to play, the other could not do any better than using a coin flip. This is the Nash equilibrium for this game. It is called a "mixed-strategy equilibrium" since players use a probabilistic mix of each of their decisions. In contrast, an equilibrium in which no strategies are random is called a "pure-strategy" equilibrium, since all of the strategies are played with probability 1.

In a Nash equilibrium, neither player can do better by deviating, so any Nash equilibrium must be on the best-response lines for both players. The only intersection of the solid and dashed lines in figure 10.1 is at probabilities of 0.5 for each player. Note that it is important for each player to be unpredictable, which could be accomplished by basing the decision on a feeling or impulse about what might work best in that moment. The key is that whatever motivates one's decision should be unobservable by the other player. So going back to the initial example, a penalty kicker might base a kick direction decision on a feeling of strength in one leg or on breezes that are not felt by the goalie.

10.2 Battle of the Sexes

Next consider a game with two equilibria in non-random strategies, the payoffs for which are shown in table 10.2. This is a game where two friends who live on opposite sides of Central Park wish to meet at one of the entrances, i.e., on the East side or on the West side. It is obvious from the payoffs that Column wishes to meet on the West side, and Row prefers the East side. But notice the zero payoffs in the mismatched (West and East) outcomes, i.e., each person would rather be with the other than to be on the preferred side of the park alone. This is known as a "battle-of- sexes" game, since the motivating examples were based on different preferences of men and women.

If the game is repeated, most people would take turns, which is what tends to happen when two people play the game repeatedly in controlled experiments with the same partner. In fact, coordinated switching often arises even when explicit communication is not permitted. Table 10.3 shows a decision sequence for a pair who were matched with each other for six rounds of a classroom experiment, using payoffs in table 10.2. There is a "match" on East in the first round,

Table 10.2. A Battle-of-Sexes Game (Row's Payoff, Column's Payoff)

| | Column: | |
Row:	West	East
West	1, 4	0, 0
East	0, 0	4, 1

Table 10.3. Alternating Choices for a Pair of Subjects in a Battle-of-Sexes Game with Fixed Partners

	Round 1	Round 2	Round 3	Round 4	Round 5	Round 6
Row Player	East ($4)	East ($0)	West ($1)	East ($4)	West ($1)	East ($4)
Column Player	East ($1)	West ($0)	West ($4)	East ($1)	West ($4)	East ($1)

and Column then switches to West in round 2, which results in a mismatch and earnings of $0 for both. Row switches to West in round 3, and they alternate in a coordinated manner in each subsequent period, which maximizes their joint earnings. Four of the six pairs alternated in this manner, and the other two pairs settled into a pattern where one earned $4 in all rounds.

The problem is harder if the game is played only once, without communication, or if there is repetition with random matchings. Consider the battle-of-sexes game in table 10.4. The payoffs in this table were used in a classroom experiment conducted at the College of William and Mary. There were 30 students located in several different computer labs, with random matchings between those designated as row players and those designated as column players. Table 10.5 shows the percentage of times that players chose the preferred location (Top for Row and Right for Column). Intuitively, one might expect the percentage of preferred-location (Top for Row, Right for Column) choices to be above one-half, and in fact this percentage converges to 67%. This mix of choices does not correspond to either equilibrium in pure strategies.

Instead of looking for mathematical coincidences, it is useful to calculate some expected payoff expressions as was done in the previous section. First consider Row's perspective when Column is expected to play Right with probability p. Consider the top row of the payoff matrix in table 10.4, where Row thinks the right column is relevant with probability p. If Row chooses Top, Row gets 2

Table 10.4. A Battle-of-Sexes Game (Row's payoff, Column's payoff)

Row:	Column:	
	Left	Right
Top	2, 1	0, 0
Bottom	0, 0	1, 2

Table 10.5. Percentage of Preferred-Location Decisions for the Battle-of-Sexes Game in Table 10.4

Round	Row Players	Column Players	All Players
1	80	87	83.5
2	87	93	90
3	87	60	73.5
4	67	67	67
5	67	67	67
Nash	67	67	67

when Column plays Left (expected with probability $1 - p$) and Row gets 0 when Column plays Right (expected with probability p). When Row chooses Bottom, these payoffs are replaced by 0 and 1. Thus Row's expected payoffs are:

- Row's expected payoff for Top = $2(1 - p) + 0(p) = 2 - 2p$,
- Row's expected payoff for Bottom = $0(1 - p) + 1(p) = 0 + p$.

The expected payoff for Top is higher when $2 - 2p > p$, or when $p < 2/3$. Obviously, the expected payoffs are equal when $p = 2/3$, and Top provides a lower expected payoff when $p > 2/3$. Since Row's expected payoffs are equal when Column's probability of choosing Right is 2/3, Row would not have a preference between the two decisions and would be willing to choose randomly. At this time, you might just guess that since the game looks symmetric, the equilibrium involves each player choosing their preferred decision with probability 2/3. This guess would be correct, as we shall show with figure 10.2. As before, the solid line represents Row's best response that we analyzed above. If the horizontal axis represents Row's beliefs about how likely it is that Column will play Right, then Row would want to play top as long as Column's probability p is less than 2/3. Row is indifferent when $p = 2/3$, and row would want to "go to the bottom" when $p > 2/3$.

So far, we have been looking at things from Row's point of view, but an equilibrium involves both players, so consider Column's decision where the vertical axis now represents Column's belief about what Row will do. At the top, Row is expected to play Top, and Column's best response is to play Left in order to be in the same location as Row. Thus Column's dashed best-response line starts in the upper-left part of the figure. This line also ends up in the lower-right part,

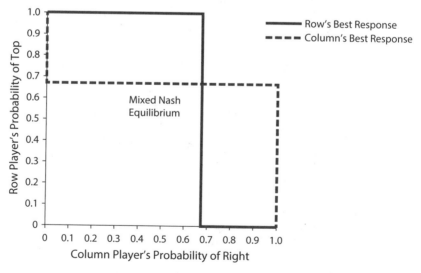

Figure 10.2. Best Responses for the Battle-of-Sexes Game in Table 10.4

since when Row is expected to play Bottom (coming down the vertical axis), Column would want to switch to the Right location that is preferred. It can be verified by simple algebra that the switchover point is at 2/3, as shown by the horizontal segment of the dashed line.

Since a Nash equilibrium is a pair of strategies such that each player cannot do better by deviating, each player has to be making a best response to the other's strategy. In the figure, a Nash equilibrium will be on both Row's (solid) best-response line and Column's (dashed) best-response line. Thus the final step is to look for equilibrium points at the intersections of the best-response lines. There are three intersections. First, consider the upper-left corner of the figure where both coordinate on Row's preferred outcome (Row earns 2, Column earns 1). Second, the lower-right intersection shows an equilibrium at Column's preferred outcome (Row earns 1, Column earns 2). We already found these equilibria by looking at the payoff matrix directly. The third intersection point in the interior of the figure is new. At this point, players choose their preferred decisions (Top for Row and Right for Column) with probability 2/3, as seen in the data.

The graph shows the equilibria clearly, but it is useful to summarize how the random strategy equilibrium would be found with only simple algebra, since the graphical approach will not be possible with more players or more decisions.

Step 1. First we need to introduce notation:

p = probability that Column chooses Right
q = probability that Row chooses Top

Step 2. Calculate expected payoffs for each decision. As noted above, Row's payoffs in the Top row are either 2 or 0, and the expected payoff for Top is shown below. The other expected payoffs are calculated similarly:

Row's Expected Payoff for Top = $2(1-p) + 0(p) = 2 - 2p$
Row's Expected Payoff for Bottom = $0(1-p) + 1(p) = 0 + p$
Column's Expected Payoff for Right = $0(q) + 2(1-q) = 2 - 2q$
Column's Expected Payoff for Left = $1(q) + 0(1-q) = q + 0$

Step 3. Calculate the equilibrium probabilities.

Equate Row's expected payoffs to determine p.
Equate Column's expected payoffs to determine q.

Nash Equilibrium in Mixed Strategies: *In order to randomize willingly, a person must be indifferent between the decisions, and indifference is found by equating expected payoffs. The tricky and unintuitive part is that equating one player's expected payoffs pins down the other's probability, and vice versa.*

The data for the battle-of-sexes game in table 10.5 are atypical in the sense that such sharp convergence to an equilibrium in randomized strategies is not always observed. Often there is a little more bouncing around the predictions, due to noise factors. Remember that each person playing that game (with random matching) was seeing a series of other people, so people have different experiences, and hence different beliefs. This raises the issue of how people learn after observing others' decisions. The learning models from chapter 7 could be applied to this setting to simulate adjustment patterns. Finally, the battle-of-sexes game discussed here was conducted under very low payoff conditions, with only one person of 30 being selected ex post to be paid their earnings in cash. High payoffs might cause other factors like risk aversion to become important, especially when there is a lot more variability in the payoffs associated with one decision than with another. If risk aversion is a factor, then the expected payoffs would have to be replaced with expected utility calculations.

10.3   The Paradox of Misaligned Profiling

Anti-terrorist strategies have received a lot of attention since the 9/11 surprise terrorist attack. Security officials have limited resources and must choose which people or types of people to subject to more extensive checks. The other side of the situation is that terrorist groups and drug smugglers must decide which types of people to use in attempts to penetrate defenses. Targeting defenses to particular demographic groups, or "profiling," is often controversial, since discrimination is sometimes evident from patterns of traffic searches or other detentions. In terrorist and smuggling settings, however, the demographics are *endogenous*, when there is a choice of which types of people to use in an attack and which types of people to stop and search. For example, Osama bin Laden once exhorted his followers to explore ways to recruit non-Muslims.

Holt et al. (2016) conducted an experiment to evaluate profiling decisions in attacker-defender games in which randomization is a natural feature of equilibrium play. The setting they considered involved one attacker, who must choose which "type" of person to send on a mission, and one defender, who must decide which "type" to search. The simplest setup involved only two possible attacker types, with one being more effective than the other. Think of the more effective type as a person who is young, athletic, ideologically trained, and the less effective type as not having those attributes and hence being less likely to succeed if not subjected to search. If the attacker sends high-reliability types, the defender will defend against those, which provides the attacker with an incentive to send the type that does not fit the profile. This cycle is reminiscent of the sequence of responses in a matching pennies game.

It helps to get a clear picture of the Nash calculations when the description is based on general notation, before the numerical parameter values used in the experiment are used. The two types, 1 and 2, have reliabilities represented by r_1 and r_2, which are the probabilities of mounting a successful attack if not detected by the defense. In this case: $r_1 > r_2$. In contrast, the defense is assumed to be 100% reliable in the sense that defending against a particular type will always block an attack from that type. But even if an attacker is not stopped by the defender, the probability of success is determined by the attacker reliability, r_1 or r_2. The attacker's payoff is a gain G in the event of a successful attack, and in that case the defender loses an amount L. This will be a zero-sum game if the attacker's gain equals the defender's loss: $G = L$.

In the equilibrium to be determined, the attacker chooses types 1 and 2 with probabilities a_1 and a_2 respectively, where these sum to 1. Consider the defender's expected payoffs first. If the defender defends against type 2, the defender will suffer a loss if the following two conditions are met: (1) the attacker sends the other type (with probability a_1), and (2) is successful (with probability r_1). So the loss $-L$ is multiplied by the product of a_1 and r_1, or $-La_1r_1$. Similarly, the expected loss from defending against type 1 is $-La_2r_2$. For randomization to occur, the attack probabilities must make the defender indifferent, and equating the defender payoffs yields an equation:

(10.1) $-La_1r_1 = -La_2r_2$ (equality of expected payoffs for defending).

Since $a_2 = 1 - a_1$, equation (10.1) can be written as an equation with a single variable, a_1, that can be solved:

(10.2) $$a_1 = \frac{r_2}{r_1 + r_2} < 1/2.$$

Given that $r_1 > r_2$, equation (10.2) implies $a_1 < a_2$, i.e., the attacker uses the more reliable type 1 *less often*. Moreover, an increase in r_1, the reliability of type 1, results in that type being used less often.

Next consider the attacker's expected payoffs when the defender defends against types 1 and 2 with probabilities d_1 and d_2, which sum to 1. An attack with type 1 is successful and earns the gain G if it is not defended (with probability d_2) and if the undetected attacker is successful (with probability r_1). So the expected payoff for an attack with type 1 is Gd_2r_1. The analogous reasoning applies for the other type, and equating the two expected payoffs yields:

(10.3) $Gd_2r_1 = Gd_1r_2$ (equality of expected payoffs for attacking).

Since $d_2 = 1 - d_1$, equation (10.3) can be written as an equation in only d_1, and solved:

(10.4) $$d_1 = \frac{r_1}{r_1 + r_2} > 1/2.$$

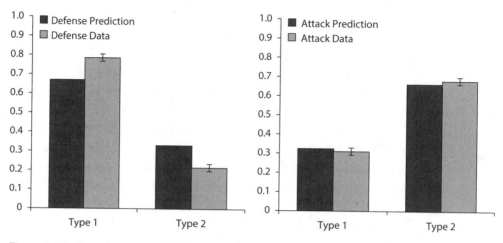

Figure 10.3. Data Averages and Theoretical Predictions

Since $r_1 > r_2$, equation (10.4) implies the defender's search targets the *more reliable* type more often in equilibrium, even though the attacker uses the *less reliable* type more often. In this sense, the attack and defense profiles are misaligned.

The experiment was run with $r_1 = 2/3$ and $r_2 = 1/3$. As a result, the predictions were that the defender would defend against the more reliable type 1 two-thirds of the time, and the attacker would attack with the less reliable type two-thirds of the time. Subjects interacted for 50 rounds with *fixed matchings*, which accentuated their incentive to be unpredictable. The gain and loss amounts, G and L, were set at $1, and subjects received private incomes each round, of $1 for defenders and $0.60 for attackers. Unannounced private incomes have the dual purpose of equalizing earnings and masking earnings to block out any fairness concerns, which would be inappropriate for an attacker/defender scenario. Subjects were paid a standard $6 show-up fee plus half of their total earnings from the 50 rounds of interactions, for a total that averaged about $26 for a 30-minute experiment. There were 72 subjects in this particular treatment. Figure 10.3 shows the theoretical predictions (dark bars) and average choice percentages (light bars) for attackers and defenders.

There was a slight tendency to over-defend against the more reliable type 1, but in general, the data are close to the theoretical predictions. An interesting feature of the data is that subjects often engaged in trying to "set up" a play, by choosing one decision several times and then switching unexpectedly. In debriefings, attackers expressed regret when a less reliable attacker was not searched but failed anyway. A useful exercise after a classroom experiment is to look for patterns of behavioral responses to most recently observed success or failure outcomes. Finally, it is interesting that the choice percentages in the very first period, with no experience, were quite close to the overall averages after 50 rounds. To summarize:

> **Paradox of Misaligned Profiling:** *In a setting in which an attacker decides whether to send a more reliable type or a less reliable type, and the defender decides which type to defend against, the Nash equilibrium involves the attacker sending the less reliable type more often and the defender defending against the more reliable type more often. The choice proportions in the experiment are close to theoretical predictions, even in the first round. These predictions explain the misaligned profiling that is observed in an experiment with fixed matchings.*

It is useful to think about how mixed-strategy equilibria are calculated when there are more than two decisions. The profiling game could be generalized to have three or more attacker types, with declining r_i parameters for type i. Some types are so ineffective that they would never be used and would never be defended against. In that case, the expected payoffs for sending the various attacker types that are actually used would all have to be equal, but the expected payoffs for those not used would be lower (and not necessarily equal).

Analogously, there can be situations in which two firms randomize over price bids in order to avoid being underbid by the other. In such cases, there would be a range of prices over which randomization occurs, and all prices in that range would have the same expected payoff. Prices outside of that range that are not selected would have lower expected payoffs. On the other hand, if firms observe private costs that change randomly and that are not observed by other firms, then they can use the randomness in their own private costs as a device to make their own bids random from the other firms' point of view. Some models of this randomized pricing and of bidding based on privately observed random values will be considered in later chapters on market power and private value auctions.

10.4   Noisy Behavior in Highly Asymmetric Games

The best-response lines in figures 10.1 and 10.2 have sharp corners, representing immediate crossovers whenever one decision offers even a very slight expected payoff advantage over the other. In other words, a decision with a payoff advantage of only a penny would be used with probability 1 and the other decision would never be used. These sharp responses, while convenient in terms of theory, are likely to be contradicted in experiments. For example, the P punishment decision was more likely to be used in the two-stage game examples in the previous chapter when the cost to the second mover was a matter of pennies. Noisy behavior of this type was noticed early on by psychologists, and it results in smoothed "better response" functions instead of the sharp best-response functions for perfectly rational responses to payoff differences, however small.

Look back at the left panel in figure 10.1, where the *curved line* represents a noisy response by the Row player as payoff differences change due to increases in the belief p that Column will choose Right. If Row were perfectly rational (and responded to arbitrarily small expected payoff differences), then Top would be played whenever p is even a little less than 1/2. The curved line shows some departure from this rationality. Notice that the probability that Row plays Top is close to 1 but not quite there when p is small, and that the curved line deviates more from the best-response line as p approaches 1/2 and payoff differences are small.

The curved "noisy best-response line" for Row still intersects Column's dashed best-response line in the center of figure 10.1, so a relaxation of the perfect rationality assumption for Row will not affect the equilibrium prediction. Similarly, suppose that we allow some noise in Column's best response, which will "smooth off" the sharp corners for Column's best-response line, resulting in an analogous curved dashed line (not shown). This would start in the lower-left corner, rise with a smooth arc as it levels off at 0.5 in the center of the graph before curving upward along the upper-right boundary of the graph. Since this line will intersect Row's downward-sloping noisy-response line in the center of the graph, we see that the "50-50" prediction would not be affected if we let each player's decision be somewhat noisy. When the game is symmetric, the best-response lines intersect in the center, and the curved noisy-response lines, which would look like a propeller, would intersect at the same point with probabilities of 0.5. Another way to think of this invariance is that if the prediction with sharp responses is at 1/2, and if noise pulls you toward 1/2, then the prediction will be unchanged. This invariance to noise effects will not hold with asymmetric games, to be considered next.

An unbalanced payoff structure is shown in table 10.6, where the Row player's payoff of 72 in the Top-Left box has been increased by a factor of five to 360. Recall that the game was balanced before this change, and in that case the choice proportions should be 1/2 for each player. The increase in Row's Top-Left payoff from 72 to 360 would make Top a more attractive choice for a wide range of beliefs, i.e., Row will choose Top unless Column is almost sure to choose Right. Intuitively, one would expect this change to move Row's choice proportion for Top up from the 1/2 level that is observed in the balanced game. This intuition is apparent in the choice data for an experiment done with Veconlab software,

Table 10.6. An Asymmetric Matching Pennies Game (Row's payoff, Column's payoff)

Row Player:	Column Player:	
	Left	Right
Top	360, 36 $\Rightarrow$	$\Downarrow$ 36, 72
Bottom	$\Uparrow$ 36, 72	72, 36 $\Leftarrow$

where the payoffs were in pennies. Each of the three sessions involved 10–12 players, with 25 periods of random matching. The proportion of Top choices was 67% for the "360 treatment" game in table 10.6.

This intuitive "own payoff effect" of increasing Row's Top/Left payoff is *not* consistent with the Nash equilibrium prediction. First, notice that there is no equilibrium in non-random strategies, as can be seen from the arrows in table 10.6, which go in a clockwise circle. To derive the mixed-equilibrium prediction, let p denote Row's beliefs about the probability of Right, so that $1-p$ is the probability of Left. Thus, Row's expected payoffs are:

Row's Expected Payoff for Top $= 360(1-p) + 36(p) = 360 - 324p$.
Row's Expected Payoff for Bottom $= 36(1-p) + 72(p) = 36 + 36p$.

It follows that the difference in these expected payoffs is: $(360 - 324p) - (36 + 36p)$, or equivalently, $324 - 360p$. Row is indifferent if the expected payoff difference is zero, i.e., if $p = 324/360 = 0.9$. Therefore, Row's best-response line stays at the top of the left panel in figure 10.4 as long as the probability of Right is less than 0.9. The striking thing about this figure is that Column's best-response line has not changed from the symmetric case; it rises from the Bottom-Left corner and crosses over when the probability of Top is 1/2. This is because Column's payoffs are exactly reflected (36 and 72 on the Left side, 72 and 36 on the right side). In other words, the only way that Column would be willing to randomize is if Row chooses Top and Bottom with equal probability. The Nash equilibrium for the asymmetric game in table 10.6 requires that Top and Bottom be played with the same probabilities (1/2 each), as was the case for the balanced payoffs in table 10.1. The reason is that Column's payoffs are the same in both games, and Row must essentially use a coin flip in order to keep Column indifferent.

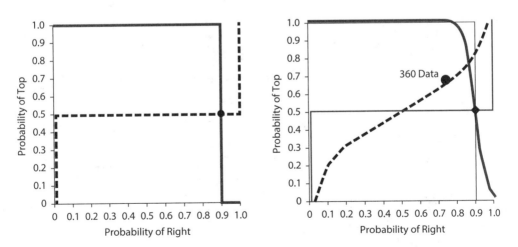

Figure 10.4. The 360 Treatment: Best Responses (left side) and Noisy Best Responses (right side)

The previous treatment change made Top more attractive for Row by raising Row's Top-Left payoff from 72 to 360. In order to produce an imbalance that makes Top *less* attractive, a second treatment reduced Row's Top/Left payoff in table 10.1 from 72 to 40. Thus, Row's expected payoff for Top is: $40(1 − p) + 36(p)$ $= 40 − 4p$, and the expected payoff for Bottom is: $36(1 − p) + 72(p) = 36 + 36p$. These are equal when the probability of Right, p, is 4/40, or 0.1. What has happened in figure 10.5 is that the reduction in Row's Top-Left payoff has pushed Row's best-response line to the bottom of the figure, unless Column is expected to play Right with a probability that is less than 0.1. This downward shift in Row's best-response line is shown on the left side of figure 10.5. The result is that the best-response lines intersect where the probability of Top is 1/2, which was the same prediction obtained from the left side of figure 10.4. Thus a change in Row's Top-Left payoff from 40 to 72 to 360 does not change the Nash equilibrium prediction for the probability of Top. The mathematical reason for this result is that Column's payoffs do not change, so Row must choose each decision with equal probability. If Row were to choose one decision with a higher probability, then Column could take advantage of the situation by not choosing randomly.

These invariance predictions are not borne out in the data from experiments reported here. Each of the three sessions involved 25 periods for each treatment, with the order of treatments being alternated. The percentage of Top choices increased from 36% to 67% when the Row's Top-Left payoff was increased from 40 to 360. The Column players reacted to this change by choosing Right only 24% of the time in the 40 treatment, and 74% of the time in the 360 treatment.

The qualitative effect of Row's own-payoff effect is captured by a noisy-response model where the sharp best-response functions on the left sides of figures 10.4 and 10.5 are rounded off, as shown on the right sides of the figures.

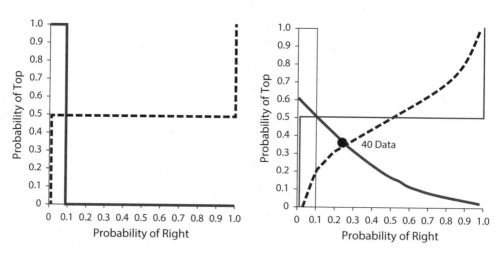

Figure 10.5. The 40 Treatment: Best Responses (left side) and Noisy Best Responses (right side)

Notice that the intersection of the curved lines implies that Row's proportion of Top choices will be below 1/2 in the 40 treatment and above 1/2 in the 360 treatment, and the predicted change in the proportion of Right decisions will not be as extreme as the movement from 0.1 to 0.9 implied by the Nash prediction. The actual data averages are shown by the black dots on the right sides of figures 10.4 and 10.5. These data averages exhibit the strong own-payoff effects that are well predicted by the curved best-response lines, especially for the 40 treatment. To summarize:

Own Payoff Effects in Asymmetric Matching Pennies: *In a mixed-strategy Nash equilibrium, a change in a payoff for one player, e.g., Row, is not predicted to affect that player's own equilibrium choice probability. This is because Column's payoffs have not changed, and to be indifferent between the two decisions over which randomization occurs, it must be the case that Row player's choice probabilities do not change. Laboratory experiments, however, do exhibit "own-payoff effects," in which a change in a player's payoff causes that player's choice probabilities to change. These effects can be modeled by replacing the sharp best-response functions with curved (noisy) response functions.*

The quantitative accuracy of these predictions is affected by the amount of curvature that is put into the noisy-response functions, and is a matter of estimation. Standard estimation techniques are based on writing down a mathematical function with a "noise parameter" that determines the amount of curvature and then choosing the parameter that best fits the data.

Finally, it is instructive to revisit the sharp best-response lines for the coordination games in figures 8.2 and 8.3 at the end of chapter 8. The center vertical lines in those figures are located where the expected payoff difference is 0, i.e., where a player would be indifferent between the two decisions. Thus, the probability p for which the dashed expected payoff difference line crosses the center vertical line is probability that equates the expected payoffs. Therefore, this crossing point represents a mixed-strategy equilibrium, which was labeled with a Nash equilibrium diamond symbol. Look at the thick arrows in these figures to convince yourself that these mixed-strategy equilibria are unstable, since the arrows point away from the equilibrium probability toward the upper or lower boundaries. When you think about it, would be silly to randomize when you are in a situation where you are trying to coordinate with someone else. Finally, notice that the thick "best-response" lines in figures 8.2 and 8.3 are drawn with "sharp" corners, indicating that the best-response probability of choosing high effort is 1 when its expected payoff is higher (right side of the vertical line), and the best-response probability is 0 when its expected payoff is lower (left side). If the best-response lines are replaced with noisy, "better-response" lines with slightly curved corners, then the resulting intersections with the expected pay-off difference line will be close to a probability of 0 at the intersection on the left, and close to 1 at the intersection on the right. In this case, adding a little noise will not have much of an effect on the high-effort and low-effort equilibria for the coordination games in those figures.

10.5 Quantal Response Equilibrium

The curved response lines used in this chapter were constructed from logit probabilistic choice functions that were discussed in the previous chapter, using the same precision of $\lambda = 0.1$ that was used previously:

$$(10.5)\;\; \Pr(D_1) = \frac{\exp(\lambda \pi_1)}{\exp(\lambda \pi_1) + \exp(\lambda \pi_2)} \;\text{ and }\; \Pr(D_2) = \frac{\exp(\lambda \pi_2)}{\exp(\lambda \pi_1) + \exp(\lambda \pi_2)} \;\text{(logit)}$$

For example, D_1 and D_2 could represent Left and Right for the column player or Top and Bottom for the row player. These functions have the property that an increase in the expected payoff for one decision will raise the choice probability for that decision. Moreover, if the expected payoffs for two decisions are equal, then the choice probabilities will be equal.

The expected payoffs on the right side of equation (10.5) depend on player's belief probabilities. These expected payoffs for each decision, in turn, determine choice probabilities $\Pr(D_1)$ and $\Pr(D_2)$. For example, the row player's expected payoffs will depend on Row's beliefs about the probability that Column will choose right. When the beliefs match the decision probabilities, the result is an equilibrium with no incentive to change, which corresponds to an intersection of the smoothed-response probabilities in the right panels of figures 10.4 and 10.5. This is known as a *quantal response equilibrium* (McKelvey and Palfrey, 1995; Goeree, Holt, and Palfrey, 2016), since the sharp best-response lines used in a Nash equilibrium analysis have been replaced by curved "quantal" responses that determine an equilibrium with some bounded rationality. To summarize:

Quantal Response Equilibrium: *The intersections of curved noisy-response lines correspond to a situation in which belief probabilities match choices in a probabilistic sense. This approach is a generalization of the notion of a Nash equilibrium that is determined by intersections of sharp best-response lines instead of curved quantal response lines. In each case, the equilibrium choice probabilities match beliefs that generate those choices.*

The logit form in equation (10.5) is flexible, since the degree of curvature is captured by a parameter that can be estimated. Logit functions of this type have been used extensively in econometric studies of binary decisions, e.g., deciding whether to take a freeway or whether to enroll in a training program. Since randomness is pervasive in some laboratory experiments, such functions are also useful in terms of providing a stochastic model that can be used for estimating other parameters of interest, e.g., risk aversion or loss aversion. *Theoretical predictions are almost never perfect due to omitted factors, so a specification of errors is needed for estimation of any parameter. In a strategic game, it is natural to "build" the errors into the analysis, using probabilistic choice response functions.*

10.6 Extensions: "Stripped Down Poker"

The insights developed in this chapter can be applied to the study of strategy in situations where it is essential to be unpredictable, e.g., in an incentivized soccer kick experiment (Composti, 2003). Another application is in terms of setting prices to avoid being "undercut" (Holt and Solis-Soberon, 1992). This section will focus on an application motivated by the game of poker: If one player only raises with a strong hand and never "bluffs," then the other player will tend to "fold" in the face of a raise. But this gives the player with a strong hand an incentive to bluff by raising with a weak hand if the bluff is unlikely to be "called." Most good poker players will bluff some, but not all of the time, which is consistent with a randomized strategy.

These observations can be illustrated with a game that begins with each of two players putting in a stake of $1. Only one person sees the draw of a single card, king or ace, and decides whether to fold and lose their dollar stake, or to raise by adding a second dollar. The second player does not see the first player's card, and in the face of raise, decides whether to fold and lose their $1 stake. To call will either win the other's $2 if that card turns out to be a king, or lose their stake of $2 if the card turns out to be an ace. This "stripped down poker" game can be done in class by letting the instructor view the card and play against one of the class members (Reiley, Urbancic, and Walker, 2008). It can also be played with the Veconlab Signaling Game (Poker setup option, fixed matchings to add interest, not letting groups go at own their own pace to prevent inferences from time delays).

Suppose that Ace and King are equally likely. Since an Ace will win if challenged, the first mover with an Ace should never fold and always raise, which will generate a gain of $2 if it is called and $1 if not. Next consider the first player's decision if a King is drawn. A fold will lose $1 for sure. A raise with a King will lose $2 if called, and will gain $1 if the other player folds. If the probability of being "called" is γ, the expected payoff for raising is $-2\gamma + 1(1 - \gamma)$, which equals the payoff from folding when

$$(10.6) \qquad\qquad -1 = -2\gamma + 1(1 - \gamma),$$
$$\text{(payoff if fold)} \quad \text{(expected payoff if bluff)}$$

or equivalently, when $\gamma = 2/3$. With a 2/3 call rate, the first mover is indifferent and would be willing to randomize between bluffing and folding with a King.

Similarly, the bluff probability β is determined by an analysis of the second player's decision of whether or not to call when the first player raises. This calculation uses Bayes' rule (chapter 5), but it can be explained by analogy to the idea that the probability that a patient actually has a disease after a positive test result is the ratio of "true positives" to the rates of all positive test results, either true or not. In particular, suppose that the first mover always raises with an Ace, but only raises with probability β with a King, i.e., "beta" is the bluff rate. If the second player encounters a raise, then it could be an Ace (a "true positive" in the medical analogy) or it could be a King (a "false positive"). The Ace is drawn with probability 1/2, and in that case there is a raise for sure, so the true positive rate is 1/2, as shown in the numerator of equation (10.7). The King is also drawn with probability 1/2, and if it is, a bluff occurs with probability β, so the rate of false positives is $\beta/2$. The probability of encountering an ace is the ratio of this true positive rate (1/2) to the rate of all positives (either true or false), which is $\frac{1}{2} + \beta/2$ in the denominator:

$$(10.7) \qquad \Pr(\text{Ace}|\text{Raise}) = \frac{\text{rate of true positives}}{\text{rate of all positives}} = \frac{\frac{1}{2}}{\frac{1}{2} + \frac{\beta}{2}} = \frac{1}{1 + \beta}.$$

In a mixed-strategy equilibrium, the second player who sees a raise must be indifferent between calling and folding. A decision to fold will lose the $1 stake, as shown by the -1 on the left side of equation (10.8). The expected payoff for calling is shown on the right side. The second player who calls and encounters an Ace will lose 2, as indicated by the first term, and will earn the other player's $2 stake if the other player is bluffing and has a King.

$$(10.8) \qquad -1 = -2\Pr(\text{Ace}|\text{Raise}) + 2\Pr(\text{King}|\text{Raise})$$
$$\text{(payoff if fold)} \quad \text{(expected payoff if call)}$$

Note that $\Pr(\text{King}|\text{Raise}) = 1 - \Pr(\text{Ace}|\text{Raise}) = \beta/(1 + \beta)$ from (10.7), so the expected payoff equality in (10.8) holds when

$$(10.9) \qquad -1 = -2\frac{1}{1+\beta} + 2\frac{\beta}{1+\beta}$$

or equivalently, when $\beta = 1/3$. In words, the second mover who sees a Raise is indifferent between folding and calling when the first player's bluff rate is 1/3.

To summarize, a mixed-strategy equilibrium for this simple poker game involves a bluff rate of 1/3 and a call rate of 2/3. In Veconlab class experiments, first movers tend to bluff too much at first, but bluff rates end up being near a half after first movers have been disciplined by higher call rates. Popova (2006)

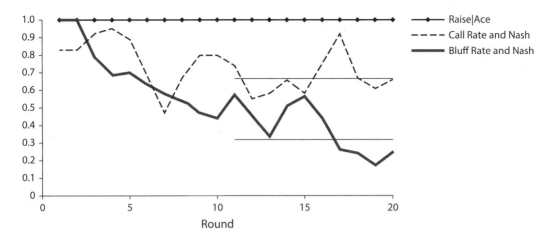

Figure 10.6. Bluff and Call Proportions for a "Stripped Down Poker" Experiment. *Source*: Popova (2006).

reports similar tendencies in a research experiment done for a senior thesis. The data averages for one session are shown in figure 10.6, where the thin dashed line tracks the call rates and the thick dark line tracks the bluff rate.

Chapter 10 Problems

1. Consider a game with playing cards where the people on the left side of the room have a red 8 (hearts or diamonds) and a black 2 (clubs or spades), and those on the right side have a black 8 and red 2. Each person is paired with someone on the other side and chooses a card to play. If the colors of the cards do not match (red and black), each earns nothing. If the colors match, earnings for each person are equal to the number on the card that the person played. (a) Write out the payoff matrix for this game. (b) Find all equilibria in non-random strategies. (c) Find expressions for left side player's expected payoffs, one for each decision, where p denotes the probability that the other player chooses their high number card. (d) Find expressions for the right side player's expected payoffs for each decision, black 8 or red 2. (e) Find the equilibrium with randomized strategies.

2. Suppose that the payoffs in table 10.4 are changed by raising the 2 payoff to a 3, for both players. Answer (c), (d), and (e) of problem 1 for this game.

3. Find the Nash equilibrium in mixed strategies for the coordination game in table 8.4 in chapter 8, and illustrate your answer with a graph.

4. The data in the table below were for the battle-of-sexes game in table 10.2. The 12 players were randomly matched, and the final five periods of play are shown. Calculate the percentages of "preferred" choices (East for Row and West for Column), and average these two numbers. What is the mixed-strategy Nash equilibrium?

Round	6	7	8	9	10
Row	6 East	3 East, 3 West	5 East, 1 West	5 East, 1 West	6 East
Column	2 East, 4 West	2 East, 4 West	1 East, 5 West	3 East, 3 West	2 East, 4 West

5. Discuss the pros and cons of using fixed versus random matchings in the attacker defender game in section 10.3. Can you imagine a situation in which the defender is stationary, but attackers come from random directions? How might this be implemented?

6. Explain why the diamond on the center vertical line in figure 8.2 from chapter 8 represents a mixed-strategy Nash equilibrium. Use the figure to explain why this mixed strategy is not stable.

7. Graph the best response functions for the simplified poker game with β on the vertical axis and γ on the horizontal axis. Explain the locations of these lines using intuition about when it is best to bluff and when it is best to call.

8. Suppose that each dollar lost is perceived as $-\$L$, where $L > 1$. Recalculate the bluff and call probabilities for the stripped down poker game in terms of loss aversion L. What is the effect of loss aversion on bluff and call rates?

11

Choice of Partners, Social Dilemmas

In games with a final period, one approach is to consider decisions that would be made in the final period and then work backwards via a process of backward induction, as discussed in chapter 9. Many social interactions, however, have no fixed final period, but rather, a probability of continuation at each point. In such settings, a self-serving "defection" in an early period may have negative consequences later if the other player retaliates. This suggests that the degree of cooperation should increase with higher continuation probabilities. Prisoner's dilemma experiments with random stopping will be considered in the first section of this chapter.

The two players in the repeated prisoner's dilemmas are stuck with each other. In contrast, binary business and social relationships can be more fluid, as we all know from experience! For example, businesses can terminate purchases from suppliers that cut corners on cost. In laboratory experiments, the ability to switch suppliers in this manner can discipline a market and enhance performance. The second section of this chapter is a prisoner's dilemma experiment in which subjects can exit and form new links with other players in a controlled manner. Moreover, this "choice of partners" game permits players to increase the scale of the interaction if mutually agreed on, e.g., payoffs and losses are doubled if scale is doubled. In effect, the possibility of unilateral exit introduces a third decision into the prisoner's dilemma, and the "scale-up" option adds another dimension in which players can reward each other if trust is established.

The third game to be considered is an N-player social dilemma with a richer set of decisions that replace the cooperate-defect dichotomy. This game, known as a "traveler's dilemma," involves two travelers who lose identical pieces of luggage and must make independent loss claims. If the claims are equal, then the claims are paid by the airline. Since the lost items happen to be identical, the airline knows the claims should be the same, but does not know exactly what the purchase price was. So the airline announces a policy that if the claims are unequal, each person only receives the *lower* of the two claims, with a penalty subtracted from the payment to the high claimant and added to the payment to the low claimant. Both claimants would be better off making identical high claims, but each has an incentive to "undercut" the other to obtain the reward for having the lower claim. This game is similar to a prisoner's dilemma in that the unique equilibrium involves payoffs that are lower than can be achieved with cooperation (submitting the maximum permitted claim). But recall that in a single-period prisoner's dilemma, the best response to any decision made by the other player is to defect, and therefore, the best response does not depend on whether the other is expected to cooperate or defect. In contrast, the optimal decision in the traveler's dilemma *does* depend on what the other person is expected to do. Thus the traveler's dilemma is more sensitive to interactions of imprecise beliefs about others' decisions. These interactions can cause data patterns to be quite far from Nash predictions, as will be evident from the experiment results. Even though the traveler's dilemma backstory is artificial, the game provides an important paradigm, and it has close similarities to some Bertrand price competition games in markets where the firm with the lowest price sells more. The traveler's dilemma is one of the author's favorite games, and the choice of partners game is another!

Notes for the Instructors: A basic prisoner's dilemma game is one of the setup options for the Matrix Game program on the Veconlab Games menu, and the Choice of Partners game is also listed on this menu. It works best with at least 12 participants so that people have enough new partners to meet. Since it takes time to choose partners and scales in each round, it is best to run the game for no more than nine rounds. The Traveler's Dilemma game is also found on the Games menu. Alternatively, it is possible for students to play an online demonstration version of the Traveler's Dilemma, where the "other decision" is taken from a database of stored decisions made by University of Virginia law students in a behavioral game theory class. This online game only takes about ten minutes to do by yourself, and it can be accessed at http://veconlab.econ.virginia.edu/tddemo.htm. Try it; you can compare your earnings with the law student team that encountered the same series of other partners that you do in this demo.

11.1 Repeated Prisoner's Dilemma Games with Random Termination

The original game run that motivated the prisoner's dilemma story was devised at RAND and used in an experiment with 100 repeated plays with the same partner. The idea was to show that defection was not the predominant decision, and in that sense, the experiment succeeded. From a theoretical perspective, the final period could be analyzed as a one-period game, which would yield mutual defection, and given that, mutual defection in the second to last period, and so forth, in a process of backward induction. In actual economic interactions, the final period may not be known, e.g., firms may go out of business, buyers may relocate to a different city, government policies are not renewed, and so forth.

Uncertainty about the future can be implemented by using a fixed probability of termination, which could be done by throwing dice or using a sequence of dice throws done in advance to announce continuation or stopping. The sequence of dice throws done in advance has the property that different groups of subjects will be exposed to the same number of plays of the game (Holt, 1985). Even though the probability of continuation was not varied in that experiment, a higher probability of continuation would be expected to make cooperation more likely, since there is a longer sequence of future benefits of continued cooperation. In other words, defection in a prisoner's dilemma generates an immediate benefit, but the lost benefits of future cooperation can dominate if the probability of continuation is high. Consider the prisoner's dilemma game shown in table 11.1 below. Notice that the immediate gain from defection is 2. But what if the other player responds in the next round with an anticipated defection, so that payoffs go down to 3 each? In this case the cost is 5, which exceeds the benefit to the initial defector. Now suppose that the probability of continuation is δ, and the probability of a random stop after the current and each future round is $1 - \delta$. For example, if a six-sided die is to be thrown, with a 1 indicating stop, then $\delta = 5/6$. A *grim trigger strategy* for the repeated game would be to cooperate unless the other defects, and then defect *forever after* in response to a defection. This extreme punishment would provide the strongest incentive not to defect.

In order to calculate the incentive to defect, we compare the immediate benefit of 2, calculated as $10 - 8$ from table 11.1, with the cost of 5, calculated as $8 - 3$. If the other player uses a grim trigger strategy (of responding to an initial defection with defection every subsequent period), then the initial defector incurs a cost of 5 in each subsequent period until the process stops. The duration of this cost is not known for sure, but the expected cost contains a term that is 5δ, since δ is the probability that play continues for 1 period and the payoff reduction of 5 is realized. There is also a chance that play continues for 2 periods, which happens with probability δ^2, so there is a term $5\delta^2$ in the expected cost calculation. Reasoning in this way, the expected cost of defection is $5(\delta + \delta^2 + \delta^3 + \ldots)$, which can be written as $5\delta(1 + \delta + \delta^2 + \ldots)$. A standard formula for an infinite series can be used to express the sum in parentheses as $1/(1 - \delta)$, which

Table 11.1. A Prisoner's Dilemma (Row's Payoff, Column's Payoff)

Row Player:	Column Player:	
	Left	Right
Top	8, 8	0, 10
Bottom	10, 0	3, 3

is actually easy to derive (problem 7). Therefore, the expected cost of defection is $5\delta/(1-\delta)$. When the probability of continuation is 5/6, as would be the case with the die throw example, then the expected cost of defection is $5\frac{\frac{5}{6}}{1-\frac{5}{6}} = 5\frac{\frac{5}{6}}{\frac{1}{6}} = 5(5) = 25$, which is much higher than the immediate benefit of 2.

The calculations used in the previous paragraph suggest an experiment in which the incentive to defect and the probability of continuation are changed systematically to see whether these affect cooperation rates in infinitely repeated prisoner's dilemma games. Dal Bó and Fréchette (2011) ran such an experiment, using either continuation probabilities of 1/2 or 3/4, and payoffs shown in table 11.2. Note that the treatments include three alternative payoff pairs for mutual cooperation in the top-left box, which provides three different incentives to defect. As cooperation payoffs are increased from 32, to 40, to 48, the incentive to defect declines from 18, to 10, to 2. Therefore, there are $3 \times 2 = 6$ treatments for the 3 payoff variations and 2 continuation probabilities. The only treatment that does not support full cooperation with a grim strategy punishment is the combination of the 32 payoff and the continuation probability of 1/2. With a cooperation payoff of 32, the incentive to defect is 18, and the loss from future cooperation is $32 - 25 = 7$. When $\delta = \frac{1}{2}$, the expected value of this loss over future periods is $7\delta/(1-\delta) = 7$, which is less than the immediate benefit of 18 from defection. In all of the other treatments, cooperation *is* supported by a grim punishment threat. For example, with a cooperation payoff of 40, the gain from defection is 10 ($= 50 - 40$), which is lower than the expected cost in terms of lost cooperative payoffs, calculated as $(40 - 25)\, \delta/(1-\delta) = 15$ when $\delta = \frac{1}{2}$.

The experiment was run with computer-generated stops. After a stop, people were rematched randomly and a new sequence would begin. Table 11.3 shows the percentages of cooperative decisions over all rounds. The main qualitative feature of the results is that cooperation is increasing in both the continuation probability and the cooperative payoff. These differences are all statistically significant, with the strongest effects associated with increasing the continuation probability and going from a cooperation payoff of 40 to 48.

Table 11.2. A Prisoner's Dilemma with Variable Incentives to Defect Determined by the Three Possible Payoff Pairs for Mutual Cooperation in the Top-Left Box

Row Player:	Column Player:	
	Left	Right
Top	32, 33 (40, 40) (48, 48)	12, 50
Bottom	50, 12	25, 25

Table 11.3. Percentage of Cooperative Decisions in All Rounds by Treatment

	Cooperative Payoff		
	32	40	48
$\delta = \frac{1}{2}$	10%	18%	35%
$\delta = \frac{3}{4}$	20%	59%	76%

Source: Dal Bó and Fréchette (2011).

Note that cooperation is only 10%, in the cell for which cooperation would not be supported by a grim punishment strategy. In all of the other cells, cooperation could be supported by grim punishments, but in three of those cells the rate of cooperation is still below 50%. Cooperation is highest in the lower-right cell, where the cooperation payoff is 48 and the immediate gain from defection is only $50 - 48 = 2$ and the loss from cooperation is $(48 - 25) = 23$. Assuming a grim trigger strategy response to the initial defection, this loss must be multiplied by a factor of $\delta/(1 - \delta) = (3/4)/(1/4) = 3$ when the continuation probability is 3/4.

To summarize, even though the immediate gain from defection is only 2, the expected value of the loss of cooperative payoffs is $3 \times 23 = 69$ with a grim punishment. The bottom-right number in table 11.3 indicates that the rate of cooperation is only 76% in this treatment. One likely explanation for the observed patterns is that punishments may not always be so grim, due to randomness or something else that lets players restore trust.

Cooperation in Infinitely Repeated Prisoner's Dilemma Experiments: *Cooperation rates are low when it cannot be supported by a grim punishment (defect forever after). Cooperation rates are increasing in the probability of continuation and the cooperative payoff, as expected. But cooperation may not be very high even in some treatments for which, in theory, it can be supported by a grim punishment.*

11.2 Choice of Partners Games

One artificial aspect of a prisoner's dilemma is that there is no exit option that is equivalent to refusing to buy or sell in a market. An early experiment run by political scientists introduced the option of exiting, with the remaining participants being randomly matched in prisoner's dilemma games. The hypothesis was that cooperation might decline if people who are prone to cooperate are more likely to become frustrated and exit. Surprisingly, the opposite pattern was observed, leading the authors to conclude that cooperators were less likely to exit.

It is easy to specify multi-person social dilemmas in which each person enjoys high earnings if all cooperate, but each has a private incentive to defect. In societies with communal production, for example, each person only receives a per capita share of the joint harvest, say $1/N$, so it is better from a selfish perspective to work hard in the evening in one's own garden rather than exert a lot of effort during the day in the common field. This was certainly the case in Jamestown in 1611. When the new governor arrived, he found starved, emaciated men bowling in the streets while fields were unplanted. A similar but less dire situation developed in the New England Plymouth colony, where Dutch investors had decided in advance to prohibit private ownership of houses and gardens.

One realistic feature of exit from multi-person groups is that people have to decide where to go next (not an option in Jamestown, where colonists were prohibited from leaving). Consider a situation in which the sizes of various groups depend on who joins and leaves and where there is public information about the overall level of cooperation in each group. This setup, when implemented in the lab, generated "cooperation chasing" as defectors switched to more cooperative groups, which caused the cooperators in those groups to become discouraged.

The lessons from the cooperation-chasing study were reinforced with a subsequent experiment in which people were divided into groups based on previous tendencies to cooperate. Participants were regrouped after each round of play according to cooperativeness in the previous round, so that the initial cooperators ended up being kept together, protected from defectors. This sorting procedure, which was not announced, resulted in higher overall levels of cooperation since it shielded the cooperators from discouragement and regret. The next step in this line of research was to let participants exclude defectors. Exclusion, e.g., by voting, turned out to provide strong psychological incentives for cooperation. The effects of exclusion will be revisited in a later chapter on the management of "common pool" shared natural resources, like hunting grounds or water reserves.

Another manifestation of exclusion is when competition provides one person with two or more potential partners. One early paper involved a single buyer and two competing sellers. Each round began with the buyer selecting one of the sellers, who would then decide whether to deliver high or low quality. In a single interaction, it is in the seller's private interest to lower cost and cut quality. With a two-round horizon, a dissatisfied buyer can switch to the other seller in the second round. This switching was common enough, but overall average quality levels were much higher in ten-round sequences, in which the buyer could switch back and forth, and sellers could learn to expect to lose sales after a low-quality delivery.

The experiment to be discussed in this section was done using an even simpler structure, with the prisoner's dilemma payoffs in table 11.4, which has a unique equilibrium that provides each player with a negative payoff of -50

Table 11.4. A Prisoner's Dilemma with an Unfavorable Equilibrium

Row Player:	Column Player:	
	A	B
A	1, 1	−1.5, 2
B	2, −1.5	−0.5, −0.5

cents. This outcome is clearly worse than not playing the game at all, since a unilateral exit decision by either player would result in 0 payoffs for each.

In many business settings, it may be advisable to order a small quantity and then scale up orders on connections that are profitable. This scale dimension can be implemented for the game in table 11.4 by letting each of two potential players select a proposed scale, with the minimum of the proposed scales being implemented. In the experiment, a scale of 0 would terminate the connection, a scale of 1 was followed by a play of the prisoner's dilemma shown in the table, and scales of 2 or 3 resulted in games with doubled or tripled payoffs, respectively. Notice that scaling up the payoffs also increases the possible losses, so the high scale is a risky position that requires mutual trust to maintain. The logic behind using the minimum proposed scale is that it takes two people to trade, and either one can break off a trading agreement.

The baseline treatment involved random re-matchings in each round, for a pre-announced total of nine rounds, so there was no opportunity to scale up connections on profitable links and terminate unprofitable ones in this "strangers" treatment. Cooperation levels were mixed in this baseline, as shown by the gray dashed line in figure 11.1 that declines steadily in later rounds.

The other two treatments with partner choice, exit, and scale opportunities resulted in sustained high rates of cooperation, even though standard backward induction arguments would predict uniform defection. The cooperation rates for active links in these treatments, shown by the solid lines at the top of figure 11.1, were basically flat and did not decline sharply until the pre-announced final round.

In the "assigned partners" treatment (dark line with empty circles), each subject was matched with one person in the first round, with scale choices followed by cooperate (A) or defect (B) choices after the scale was determined. Each person was matched with the same person again in round 2, and this time they had more information on which to base a scale decision. In addition, each met one new person with no visible history, and they also had to propose a scale for this second link. As before, A or B choices were made for each link after the scales had been determined. In round 3, each person met a third person, again with no visible history, and they could propose scales (exit, 1×, 2×, or 3×) for each of the three possible links. There were no new links in subsequent rounds.

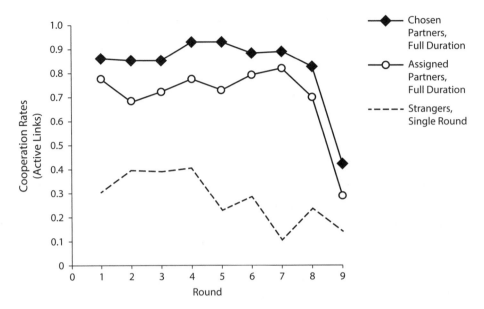

Figure 11.1. Cooperation Rates with Exit and Scale Options. *Source*: Holt, Johnson, and Schmidtz (2015).

Consider the sequence for one of the subjects, ID2, which is shown in table 11.5. In the top row on the left, ID2 was matched with ID7, and the minimum of their proposed scales was 1×. They both cooperated, as indicated by the "A A" outcome, and earned $1 each on this link. Moving to the right in the top row, we see that the scale for this link went up to 2× in round 2 and 3× in round 3, where it stayed, with highly profitable mutual cooperation ($3 each) until round 8 when one person defected. The link went inactive in the final round. The second row of the table shows the link with ID8, which began in round 2 at 2× scale, but became inactive immediately after ID8 defected. The third link was even more of a disaster; it also went inactive immediately after an initial defection by ID12, but the two people managed to restart the link at a low 1× scale, only to defect on each other in subsequent rounds. This sequence indicates that it is difficult to reestablish trust once it is gone. And even though the person who defects makes $1 more than they would by cooperating (or $3 more at 3× scale), this short-term gain is very costly. *An analysis of the data indicated that those who defected the most tended to earn the least.*

In the "chosen partners" treatment, each person met 1 new person in each round, but after round 3 they could only propose positive scales (1×, 2×, or 3×) for at most three links. So to add a link after round 3, a person would have to sever another link. If a link does go inactive, this process lets the person try to establish a good working relationship with someone else. This treatment also resulted in sustained high cooperation until the end, as indicated by the top line in figure 11.1. People with inactive links were able to establish new links in later rounds,

Table 11.5. History of Decisions for ID2 (Listed First in Each Pair of Decisions) with Three Assigned Partners and with the Scale (1×, 2×, or 3×)

Round:	1	2	3	4	5	6	7	8	9
ID 2, ID 7:	A A 1x	A A 2x	A A 3x	A A 3x	A A 3x	A A 3x	A A 3x	A B 3x	Inactive
ID 2, ID 8:		A B 2x	Inactive	Inactive	Inactive	Inactive	Inactive	Inactive	Inactive
ID 2, ID 12:			A B 1x	Inactive	B A 1x	A B 1x	A A 1x	A B 1x	B B 1x

which sometimes worked fine, and other times failed, since some of these new partners were people who defected previously, and others were people who had experienced defection and were probably cautious as a result. In other words, the "pool" of available people was tainted. For both the chosen and assigned partners treatments, a common pattern was to increase the scale of profitable links, with a somewhat sharper rate of increase for the chosen partners treatment, where most active links reached the 3× scale after several rounds. To summarize:

> **Prisoner's Dilemma with Choice of Partners:** *Even with a fixed, known number of rounds, sustained high cooperation is observed in prisoner's dilemma experiments with the option to exit from uncooperative relationships and to scale up the intensity of cooperative relationships. Scales tend to increase on profitable links, and those who do defect almost invariably earn less.*

The choice of partners experiment provides important insights for students who might otherwise perceive everyday social and business interactions in terms of being risky prisoner's dilemmas with opportunities to benefit from defection and free riding. Another notable aspect of the experiment is the presence of strong pro-social feelings that developed on mutually profitable links. A sense of excitement and satisfaction was "in the air" and might have been even more powerful in a setting with no fixed final period, e.g., with the random stopping rule discussed earlier. There was no official debriefing, except in a couple of software test runs, but some subjects were eager to talk and stayed around after being paid. They had lots of questions, including: "Who can I talk to, who can I email?" "How can I work in the lab?" "Can you help me, I'm worried about the feelings of the other person?" In some cases, students continued to seek out opportunities to discuss and reflect on the choice of partners experiment a year or more later.

11.3 The Traveler's Dilemma

The version of the traveler's dilemma game described in the introduction was first introduced by Basu (1994). He viewed the game as more of a dilemma for the game theorist than for the players, who tend to behave in a manner that he found reasonable. Recall that the game specifies a penalty/reward, which is taken away from the high claimant and given to the low claimant. With a very low penalty rate, there is little risk in making a high claim, and yet each person has an incentive to "undercut" any common claim level. For example, suppose that the upper limit on permitted claims is $200, and the amount paid is minimum of the two traveler's claims, minus a penalty of $5 that is taken from the high claimant and given to the low claimant. So if the claims are $100 and $200, the minimum is $100, and the person who claimed $100 would receive $105, with the other person only receiving $95 after the penalty is deducted from the minimum. Once in separate rooms, each person would surely consider the full claim of $200, but each might wonder about a deviation to $199 that would reduce the minimum by $1, which is more than made up for by the $5 reward for being low. In fact, there is no possible belief about the other's claim that would make one want to claim the upper limit of $200. If one expects any lower claim, then it is better to undercut that lower claim as well, which suggests that the equilibrium would be at the lowest possible claim of $80.

The argument in the previous paragraph can be more precise, based on an assumption that each person knows that the other is perfectly rational. Recall that there is no belief about the other's claim that would justify a claim of $200. If they each know that the other will never claim $200, then the upper bound has shifted to $199, and there is no belief that justifies a claim of $199. To see this, suppose that claims must be in integer dollar amounts and note that $199 is a best response to the other's claim of $200, but it is not a best response to any lower claim. Since $200 will not be claimed by any rational person, it follows that $199 is not a best response to any belief about the other's decision. Reasoning in this manner, we can rule out all successively lower claims except for the very lowest feasible claim. Similar reasoning shows that no configuration with unequal claims can be an equilibrium. This argument based on common knowledge of rationality is called "rationalizability," and the minimum possible claim in this game is the unique *rationalizable* equilibrium. Despite the persuasive adjective, this theory will provide poor predictions for about half of the treatments.

The dilemma for the theorist is that the Nash equilibrium is not sensitive to the size of the penalty/reward level, as long as this level is larger than the smallest possible amount by which a claim can be reduced. For example, the unilateral incentive to deviate from any common claim in the traveler's dilemma is not affected if the penalty/reward rate is changed from $5 to $2. If both players were planning to choose a claim of $200, then one person's deviation to $199 would reduce the minimum to $199, but would result in a reward of $2 for the deviator. Thus the person deviating would earn $199 + $2 instead of the $200 that would be obtained if they both claim $200. The same argument can be used to show that there is an incentive to undercut any common claim, whether the

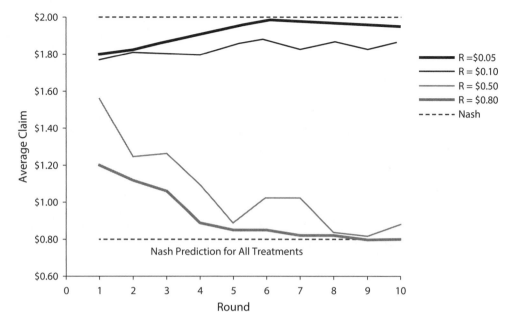

Figure 11.2. Data for the Traveler's Dilemma. *Source*: Capra et al. (1999).

penalty/reward rate is $2 or $200. Thus, the Nash equilibrium is not sensitive to this penalty/reward rate as long as it exceeds $1, whereas one might expect observed claims to be responsive to large changes in this payoff parameter.

The Capra et al. (1999) experiment was motivated by this expectation that the penalty/reward parameter would have a strong effect on actual claim choices. Such a result would contradict the unique Nash prediction that is independent of changes in this parameter. Each session involved about 10–12 subjects, who were randomly paired at the start of each round, for a series of ten rounds with the same penalty/reward parameter. Claims were required to be between $0.80 and $2.00. There were six sessions, each with a different penalty/reward parameter. The data averages for four of the sessions are plotted in figure 11.2, where the horizontal axis shows the round number. The penalty/reward parameter is denoted by R, which ranges from 5 cents to 10 cents (top two dark lines the figure), to higher levels of 50 cents and 80 cents (bottom two lines).

With a high penalty-reward parameter of $0.80, the claims average about $1.20 in round 1 and fall to levels approaching $0.80 in the final four rounds, as shown by the thick solid line at the bottom of the figure. The data for the $0.50 treatment start somewhat higher but also approach the Nash prediction in the final rounds. In contrast, the round 1 averages for the $0.10 and $0.05 treatments, plotted as dark lines, start at about a dollar above the Nash prediction and actually rise slightly, moving *away* from the Nash prediction. The data for the intermediate treatments ($0.20 and $0.25) are not shown, but they stayed in the middle range ($1.00 to $1.50) between the top and bottom pairs of lines, with some more variation and crossover between the two intermediate treatments. To summarize:

> **Traveler's Dilemma Experiment Results:** *The most salient aspect of the data, the sensitivity to the size of the penalty/reward rate, is not explained by the Nash equilibrium. The equilibrium prediction is that players will choose the lowest permitted claim, a prediction that is unaffected by changes in the penalty/reward rate. In traveler's dilemma games with low penalty/reward rates, the claims tend to rise above the Nash prediction toward the opposite end of the set of feasible claim amounts. But the data do converge to the Nash equilibrium for treatments with high penalty/reward rates.*

The online version of the traveler's dilemma mentioned in the introductory paragraph of this chapter is structured so that the "other decisions" are claims saved in a database from an experiment involving Virginia law students from a behavioral game theory class. After finishing five rounds, the person playing the demo is told how their total earnings compare with the total earned by the law student who faced the same sequence of other claims that they themselves saw. The average claims for the setup with a penalty/reward of only 10 are fairly high, in the 180 range, so someone who stays closer to the Nash equilibrium of 80 will typically earn less than the law student earnings benchmark. In fact, this was the case when a Nobel Prize–winning economist started with relatively low claims and raised them somewhat after encountering high claims by others. In the end, he earned about 25% less than the law student who had faced the same sequence of other claims.

11.4   Learning and Experience in the Traveler's Dilemma

Notice that the most salient feature of the traveler's dilemma data, the strong effect of the penalty/reward parameter, is not predicted by the Nash equilibrium. One might dismiss this game on the grounds that it is artificial. There is some truth to this, although many standard market economic games do involve payoffs that are higher for the firm with the lowest price. The traveler's dilemma game, however, is not intended to model a specific type of price competition. Instead, it involves an intentionally abstract setting, which serves as a paradigm for particular types of strategic interactions. *The traveler's dilemma is no more about lost luggage than the prisoner's dilemma is about actual prisoners.* If standard game theory cannot predict well in such simple situations, then some rethinking is needed. At a minimum, it would be nice to have an idea of when

Table 11.6. Traveler's Dilemma Data for a Classroom Experiment with $R = 10$

Round	Average (10 claims)	SuzSio	K Squared	Kurt/ Bruce	JessEd	Stacy/ Naomi
1	137	100 (133)	080 (195)	139 (140)	133 (100)	150 (135)
2	131	095 (191)	098 (117)	135 (140)	80 (130)	127 (200)
3	144	125 (135)	096 (117)	135 (100)	199 (199)	134 (200)
4	142	115 (125)	130 (100)	125 (115)	198 (115)	150 (134)

Key: Own Claim (Other's Claim).

the Nash equilibrium will be useful and when it will not. Even better would be a theoretical apparatus that explains both the convergence to Nash predictions in some treatments and the divergence in others. The rest of this chapter pertains to several possible approaches to this problem. We begin with an intuitive discussion of learning from an actual class experiment.

Behavior in an experiment with repeated random pairings may evolve from round to round as people learn what to expect. For example, consider the class data in table 11.6. Claims were required to be between 80 and 200 cents, and the penalty/reward parameter was 10 cents. There were 20 participants, who were divided into 10 pairs. The table shows the decisions for five of the teams during the first four rounds. The round is listed on the left, and the average of all ten claims is shown in the second column. The remaining columns show some of the team's own decisions, which are listed next to the decision of the other team for that round (shown in parentheses).

First consider the round 1 decision for "Stacy/Naomi" on the right side of the table, who claimed 150. The other team was lower, at 135, so the earnings for Stacy/Naomi were the minimum, 135, minus the penalty of 10 cents, or 125. "SuzSio" began lower, at 100, and encountered a claim of 133. This team then cut their claim to 95, and encountered an even higher claim of 191 in round 2. This caused them to raise their claim to 125, and they finished round 10 (not shown) with a claim of 160.

The team "K Squared" (Katie and Kari) began round 1 with a decision of 80 cents, which is the Nash equilibrium. They had the lower claim and earned the minimum, 80, plus the 10 cents for being low. After observing the other's claim of 195 for that round, they raised their claim to 98 in round 2, and eventually to 120 in round 10. The point of this example is that the best decision in a game is not necessarily the equilibrium decision; it makes sense to respond to and anticipate the way that the others are actually playing. If "K Squared" had played the Nash equilibrium decision of 80 in the first four rounds, they would have earned 90 in each round, for a total of 360. They actually earned 409 by adapting to observed behavior and taking more risk. It turns out that a claim of 200 in all rounds would have been even more profitable.

Teams who were low relative to the others' claims tended to raise their claims, and teams that were high tended to lower them. This qualitative adjustment rule, however, does not explain why "SuzSio's" claim was lowered even after it had the lower claim in round 1. This reduction seems to be in anticipation of the possibility that other claims might fall. In the final round, the claims in this class ranged from 110 to 200, with an average of 146 and a lot of dispersion. Remember, the unique Nash equilibrium is 80.

One way to describe the outcome is that people have different beliefs based on different experiences, but by round 10 most expect claims to be about 150 on average, with a lot of variability. If claims had converged to a narrow band, say with all at 150, then the "undercutting" logic of game theory might have caused them to decline as all seek to get below 150. Nevertheless, the variability in claims did not go away, perhaps because people had different experiences and different reactions to those experiences. This variability made it harder to figure out the best response to others' claims. In fact, claims did not diminish over time; if anything there was a slight upward trend. The highest earnings were obtained by the "JessEd" team, which had a relatively high average claim of 177.

The first ten rounds of this classroom experiment were followed by ten more rounds with a higher payoff parameter (50 cents). This created a strong incentive to be the low claimant, and claims did decline to the Nash equilibrium level of 80 after the first several rounds. Thus, convergence to a Nash equilibrium seems to depend on the *magnitudes* of incentives, not just on whether one decision is slightly better than another. At an intuitive level, this is what you need to take away.

Variability and Adaptive Adjustment in the Traveler's Dilemma: *With random matching, different people have different experiences, and the resulting variability makes it difficult to apply sharp "undercutting" strategies. People tend to raise their claims after encountering high claims and vice versa, with overall levels being affected by the size of the penalties and rewards.*

11.5   Iterated Rationality and Convergence to Equilibrium

The analysis of the matching pennies game in the previous chapter involved looking at the intersection of the curved "stochastic best response" lines, in a manner that is analogous to looking at supply and demand intersections. The intersections of stochastic response lines have the property that probabilities corresponding to Row and Column player beliefs are equal to the choice probabilities of the other player. Such an intersection is a stochastic ("quantal response") equilibrium in the sense that Row's beliefs about Column's decisions match Column's choice probabilities and vice versa.

Instead of having just two decisions like Top or Bottom, the traveler's dilemma has many possible claim choices. For the simple example used in this section, only three amounts will be considered: 80, 90, and 100. A quantal response equilibrium will be a set of three probabilities with an equilibrium property that beliefs match choice probabilities. The process of finding the equilibrium used here will be iterative; we start with a set of beliefs and generate a set of choice probabilities. If the choice probabilities differ from the original set of belief probabilities, then this is not an equilibrium, and we iterate again. This iterative approach has the added advantage of being related to the discussion in chapter 8 of iterated rationality for the guessing game (levels 0, 1, 2, etc.).

For the traveler's dilemma game with three possible claims, the initial "flat" beliefs are 1/3 for each of the three possible claims, which corresponds to level 0. By using the expected payoffs for these flat prior beliefs, one can find the claim, 80, 90, or 100, that has the highest expected payoff. A *best response* to level 0 beliefs would be to use the claim with the highest expected payoff with probability 1. This would represent "best-response" level 1 thinking, etc. The expected payoff calculations are shown in table 11.7, which has a row for each of the three possible claim amounts. Since the possible claims are 80, 90, and 100, the penalty parameter has to be larger than the minimum claim increment (10), so we begin with a large penalty of 30.

The left column of table 11.7 shows the level-0 belief probabilities (1/3) for the three possible claim amounts listed in the second column. The third column shows the formulas for finding expected payoffs as sums of products of belief probabilities and associated payoffs. The top row for claim 80 is for the case where you choose 80, so the minimum claim will be 80 for sure. The first term (1/3)*(80) is for the case where the other person also chooses 80, so there is no reward. The second term, (1/3)*(80 + 30), is for the case in which the other person chooses 90, so you earn a reward of 30 for being low at 80. The third term is analogous. The sum of these products is 100, as shown in the top row, right column. Looking over the rest of the table, notice that there is no penalty or reward along the diagonal coming down from left to right, since these are the

Table 11.7. Expected Payoffs for a Three-Decision Traveler's Dilemma, $R = 30$

Level 0 Belief	Claim	Your Expected Payoff for Each Claim (Calculated with Level 0 Beliefs)	Expected Payoff
1/3	80	(1/3)*(80) + (1/3)*(80+30) + (1/3)*(80+30)	100
1/3	90	(1/3)*(80−30) + (1/3)*(90) + (1/3)*(90+30)	86.67
1/3	100	(1/3)*(80−30) + (1/3)*(90−30) + (1/3)*(100)	70

terms where the other person's decision matches yours (no penalty or reward). Above the diagonal, the other claim is higher and you receive a reward of +30, and you incur a penalty of −30 below the diagonal.

Notice from the right column that the highest expected payoff is for a claim of 80, and this would be the level 1 best response. Also, notice that the expected payoff for the 90 claim is only about 15% lower, which suggests that the percentage of people choosing 90 might not be much lower than the percentage choosing 80. Differences would be even smaller for lower penalties, e.g., for $R = 11$ instead of R = 30. One way to model this is to use "better responses" or "quantal responses" instead of "best responses." As indicated in chapter 10, a standard "logit" approach is to use ratios of exponential functions with denominators structured to ensure that probabilities sum to 1. Let the expected payoffs for the three claims be represented by π_{80}, π_{90}, and π_{100}, so that the logit formulas analogous to equation (10.5) are:

(11.1)

$$Pr(80) = \frac{\exp(\lambda \pi_{80})}{\exp(\lambda \pi_{80}) + \exp(\lambda \pi_{90}) + \exp(\lambda \pi_{100})}$$

$$Pr(90) = \frac{\exp(\lambda \pi_{90})}{\exp(\lambda \pi_{80}) + \exp(\lambda \pi_{90}) + \exp(\lambda \pi_{100})}$$

$$Pr(100) = \frac{\exp(\lambda \pi_{100})}{\exp(\lambda \pi_{80}) + \exp(\lambda \pi_{90}) + \exp(\lambda \pi_{100})}$$

First, notice that these ratios in (11.1) only differ in terms of which expected payoff was used in the numerator, and all three choice probabilities sum to 1. The precision λ is a parameter that determines the amount of noise. In the extreme case of complete randomness, $\lambda = 0$, and all of the resulting exp(0) terms have values of 1, so the choice probabilities are 1/3 each. As the precision increases, the probabilities of decisions with higher expected payoffs tend to increase. The figures for the matrix games (with payoffs expressed in pennies) in the previous chapter were drawn with $\lambda = 0.1$, and that is the precision that will be used here, too.

The next step is to use the expected payoff formulas in table 11.7 to set up a spreadsheet for calculations. In addition, a column is added that calculates the exponential expressions in (11.1), sums them, and takes ratios to determine choice probabilities, which we will refer to as "noisy Level 1" probabilities. Then we can use those probabilities (instead of the initial 1/3 beliefs) to determine noisy Level 2 probabilities. It will turn out that this process converges after just three or four iterations, which is an equilibrium of the noisy process. (Of course, convergence does not imply that the equilibrium is unique, although advanced methods can be used to show that it is unique in this case.)

Table 11.8 shows the layout of the spreadsheet. The reason to do this in a spreadsheet is that you can set the penalty parameter in a designated cell with a fixed cell reference, generate the predictions, and then see how the predictions

Table 11.8. Logit Responses ($\delta = 0.1$) for a Three-Decision Traveler's Dilemma

A	B	C	D	E	F
Penalty 30	Level 0 Belief	Claim	Expected Payoffs	exp($\lambda \pi$)	Level K+1 Belief
Cell A8	=1/3	80	=(B8)*(C8) + (B9)*(C8+$A$7) + (B10)*(C8+$A$7)	=exp(0.1*D8)	=E8/E11
Cell A9	=1/3	90	=(B8)*(C8–$A$7) + (B9)*(C9) + (B10)*(C9+$A$7)	=exp(0.1*D9)	=E9/E11
Cell A10	=1/3	100	=(B8)*(C8–$A$7) + (B9)*(C9–$A$7) + (B10)*(C10)	=exp(0.1*D10)	=E10/E11
Cell A11				=E8+E9+E10	

change by changing the penalty number in that designated cell. You should set up a spreadsheet as you read so that you can play with it later (if you are the type of person who likes to see how things work). Otherwise, you can just skim the four steps that follow.

Step 1: Create an Excel spreadsheet and insert a penalty parameter value of 30 in cell A7 (the 7th row of the A column). This cell will be referenced as $A$7 in Excel commands, where the $ signs will keep the reference fixed if you copy and paste a block of text that refers to this cell. The spreadsheet columns are listed as A, B, C, etc. Beliefs will be in column B, and Claims will be in column C, so that references to these cells will be easy to remember as beliefs and claims. To further enhance readability, use rows 8, 9, and 10 for claims of 80, 90, and 100.

Step 2: Next you will specify initial beliefs and expected payoffs. The initial Level 0 beliefs are shown in column B as all being equal, with the Excel command "=1/3" to put 0.333 in each of those cells. (The "=" sign signals the program to do the calculation.) Then beliefs and claims are used in the commands in column D, again with "=" signs. The notation there matches the formulas in table 11.7. So in row A8 for a claim of 80, the minimum claim will be 80, and hence there is a C8 in each of the three terms. Note carefully the dollar signs for the reference to the penalty $A$7, which refers to the number 30 in cell A7. After typing in those commands, you will see expected payoff values in the D column; which should match those from the right column of table 11.7: 100, 86.67, and 70 (unless you made an error and need to recheck your formulas).

Step 3: The next task is to use the logit probabilistic choice formulas to determine the responses to the initial beliefs. The exponential commands like "=exp(0.1*D8)" calculate exponentials of precision-weighted expected payoffs with $\lambda = 0.1$. (You could have even used a fixed cell reference for the precision in order to experiment with different degrees of noise.) These exponentials are summed in cell E11, which is then the denominator for the logit formula, in column F. In other words, column F will show the choice probabilities that are noisy responses to the initial level 0 beliefs.

Step 4: Finally, you can repeat the noisy logit response process for higher levels of noisy responses, e.g., Level 2, Level 3, etc. Iteration is easy: just highlight the block of text ***shown in gray*** plus column headings (C7:F11) and copy it into a block with G7 as the upper left corner, i.e., into the block (G7:J11). You should be able to just put the cursor into G7 and paste. Then the beliefs that show up in column J will be for noisy Level 2. Copy the block one more time and paste to cell K7 to get Level 3 beliefs that show up in column N. You may want to adjust the column labels for the different "K+1" levels of beliefs so that they are correct for levels 1, 2, 3, and 4.

Table 11.9 shows the iterated beliefs for the claims for the various levels of iterated thinking, beginning with the Level 0, equal probabilities for each decision, shown on the left side. Level 1 probabilities are not equal, and are much higher for the lower claims of 80 and 90. Notice that the Level 4 choice probabilities essentially match those of Level 3, so the process converged quite quickly in this case. The second-to-last row of the table shows the average claims for $R = 30$, which are calculated as sums of products of probabilities and associated claim amounts. These averages fall from 90 (an average of 80, 90, and 100) on the left to about 81 on the right. The probabilities on the right are a stochastic (quantal

Table 11.9. Choice Probabilities for a Logit Function with $\lambda = 0.1$ and $R = 2$

Claim	Level 0 Probabilities	Level 1 Probabilities	Level 2 Probabilities	Level 3 Probabilities	Level 4 Probabilities
80	1/3	0.76	0.90	0.90	0.90
90	1/3	0.20	0.06	0.05	0.06
100	1/3	0.04	0.03	0.04	0.04
Avg. Claim ($R = 30$)	90	82.3	81.3	81.4	81.4
Avg. Claim ($R = 11$)	90	88.5	88.0	87.8	87.8

response) equilibrium in the sense that with those as beliefs, the noisy best responses match the beliefs, so there would be no tendency for change. Notice that the quantal response equilibrium of 81 with a penalty reward of 30 is quite close to the Nash prediction (no noise) of 80.

The bottom row of table 11.9 shows average claims after the penalty in the $A$7 cell of the spreadsheet is reduced from 30 to 11 (the lowest integer it can be and still have a Nash prediction of the claim of 80). This change does not alter the unique Nash prediction, but the logit model predicts that claims will be increased to about 88, which is close to the midpoint of the two extreme claims of 80 and 100. The logit model can explain why claims might be close to Nash levels for a game with a high penalty parameter, and why claims might be higher with a low penalty parameter. This sensitivity to the penalty parameter of the game is intuitive, since there is less risk of raising one's claim with a low penalty.

Finally, you might wonder what all of this means for the traveler's dilemma experiment data shown in figure 11.2. To accomplish this, an expanded version of the spreadsheet, with a row for each claim in the interval from 80 to 200, was used to calculate logit response probabilities for each level, again using the parameter $\lambda = 0.1$. The resulting claim distributions for each level are plotted in figure 11.3, with claims from 80 to 200 on the horizontal axis, and choice probabilities on the vertical axis. The flat dashed line represents level 0, with equal probabilities for all claims in this interval.

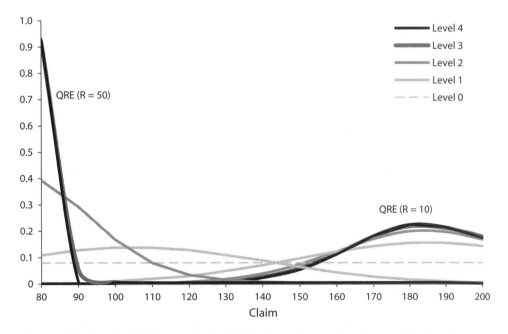

Figure 11.3. Traveler's Dilemma Claim Distributions for Levels of Iterated Rationality

When $R = 10$, a relatively low penalty for the expanded game, the logit responses form gentle hill shapes that converge to the dark curve on the right side of the Figure that peaks above 180. This prediction is close to the average claim of 186 that was observed in the final five rounds of the experiment treatment for $R = 10$. For a high penalty of 50, in contrast, the probabilities pile up near the Nash prediction of 80 on the left, which is consistent with the low claims observed in the experiment for the two treatments with high penalty parameters. The same precision parameter was used for both sets of lines, so the same theory that predicts high claims in one treatment also predicts low claims in the other. The black line on each side lies over the top of the level 3 line, i.e., the level 3 and level 4 lines are virtually identical. Thus, these are claim distributions that get mapped into the same distributions, i.e., they each represent a *quantal response equilibrium*. Therefore, the black lines are given "QRE" labels for each treatment.

> **Iterated Rationality and Stochastic Equilibrium in the Traveler's Dilemma:**
> *The sequence of noisy better responses generates associated beliefs that start with level 0 (equal probabilities) and converge to an equilibrium list of beliefs that are the same as the noisy responses to those beliefs. This model can explain why observed claims in the experiment (with a claim range from 80 to 200) converge to near-Nash levels (near 80) in treatments with high penalties, and why claims tend to cluster at much higher levels when penalties are relatively low. In this game, behavioral game theory provides a clear explanation for intuitive deviations from stark Nash equilibrium predictions.*

11.6 Extensions

In infinitely repeated games, cooperation can be enhanced by increasing the probability of continuation and reducing the immediate payoff from defection. The effects of cooperation in such situations might be to generate prosocial attitudes toward others in the group (*reciprocity*) or even toward total "third party" strangers (which is termed *indirect reciprocity*). The result could be increases in cooperative tendencies that are even higher than predicted by payoff and probability considerations. We will return to this idea after some ways of measuring prosocial behavior in bargaining and trust games are developed in chapters 14 and 15.

The original experimental papers on prisoner's dilemma with exit (Orbell, Schwartz-Shea, and Simmons, 1984, and Orbell and Dawes, 1993) were done in leading political science and sociology journals. The possibility of exit tended to increase cooperation, but not by much. Davis and Holt (1994c) ran the experiment in which supplier switching promoted high-quality deliveries, as compared with low-quality deliveries that were prevalent in one-period match-ups. Much of the recent work in this area pertains to the effect of flexibility in terms of adding and breaking links in network experiments, e.g., Fehl, van der Post, and Semmann (2011). This literature will be discussed in more detail in chapter 16 on public goods (voluntary contributions) games.

An interesting one-shot traveler's dilemma experiment is reported by Becker, Carter, and Naeve (2005), who invited members of the Game Theory Society to submit claims or mixed strategies. The range of allowed claims was the interval [2, 100], and the penalty/reward rate was 2. Participants were also required to submit a probability distribution on this interval representing their beliefs about other decisions. Two of the entrants into the contest were selected at random, and were paid a monetary prize. One person was paid 20 times their earnings as determined by playing against other submitted strategies, and another person was paid a prize that was related to the accuracy of that person's submitted beliefs. In total, 26 of the 45 pure strategies submitted were above 90, and only 3 corresponded to the unique Nash equilibrium, 2. The modal decision was 100 (10 cases). The submitted beliefs looked somewhat like the actual distributions of decisions from the experiment. These results indicate that experts in game theory recognize a situation where the Nash equilibrium prediction will not be a good guide for what to expect.

There are several other games with a similar payoff structure that depends on the minimum of all decisions. For example, the shopping behavior of un-informed consumers may make firms' profits sensitive to whether or not their price is the minimum of the prices offered in the market. Capra et al. (2002) provide experimental data for a price competition game with this general structure. If one firm sets a lower price than the other, then it will obtain the larger market share, and the high-price firm will have to match the other's price to get any sales at all. This is like the traveler's dilemma in that earnings are determined by the minimum price, with a penalty for having a higher price. In particular, each firm earns an amount that equals the minimum price times their sales quantity, but the firm that had the lower price initially will have the larger market share by virtue of picking up sales to the informed shoppers.

The Capra et al. (2002) price competition game also has a unique Nash equilibrium price at marginal cost, since at any higher price there is a unilateral incentive to cut price by a very small amount and pick up the informed shoppers. The Nash equilibrium is independent of the proportion of informed buyers who respond to even small price differences. Despite this independence property, it is intuitively plausible that a large fraction of buyers who are informed about price differences would provide sellers with less power to raise prices. This intuition was confirmed by the results of the experiment. As with the traveler's dilemma, data averages were strongly affected by the parameter that determines the payoff differential for being low, even though this parameter does not affect the unique Nash equilibrium at the lowest possible decision.

The iterated introspection approach in section 11.5 is based on a small literature on how to model games played only once. As mentioned, in chapter 8, Stahl and Wilson (1995) and Nagel (1995) were influenced by one-shot guessing game data to consider one or more levels of iterated strategic thinking. Camerer, Ho, and Chong (2004) incorporate heterogeneity by allowing different people to have different levels of strategic thinking, with a one-parameter distribution that determines the probability that a person will be of each different level. This *cognitive hierarchy* model has the intuitive property that people assume that others have lower levels of rationality than they do. For example, a level 2 player would view the world as being composed of level 0 and level 1 players. When confronted with the data for 1-shot games, this approach suggests that people engage in about 1.5 steps of this reasoning process.

Goeree and Holt (1999b, 2001, 2004) introduce stochastic elements into the iterated introspection process. In their *noisy introspection* model, the amount of "noise" is higher for higher levels of iterated reasoning in the sequence from "he thinks" to "he thinks I think" to "he thinks I think he thinks," etc. The insight here is that one person may have some idea about what the other will do, but has a more dispersed idea about what the other thinks the first person will do, and an even more dispersed idea about what the second person thinks the first person thinks, etc. For a related model with two different error parameters, see Weizsäcker (2003) and Kübler and Weizsäcker (2004).

Chapter 11 Problems

1. Show that unequal claims cannot constitute a Nash equilibrium in a traveler's dilemma.

2. What is the Nash equilibrium for the traveler's dilemma game where there is a $5 penalty and a $5 reward, and claims must be between −$50 and $50? (A negative claim means that the traveler pays the airline, not the reverse.)

3. (non-mechanical) Consider a traveler's dilemma with N players who each lose identical items, and the airline requests claim forms to be filled out with the understanding that claims will be between $80 and $200. If all claims are not equal, there is a $5 penalty for anyone whose claim is not the lowest and a $5 reward rate for the player(s) with the lowest claim(s), as long as there is at least one player with a higher claim. Speculate on what the effect would be on average claims of increasing N from 2 to 4 players in a setup with repeated random matchings.

4.   Calculate what the earnings would have been for the "K-squared" team if they had chosen a claim of 200 in each of the first four rounds of the game summarized in table 11.6.

5.   Explain why the Nash equilibrium (with sequential rationality) for the Choice of Partners game in section 11.2 must involve exit in the final round. Then what would backward induction imply about the second to last round?

6.   Take the spreadsheet in table 11.8 and double the precision in column D from 0.1 to a much higher level, say 1.0. First conjecture on how this increased precision would change the average claim amounts for the $R = 11$ row of table 11.9. Then recalculate those average claims for $R = 11$ and this higher precision. Explain the intuition for the effects of higher precision.

7.   To derive the "present value" formula used in the first section of this chapter: $1 + \delta + \delta^2 + \ldots = 1/(1 - \delta)$, define the sum $V = 1 + \delta + \delta^2 + \ldots$. It follows from this definition that $V = 1 + (\delta + \delta^2 + \ldots) = 1 + \delta(1 + \delta + \delta^2 + \ldots)$. The next step is up to you.

12

Contests and Rent Seeking

Administrators and government officials often find themselves in a position of having to distribute a limited number of prized items (locations, licenses, etc.). Contenders for these prizes may engage in lobbying or other costly activities that increase their chances for success. No single person would spend more on lobbying than the prize is worth, but with a large number of contenders, expenditures on lobbying activities may be considerable. This raises the disturbing possibility that the total cost of lobbying by all contenders may "dissipate" a substantial fraction of the prize value, as noted 50 years ago by Gordon Tullock, who had observed rampant rent seeking while serving as a diplomat in China before becoming an academic. In economics jargon, the prize is an "economic rent" that is the difference between the value of an object or license and the opportunity cost of using that object productively in the next best available use. Real effort costs associated with non-market competition for a prize are referred to as *rent seeking*. Such activities are thought to be more prevalent in developing countries, where some estimates are that non-market competition consumes a significant fraction of national income, as documented by Ann Krueger (1974), who coined the phrase. She provided major insights into why some nations "fail." One only has to serve as a department chair in a large US university, however, to experience the frustrations of effort-based competitions between large numbers of departments for prizes that are small, especially in "lean" budget years.

The stylized lottery game considered in this chapter is one for which the probability of obtaining the prize is equal to one's share of the total lobbying expenditures by all contenders. Explicit lotteries are sometimes used by public agencies to allocate commodities or licenses, e.g., auto license plates in Beijing. The lottery game, however, provides a paradigm for understanding the social costs of lobbying competitions more generally, e.g., "beauty contest" competitions for TV broadcast licenses awarded by government agencies. The nature of the Nash equilibrium for this game is quite intuitive and is illustrated with discrete examples.

The results of rent-seeking experiments are important because they highlight the potential costs of administrative (non-market) allocation procedures, especially in comparison with efficient market-based allocations. It is, of course, inappropriate to sell some things, like access to school slots, dorm rooms, or medical residency programs. Lotteries can be used in such situations. A better solution is to let people submit ranked preference lists, and such "matching mechanisms" will be discussed in the final chapter of this book.

> Note to the Instructor: The lottery version of the rent-seeking game can be implemented with playing cards in an engaging game, as indicated in the instructions appendix for this chapter. The hand-run version in the appendix has the advantage of permitting a comparison with an auction-based allocation, but the web-based Veconlab Rent Seeking game (on the Public menu) permits additional options, like eliciting an ex ante assessment of the chances of winning.

12.1 Government with "a Smokestack on Its Back"

One of the most spectacular successes of government policy has been the auctioning of bandwidth used to feed the exploding growth in the use of cell phones and wireless devices. The licenses specify a particular frequency band and geographic area. The US Federal Communications Commission (FCC) has allocated major licenses with a series of auctions that have raised more than $100 billion without adverse consequences. European countries followed with similarly successful auctions, which often raised many more billions than expected.

Auctions can bring together large numbers of potential competitors to allocate commodities quickly to those with the highest valuations, as discussed in later chapters in this book. In the United States, much of this bandwidth was originally reserved for the armed forces, but it was underused at the end of the Cold War. Fortunately, the transfer created a large increase in economic wealth, without significant administrative costs, although it could have easily been otherwise. Radio and television broadcasting licenses were traditionally allocated via administrative proceedings. The successful contender would have to convince the regulatory authority that service would be of high quality and that community social values would be protected. This often required establishing a technical expertise and an effective lobbying presence. The lure of extremely high potential profits over many decades was strong enough to induce large expenditures by aspiring providers. Those who had acquired licenses in this manner were, understandably, opposed to market-based allocations that forced the recipients to pay for the licenses. Similar economic pressures may explain why "beauty contest" allocations based on lobbying and administrative proceedings continue to be prevalent in many countries.

The first crack in the door appeared in the late 1980s, when the FCC decided to skip the administrative proceedings in the allocation of hundreds of local cell phone licenses. The forces opposed to the pricing of licenses did manage to block an auction, and the licenses were allocated by lottery instead. There were about 320,000 applications for 643 licenses. Each application involved significant paperwork (legal and accounting services), and firms specializing in providing completed applications began offering this service to interested competitors. The resources used to provide this service have opportunity costs. Hazlett and Michaels (1993) report that accounting firms were charging about $600 to fill out the paperwork and organize the required documentation for each lottery application. Using this average fee, the authors estimated the total cost of all submitted applications. The lottery winners often sold their licenses to more efficient providers, and resale prices were used to estimate the total market value of the licenses. Each individual lottery winner earned very large profits on the difference between the license value and the application cost, but the costs incurred by others were lost. The total cost of the transfer of this property was estimated to be roughly *40%* of the market value of the licenses. There are, of course, the indirect costs of subsequent transfers of licenses to more efficient providers, a process of consolidation that may take many years. In the meantime, inefficient provision of the cellular services may have created ripples of inefficiency in the economy. Episodes like this caused Milton and Rose Friedman (1989), in a discussion of the unintended side effects (negative externalities) of government policies, to remark: "Every government measure bears, as it were, a smokestack on its back." In this case, the massive paperwork that was received caused the floor of a government warehouse to collapse. Nobody died, but this event adds color to the phrase "smothered in paperwork."

In a classic paper, Gordon Tullock (1967) pointed out that the real costs associated with non-market competitions may destroy or "dissipate" much of the value of those rents in the aggregate, even though the winners in such contests may earn large profits. This destruction of value is often invisible to those responsible, since the contestants participate willingly, and the administrators often enjoy the process of being wined and dined. Moreover, some of the costs of activities like waiting in line and personal lobbying are not directly priced in the market. These costs can be quite apparent in a laboratory experiment of the type to be described next. The costs of this non-market rent-seeking activity are particularly disturbing in light of the fact that such costs can be avoided with auctions that use market prices instead of costly efforts to determine the winners. To summarize:

Rent Seeking: *The duplicated "all-pay" competitive efforts in non-market allocations of economic rents can be very detrimental for the economy as a whole if alternative market-based allocations (e.g., auctions) are available.*

This lesson should not be forgotten by a reader who becomes immersed in the chapter's presentation of various factors that increase or decrease the proportion of the prize value that is dissipated in rent-seeking competitions and tournaments.

12.2 Rent Seeking in the Classroom Laboratory

In many administrative (non-market) allocation processes, the probability of obtaining a prize or monopoly rent is an increasing function of the amount spent in the competition. In the FCC lottery, the chances of winning a license were approximately equal to the applicant's efforts as a proportion of the total efforts of the other contestants. This provides a rationale for a standard mathematical model of "rent seeking" with N contestants, which was first proposed by Tullock (1980). The effort for person i is denoted by x_i for $i = 1, ... N$. The total cost of each effort is a cost c times the person's own effort, i.e., cx_i. In the simplest symmetric model, the value of the rent, V, is the same for all, and each person's probability of winning the prize is equal to their own effort as a fraction of the total effort of all contestants. Here the expected payoff is the probability of success, which is the fraction of total effort, times the prize value V, minus the cost of effort:

$$(12.1) \qquad \text{expected payoff} = \frac{x_i}{\sum_{j=1...N} x_j} V - cx_i.$$

The effort cost on the right is not multiplied by any probability since it must be paid whether or not the prize is obtained.

Goeree and Holt (1999b) used the payoff function in (12.1) in a classroom experiment with $V = \$16,000$, $c = \$3,000$, and $N = 4$. Each of the four competitors consisted of a team of students. Efforts were required to be integer amounts. This requirement was enforced by giving 13 playing cards of the same suit to each team. Rent-seeking effort was determined by the number of cards that the team placed in an envelope. Each team incurred a cost of $3,000 per card played, regardless of success or failure. The cards played were collected, shuffled, and one was drawn to determine which contender would win the $16,000 prize. The number of cards played varied from team to team, but each team played about three cards on average. Thus a typical team incurred 3 × $3,000 in expenses, and the total lobbying cost for all four teams was over $36,000, all for a prize worth only $16,000!

Similar results were obtained in a separate classroom experiment using the Veconlab setup. The parameters were the same (four competitors, a $16,000 value, and a $3,000 cost per unit of effort), and the 12 teams were randomly put into groups of 4 competitors in a series of rounds, with an initial capital of $100,000. The average number of lobbying effort units was 3 in the first round, which was reduced to a little over 2 in rounds 4 and 5. Even with two units expended by each team, the total cost would be 2 (number of effort units) times 4 (teams) times $3,000 (cost of effort), for a total cost of $24,000. This resulted in overdissipation of the rent, which was only $16,000.

12.3   Nash Equilibrium in Contests

A Nash equilibrium for this experiment is a lobbying effort for each competitor such that nobody would want to alter their expenditure given that of the other competitors. First, consider the case of four contenders, a $16,000 value, and a $3,000 effort cost. Note that efforts of 0 for all players cannot constitute an equilibrium, since any person could deviate to an effort of 1 and obtain the $16,000 prize for sure at a cost of only $3,000. Next suppose that each person is planning to choose an effort of 2 at a cost of $6,000. Since the total effort is 8, the chances of winning are 2/8 = 1/4, so the expected payoff is 16,000/4 − 6,000, which is *minus* $2,000. This cannot be a Nash equilibrium, since each person would have an incentive to deviate to 0 and earn nothing instead of losing $2,000. Next, consider the case where each person's strategy is to exert an effort of 1, which produces an expected payoff of 16,000/4 − 3,000 = 1,000. To verify that this is an equilibrium, we have to check the profitability of unilateral deviations. A reduction to an effort of 0 with a payoff of 0 is obviously bad. A *unilateral* increase to an effort of 2, when the others maintain efforts of 1, will result in a 2/5 chance of winning, for an expected payoff of 16,000(2/5) − 6,000 = 400, which is also worse than the payoff of 1,000 obtained with an effort of 1. These calculations are summarized in table 12.1, which shows the payoffs for a player with efforts of 0, 1, or 2, as indicated in the left column. The remaining columns are for the case where the three others choose efforts of 0, 1, or 2. In the symmetric equilibrium, the effort of 1 for each of the four contestants results in a total cost of 4(3,000) = 12,000, which dissipates three-fourths of the 16,000 rent. This raises the issue of whether a reduction in the cost of rent-seeking efforts might reduce the extent of wasted resources. In the context of the FCC lotteries, for example, this could correspond to a requirement of less paperwork for each separate application. Suppose that the resource cost of each application is reduced from $3,000 to $1,000. This cost reduction causes the Nash equilibrium effort level to rise from 1 to 3 (see problem 1). Thus, cutting the cost to one-third of its original level is predicted to triple the amount of rent-seeking activity, so the social cost of this activity would be unchanged.

The Nash equilibrium calculations done in table 12.1 were based on considering a particular level of rent-seeking activity, common to each person, and showing that a deviation by one person alone would not increase that person's expected payoff. This trial-and-error approach is straightforward, but tedious, and it has the additional disadvantage of not explaining how the candidate for a Nash equilibrium was found in the first place. The appendix at the end of the

Table 12.1. Rent-Seeking Payoff Calculations for a Prize Worth $16,000 and an Effort Cost of $3,000, with Three Other Competitors

	3 Others Choose 0	3 Others Choose 1	3 Others Choose 2
Own effort = 0	earn $0	earn $0	earn $0
Own effort = 1	earn $13,000	earn $1,000 (Nash)	earn −$714
Own effort = 2	earn $10,000	earn $400	earn −$2,000

chapter offers a simple calculus derivation of a formula for the Nash equilibrium rent-seeking effort, assuming that players are risk neutral:

(12.2)
$$x^* = \frac{(N-1)}{N^2} \frac{V}{c}.$$

For example, if $V = \$16,000$, $c = \$3,000$, and $N = 4$, then the equilibrium effort is $(3/16)(16/3) = 1$, as was verified previously with the method of checking deviations. When the cost is reduced to $1,000, the equilibrium effort is $(3/16)(16/1) = 3$. Since cutting the cost by one-third causes a tripling of the equilibrium effort, the predicted effect of a cost reduction is that the total amount spent on rent-seeking activity is unchanged, i.e., invariant to the effort cost. This conclusion would need to be modified if the effort cost were a nonlinear function of the effort. The linearity assumption used here corresponds to the case of a lottery in which accounting firms charge a flat rate for each application processed.

Finally, consider the total expenditures on all rent-seeking activity, as measured by the product of the number of contenders, the effort per contender, and the cost per unit of effort: Nx^*c. It follows from (12.2) that this total expenditure is: $(N-1)/N$ times the prize value, V. Thus the proportion of V that is dissipated is a fraction, D, which is an increasing function of the number of contenders.

(12.3) $D = \frac{N-1}{N}$ (equilibrium rent-seeking expenditures as a fraction of value).

With two contenders, half of the value is dissipated in a Nash equilibrium, which increases to two-thirds with 3 contenders, three-fourths with 4 contenders, as it approaches 1 (full dissipation) with large numbers. Competition, which is normally good for market efficiency, is bad in non-market allocations. To summarize:

Nash Equilibrium Rent Dissipation in a Lottery Contest: *In equilibrium, individual effort is proportional to the ratio of prize value to effort cost, with the proportionality factor that is decreasing in group size, N. Rent dissipation, D, measured as the ratio of total expenditures on rent-seeking activity to prize value, $(N-1)/N$, which is increasing in group size and independent of the effort cost. In theory, the* *social cost of rent seeking is large with only two contenders, and only gets larger with more competition. But a decrease in the per unit cost of effort is predicted to be exactly offset by increases in equilibrium effort.*

12.4 Economic Effects of Changes in Cost and Group Size

The experiments discussed in this section are motivated by two predictions: First, although an increase in the number of competitors is predicted to reduce the amount of lobbying effort for each one, the total social cost of lobbying is predicted to be higher with more competitors. Second, a reduction in the cost of rent-seeking effort generates an offsetting increase in this effort, so the fraction of rent dissipation is predicted to be independent of the cost, c. For example, recall the previous setup with four contenders in which the reduction in c from \$3,000 to \$1,000 raised the Nash equilibrium lobbying effort from 1 to 3, thereby maintaining a constant total lobbying expenditure.

These predictions were initially evaluated in a class experiment with a "2 × 2" design: high versus low lobbying costs, and high versus low numbers of competitors. There were 20 rounds (5 for each treatment), with no balancing controls for sequence effects. The prize value was \$16,000 in all rounds, and each person began with an initial cash balance of \$100,000. One person was selected at random ex post to receive a small percentage of earnings. The treatments involved a per-unit cost of either \$500 or \$1,000, and either two or four contenders. All of the observed effort levels were higher than the Nash equilibrium predictions, and the total cost of the rent-seeking activity, as shown in table 12.2 (rounded to the nearest \$1,000), was a high fraction of the total prize value in all cases. With four competitors, the rent was either fully dissipated or overdissipated, as can be seen from the numbers in the bottom row. Next, consider the effort cost comparisons across each row of the table. The total rent-seeking expenditures, on average, increase, moving from left to right in table 12.2. Although the doubled effort cost did result in lower efforts (not shown), those efforts did not fall by half as predicted, so total rent-seeking expenditures increased.

There are reasons to be skeptical of the results of a single class experiment, with minimal incentives, no balancing of the order of treatments, and the

Table 12.2. Total Rent-Seeking Expenditures (Rounded Off) in a Classroom Experiment with a Prize Value of \$16,000

	Low-Effort Cost (\$500)	High-Effort Cost (\$1,000)
$N = 2$	\$10,000	\$12,000
$N = 4$	\$16,000	\$24,000

typical semi-social classroom setting. In this case, however, the class results are roughly consistent with data from an experiment with 16 sessions in a 2×2 design that varies group size and effort cost. The parameters are basically scaled-down versions of those used in the class experiment, with a prize value of $1.60 and an effort cost of $0.10. Participants were recruited in cohorts of 12, with random matchings for 20 rounds of play. Each cohort only participated in a single group size treatment, 2 or 4, with either high or low effort cost (10 cents or 5 cents per unit effort). Four separate sessions (groups of 12 people) were run for each of the four treatments in this design. Subjects were given endowments of $1.30 at the start of each round, but they could pay for higher efforts out of prior earnings and were permitted to spend up to $1.90 per round (efforts from 1 to 19 in the $0.10 cost treatment and from 1 to 38 in the $0.05 cost treatment). The motivation for letting people spend more on effort than the prize is worth was to avoid a "pile-up" at the highest effort, and to let people express extreme desires to try to win at all costs.

Figure 12.1 shows the average efforts for each group size in the high-cost treatment, with $c = 10$ cents. Notice that both data sequences are above the Nash equilibrium predictions calculated from equation (12.2), which are shown by the relevant horizontal lines (the dashed line at 4 for $N = 2$ and the gray line at 3 for $N = 4$). The data lines have markers for data averages, but otherwise have the same gray or dark features as the Nash predictions. As before, we see over-dissipation in the four-person design, since the average efforts of about 5 yield a total effort cost of about $4 \times 5 \times \$0.10 = \$2$, as compared with the $1.60 prize value. There is a downward trend in the effort averages for the $N = 2$ dashed

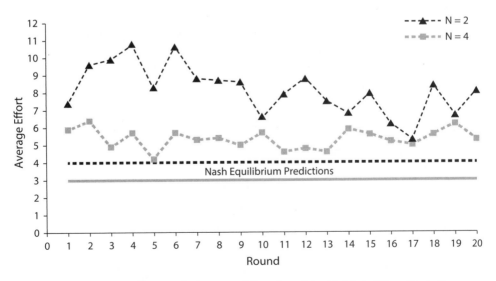

Figure 12.1. Average Efforts with Groups of Size 2 and 4 with High-Effort Cost. *Source:* Holt and Smith (2017).

Table 12.3. Rent Dissipation Percentages by Session: Cost and Numbers Effects

	Low-Effort Cost	High-Effort Cost
$N = 2$	77, 67, 62, 37 **Avg. = 61%**	91, 85, 85, 94 **Avg. = 89%**
$N = 4$	77, 83, 130, 81 **Avg. = 93%**	132, 111, 93, 87 **Avg. = 106%**

Source: Holt and Smith (2017).

line, and therefore, data averages for the final ten rounds (last half) were used in statistical comparisons. Recall that the Nash prediction from equation (12.3) is that the dissipation ratio of total rent-seeking expenditures to prize value be $(N − 1)/N$, which is 50% for $N = 2$ and 75% for $N = 4$. The actual rent dissipation percentages by session are shown in table 12.3, with treatment averages in bold. As was the case with the class experiment, rent dissipation goes up with higher costs and larger groups, with average dissipation ratios a little above 100% (expenditures above value) for the large groups with high costs (bottom right cell in the table).

Before proceeding, it is useful to summarize results, although we will have more to say about underlying statistical support for these results later.

Rent-Seeking Experiment Results

- *Rent Dissipation: Total expenditures on rent seeking are greater than 50% of prize value, and are greater than Nash predictions in all treatments.*
- *Group Size: An increase in the number of contenders tends to increase the total expenditures on rent seeking, which is directionally consistent with Nash predictions.*
- *Cost Effects: An increase in the cost of effort tends to raise total expenditures on rent seeking, which is inconsistent with the Nash invariance prediction that changes in effort costs will be fully offset by changes in effort in the opposite direction.*

12.5   A Preview of Nonparametric Statistical Analysis

An examination of the individual session averages in table 12.3 shows some variability from one session to another within the same treatment, especially the session with extreme rent dissipation of 130% in the lower left cell. Statistical tests provide a method of distinguishing directional patterns from randomness. The intuition is conveyed by a simple example. Suppose that you want to test to see whether a coin is fair. Instead of using a coin, we could use a small rubber "piggie" taken from a simple child's game in Denmark. (This game was presented to the author after he once taught a short experimental workshop to Danish students, using six- and ten-sided dice many times.) The pig can land in various positions, on its head, back, legs, or side; this is clearly a situation of ambiguity. Consider the hypothesis that the probability of a side landing is 1/2. This hypothesis loses all credibility if one flips the pig and observes six side landings in a row. Assuming the equal probability hypothesis were true, the probability of seeing an outcome this extreme is (1/2)(1/2)(1/2)(1/2)(1/2)(1/2) = 1/64 = 0.0156. This probability of seeing something as extreme or more extreme, given that the null is true, is referred to as a "*p* value." In this case, there is no reason to anticipate deviations in one direction or the other, so the alternative hypothesis allows a difference in either direction. Since six non-side landings would also have a probability of 0.0156 under the null, the *p* value to use is twice that amount, or about 0.03. A statistical test sets a standard cutoff probability, e.g., 0.05, for which the null hypothesis is rejected if the *p* value is below that cutoff. Some of the statistical tests commonly used by experimentalists are discussed in the methodology chapter that follows. But the intuitive framework will be sketched here to provide a preview and a quick perspective on group size and effort cost effects.

First, remember that each of the session averages in table 12.3 is an average for 12 people for the final ten rounds. The 120 observations for a given session are *not independent* observations, since people are competing with each other and there can be group effects. The session with a dissipation percentage of 130%, for example, may have evolved with some people being very aggressive in terms of effort expenditures, and others exerting more effort to stay competitive. There are econometric methods to deal with "clustered" sets of observations if one is willing to accept the underlying assumptions, but most experimental economists prefer the conservative approach of taking each session average as a single observation.

Next look at the top row of table 12.3, where all four session averages with low costs on the left are lower than the four session averages with high costs on the right. If there really were no effort-cost effect, then the ranks of these numbers should be mixed up a bit, instead of having the highest numbers on the right. (We are still focused on the top row only.) In fact, there are "8 take 4" = $\frac{8!}{(4!)(4!)}$ = 70 ways that eight session averages in the top row can be divided into two groups of 4, and of these, we see the most extreme order with all four of the numbers on the right side with high cost being higher. Under the null hypothesis of no effort-cost effect, each of these 70 groupings or "permutations" would be equally likely, so the chances of seeing something as extreme (or more) as what we see in the top row is 1/70 = 0.014. But there is another grouping that is just as extreme in the other direction, with the four smallest numbers on the right. So the chances of seeing something as extreme or more so *in either direction* is 0.028 under the null hypothesis of no effect. This is very unlikely, less than 3%, so in this sense we can reject the null hypothesis at standard significance levels (a 5% cutoff is often used).

The analysis thus far only pertained to the data in the top row of table 12.3. One way to use both rows is to consider regroupings of the data within each row but at the same time, thus holding the group size constant. A procedure for using all of the data, holding group size constant, will be discussed in the next chapter. This joint analysis based on all 16 session averages will provide reasonable statistical support for both group size and effort cost effects.

An alternative, more standard way to use all of the data would be to run a multiple regression (16 observations) with independent variables both for group size ($G = 0$ for group size 2, $G = 1$ for group size 4) and for effort cost ($EC = 0$ for low and $EC = 1$ for high). The variable to be explained ("dependent variable") is the rent dissipation percentage, D_i, for session i. Most students will have encountered regression in other courses, but the idea is to fit a model of the form: $D_i = \alpha + \beta G_i + \gamma EC_i$, where the Greek letters are parameters to be estimated. It turns out that the group size and effort cost coefficients are both positive and significant at the 5% level in such a regression. So the regression and the results of permutation tests hinted at previously tell the same story, in support of the existence of both group size and effort cost effects on rent dissipation.

12.6 Demographic and Individual Preference Effects in Contests

The first rent-seeking experiment was run by Millner and Pratt (1989). Instead of having simultaneous effort choices, they let people raise efforts at will during a fixed decision period. In the 1980s, computerized interfaces were first being used, and the subjects could see the other's effort choice at any point in time on their screens. Given the effort cost of 1, and the group size of 2, the predicted effort level (for a simultaneous choice model) was $(N-1)/N^2$ times the value of 8, which equals 2. The average effort expenditure in the experiment with continuous choices was slightly above this level, at $2.24. In a companion paper, Millner and Pratt (1991) used a lottery-choice decision to separate people into groups according to their risk aversion. Their risk task involved a choice between a sure $12 and a gamble between $18 and $8, with a switch point determined by choices made in a menu with increasing probabilities of winning in the gamble. Groups with more risk-averse individuals tended to spend less on rent seeking, which is sensible since a high expenditure exposes one to a larger cost if the effort is not successful. This procedure of sorting people by a measured preference trait was quite innovative and will be encountered again in a subsequent chapter on auctions with private values.

In theory, the effect of risk aversion on effort choice in a contest is more complicated than the simple intuition in the previous paragraph suggests. But the finding that risk-averse individuals exert less effort has been reported in a number of other papers as well, e.g., see Sheremeta (2013) or the survey papers mentioned in the extensions section at the end of the chapter. Anderson and

Freeborn (2010) also find that more risk-averse subjects (based on a Holt-Laury measure) have lower rent-seeking expenditures, an effect that is particularly strong for female subjects.

Price and Sheremeta (2015) use multiple regression to separate the effects of individual characteristics, such as gender and risk aversion, from treatment variables like whether initial cash reserves were endowed or earned in a pre-experiment task. The estimated regression coefficients show that more risk-averse people choose lower rent-seeking efforts, but when a gender term is added to the regression, the estimated gender coefficient indicates that women tend to choose *higher* efforts. Surprisingly, rent-seeking efforts are about 25% higher for women than for men, and the gender effect is stronger than the treatment effects. Another variable that had about the same effect as gender, but in the opposite direction (26% lower), was a self-reported measure of religiosity (1 if religion is very important). Being an economics major had no effect. A particularly interesting finding was that requiring subjects to first earn the cash balances tended to reduce efforts by about 10–15% below levels observed with exogenous cash endowments. But even with earned cash balances, efforts were still much higher than Nash predictions.

Price and Sheremeta also used a post-experiment task in which subjects were given a chance to bid against each other for a "prize" with a *zero* value. The bids involved real costs, and yet a significant fraction of subjects made bids in which the only prize was the joy of winning. Those subjects who exhibited a joy of winning in the final task tended to exert higher efforts in the standard lottery contest rounds that preceded.

Others have also found that women compete more aggressively in contests and earn less as a result, and that given the choice between contest (one wins) and piece rate (steady pay for effort) that women tend to avoid contests relative to men (Niederle and Vesterlund, 2007). To summarize up to this point:

> **Gender and Risk Aversion Effects in Contests:** *Risk-averse subjects tend to choose lower levels of rent-seeking efforts in contests. Women tend to prefer piece rate payment schemes to contest-based payoffs. Women in a contest tend to choose higher rent-seeking efforts than men.*

The risk aversion choice menu used by Price and Sheremeta paper is shown in table 12.4 (with references to bingo cages replaced by dice). This menu also has a sure payoff on the safe side, which is $1 in all rows for option A. The presence of a sure option is thought to generate gender differences that do not show up in standard Holt-Laury menus for which the safe option is a gamble, not a sure payment. As noted in chapter 3, a safe option may establish a reference point, so that a $0 payoff for the risky option might be perceived as a

Table 12.4. The Price-Sheremeta (2015) Risk Aversion Task

Row	Option A	Option B		A or B
1	$1	$3 never	$0 if 1,2,3,4,5,6,7,8,9,10,11,12,13,14,15,16,17,18,19,20	
2	$1	$3 if 1 is throw of die	$0 if 2,3,4,5,6,7,8,9,10,11,12,13,14,15,16,17,18,19,20	
3	$1	$3 if 1 or 2	$0 if 3,4,5,6,7,8,9,10,11,12,13,14,15,16,17,18,19,20	
4	$1	$3 if 1,2,3	$0 if 4,5,6,7,8,9,10,11,12,13,14,15,16,17,18,19,20	
5	$1	$3 if 1,2,3,4	$0 if 5,6,7,8,9,10,11,12,13,14,15,16,17,18,19,20	
6	$1	$3 if 1,2,3,4,5	$0 if 6,7,8,9,10,11,12,13,14,15,16,17,18,19,20	
7	$1	$3 if 1,2,3,4,5,6	$0 if 7,8,9,10,11,12,13,14,15,16,17,18,19,20	
8	$1	$3 if 1,2,3,4,5,6,7	$0 if 8,9,10,11,12,13,14,15,16,17,18,19,20	
9	$1	$3 if 1,2,3,4,5,6,7,8	$0 if 9,10,11,12,13,14,15,16,17,18,19,20	
10	$1	$3 if 1,2,3,4,5,6,7,8,9	$0 if 10,11,12,13,14,15,16,17,18,19,20	
11	$1	$3 if 1,2,3,4,5,6,7,8,9,10	$0 if 11,12,13,14,15,16,17,18,19,20	
12	$1	$3 if 1,2,3,4,5,6,7,8,9,10,11	$0 if 12,13,14,15,16,17,18,19,20	
13	$1	$3 if 1,2,3,4,5,6,7,8,9,10,11,12	$0 if 13,14,15,16,17,18,19,20	
14	$1	$3 if 1,2,3,4,5,6,7,8,9,10,11,12,13	$0 if 14,1,16,17,18,19,20	
15	$1	$3 if 1,2,3,4,5,6,7,8,9,10,11,12,13,14	$0 if 15,16,17,18,19,20	

loss, which is thought to trigger a gender effect. The innovative feature of the Price-Sheremeta risk elicitation menu is that the possible random outcomes (explained as throws of a 20-sided die) are listed in a row, so that the width of the list gives a rough perspective of the overall chances of each possible payoff amount, $3 or $0, for option B.

12.7 Alternative Explanations for "Overbidding"

There is no general agreement about what is causing effort expenditures to exceed Nash equilibrium predictions in these experiments. Figure 12.2 shows the frequency distribution counts for the $N = 2$ treatments (last half). The dark bars are for the high-effort cost sessions, in which each unit of effort costs 10 cents, and the Nash equilibrium is at 4 such units. With high costs, the maximum permitted effort of 19 would result in an expenditure of $1.90. For the low-cost setup, effort only costs 5 cents, and effort choices were permitted to range from 0 to 38, so that the maximum expenditure is held constant at 38 times $0.05 = $1.90. In order to facilitate comparisons, the horizontal axis is the number of 10-cent

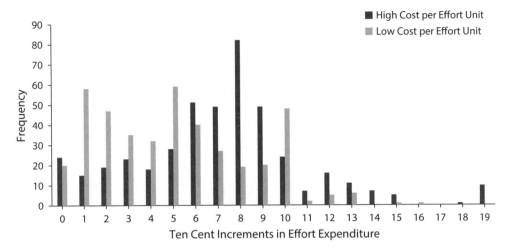

Figure 12.2. Frequency of Effort Expenditure Decisions (Bids) for N = 2 Sessions, with High and Low Costs Per Effort Unit. *Source*: Holt and Smith (2017).

effort expenditure units, which will be referred to as "bids" since the units are monetary. In the high-cost treatment, the bids equal the number of effort units that cost 10 cents each. In the low-cost treatment, the effort numbers are essentially divided by 2 and rounded off, so that they also range from 0 to 19, so each bid unit on the horizontal axis is for a 10-cent expenditure (two 5-cent effort units). This normalization is particularly convenient for comparisons since the Nash equilibrium prediction is for the rent-seeking expenditure to be the same, at $0.40 in both treatments. This is because when the effort cost is halved, the predicted effort level is doubled from 4 to 8 units, so that the predicted *expenditure* stays the same at four 10-cent expenditure units. The figure shows clearly how strong the tendency is to bid above Nash predictions. In addition, effort expenditures are larger in the high-cost treatment, in sharp contradiction of the Nash invariance prediction.

The other striking aspect of the figure is the wide spread of decisions, which is particularly notable since the data are for the final ten rounds and it can be assumed that learning has occurred in prior rounds. There is, of course, heterogeneity across people. The dark bar at 19 on the far right was for a single person who chose the maximum effort in all 20 rounds. That person won the contest in nine of the first ten rounds, earning the $1.30 endowment plus the $1.60 prize value minus the $1.90 effort expenditure, for a net gain of $1.00 in those rounds! In fact, there is a documented tendency for subjects who win in one round to bid higher in the next. The positive reinforcement associated with winning may be what is driving the higher subsequent bids, which increases variability for the same person (Sheremeta and Zhang, 2010). In a sense, the spread is not so surprising given the large effects reported for individual characteristics (gender, risk aversion, religiosity, and joy of winning).

Variability can reinforce overbidding, since the Nash equilibrium in figure 12.2 is on the left side of the range of possible bids. As noted in Sheremeta (2015), the effects of decision "noise" in models of bounded rationality (like "quantal response") are to spread decisions out to some extent over the set of all possible decisions, and such spreading would pull the average bids up if the Nash equilibrium is low in the range of feasible bids. This observation is supported by contest experiments that alter the range of decisions, holding everything else constant. Overbidding is more prevalent when the upper limit of the set of possible efforts is increased.

Overbidding is probably caused by a combination of factors. However, one that can be ruled out is risk aversion, since more risk-averse subjects tend to bid lower in contests. On the other hand, the inverse-S shaped probability weighting functions would tend to enhance the perceived importance of low win probabilities, which would be most relevant in contests with more bidders. The probability weighting conjecture is supported by contest experiments, summarized in Sheremeta (2015), in which subjects are awarded a proportion of the prize, instead of awards to a random winner. With proportional awards, probabilities are no longer as relevant, so the effects (if any) of probability weighting might be reduced. Another factor that could work in the opposite direction is that high efforts intended to raise the win probability will push it into the range where it would generally be thought to be underweighted.

Another possible explanation considered by Sheremeta (2015) is the sheer joy of winning, which was observed for about 40% of the subjects in the Mago, Samek, and Sheremeta (2016) experiment that involved bidding for a prize of value 0. At this point, the consensus seems to be that the overbidding may be due to several behavioral factors, and experiments offer some support for several explanations. Any estimation that includes joy of winning, probability weighting, and bounded rationality would require a model that incorporates errors. To summarize:

> Overbidding in Contest Experiments: *There are three factors—bounded rationality, joy of winning, and probability weighting—that have been identified as possible causes of overbidding, using treatments that manipulate the effects of each factor. Regardless of the cause, the massive wastes associated with effort-based competition pose a major problem, and it is important to look for ways to reduce the costs or substitute market allocations like auctions whenever appropriate.*

12.8 Extensions and Further Reading

There have been a number of research experiments using the payoff structure in equation (12.1), and overdissipation is the typical result, e.g., Potters, de Vries, and van Winden (1998). Anderson and Stafford (2003) evaluated predictions of a generalization of the basic model, by giving different players different cost levels and allowing an initial decision of whether or not to compete at all. The total expenditures on rent seeking exceeded theoretical predictions in all treatments they considered, and the rent was overdissipated in most treatments. The high effort levels reduced profitability for entrants, so the rate of participation in the rent-seeking contest was lower than predicted in theory. An extensive recent survey of the literature can be found in Dechenaux, Kovenock, and Sheremeta (2015).

Rent-seeking models have been used in the study of political lobbying (Hillman and Samet, 1987), which is a natural application because lobbying expenditures often involve real resources. In fact, any type of contest for a single prize may have similar strategic elements, e.g., political campaigns or research and development contests (Isaac and Reynolds, 1988). Of course, some lotteries involve simple ticket purchases instead of socially costly efforts. In this case, the expenditures are not socially wasteful, but rather are simple transfer payments of cash. Davis and Reilly (1998) use this observation to compare revenues raised by lotteries with revenues from auctions.

In all but one of the experiments reviewed in this chapter, the prize is the same for each contestant. In contrast, when non-price competition is costly, rent seeking may have a role to play if it helps to select contestants with higher private values. In the case of a binding price floor, for example, there is an excess of eager sellers who are not permitted to compete by lowering price, as discussed in chapter 2. Even though some sellers may have lower costs, the high-cost sellers may make sales if allocations are determined by random events like who arrives first or who buyers call first. In this case, seller efforts (like putting up a larger sign or arriving early to get a prime location at a farmers' market) may permit more efficient sellers to enhance their chances of making a sale. Of course, efficiency consequences of eliminating the price floor are much more favorable than having sellers compete in non-price dimensions, even if that competition improves seller selection (Finley et al., 2018).

Appendix. Mathematical Derivation of the Equilibrium

A calculus derivation is relatively simple, and is offered here as an option for those who are familiar with basic rules for taking derivatives. First, consider the expected payoff function in (12.1), modified to let the decisions of the $N-1$ others be equal to a common level, x^*.

$$(12.4) \qquad \text{expected payoff} = \frac{x_i}{x_i + (N-1)x^*} V - c x_i.$$

This will be a "hill-shaped" (concave) function of the person's own rent-seeking activity, x_i, and the function is maximized "at the top of the hill" where the slope of the function is zero. To find this point, the first step is to take the derivative and set it equal to zero. The resulting equation will determine player *i*'s best response when the others are choosing x^*. In a symmetric equilibrium with equal rent-seeking activities, it must be the case that $x_i = x^*$, which will yield an equation that determines the equilibrium level of x^*. This analysis will consist of two steps: setting the derivative of player *i*'s expected payoff equal to zero, and then using the symmetry condition that $x_i = x^*$.

Step 1. In order to use the rule for taking the derivative of a power function, it is convenient to express (12.4) as a power function:

(12.5)   expected payoff $= x_i [x_i + (N - 1)x^*]^{-1}V - cx_i$.

The final term on the right side of (12.5) is linear in x_i, so its derivative is $-c$. The first term on the right side of (12.5) is the product of x_i and a power function that contains x_i, so the derivative is found with the product rule (derivative of the first function times the second, plus the first function times the derivative of the second), which yields the first two terms on the left side of:

(12.6) $[x_i + (N - 1)x^*]^{-1}V - x_i[x_i + (N - 1)x^*]^{-2}V - c = 0.$

Step 2. Next use the fact that $x_i = x^*$ in a symmetric equilibrium. With this substitution into (12.6), one obtains a single equation in x^*:

(12.7) $[x^* + (N - 1)x^*]^{-1}V - x^*[x^* + (N - 1)x^*]^{-2}V - c = 0,$

which can be simplified to:

(12.8) $\dfrac{V}{Nx^*} - \dfrac{V}{N^2 x^*} - c = 0.$

Finally, we can multiply both sides of (12.8) by $N^2 x^*$, which yields an equation that is linear in x^* that reduces to the formula given previously in the main text in equation (12.3).

Chapter 12 Problems

1.   Suppose that the cost per unit of effort is reduced from \$3,000 to \$1,000, with four competitors and a prize of \$16,000. Show that a common effort of 3 is a Nash equilibrium for the rent-seeking contest. Please do this by checking the profitability of deviations to efforts of 2 or 4, not by referring to a calculus-based formula.

2.   Now suppose that the effort cost in problem 1 is reduced to \$500, with a prize value of \$16,000 and four bidders as before. Show that a common effort of 6 is a Nash equilibrium (consider unilateral deviations to 5 and 7).

3. Calculate the total amount of money predicted to be spent in rent-seeking activities for the setup in problem 1. Is the reduction in the per-unit (marginal) cost of rent seeking from $3,000 to $1,000 predicted to reduce *total* expenditures on rent-seeking activity? Explain.

4. Would a reduction in competitors from four to two in problem 2 reduce the predicted extent of rent dissipation? (Rent dissipation is the total cost of all rent-seeking activity as a proportion of the prize value.)

5. (non-mechanical) In what sense are the lengths of the lists of numbers in table 12.4 biased in terms of the impressions they convey about probabilities? How might this bias be reduced?

6. (non-mechanical) Speculate about whether prospect theory might provide an explanation for overbidding in contests, even when endowments are given that protect subjects from negative earnings if they are unsuccessful in the contest.

7. (non-mechanical) One procedure sometimes used by the US National Science Foundation to reduce paperwork expenses is to require 3-page pre-proposals to be submitted by applicants competing for extremely large (low-probability) grants. Then only a subset of the pre-proposals that are deemed the best will be invited to submit full proposals. What might be the pros and cons of such an approach?

13

Methodology

In order to evaluate the experiment results with a careful, skeptical eye, it is essential to understand underlying methodological considerations. The development of good design instincts typically comes from actually running experiments and reading others' work. This is because readers tend to glaze over pages of abstract discussions of design and statistical tests. A consideration of methodology has been delayed until this point so that examples with decision and game experiments can be used. The first part of this chapter pertains to general considerations for designing the treatment structures of experiments, and for dealing with questions of appropriate incentives and phrasing of instructions. This section ends with a list of "fatal flaws" in design that should be avoided. The second part of the chapter pertains to the nonparametric statistical tests that are most commonly used for the relatively small data sets encountered in economics experiments. Each test is explained in the context of a specific experimental data set.

The standard nonparametric tests involve transforming data into ranks (1 for smallest value, 2 for second smallest, etc.). One advantage of this transformation is that distributions of ranks of continuous measures are carefully documented and available in tabular form or statistical software programs for standard tests. For each of the tests considered in the second section, there is a test statistic that is computed by comparing the data ranks for one treatment with those for another treatment. If the treatment change has no effect, then the ranks will be "mixed up" with some low ranks and some high ranks for each treatment. The *null hypothesis of no treatment effect* is that the ranks are drawn from the same distribution. But if the data ranks for one treatment are generally higher than for another, then the data do not offer much support for the null hypothesis.

For example, suppose that there are four separate sessions (groups of subjects) in each of two treatments, and that the outcome for each session is an overall average (average price, bid, effort, etc.). These eight session averages can be ranked from low (1) to high (8). If the session ranks are draws from the same probability distribution for each treatment, then it would be very unlikely that the four lowest ranks occur in one treatment and the four highest ranks occur in the other. This statement is true, regardless of whether the unranked data are draws from a normal distribution (with mean and variance) or from some other distribution with only a single parameter, e.g., uniform, or even from a distribution with many parameters. Extreme values of the test statistic (based on ranks) will justify rejection of the null hypothesis of no effect, and standard tables for such calculations will be provided for each test. These rank-based tests are *non-parametric* since the underlying distributions of *ranks* across treatments that are used to construct relevant statistical tables do not depend on specific parameters of the *data* distributions.

13.1 Experiment Design and Procedures

Treatment Structure

A treatment is a completely specified set of procedures, which includes instructions, incentives, subjects, rules of play, etc. This notion is analogous to the case of a medical trial, where the treatment would include details about dosage, timing, subject selection, measurement, and all of the features that would be necessary for someone else to replicate the trial at a later time. The description of a treatment in an economics experiment should include the detail necessary for setting up a replication study that the original authors would consider to be valid.

Just as scientific instruments need to be calibrated, it is useful to calibrate economics experiments. *Calibration* typically involves a "baseline treatment" for comparisons. For example, suppose that individuals are each given a sum of money that can either be invested in an "individual account" or a "group account." Investments in the group account have a lower return to the individual, but a higher return to all group members. If the typical pattern of behavior is to invest half in each account, then this might be attributed either to the particular investment return functions used or to "going 50–50" in an unfamiliar situation. In this case, a pair of treatments with differing returns to the individual account may be used to tease out a generalizable investment pattern. Suppose that the investment rate for the individual account is 50% in one treatment, which could be due to confusion or uncertainty, as indicated above. If we would also observe that the share of investment in the group account rises when the rate of return is increased, this would have important economic implications.

The next example involves a market experiment, in which subjects post prices and buyers are randomly selected to make purchases at those prices. If a small number of sellers have multiple units that they can sell, they might try to exercise "market power" by offering fewer units for sale in an attempt to drive up the price. High prices may be attributed to small numbers of sellers or to the way in which sellers are constrained from offering private discounts to particular targeted buyers. These issues could be investigated by changing the number of sellers, holding discount opportunities constant, or by changing the nature of allowed discounting opportunities while holding the number of sellers constant. Many economics experiments involve a *2 × 2 design* with treatments in each of the four cells, e.g., low numbers with discounting, low numbers with no discounting, high numbers with no discounting, and high numbers with discounting.

A major, but unfortunately very common, flaw is a treatment structure change that modifies more than one factor at the same time. This makes it difficult to determine the cause of any observed change in behavior. In the previous example, if there were only two treatments: low seller numbers and no discounting, and high seller numbers and discounting, then a difference could be due to either the change in the number of sellers or the discounting opportunities. However, it might sometimes be useful to change more than one treatment dimension at a time if the alternative policies being considered require this. Even then, it might be useful to go back and deconstruct the effects by changing one dimension at a time.

Another example, from chapter 3, involved an observation of no difference in risk-taking behavior with low real payoffs of several dollars and high hypothetical payoffs. This difference could be due either to the nature of payoffs (real versus hypothetical) or to the scale (low versus high). As noted in chapter 3, Holt and Laury (2002) report that scaling up hypothetical choices had little effect on the tendency to select the safer option in their choice menu, whereas scaling up the payoffs in the real (non-hypothetical) treatments to hundreds of dollars caused a sharp increase in the tendency to choose the safer option. It is interesting to note that Holt and Laury were (correctly) criticized for presenting each subject with a low-payoff lottery choice *before* they made high-payoff choices, which can confound payoff scale and order effects. To address this issue, Holt and Laury (2005) reported a second experiment in which each person only made decisions in a single payoff-scale treatment (real or hypothetical), all done in the same order. If the intensity of a causal factor is of interest, then it may be best to have multiple treatments that vary the intensity of that factor, e.g., by scaling up payoffs in lottery-choice experiments by factors of $1\times$, $10\times$, $50\times$, etc.

> **Treatment Structure:** *In order to infer the cause of a change in economic outcomes, it is necessary to assign subjects to treatments exogenously, so that results in the presence of a possible causal factor can be compared with a baseline treatment in which that factor is not present. The separate effects of two possible causal factors can be investigated by changing those factors independently in a 2 × 2 design. In other cases, it is useful to use multiple treatments that vary the intensity of the causal factor.*

"Between-Subjects" versus "Within-Subjects" Designs

Many research experiments are based on using a single treatment for each subject or group of subjects, e.g., 20 rounds of random matching for the battle-of-sexes game with a low payoff asymmetry for some groups, and 20 rounds with a high payoff asymmetry for other groups. In contrast, classroom experiments typically involve two or more treatments, or even half of the class doing one pair of treatments and the other half logging in with a different "session name code" to be exposed to a different pair of treatments. This procedure, dictated by limits on class time and available numbers of participants, can be problematic when behavior in one treatment carries over into the second.

Those kinds of sequence effects, as explained later, are avoided with a *between-subjects* design that is commonly used in research settings. The alternative is to let each person make decisions in both treatments, which puts more people into each treatment but affords less time to complete multiple rounds of decision making in each treatment. This approach is called a *within-subjects* or sequential design, since behavior for a group of subjects in one treatment is compared with behavior for the same group in the other treatment. The "between-subjects" and "within-subjects" terminology is commonly used in psychology, where the unit of observation is typically the individual subject. Subjects in economics experiments generally interact, so the unit of analysis is the group or market, which explains the use of "within-group" and "between-group" terminology.

Each type of design has its advantages. If behavior is slow to converge or if many observations are required to measure what is being investigated, then the parallel (between-subjects or between-group) design may be preferred since it will generate the most repeated observations per treatment in the limited time available. Moreover, subjects are only exposed to a single treatment, which avoids *sequence effects*. For example, sequence effects were quite strong in the traveler's dilemma experiment (chapter 11). A group that played ten rounds of a traveler's dilemma with a low penalty rate tended to make high claims, and this tendency carried over to part B when the penalty rate was raised. Conversely, those groups that played the game with a high penalty initially tended to stay with low claims in part B, even after the penalty rate was lowered. The effects of penalty rate changes were in the predicted direction, but the sequence effects were so strong that the analysis of comparisons across groups was only done with the part A data.

When students are asked to design an experiment to run on their classmates, they often come up with sequences of treatments for the same group. The most common ex post self-assessment is that "I wish I had used a single treatment on each half of the class so that we would not have to wonder about whether the change in behavior between treatments was due to prior experience." Even in research experiments, it is annoying to have to report and analyze decisions differently depending on where they were observed in a sequence. In the author's experience, the use of a within-subjects setup is often regretted, e.g., the Holt and Laury (2002) risk aversion paper and the traveler's dilemma paper. In some settings, there are large differences between groups in terms of their cooperative tendencies, and within-subjects comparisons should be considered, but with reversed orders in every other session to control for sequence effects. For example, suppose that you have a group of adults, and you want to determine whether a drink with caffeine improves running speeds. In any group of adults, performance can differ by factors of 2 or more, depending on age, weight, overall health, etc. In this case, it would be desirable to time running speeds for each person under both caffeine conditions on different days. The advantage of a sequential (within-subjects) design is that individual differences are evaluated by letting each person serve as their own control. Sometimes sequence effects are themselves the focus of the experiment, e.g., the issue of whether the imposition of a minimum wage causes worker aspiration wage levels to increase (Falk, Fehr, and Zehnder, 2006). To summarize:

Between- versus Within-Subjects Designs: *A within-subjects design (two or more treatments per group) is appealing if there is high variability across individuals or groups relative to the variability caused by sequencing. A between-subjects design (one treatment per group) is better when there is less variability across individuals or groups, and when there are sequence effects that cause behavior in one treatment to be influenced by what happened in an earlier treatment.*

Sometimes the best choice between sequential and parallel designs is not clear, and a pair of experiments, one with each method, provides a better perspective on the behavior being studied, e.g., the Holt and Laury (2002, 2005) combination of within and between designs.

Incentives

Economics experiments typically involve monetary decisions, e.g., prices, costly efforts, etc. Most economists are suspicious of results of experiments done with hypothetical incentives, and therefore real cash payments are almost always used in laboratory research. As we shall see in later chapters, incentives may sometimes not matter at all, but sometimes matter a lot. For example, people have been shown to be more generous in offers to others when such offers are hypothetical than when generosity has a real cost, and people tend to become considerably more risk averse when the stakes are very high than is the case where the stakes are hypothetical or involve only several dollars. Conversely, it is hard to imagine that scaled-up money payments would help serious students raise their GRE scores, and there is even some psychological evidence that money payments interfere with a child's test performance. In economics experiments, the general consensus is that money payments or other non-hypothetical incentives are necessary for studying markets and economic interactions involving decisions with clear consequences. Tests of cognitive abilities, however, are generally not incentivized when done in conjunction with economics experiments (Brañas-Garza, Kujal, and Lenkei, 2016).

There remains an important issue of just how large payoffs should be. Scaling up payoffs does not have much of an effect in many laboratory situations (Smith and Walker, 1993), with the notable exception of risk aversion tasks. In any case, the use of hypothetical incentives often results in more noise or nonsensical decisions, as was apparent from the discussion of belief elicitation experiments in chapter 5. To summarize:

> Incentives: *There are many documented situations in the economics and psychology literature where monetary incentives do not seem to have much of an effect. However, in other cases, using incentives may reduce errors and noise. In the absence of a widely accepted framework for identifying problematic situations with precision, it is advisable to use money incentives in economics, especially for experiments that are used to study behavior in settings that are intended to parallel markets involving highly motivated consumers or producers.*

Replication

One of the main advantages of experimental analysis is the ability to repeat the same setup numerous times in order to determine average tendencies that are relatively insensitive to random individual or group effects. Replication requires that instructions and procedures be carefully documented. It is essential that instructions to subjects be written as a script that is followed in exactly the same manner with each cohort that is brought to the laboratory. Having a set of written instructions helps ensure that unintended "biased" terminology is avoided, and it permits other researchers to replicate the reported results. For example, if the experimenters provide a number of examples of how prices determine payoffs in a market experiment, and if these examples are not contained in the written instructions, the different results in a replication may be due to differences in the way the problem is presented to the subjects. The general rule can be summarized:

> Replication and Reporting: *The experimenter should report enough detail, including instructions and examples, so that someone else could replicate the experiment in a manner that the original author(s) would accept as being valid, even if the results turn out to be different.*

Finally, one good way to get a feel for the experiment in the paper is to read the instructions appendix first. With ambiguity aversion, for example, it often seems to be the "nature of the beast" that the procedures are confusing. This is why a key paragraph from the instructions was used in chapter 6 to convey the structure of the experiment.

Control

Another advantage of experimentation is the ability to manipulate factors that may be affecting behavior, so that extraneous factors are held constant ("controlled") as the treatment variable changes. Control can be impaired when procedures distort the incentives that participants actually face. The use of biased terminology may allow participants to act on "homegrown values" that conflict with or override the induced incentives. For example, experiments pertaining to markets for emissions permits typically do not use the word "pollution," since that might cause environmentally sensitive subjects to avoid the "polluting" activity even if it enhances their earnings. This guilt effect could differ from one person to another, and this is the sense in which *homegrown values* unrelated to the experiment incentives could intervene. Moreover, if people are trading physical objects like university sweatshirts, differences in individual valuations make it hard to reconstruct the nature of demand in a market experiment. There are, of course, situations where non-monetary rewards are desirable, such as experiments designed to test whether ownership of a physical object (like a school sweatshirt) makes it more desired, which is known as an *endowment effect* (Kahneman, Knetsch, and Thaler, 1991). How much control is desirable should be judged in the context of the purpose of the experiment.

Another factor that can disrupt control is the use of deception. If subjects suspect that announced procedures are not being followed, then the specified incentives may not operate in the intended ways. This counterproductive incentive effect is probably the main reason that deceptive practices are much less common in incentive-based economics experiments than is the case in social psychology. Even if deception is hidden during the experiment, subjects may well find out afterward by sharing experiences with friends. Therefore, the careful adherence to non-deceptive practices provides a "public good" for those who run experiments in the same lab at a later time. Ironically, the perverse incentive effects of deception in social psychology experiments may be aggravated by ex post debriefings that are sometimes required by human subjects committees. The best approach is to avoid blatant deception and to use common sense in gray areas where aspects of the procedures are not explained. An example of a commonly used and generally accepted omission could be the use of gender or risk aversion tests to sort subjects into separate markets, without explaining that sorting.

Context in Laboratory and Field Experiments

An important design decision for any experiment pertains to the amount and richness of context to provide. A little economic context can be very useful. For example, it is possible to set up a market with a detailed and tedious description of earnings in abstract terminology that does not mention the word "price." But market terminology helps subjects figure out "which way is up," i.e., that sellers want high prices and buyers want low prices. Another example of compromise is the "misaligned profiling" experiment summarized in chapter 10, which models the interactions of security officials and terrorists seeking to avoid detection. The instructions used the "attack" and "defend" terminology, but did not have any references to security or terrorists.

Even though context may help subjects understand the strategic landscape, it is an accepted practice in economics experiments to strip away a lot of social context that is not an essential part of the economic theories being tested. If the alternative theories do not depend on social context, then the best approach is to hold context constant as the economic parameters are changed. This process of holding context constant may involve minimizing its unintended and unpredictable effects, e.g., by taking steps that increase anonymity during the experiment. Even then, a lot can be learned by re-introducing social context in a controlled manner, e.g., by comparing individual decisions with those made by groups of people.

Social context can sometimes be critically important, as in some politics experiments where it is not possible to recreate the "knock on the door" or phone campaign solicitation in the lab. In such cases, researchers use *field experiments* involving people in their natural environments, so that they are unaware that they are participating in an experiment. For example, Gerber and Green (2000) targeted political messages to randomly selected voters, using phone, personal contact, or mail, in order to evaluate the effects of these messages on voter turnout, which was determined by looking at precinct records after the election. Field experiments can also provide more relevant groups of subjects and can be used to avoid *experimenter demand effects*, i.e., situations in which behavior in the lab may be influenced by subjects' perceptions of what the experimenter wants or expects. There are, of course, intermediate situations where the lab setup is taken to the field to use traders from particular markets in a context that they are familiar with. For example, List and Lucking-Reiley (2000) set up auctions for sports cards at a collector's convention. Another type of *enriched laboratory experiment* involves making the lab feel more like the field, as in voting experiments where subjects are seated in a comfortable room decorated like a living room and are shown alternative campaign ads that are interspersed with clips from a local news show.

Although field experiments can introduce a more realistic social context and environment, the cost is often a partial loss of control over incentives, over measurement of behavior, or over the ability to replicate prior results under identical conditions. In one example, presented in chapter 19, the effects of alternative political ads on voting behavior in an enriched laboratory setting are measured indirectly by surveys of voters' intentions, since actual votes for one candidate or another are private information. In addition, replication may be complicated by interactions between political ads used in the experiment and the positive or negative dynamics of an ongoing political campaign. In other cases, reasonably good controls are available. For example, even though individual valuations for sports cards are not induced directly, they can be approximately controlled by using matched pairs of cards with identical book values.

To the extent that social context and target demographics are important in a field experiment, each field experiment is like a data point that is specific to that combination of subjects and context unless appropriate random selection of subjects is employed. Thus, the results from a *series* of related field experiments become more persuasive, just as do the results from series of laboratory experiments. There can also be interactions between the two approaches, as when results from the lab are replicated in the field, or when a general, but noisy, pattern discovered in diverse field situations shows up in a laboratory setting that abstracts away from the diverse field conditions. This point is made convincingly by Kagel and Roth (1995) in a study of matching markets for medical residents that will be discussed in the final chapter of this book. To summarize:

Context: *In a lab experiment, deciding on how much context to provide is typically a compromise between ensuring that subjects understand the strategic situation clearly, without introducing terminology that will trigger emotional responses or heterogeneous "homegrown values," unless such things are the focus of the experiment itself.*

Fatal Errors

Professional economists often look to experimental papers for data patterns that support existing theories or that suggest desirable properties of new theories and public policies. Therefore, the researcher needs to be able to distinguish between results that are replicable from those that are artifacts of improper procedures or confirmation bias. Even students in experimental sciences should be sensitive to procedural matters so that they can evaluate others' results critically. Those who are new to experimental methods in economics should be warned that there are some fatal errors that can render the results of economics experiments useless. As the above discussion indicates, these include:

Fatal Procedural Errors

- *using inadequate or inappropriate incentives*
- *using non-standardized or biased instructions and suggestive examples*
- *using inappropriate context that induces uncontrolled homegrown values*
- *not holding constant the effects of psychological biases across treatments*

> - *including an insufficient number of independent observations*
> - *losing control due to deception or confusing procedures*
> - *failing to provide a calibrated baseline treatment*
> - *changing more than one design factor at the same time*

13.2   Statistical Testing: Independence of Observations

In order to reach conclusions supported by standard statistical arguments, it is necessary to have enough independent observations to justify a firm conclusion.

For example, consider the hypothesis that people are more risk averse when sleep deprived, which could be assessed using a within-subjects design. Under the null hypothesis of no effect, measured risk aversion is just as likely to be higher or lower with sleep deprivation, so a single observation with higher measured risk aversion in this treatment would not offer much evidence of a treatment effect. Under the null hypothesis of no treatment effect, the chance that measured risk aversion is higher with sleep deprivation is 1/2. So if eight subjects all exhibit more risk aversion when they are sleep deprived, then the chances of an outcome this extreme would be $(1/2)^8 = 0.0004$. This probability corresponds to a less than 1% chance, so the null hypothesis could be rejected at the 1% significance level. This test, based on simple binomial calculations, depends only on the signs of differences. As shown below, a more precise ("signed rank") test is available if the magnitudes of differences can be compared.

These arguments are based on ordinal comparisons of risk aversion measures and do not depend on specific distributions of such measures (e.g., normal). For this reason, rank-based tests are *nonparametric*. The clearest presentation of the nonparametric tests commonly used by experimenters is found in Siegel (1956), which contains many examples from very early economics and psychology experiments. These statistical arguments will be elaborated in the following sections. For now, the main idea is that in order to make a statistical claim, the results have to be sufficiently strong and based on enough independent observations. Only in this manner can we conclude that the observed data patterns are unlikely to be observed under the null hypothesis of no treatment effect.

Independence of observations can be lost due to contamination. For example, if four pairs of people are negotiating over the division of $10 and if the first agreement reached is announced to the others, then the announced outcome might affect the remaining three agreements. A more subtle case of contamination is with re-matching, because people are dealing with different partners in the second round. Without announcements, there would be four independent bargaining outcomes in round 1, but the round 2 results might depend on subjects' experiences in the first round, which could result in contamination and loss of independence. If random matching is desired in order to make each round more like a single-period game, then the standard approach is to treat *each group* of people who are being re-matched in a series of rounds as a single independent observation. In this case, the experimenter would need to bring in a number of separate groups for each of the treatments (or treatment orders) being investigated. If this seems like an unnecessarily conservative approach, remember that the experimenter is often trying to persuade skeptics about the importance and generalizability of the results.

Even if people are being randomly re-matched, it is possible that they might try to "teach" others a lesson to discipline the group, in hopes of affecting behavior of future partners. One approach to this problem is to implement a *rotation matching scheme* in which subjects in different roles (e.g., proposers and responders) meet each other in a sequence that ensures that all prior partners (and their subsequent partners) are never encountered again. This is done by putting the ID numbers in two circular arrays and rotating the outer circle one notch each round, stopping before repetition occurs. Brandts and Holt (1992) implemented this matching protocol in a two-person game that involved sending costly signals to the other player. The instructions stated that the subject would never be matched with the same person twice, and would also not be matched with anyone who had been matched with someone they had encountered in a prior round. That idea was clarified by saying that "it is as if you go down the receiving line at a party and tell everyone you meet a story that they repeat to everyone they meet, etc. Even if everyone who hears the story repeats it, these people are behind you in the receiving line and you would not encounter anyone who has heard the story."

Finally, it is important to qualify this discussion of independent observations by noting that a lot might be learned from even a very small number of market trading sessions, each with many (e.g., hundreds of) participants. The analysis requires careful econometric modeling of the dynamic interactions within a session, so that the unmodeled factors can reasonably be assumed to be independent shocks. After all, the US macro-economy is a single observation with many interactions, but this does not paralyze macroeconomists or invalidate econometric macroeconomic models of interactive dynamics. If the process being studied does not require dynamic interactions of large numbers of participants, then a design with more independent observations allows the researcher to reach conclusions without relying on extra modeling assumptions.

13.3   Mann-Whitney Tests for Between-Subjects Designs

The costs of obtaining independent observations with independent sessions often lead experimental economists to work with a relatively small number of observations. Even though sample means tend to have normal distributions when there are large numbers of observations, the parametric assumptions like normality that are used in standard tests are not appropriate with small sample sizes. Nonparametric tests are useful in these situations.

The most frequently used nonparametric test is the Mann-Whitney test for a between-groups design with independent observations for each of two treatments. This test can be illustrated with data from a social dilemma game, in which each player decides whether to make a pro-social decision ("volunteer") or not. Volunteering entails a small cost of $0.20. Everyone in the group earns a benefit of $1 if at least one person in the group volunteers, e.g., it only takes one person to call for help. If nobody volunteers, they all earn only $0.20. Therefore, a player would want to volunteer if nobody else is going to, but would not want to volunteer if someone else is going to bear the cost. Hence the name "volunteer's dilemma."

With simultaneous decisions, the Nash equilibrium involves a lower probability of volunteering in larger groups, as will be discussed in chapter 17. Table 13.1 summarizes a subset of the data from a volunteer's dilemma experiment. The table shows average volunteer rates for separate sessions done under three treatment conditions: with group sizes of 2, 3, or 6. Each session involved 12–36 subjects who were randomly matched for 20 rounds. The average volunteer rates by session are shown in the second row of the table. The data ranks are shown in parentheses, with the three smallest ranks (1, 2, 3) for the group size 6 treatment on the right. Even though there is a lot to be learned from the overall pattern, with lower volunteer rates for larger groups, the discussion will begin with a restricted comparison of the $N = 3$ and $N = 6$ treatments on the right, using a Mann-Whitney test for two unmatched samples. The basic intuition for this test will later be applied to all three treatments, using a different test.

The test statistic, referred to as the Mann-Whitney U, is found by counting the number of times each of the ranks for one treatment exceeds a rank from the other. For the $N=3$ treatment in the center column, the rank 6 beats the ranks of 3, 2, and 1 for the $N=6$ treatment, so there are three binary wins so far. In addition, the rank of 5 in the center column also beats all three lower ranks in the right column, as does the rank of 4. So the binary win count is nine for $N = 3$ over $N = 6$. The binary win count in the other direction is 0, since none of the three ranks in the $N = 6$ treatment are larger than any of those for $N = 3$. The test statistic U is the *minimum of the binary win counts in each direction*, which is 0 in this case.

Table 13.1. Session Average Volunteer Rates (Ranks) for Group Sizes 2, 3, and 6

Group Size	N = 2	N = 3	N = 6
Volunteer Rate (data ranks)	0.55, 0.51, 0.49 (9, 8, 7)	0.42, 0.39, 0.38 (6, 5, 4)	0.31, 0.28, 0.20 (3, 2, 1)

To understand the intuition, consider what a low test statistic means. The most extreme outcome that could be observed is for the binary win counts to be as unequal as possible, in this case 9 and 0. A less extreme outcome would involve some wins in the opposite direction, e.g., 7 wins in one direction and 2 in the other, for a U of 2. *A higher U value indicates a less significant result.*

For the comparison of the two group sizes, there are six observations in total and 3 observations per group size treatment. Thus there are "6 take 3" possible ways that the ranks could be allocated to the two treatments. Since "6 take 3" = $\frac{6!}{(3!)(3!)}$ = 20, and all 20 possible permutations are equally likely under the null hypothesis of no treatment effect, the chance of seeing any one outcome is $1/20 = 0.05$. This also holds for the likelihood of observing the most extreme outcome. Thus, 0.05 would be the "p value" for a test of the null hypothesis when the alternative is specified in one direction (lower volunteer rates for larger groups).

Suppose that prior analysis does not suggest a direction, i.e., that large groups might have either higher or lower volunteer rates as compared with small groups. Then an outcome as extreme in the opposite direction has to be considered as well. In this case, it would be the three highest ranks for group size 6. That outcome also has a probability of 1/20 under the null hypothesis, so the p value for a two-tailed test would be $1/20 + 1/20 = 1/10$. Since it is natural to expect volunteer rates to be lower in larger groups ("diffusion of responsibility"), a one-tailed test is appropriate in this case. The general procedure is to establish a cutoff critical value, e.g., 0.05 ("5%") and reject the null if the data are so unlikely that the p value is less than or equal to this cutoff. So in this case, the null of no group size effect could be rejected at the 5% level (one-tailed test).

After the test statistic, U, is computed, the p value can be determined by statistical software or by looking at a table of cutoff p values. To summarize:

Mann-Whitney U Test Statistic

- *Combine the observations for two treatments.*
- *Rank the combined observations from 1 (lowest) to N (largest).*
- *Calculate the number of "binary wins" for the ranks in one treatment over those of the other, and then calculate binary wins in the opposite direction.*
- *The minimum of the binary win counts is the test statistic U.*
- *A low U value indicates a more unequal binary win count and a stronger rationale for rejecting the null hypothesis of no treatment effect.*

For example, table 13.2 shows the relationship between p values and U values for treatments with equal numbers of observations. The top row applies to the case just considered, with three observations, or sessions, in each treatment. With $U = 0$, the p value of $1/20 = 0.05$ is shown in the upper-left corner. For a two-tailed test, the value in the table is simply doubled. So the data that result

Table 13.2. Mann-Whitney U Test: Table of p Values for One-Tailed Test

	U=0	U=1	U=2	U=3	U=4	U=5	U=6	U=7	U=8	U=9	U=10
$n_1 = n_2 = 3$	.05	.10	.20	.35	.50						
$n_1 = n_2 = 4$	.014	.029	.057	.100	.171	.243	.343	.443			
$n_1 = n_2 = 5$	.004	.008	.016	.028	.048	.075	.111	.155	.210	.274	.345
$n_1 = n_2 = 6$	.001	.002	.004	.008	.013	.021	.032	.047	.066	.090	.120
$n_1 = n_2 = 7$	.000	.001	.001	.002	.003	.006	.009	.013	.019	.027	.036

Key: U = minimum of binary win counts

in rejection of the null at the 5% level (one-tailed) would only result in a rejection at the 10% level with a two-tailed test. In academia, outside observers (and especially reviewers) prefer to see two-tailed tests, unless there is a strong argument for anticipating results in one direction.

More extensive tables are available for different numbers of observations in each treatment. It is even easier to use statistical software packages that report both U test statistics and associated p values, usually for two-tailed tests. The emphasis here is on understanding the intuition behind the tests so that the reader can use and interpret the results of the statistical software and choose appropriate tests.

In order to solidify one's intuition, consider how the p values for table 13.2 are actually determined. For the top row with 3 observations per treatment, there are 6 ranks and 20 different ways those ranks could be assigned to the treatments. The 10 permutations that yield higher sums of ranks for Treatment A are listed in table 13.3. Specifically, the 3 ranks in the Treatment A row have a higher sum than the 3 residual ranks in the corresponding cell of the Treatment B below. The U values are determined by the minimum of the binary win counts, and are shown in the Minimum Win Count row. For example, for the second most extreme outcome (second column from the right), the rank 4 in Treatment B beats the rank 3 above it in Treatment A, so the minimum binary win count is 1, which is the U value. There are two outcomes as extreme or more so, so the p value for $U = 1$ is 1/20 + 1/20 = 0.1, as shown in the bottom row of the $U = 1$ column of table 13.3. The other p values in this row are calculated similarly, by adding up probabilities of outcomes *as extreme or more extreme* in each case. The key insight is that the Mann-Whitney test is a permutation test done on data ranks, where the p value probability is determined by ratios of permutation counts. To summarize:

Mann-Whitney Test: *The p value provided by test statistic tables is calculated as a ratio. The numerator is the number of permutations of ranks between treatments that yield an outcome as extreme or more so than what was observed in the data, and the denominator is the total number of possible permutations.*

Table 13.3. Permutations of Ranks for Three Observations in Each Treatment Yielding a Higher Rank Sum for Treatment A

Permutation	1	2	3	4	5	6	7	8	9	10
Treatment A	5	6	6	5	6	6	6	6	6	6
	4	3	4	4	4	5	4	5	5	5
	2	2	1	3	2	1	3	2	3	4
Treatment B	6	5	5	6	5	4	5	4	4	3
	3	4	3	2	3	3	2	3	2	2
	1	1	2	1	1	2	1	1	1	1
Minimum Rank Sum	10	10	10	9	9	9	8	8	7	6
Minimum Win Count	$U=4$	$U=4$	$U=4$	$U=3$	$U=3$	$U=3$	$U=2$	$U=2$	$U=1$	$U=0$
p Value for one-tailed test	$p=.50$			$p=.35$			$p=.20$		$p=.10$	$p=.05$

A less important insight pertains to a different (but more commonly encountered) method of calculating the Mann-Whitney U test statistic. For each permutation, take the sum of the ranks for Treatment A and the sum of the ranks for Treatment B, and find the minimum of those sums, which is shown in the Minimum Rank Sum row. The lowest possible rank sum is then subtracted from this minimum to determine the test statistic U. In this case, the lowest possible sum of ranks for 3 observations $(1 + 2 + 3)$ is 6. Note that if you subtract 6 from the entries in the Minimum Rank Sum row, you obtain the U value in the Minimum Win Count row. This works for all rows and is true in general: $(n + 1)n/2$ to determine the sum of the first n integers, which is the lowest possible rank sum. This explains the intuition for the typical Mann-Whitney U formula when it is presented as the minimum of the rank sums for the data observed, minus the lowest possible rank sum for that sample size. Of course, the binary win count method is easier, since it does not require conversion to ranks (binary wins can be calculated with either original data or ranks).

Example 13.1

Table 13.4 reproduces the results of a rent-seeking experiment that was discussed at length in the previous chapter. This was a 2 × 2 design with changes in effort cost (low on the left and high on the right) and changes in the group size (group size 2 in the top part and group size 4 in the bottom part). There were 16 sessions in total, with 4 session results in each of the treatment cells of the table.

Table 13.4. Rent Dissipation Percentages by Session: Cost and Numbers Effects

	Low-Effort Cost	High-Effort Cost
$N = 2$ (data)	77, 67, 62, 37	91, 85, 85, 94,
$N = 2$ (ranks)	4, 3, 2, 1	7, 5.5, 5.5, 8
$N = 4$ (data)	77, 83. 130, 81	132, 111, 93, 87,
$N = 4$ (ranks)	1, 3, 7, 2	8, 6, 5, 4

The results for each session are summarized by a percentage ("rent dissipation") that is a measure of the total effort expended to obtain a prize ("rent").

First consider the top ($N = 2$) row, where all 4 sessions with high effort costs on the right yielded higher dissipation percentages than the 4 sessions with low effort costs in that row. This is also a most-extreme case, where there are no binary wins in one direction, so $U = 0$. The $U = 0$ column for the second row of table 13.2 with 4 observations per treatment shows a p value of 0.014. This p value can be explained in terms of permutations. With 8 observations in the top row, there are "8 take 4" $= \frac{8!}{(4!)(4!)} = 70$ possible permutations of the ranks. (See problem 4 for the intuition for the ratio of factorial expressions.) Under the null hypothesis of no effect, each of these permutations would be equally likely. The most extreme outcome was observed, which yields a p value of 1/70.

In contrast, there is some overlap in the data with group size 4, as shown in the bottom row of table 13.4: the session averages were (77, 83, 130, 81) with low effort cost on the left side, and the 130 is higher than three of the percentages (111, 93, and 87) on the right side. You should convince yourself that the *minimum* binary win count is 3, so $U = 3$. Then use table 13.2 to determine the p value for a one-tailed test. You can check the test statistic calculation by using the data ranks to compute rank sums. Then take the minimum rank sum and subtract 10 (the lowest possible rank sum for 4 integers) to verify that the test statistic is also 3 for this method.

This effort cost example is useful in the sense that it illustrates both an extreme outcome and an outcome with some overlap in the ranks. Instead of running tests for effort cost effects separately for the two group sizes, it is possible to conduct a single test based on simultaneous "left-right" permutations of ranks within each row. The intuitive idea is to determine the proportion of those simultaneous permutations that result in an outcome as extreme or more so than what is observed in the table. This combined test, which uses all of the data, is based on the same permutations of high and low effort cost labels that are used in the separate tests, while keeping the data "stratified" into separate rows for different group sizes. This *stratified permutation test* will be considered in a later section.

13.4 Wilcoxon Tests for Matched Pairs (Within-Subjects Designs)

Suppose each subject (group) is exposed to two treatments in a between-subjects (between-groups) design, perhaps in an alternating order to control for order effects. Therefore, each person or group generates a "matched pair" of observations. The idea is to look at treatment differences for each observation and rank them in terms of absolute value. (If an observation shows a 0 treatment difference, it is ignored and not included in the sample size N.) If the treatment differences in one direction are larger in absolute value than in the other direction, then the null hypothesis is called into question. This matched-pairs test involves computing (signed) treatment differences, converting them into absolute values, and then comparing the sum of absolute values of treatment differences in one direction with those in the other direction. The computational trick used in this comparison is accomplished by reattaching the original sign of the treatment difference to the rank of the absolute value of the difference, and hence, this is known as a *Wilcoxon signed rank test*. The test statistic is the sum of signed ranks. But it is best to explain this procedure by example.

Consider the data in table 13.5 with 6 matched pairs. There are 6 separate groups of subjects. Each group did two treatments, with either a 5-seller market or a 3-seller market (with two subjects sitting out for that treatment). The treatment differences in the third row are generally positive, in conformity with the intuition that prices could be higher with fewer sellers. The one deviation is the price of 479 in Session 2, in which the average price was higher with 5 sellers. The fourth row shows the absolute values of the differences in the row above it. These absolute values are then converted into ranks in the fifth row. Notice that the difference of -9, which is the lowest number in the Difference row, turns out to be the second lowest in the Absolute Value of Difference row, so it receives a rank of 2. After the absolute values of the differences are ranked, the sign of the original difference is attached to the rank of the absolute difference. For example, the signed rank for Session 2 is -2, as shown in the bottom row.

Table 13.5. Average Prices with Matched Pairs (* Treatment First in Sequence)

Session	S1	S2	S3	S4	S5	S6
3 Seller Price Averages	425	470*	408	436*	424	517*
5 Seller Price Averages	415*	**479**	392*	401	392*	512
Treatment Differences	10	−9	16	35	32	5
Absolute Values of Differences	10	9	16	35	32	5
Ranks of Absolute Values	3	2	4	6	5	1
Signed Ranks	3	−2	4	6	5	1

Table 13.6. Wilcoxon Signed Rank Test: Critical Values for ± W

	Level of significance for a one-tailed test (double for two-tailed)			
	0.05	0.025	0.01	0.005
N = 5 pairs	15	—	—	—
N = 6 pairs	17	21	—	—
N = 7 pairs	22	24	28	—
N = 8 pairs	26	30	34	36
N = 9 pairs	29	35	39	43

Key: W = sum of signed ranks of absolute differences

The most extreme outcome would be for all of the differences to be positive, in which case the sum of signed ranks would be 21. For the observed outcome, the sum of signed ranks in the bottom row of the table is 17, so $W = 17$ is the test statistic to be used in looking at a table of critical values to decide whether to reject the null hypothesis of no effect. Table 13.6 shows critical values for W for various numbers of matched pairs and significance levels, for a one-tailed test. For $N = 6$ (matched pairs) in the second row, the critical value is 17 for $p = 0.05$, so the null hypothesis can be rejected at the 5% level. To summarize the procedure:

Wilcoxon Matched Pairs Test

- *Calculate the treatment difference for each matched pair.*
- *Determine the absolute value of each treatment difference.*
- *Rank the absolute values of all non-zero differences between pairs.*
- *Attach signs to the ranks of absolute values to obtain signed ranks.*
- *Calculate the test statistic W = sum of signed ranks.*
- *The null hypothesis is rejected if W is greater than or equal to a critical value in a table based on the significance level and number of pairs.*

Notice that, for each signed rank, the are two possible signs, positive or negative. The intuition behind the test can be understood by noting that there are $2^6 = 64$ possible ways that the signs of the 6 ranks can be permuted. Under the null hypothesis, either sign is equally likely, so there are 64 equally likely ways that the differences could have been signed. Of these, there are 3 outcomes as or more extreme than what was observed: the actual outcome shown in the bottom row of table 13.5 with a −2 reversal, or a case of a single −1 reversal, or a case of no reversals (all positive treatment differences). This probability is $3/64 = 0.047$, which is consistent with the recommendation from table 13.6 as indicated previously.

The point is that the Wilcoxon signed rank test is equivalent to a permutation test done by randomly permuting the signs of the absolute values of ranks of treatment differences, and that is the way that the critical values for the Wilcoxon test are determined. The critical value tables, however, make this process easier in the sense that the procedure is reduced to finding a sum of signed ranks instead of considering all 2^n possible permutations.

Example 13.2

The data set in table 13.5 is taken from Davis and Holt (1994a), but the boldfaced number 479 in session S2 was changed to provide an example of how the lowest difference (-9) might not be the lowest absolute value of the difference. The actual number for the 5-seller treatment of session S2 was 471 (instead of 479), so the difference was -1, which would receive a rank of 1 and a signed rank of -1. To test your understanding, you should determine the sum of signed ranks for the original data with the 471 average. What would be the probability of seeing something as extreme or more so (expressed as a ratio of some integer over 64)?

13.5 The Jonckheere Test of a Directional Effect of Treatment Intensity

Many experiments involve more than two treatments, with the stimulus factor being changed in intensity. For example, payoff scales can be increased from $1\times$ to $10\times$ to $20\times$. The data in table 13.1 provided volunteer rates by sessions for group sizes of 2, 3, and 6. The initial analysis was based on a Mann-Whitney 2-sample test, used to reject the null hypothesis of no treatment effect going from 3 to 6 subjects per group. A similar test with a similar outcome could be done comparing group sizes 2 and 3. In each case the p value is $1/20 = 0.05$ for a one-tailed test. It is advisable to avoid running multiple statistical tests when it is possible to evaluate the overall effect with a single test. There is a directional test, known as the *Jonckheere test*, that can be used when the alternative hypothesis is directional. In this case, the directional alternative hypothesis would be that volunteer rates are lower for larger group sizes, i.e., that the distributions of volunteer rates have medians that are lowered as group size increases from 2 to 3 to 6.

The test procedure is similar to the binary win calculations done for the Mann-Whitney test. First arrange the data in categories of increasing predictions to the right: (20, 28, 31) for $N = 6$, (38, 39, 42) for $N = 3$, (55, 51, 49) for $N = 2$. (This initial arrangement makes it unnecessary to do win counts in each direction and use the minimum, as was done for the Mann-Whitney test.) Begin with 20 on the left. All six of the observations in the two categories to the right are larger than the 20. So the binary win count is 6 so far. There are another 6 larger observations to the right for the 28, and another 6 for the 31, and the win

Table 13.7. Table of Critical Values for the Jonckheere J Test Statistic

Groups								2	3	4	5	6
and	2	3	4	5	6	7	8	2	3	4	5	6
Sample	2	3	4	5	6	7	8	2	3	4	5	6
Sizes	2	3	4	5	6	7	8	2	3	4	5	6
0.1	10	20	34	50	71	94	121	18	37	63	95	134
0.05	11	22	36	54	75	99	127	19	39	66	100	140
0.01	12	24	40	59	82	109	139	21	43	72	109	153
0.005	—	25	42	62	86	113	144	22	45	76	113	158

Key: J = sum of counts of observations in categories to the right that are larger

count is 18 at this point. Then move to the middle category, where each number is smaller than all 3 numbers in the right side category. There is no need to run win counts for the far right category, since there is no category to its right. So the sum of all of these binary comparisons is $6 + 6 + 6 + 3 + 3 + 3 = 27$. The test statistic is the sum of the counts, in this case, $J = 27$.

By checking a table of critical values in table 13.7, one can verify that this test statistic is higher than the relevant cutoff of 25 for $p = 0.005$ (3 categories, 3 observations), so the null hypothesis can be rejected at the 1% level. Again, a sharper, more reliable result is obtained by using all of the data for a single test if an ordered alternative is appropriate. The Jonckheere test is, relatively speaking, underutilized in experimental economics, given that the increasing intensity treatment structures are quite common. To summarize:

Jonckheere Test: *With more than two treatments, the Jonckheere test is appropriate when there is a natural directional hypothesis, e.g., due to increasing the intensity of a treatment variable. The relevant steps are:*

- *Arrange the observations in columns by treatment.*
- *Arrange the treatment data columns from left (lowest predicted values) to right (highest predicted values).*
- *For each data observation in the left column, count the number of observations in columns to its right that are higher (as predicted). Repeat for each observation in the left column.*
- *Repeat this "less than" count for each observation in the other columns, with comparisons made with observations in columns to its right. There is no need to perform "less than" counts for the far right column, since there are no observations to the right of that column.*
- *The test statistic J is the sum of these "less than" counts.*
- *The numbers of observations in each column are used to find the relevant p value in the Jonckheere table 13.7 of critical values.*

13.6 Stratified Permutation Tests

Permutation tests are really quite flexible in terms of dealing with unusual configurations of data that often arise. Sometimes there is a secondary treatment or procedural variation that is unrelated to the treatment effect of primary interest. One approach is to sort the data in terms of different values of the secondary treatment and then perform a separate test for each value of the secondary treatment. The stratified permutation test permits the experimenter to run a single test using all of the data.

Stratification can be illustrated with the rent dissipation data pattern that is shown in table 13.4. Up to this point, each time the effect of effort cost (high or low) on rent dissipation has been considered, it has been for a specific group size, either $N = 2$ in the top row of table 13.4 or for the $N = 4$ case in the bottom row. First, recall that there are "8 take 4" = 70 possible permutations of the 8 ranks in each row of the table. If one considers the top row alone and uses the data ranks, the p value for a one-tailed test is $1/70 = 0.0143$, since the observed outcome with no reversals is the most extreme, with the four lowest ranks on the left side. If one only considers the bottom row, the p value is 0.10, since there are 7 possible permutations of ranks that yield rank sums on the left side that are less than or equal to the $1 + 3 + 7 + 2 = 13$ that is observed, which yields $p = 7/70 = 0.1$. (Finding the 7 permutations is tedious and should be skipped.)

Now suppose that the ranks in both rows are permuted simultaneously, while keeping the rows separate. This involves permuting the effort cost labels on the individual observations, but not changing the group size labels. There are 70 permutations within a row, so there are $70 \times 70 = 4{,}900$ possible permutations for both rows, which is a lot of possibilities to enumerate and evaluate. In this case, it is easiest to use a large number of *simulations* instead of exact permutations. Each simulation generates a random reassignment of data points between the two effort cost columns (4 on each side). This is done by simultaneously permuting the 8 ranks in the top row and the 8 ranks in the bottom row, while keeping the permuted ranks in their respective rows as they move side to side within a row. For each simulation, the calculation program computes the difference between the average of the ranks with high effort cost labels and the ranks with low effort cost labels. This simulated difference measure is compared with the observed difference between the average ranks on the right and left sides of table 13.6. The observed difference is 3.25 (with an average rank of 2.88 for low effort costs and 6.13 for high effort costs, the difference is $6.13 - 2.88 = 3.25$). The program counts up the number of simulations with treatment average differences that are as or more extreme than the 3.25 observed difference in average ranks. The p value is the ratio of the number of as-or-more-extreme simulated differences to the total number of simulations.

With 100,000 simulations, the result is that 283 simulations resulted in a difference that was more extreme than what was observed, and 570 simulations were more extreme in either direction, so the ratio is 283/100,000 = 0.00283. Thus the result is significant at a p value of 0.003 (one-tailed test) or 0.006 (two-tailed test). This procedure has the advantage of using all of the data, and thereby providing more reliable results. In addition, by avoiding the need to do two separate tests, it reduces the risk of finding a statistically significant effect by chance when there is actually no treatment effect. Notice that a matched-pairs test is a special case of stratification, with a single observation for each treatment in each matched pair (strata).

In Holt, Porzio, and Song (2017), for example, the primary treatment difference involved the gender of traders in an asset market, male or female, with a balanced set of price bubble measures for all-male and all-female markets, each run for 25 periods. There was also another balanced set of bubble measures for gender-sorted asset markets that were only run for 15 periods, and bubbles tend to be lower for both gender groupings in the short-duration markets. Instead of reporting separate tests for gender effects for short and long markets, it is better to do a single test where permutations are stratified by market length, so that the gender labels associated with short markets are permuted and the gender labels associated with the long markets are also permuted at the same time. The main idea is that the stratification controls for the secondary treatment variation (market length) while the main treatment variation is evaluated. To summarize:

> **Stratified Permutation Tests for Multiple Data Groupings:** *A stratified permutation test generalizes the Mann-Whitney analysis of two treatments to the case of two treatments with stratified data generated by changes in other variables (subject pool or secondary treatment changes). The Wilcoxon Matched Pairs test is a special case where each pair is a separate strata.*

13.7 Extensions and a Word of Caution

Harrison and List (2004) survey and categorize different types of field experiments. Levitt and List (2007) present a strong argument that field experiments are more appropriate than lab experiments for the measurement of social preferences. Camerer (2016) provides a thoughtful response and a defense of lab experiments.

For both types of experiments, it is important to think carefully about the appropriate sample sizes in advance. For example, if one treatment involves hypothetical incentives that induce high behavioral variability, and another uses cash incentives with less variability, then it would make sense to allocate more subjects to the treatment with high variability. List, Sadoff, and Wagner (2011) discuss these issues and offer some rules of thumb for sample size calculations. Advance consideration of sample sizes is essential with field experiments for which it is typically difficult or impossible to augment sample sizes at a later date. Even with lab experiments, augmenting sample sizes ex post can invalidate statistical conclusions. To protect against data mining by increasing sample sizes on the fly, it is possible to register sample sizes and hypotheses in advance on some sites.

As discussed in the chapter, one problem with running multiple statistical tests is the chance that a "statistically significant" result could be due to chance, i.e., a false rejection of a null hypothesis. The approach discussed in this chapter involves combined tests that can provide a single perspective on multiple comparisons, e.g., the Jonckheere test for an ordered alternative hypothesis, or the stratified permutation test to accommodate different strata caused by nuisance variations, e.g., different subject pools or secondary treatment variations. A different approach is to go ahead and run multiple tests, and then adjust the p values for the degree of multiplicity, a "family-wise adjustment" procedure that is standard in medical trials.

A related issue of appropriate p values has come up as a result of a failure of some studies to replicate, especially in social psychology. In particular, some widely cited studies involving *priming* effects of small social cues have not been replicated. In an email to a few colleagues that was later widely circulated, Danny Kahneman warned that priming is a "train-wreck looming" and he recommended a "daisy chain" of replications done by one lab of another lab's work. A focus on replication will probably result in lower p value standards for published research, especially in the case of a new discovery that does not fit a pattern established in related studies.

This chapter is heavily influenced by the author's joint work, Holt and Sullivan (2017), which is primarily focused on permutation tests, although parallels with standard nonparametric tests are mentioned. The classic reference on nonparametric tests is the Siegel (1956) book that was mentioned earlier in the chapter. This book is written with a clarity that comes from having encountered many of the testing issues in his own experimental research. An updated version can be found in Siegel and Castellan (1988). There are many other fine nonparametric statistics books, e.g., Gibbons and Chakraborti (2014).

Chapter 13 Problems

1. (non-mechanical) It has been observed that collusion on price may be hard to establish if it involves unequal sacrifices, but that once established, collusive agreements in experiments can be stable under some conditions. In an

experiment to evaluate the effects of communication opportunities on price levels, would you prefer to have each group of sellers subjected to a single treatment, or would you prefer to have each group experience some periods with communication and some without, perhaps alternating the order? What considerations might be relevant?

2. Consider a between-subjects design with two treatments: "Dime" and "Quarter" (maybe due to price or quantity discount treatments). If there are just two separate sessions, one with each treatment, and if the one with the lowest price outcome is listed on the left, then the two possible ranking outcomes can be represented as DQ and QD. Suppose instead that there are two sessions with each treatment, so that the four sessions can be ranked in order by average prices. For example, one of the possible rankings is DDQQ. Find the other five rankings.

3. The setup in the previous problem, with two sessions in each of two treatments, yields 6 possible rankings of the D and Q designators. Mathematically, this represents the number of ways of ranking 4 things by type, when there are two things of each type. The mathematical formula for this is: (4*3*2*1) divided by (2*1)*(2*1), or (4!)/(2!)*(2!), which yields 24/4 = 6. Now consider a setup with a total of six sessions, with three sessions for each of two treatments, D and Q. Either calculate or count the number of rankings by type, and explain your answer. (It is not permissible to simply quote the number given in the chapter.)

4. Think about the intuition behind the ratio of factorial expressions in the previous question. With 4 items, there are 4 ways to pick the first element in a sequence, 3 ways to pick from the remaining 3 elements, 2 ways to pick the third element, and only 1 element left for the final position, so the number of possible orders is 4*3*2*1. Thus the numerator of the formula (4!) is the total number of possible rankings for the sessions. Division by the factorial expressions in the denominator reduces the number in the numerator. This is done because all that is required in the statistical argument is that the sessions be identified by the treatment, e.g., Q or D, and not by the particular session done with that treatment. For example, if Q_1 is the first session done with the Q treatment, and Q_2 is the second session, then the two orderings, Q_1Q_2 and Q_2Q_1, are only counted once, since the theory does not make a prediction about order within the same treatment. Using this observation, explain why the denominator in the formula is the product of two factorial expressions.

5. Consider a "within-subjects" design in which each person (or group) is exposed to two treatments, e.g., two numerical decisions (D1 and D2) are made under differing conditions, with one decision to be chosen at random ex post

to determine earnings. If D1 > D2, then code this as a Heads for that person. A natural null hypothesis would be that the probability of Heads is 1/2, so the chances of two Heads in two trials is (1/2)(/12) = 1/4. If there are 5 subjects and Heads is observed in all cases, what are the chances that this could have occurred at random?

6. For the data in the bottom row of table 13.4, verify the claim in the text that the *minimum* binary win count is 3, so $U = 3$. Then use table 13.2 to determine the p value for a one-tailed Mann-Whitney test.

7. The original (unaltered) data for the Davis and Holt (1994a) matched-groups experiment are:

Session	S1	S2	S3	S4	S5	S6
3 Seller Price Averages	425	470*	408	436*	424	517*
5 Seller Price Averages	415*	**471**	392*	401	392*	512

Determine the sum of signed ranks for the original data shown here. Then determine the p value for a one-tailed test from table 13.6.

PART III

Social Preferences

Bilateral bargaining is pervasive, even in a developed economy, especially for large purchases and specialty commodities like automobiles and housing. Bargaining is also central in many legal and political disputes, and it is implicit in family relations, care of the elderly, etc. The highly fluid give and take of face-to-face negotiations makes it difficult to specify convincing structured models that permit the calculation of Nash equilibria. Experimental research has responded in two ways. First, it is possible to look at decisions in unstructured bargaining situations to spot interesting patterns of behavior, like the well-known "deadline effect," the tendency to delay agreements until the last moment. The second approach is to limit the timing and sequence of decisions, in order to learn something about fairness and equity considerations in a simplified setting. The ultimatum bargaining game discussed in chapter 14 is an example of this latter approach. In an ultimatum game, one person makes a proposal about how to split a fixed amount of money, and the other either accepts or rejects, in which case both earn zero. Seemingly irrational rejections in such games have fascinated economists, and more recently, anthropologists.

The simple scenarios discussed in chapter 15 also have alternating decision structures, but the focus is on manipulations that highlight issues of fairness, trust, and reciprocity. The "trust game" begins when one person is endowed with some cash, of which part or all may be passed to the other person. The money passed is augmented, e.g., tripled, and any part of the resulting amount may be passed back to the original person. A high level of trust would be indicated if the first person passes most of the initial cash stake, expecting the second person to reciprocate and pass back even more. The second setup that is considered, the "reciprocity game," has more of a market context, but the underlying behavioral factors are similar. Here, the employer announces a wage, and seeing this, the worker chooses an effort level, which is costly for the worker but which benefits the employer. The issue is whether fairness considerations, which are clearly present in bilateral negotiations, will have an effect in more impersonal market settings.

Many public programs and policies are designed to remedy situations where the actions taken by some people affect the well-being of others. The classic example is that of the provision of a "public good" like national defense, which can be consumed freely by all without crowding and the possibility of exclusion. Chapter 16 introduces a model of voluntary contributions to a public good, where the private cost to each person is less than the social benefit, and yet the cost is greater than the private benefit from contribution. The result is that individuals have an incentive to "free ride" on others' contributions. Voluntary contributions experiments (and the associated tendency to free ride) provide a workhorse for the analysis of social preferences, altruism, reciprocity, punishment, etc. Associated field experiments manipulate matches and other nudges to trigger charitable contributions from groups of people who are not aware that they are in an experiment.

A similar situation occurs when the provision of a public good requires that a certain target or threshold of contributions be reached. A volunteer's dilemma is a case in which the threshold is one, i.e., when it only takes one "volunteer" to provide the good outcome preferred by all. The "volunteer's dilemma" discussed in chapter 17 is whether to incur the private cost to achieve this good outcome for all, at the risk of duplication. This game differs from a standard public goods setup in that the private benefit to the volunteer exceeds the cost.

External effects on others' well-being may arise in a negative sense as well, as is the case with pollution or overuse of a common resource. The "common pool resource" game in chapter 18 is a setting where each person's efforts to secure benefits from a shared resource tend to diminish the value that others derive from their efforts, as might happen with excessive harvests from a fishery. The resource is like a public good in the sense that the problem arises from non-exclusion, but the difference is the presence of congestion effects, i.e., when one person's use reduces the value for others.

Public expenditures are typically influenced by voting and other political processes, and the outcomes of such processes can be very sensitive to the rules that govern voting. Chapter 19 provides an introduction to laboratory voting experiments and related field experiments that are motivated by political science considerations.

14

Bargaining

The simplest bargaining scenario provides one person with the ability to propose a final offer that must be accepted or rejected, just as a seller may post a price on a take-it-or-leave-it basis. Although rejections of positive amounts of money may surprise some, they will not surprise anyone who has participated in this type of bargaining.

> **Note to the Instructor:** The Veconlab Bargaining Games program has setup options for dictator, ultimatum, 2-stage, and conflict bargaining situations.

14.1 Strategic Advantage and Ultimatums

Suppose that a local monopolist can produce a unit of a commodity for $5. The sole buyer needs the product and is willing to pay any amount up to $15 but not a penny more, so $15 is the buyer's value. Then the "surplus" to be divided is $10, i.e., the difference between the value and the cost. The buyer knows the seller's cost, and hence knows that a price of $10 would split the surplus. A higher price corresponds to offering a lower amount to the buyer. The ability to make a take-it-or-leave-it offer confers a strategic advantage, at least in theory. This setup is called an "ultimatum game" since the proposer (seller) makes a single offer to the responder (buyer), who must either accept the proposed split of the surplus (as determined by the price) or reject, which results in zero earnings for both.

The ultimatum game was introduced by Güth, Schmittberger, and Schwarze (1982) in response to the observation of people "leaving money on the table" by rejecting offers in complicated multi-round negotiations. As noted in the Selten quote in chapter 1, this observation did not surprise psychologists. The reaction of the economists was to consider a highly simplified game with a single offer, because it highlights an extreme conflict between the dictates of selfish, strategic behavior and notions of fairness. If people only care about their own earnings, and if more money is preferred to less, then the proposer should be able to get away with offering a very small amount to the responder. The monopolist in the previous paragraph's example could offer a price of $14.99, knowing that the buyer would rather pay this than reject the offer and earn zero. The $14.99 price is unfair in the sense that $9.99 of the available surplus goes to the seller, and only a penny goes to the buyer. Even so, the buyer who only cares about getting the lowest possible price should accept any price offer below $15.00, and hence the seller should offer $14.99.

It is easy to think of situations in which a seller might hesitate to exploit a strong strategic advantage. Getting a reservation for dinner after graduation in a college town is the kind of thing one tries to do six months in advance. Restaurants seem to shy away from allocating the scarce table space on the basis of price, which is usually not raised on graduation day. Some moderate price increases may be hidden in the form of requiring the purchase of a special graduation meal with multiple courses, but this kind of price premium is nowhere near what would be needed to remove excess demand, as evidenced by the long lead time in accepting reservations. A possible explanation is that an exorbitant price might be widely discussed and perceived as being unfair, with a backlash that might harm future business. The higher the price, the lower the cost of rejecting the deal. In the monopoly example discussed above, a price of $14.99 would be rejected if the buyer is willing to incur a one-cent cost in order to punish the seller for charging such a price.

Several variants of the ultimatum game have been widely used in laboratory experiments because this game maximizes the tension between fairness considerations and the other extreme where people only care about their own earnings. Moreover, in the lab it is possible to set up a "one-shot" situation with enough anonymity to eliminate any considerations of reputation, reward, and punishment. The next section describes an ultimatum experiment where laboratory control was a particularly difficult problem.

14.2 Bargaining in the Bush

Jean Ensminger (2004) conducted an ultimatum experiment in a number of small villages in East Africa. All participants were members of the Orma clan. The Orma offer an interesting case where there is considerable variation in the extent of integration into a market economy, which may affect attitudes toward fairness. The more nomadic families raise cattle and live largely on the milk and other products, with very little activity in the way of market purchases or sales. Although some nomadic families have a high level of wealth, which is kept in the herd, they typically have low incomes in terms of wage payments. Other Orma, in contrast, have chosen a more sedentary lifestyle for a variety of reasons, including the encroachment of grazing lands. Those who live sedentary lives in villages do purchase food with money income obtained as wages or crop revenues. Thus, exposure to a market economy is quite variable and is well measured as direct money income.

One can imagine two plausible conjectures about the effect of market integration on fairness attitudes. The first possibility is that interactions among nomadic people involve trust and reciprocity, whereas a market with more anonymous arrangements may make people more selfish. The opposite conclusion could be reached by reasoning that many bargaining situations in a market context end up with individuals agreeing to "split the difference," so that fair outcomes in an ultimatum game would be related to the degree of market exposure.

In Ensminger's experiment, at least one adult was recruited from each household to play "fun games for real money." The experiments were conducted in grass houses, which enabled her to isolate groups of people during the bargaining process. A "grand master," who was known to all villagers, read the instructions. This person would turn away to avoid seeing decisions as they were made. The amount of money to be divided between the proposer and responder in each pair was set to be approximately equal to a typical day's wages (100 Kenyan shillings at the time), and the game was only played once. Each proposer would make an offer by moving some of the shillings to the other side of the table before leaving the room while the responder was allowed to accept or reject this offer. Ensminger reports that the people enjoyed the games, despite some amusement at the "insanity" and "foolishness" of Western ways. She tried to move from one village to another before word of results preceded her arrival.

The data patterns for 56 bargaining pairs are shown in figure 14.1. The average offer was 44% of the stake, with a clear mode at 50%. The lowest offer was 30%, and even these unequal splits were rarely rejected (as indicated by the light part at the bottom of the bar). If people had foreseen that the 50% offers would all be accepted, then they might have lowered their offers. It is clear that the modal offer was not optimal in the sense of expected-payoff maximization against the actual ex post rejection pattern. People who made these generous

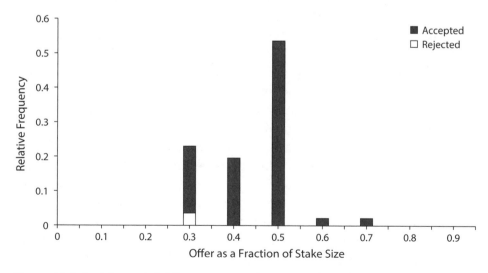

Figure 14.1. Distribution of Offers to Responders in an Ultimatum Game with the Orma (56 pairs, one-day wage stake). *Source*: Ensminger (2004).

offers almost always mentioned fairness as the justification in follow-up interviews. Ensminger was suspicious, however, and approached some reliable informants who revealed a different picture of the "talk of the village." The proposers were apparently *obsessed* with the possibility that low offers would be rejected, even though rejections were thought to be unlikely. Such obsessions suggest that expected-payoff maximization may not be an appropriate assumption for large amounts of money, e.g., a day's income. This conjecture would be consistent with the payoff-scale effects on risk aversion discussed in chapter 3.

In addition, it cannot be the case that individuals making relatively high offers were doing so *solely* out of fairness considerations, since offers were much lower in a second experiment in which the proposer's split of the same amount of money was automatically implemented. This game, without any possibility of rejection, is called a "dictator game." The average offer fell from 44% in the ultimatum game to 31% in the dictator game with no possibility of rejection. Even though there seems to be some strategic reaction by proposers to their advantage in the dictator game, the modal offer was still at the "fair" or "50–50" division, and less than a tenth of the proposers kept all of the money.

Ensminger used multiple regression to evaluate proposer offers in the ultimatum and dictator games. In both cases, the presence of wage income is significantly related to offers; those with such market interactions tend to make higher, more generous offers. Her conjecture is that people exposed to face-to-face market transactions may be more used to the notion of "splitting the difference." Variables such as age, gender, education, and wealth (in cattle equivalents) are not significant.

The unimportance of demographic effects relative to market exposure measures is also evident in a cross-cultural study involving 15 small-scale societies on five continents (Henrich et al., 2001). This involved one-shot ultimatum games with comparable procedures that were conducted by anthropologists and economists. There was considerable variation, with the average offer ranging from 0.26 to 0.58 of the stake. The societies were ranked in two dimensions: the extent of economic cooperation and the extent of market integration. Both variables were highly significant in a regression, explaining about 61% of the variation, whereas individual variables like age, sex, and relative wealth were not. The lowest offers were observed in hunter-gatherer societies where very little production occurred outside of family units (e.g., the Machiguenga of Peru). The highest offers were observed in a society where production involved joint effort. In particular, offers that provided more than half of the stake to the other person were common for the Lamelara of Indonesia, who are whale hunters in large sea-going canoes. Production in that context requires assembling large groups of men and dividing up the whale meat after the hunt.

A one-shot ultimatum game, played for money, is a strange new experience for these people, and behavior seemed to be influenced by parallels with social institutions in some cases. The Ache of Paraguay, for example, made generous offers, above 0.5 on average. The proposed sharing in the ultimatum game has some parallels with a practice whereby Ache hunters with large kills will leave them at the edge of camp for others to find and divide. To summarize:

> **Ultimatum Bargaining in the Field:** *In bargaining experiments done in the field, but under somewhat controlled lab-like conditions, the typical offer is about 40–50% of the stake. Tendencies to make fair offers and reject unfair ones are more pronounced in societies with a higher degree of market integration, and with more jointness and division in everyday production.*

14.3 Bargaining in the Lab

Ultimatum games have been conducted with more standard, student subject pools in many developed countries, and with stakes that are usually about $10. The mean offer to the other person, as a fraction of the stake, is typically about 0.4, and there seems to be less variation in the mean offer than was observed in the small-scale societies. Roth et al. (1991) report ultimatum game experiments that were run at four universities, each in a different country. The modal offer was 0.5 in the United States and Slovenia, as was the case for the Orma and for other studies in the United States (e.g., Forsythe et al., 1988). The modal offer was somewhat lower, 0.4, in Israel and Japan. Rejection rates in Israel and Japan were no higher than in the other two countries, despite the lower offers, which led the authors to conjecture that the differences in behavior across countries were due to different cultural norms about what is an acceptable division. These differences in behavior reported by Roth et al. (1991) were, however, lower than the differences observed in the 15 small-scale societies, which are less homogeneous in the nature of their economic production activities.

Recall that the Orma made no offers below 0.3. In contrast, some student subjects in developed countries make offers of 0.2 or lower, and these low offers are rejected about half of the time. For example, consider the data in figure 14.2, which is for a one-shot ultimatum game that was done in class, but with full money payments and a $10 stake. The mean offer was about 0.4 (as compared with 0.44 for the Orma), but about a quarter of the offers were 0.2 or below, and these were rejected a third of the time.

Ultimatum bargaining behavior is relatively sensitive to various procedural details, and for that reason extreme care was taken in the Henrich et al. cross-cultural study. Hoffman et al. (1994) report that the *median* offer of 5 in an ultimatum game was reduced to 4 by putting the game into a market context. The market terminology had the proposer play the role of a seller who chose a take-it-or-leave-it price for a single unit. Interactions with posted prices are

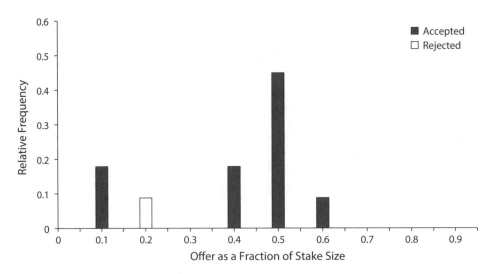

Figure 14.2. Distribution of Offers to Responders in a One-Shot Ultimatum Game (11 pairs, $10 stake, cash payments, class experiment)

more anonymous than face-to-face negotiations, and market price terminology in the laboratory may stimulate less generous offers for this reason. The median offer fell again, to 3, when high scores on a trivia quiz were used to decide which person in each pair would play the role of the seller in the market. Presumably, this role-assignment effect is due to the fact that people may be more willing to accept an aggressive (low) offer from a person who earned the right make the offer.

Ultimatum Bargaining in the Lab: *Even under controlled laboratory conditions with financial incentives, proposers in ultimatum games do not come close to making aggressively low offers that are predicted by economic rationality with selfish preferences. Low offers are more common when the game is framed with market terminology, or when the proposer has won the right to play that role in an ex ante trivia quiz.*

Many business bargaining decisions are made by teams of managers or union officials, and several laboratory experiments have examined the offers made by groups of subjects in the laboratory. The general result is that groups tend to make lower offers to other groups, i.e., groups appear to be less fair-minded, or at least they expect other groups to be less fair-minded, as compared with individuals facing other individuals (Bornstein and Yaniv, 1998). There is no clear consensus about what social dynamic is driving differences between individuals and groups, but experiments can shed some light on this process. Pallais (2005) reports an ultimatum experiment involving 220 subjects: 10 pairs of individuals, 10 paired groups of size 3, and 10 paired groups of size 7. The pie size was $10 for the individuals, $30 for the 3-person groups, and $70 for the 7-person groups. Each decision unit (group or individual) was placed in a separate classroom, and there was no imposed time limit for these one-shot ultimatum game decisions. Group discussions were not monitored, although all participants were asked to fill out ex post survey forms. The average offer made by the proposer was $4.40 for individuals, which was significantly higher than the average offers of $3.50 and $3.60 made by groups of size 3 and 7 respectively. In ex post surveys, the people in the individual treatment were much more likely to mention equity considerations. In addition, those in the individual treatment indicated more concern about the possible rejection of uneven offers than those in the group treatments did. This is a fine example of a research project done by a second-year undergraduate (now a professor at Harvard), with careful attention to procedural detail. For example, the decision was made not to use computers because of a fear that the person at the keyboard would control the discussion and thereby diminish group size effects.

14.4 Dictator Games and Procedural Variations

The perplexing results of ultimatum game experiments stimulated a round of experiments with an even simpler game—the dictator game, in which the proposer unilaterally decides on a division of the available money. The other player has no active role to play in this game. Dictator experiments were run with instructions stating that the subject and their anonymous counterpart in another room have been "provisionally allocated" an amount, $10, and that their task is to decide how to "divide" it. Hoffman et al. (1994) replicated dictator game experiments with this wording and room separation, with the result that only about 18% of the proposers passed $0 and about a third passed $4 or more to the other player. These significant deviations from the rational decision for a selfish subject to pass $0 have been attributed to fairness. The authors note that there are many repeated-game situations in which people do share. In hunter-gatherer societies, large-game hunting is risky, with each hunter having a less than 50-50 chance of success on a given trip. As a result, such large-game kills tend to be widely shared, which benefits all by spreading risk, but affords the general advantage of high-value hunting successes. In contrast, small game and gathered foods are typically less variable and are not shared. In the anthropology literature, Hawkes (1993) refers to this behavior as "reciprocal altruism." Most readers can think of analogous interactions in their student experience. The point is that the experimenter does not shape the world that subjects have lived in, and they come into the lab with repeated-game habits and expectations of sharing.

In order to see whether these "home-grown" attitudes could be diminished, the authors ran a parallel treatment with double anonymous procedures, in which nobody, *not even the experimenter*, could ever be able to observe how much money a particular subject decided to pass. One of the several anonymity protections used was that subjects were given envelopes with 10 dollar bills and 10 slips of paper, and they could keep either dollars or slips and pass a mix of dollars and slips, so the thickness of the envelope that is turned in would not reveal the subject's actual decision. In addition, two of the envelopes only had slips of paper. The envelopes were collected by a student "monitor," so that the experimenter could not later match envelopes to people. After the instructions were read, the students in the proposer room were called out one by one to bring their personal belongings so as to make a clean exit after turning in their envelopes.

With double anonymous procedures, about two-thirds passed $0, and fewer than 10% passed $4 or more. Hoffman, McCabe, and Smith (1996b) showed that the percentage of $0 pass decisions was lowered by removing some of the double-blind elements (the monitor and the all-paper-slip envelopes), and lowered even more by switching to a single-blind procedure in which the experimenter can observe subjects' pass decisions, although other subjects cannot observe each others' decisions. The "provisionally allocated" and "divide" terminology also made a difference: removing those phrases that suggest sharing from the instructions, holding the other procedures constant (no double anonymity) resulted in a doubling of the percentage of selfish $0 pass decisions, from about 20% to about 40%. The point is that instructional changes can matter a lot when they interact with values that subjects bring to the lab, as shown in the first two columns of table 14.1.

But even the most stringent double-anonymity procedures, with student monitors, unmarked envelopes with paper stuffing, etc., still result in a situation in which about 40% of dictators do pass positive amounts to the other person. Even sharper results were obtained by Cherry, Frykblom, and Shogren (2002), who used a quiz with questions from the Graduate Management Admissions Test (GMAT) to let subjects earn either $40 or $10, depending on their scores relative to a cutoff. These subjects were then given the dictator roles, and

Table 14.1. Social Distance in Dictator Games and Selfish $0 Pass Rates

	Hoffman et al. (1996b)			Cherry et al. (2002)		
Terminology	"divide"	neutral	neutral	neutral	neutral	neutral
Anonymity	single	single	double	single	single	double
Endowment	gift	gift	gift	gift	earned	earned
$0 Pass Rate	18%	42%	65%	19%	79%	97%
$0 Pass Rate, $40 Pie				15%	70%	97%

the potential recipients in the other room were told how the money amounts had been earned. In the other "gift" treatment, the money amounts, $40 or $10, were received from the experimenter. In a baseline treatment, without unearned cash, the percentage of $0 pass decisions was 15–19%, depending on whether the amount to be divided is high or low. In the earned cash treatment, this percentage of selfish decisions increased to the 70–79% range. In each case, the pass $0 rate was a little higher on average when the available amount is only $10. When earned cash was combined with double-blind procedures, the percentages of purely selfish $0 pass decisions was in the 90–95% range. Although there are some minor procedural differences between the studies discussed, the general pattern of results shown in table 14.1 is informative and can be summarized:

> **Procedural Variations in the Dictator Game:** *Double anonymity procedures increase the incidence of perfectly selfish dictator decisions (passing $0). Changes in instructions that remove references to "divide" and "provisionally allocated" have effects of a similar magnitude. When dictators first earn the money that they can subsequently divide, almost all dictator pass decisions (with double anonymity) match the perfectly selfish $0 amounts.*

14.5 Multi-Stage Bargaining

Face-to-face negotiations are typically characterized by a series of offers and counter-offers. The ultimatum game can be transformed into a game with many stages by letting the players take turns making proposals about how to split an amount of money. If there is an agreement at any stage, then the agreed split is implemented. A penalty for delay can be inserted by letting the size of the stake shrink from one stage to the next. For example, the amount of money that could be divided in the first stage may be $5, but a failure to reach an agreement may reduce this stake to $2 in the second stage. If there is a pre-announced final stage and the responder in that stage does not agree, then both earn zero. Thus the final stage is like an ultimatum game, and can be analyzed as such.

Consider a game with only two stages, with a money stake, Y, in the first stage and a lower amount, X, in the final stage. For simplicity, suppose that the initial stake is $15, which is reduced to $10 if no agreement is reached in the first round. *Warning: the analysis in this paragraph is based on the (questionable) assumption that people are perfectly rational and care only about their own payoffs.* Put yourself in the position of being the proposer in the first stage of this game, with the knowledge that both of you would always prefer the action with the highest payoff, even if that action only increases one's payoff by a penny. If you offer the other player too little in the first stage, then this offer will be rejected, since the other player becomes the proposer in the final (ultimatum game) stage. So you must figure out how much the other person would expect to earn in the final stage when they make a proposal to split the $10 that remains. In theory, the other person could make a second-stage offer of a penny, which you would accept under the assumption that you prefer more money (a penny) to less (zero). Hence the other person would expect to earn $9.99 in the final stage, assuming perfect rationality and no concerns for fairness. If you offer less than $9.99 in the first stage, it will be rejected, and if you offer more it will be accepted. The least you could get away with offering in the first stage is, therefore, just a little more than $9.99, so you offer $10, leaving $5 for yourself, which is accepted.

In the previous example, the theoretical prediction is that the first-stage offer to the other person (responder) will equal the amount of the stake that would remain if bargaining were to proceed to the second stage. This result can be generalized. If the size of the money stake in the final stage is X, then the responder making the counter-offer in that stage "should" offer a penny to the original proposer, which will be accepted. The responder making the counter-offer in the final stage can obtain essentially the whole remaining stake, i.e., $X − $0.01. Thus the proposer making an offer in the first stage can get away with offering a slightly higher amount, i.e., $X, which is accepted. In this two-stage game, the initial proposer earns the amount by which the pie shrinks, $Y − $X, and the responder earns the amount remaining in the second stage, $X.

An experiment with this two-stage structure is reported in Goeree and Holt (2001). The size of the stake in the initial stage was $5 in both cases. In one treatment, the pie was reduced to $2.00, so the proposer should demand $3 (the shrinkage) and offer $2.00, as shown in the middle column of table 14.2. This $2 offer is what the responder could earn in the final-stage ultimatum game. The average first-stage offer was $2.17, quite close to this prediction. In a second treatment, shown on the right side of the table, the pie shrinks to $0.50, so the prediction is for the first-stage offer to be very inequitable ($0.50). The average offer was somewhat higher, at $1.62, as shown in the right column of the table. The reduction in the average offer was much less than predicted by the theory, and rejections were quite common in this second treatment. Notice that the

Table 14.2. A Two-Stage Bargaining Game Played Once

	Treatment 1	Treatment 2
Size of Pie in First Stage	$5.00	$5.00
Size of Pie in Second Stage	$2.00	$0.50
Selfish Nash First-Stage Offer	$2.00	$0.50
Average First-Stage Offer	**$2.17**	**$1.62**

Source: Goeree and Holt (2001).

cost of rejecting a low offer is low, and knowing this, the initial proposers were reluctant to exploit their advantage fully in this second treatment.

The effects of payoff inequities are even more dramatic in a second two-stage experiment (Goeree and Holt, 2000). The initial proposers made seven choices corresponding to seven different situations, with the understanding that only one of the situations would be selected afterward, at random, before the proposal for that situation was communicated to the other player. The initial pie size was $2.40 in all seven cases, but the second-stage pie size varied from $0.00 to $2.40, with five intermediate cases. In each case, the theoretical prediction is that the initial proposer offer will equal the second-stage pie size.

The two extreme treatments for this experiment are shown in table 14.3. In the middle column with full shrinkage, the pie shrinks from $2.40 to $0.00. This gives the initial proposer a large strategic advantage, so the initial offer would be a penny in a game between two selfish, rational players. This outcome would produce a sharp asymmetry in the payoffs in favor of the proposer. Adding insult to injury, the instructions for this case indicated (in a neutral manner) that the initial *proposer* would receive a fixed payment of $2.65 in addition to the earnings from the bargaining, whereas the initial responder would only receive $0.25. Only if the initial proposer were to offer the whole pie of $2.40 in the first stage would final earnings be equalized. This is listed as the "egalitarian first-stage offer" of $2.40 in the middle column. The average offer ($1.59) shown in the bottom row is well above the Nash prediction ($0.01), and is closer to the egalitarian offer of $2.40.

The other extreme treatment is shown in the right-hand column of table 14.3, which is labeled "no shrinkage." Here the pie does not shrink at all in the second stage, so the theoretical prediction is that the first-stage offer will be $2.40, which is the size of the pie in the second stage. The initial responder now has the strategic advantage, since the pie remains high in the final (ultimatum) stage when this person has the turn to make the final offer. To make matters even more asymmetric, this strategic advantage is complemented with a

Table 14.3. A Two-Stage Bargaining Game with Asymmetric Fixed Payments

	Full Shrinkage	No Shrinkage
Size of Pie in First Stage	$2.40	$2.40
Size of Pie in Second Stage	$0.00	$2.40
Proposer Fixed Payment	$2.65	$0.25
Responder Fixed Payment	$0.25	$2.65
Egalitarian First-Stage Offer	**$2.40**	**$0.00**
Selfish Nash First-Stage Offer	$0.01	$2.40
Average First-Stage Offer	**$1.59**	**$0.63**

Source: Goeree and Holt (2000).

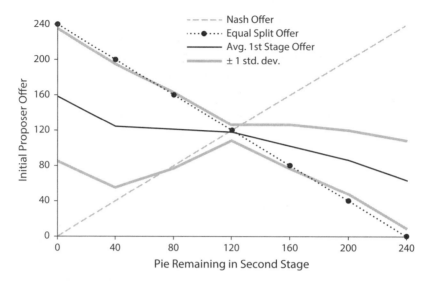

Figure 14.3. Two-Stage Bargaining Game: Average First-Stage Offer (dark line) Compared with Egalitarian Offer (dotted line) and Subgame Perfect Nash Offer (dashed line)

high fixed payment to the initial *responder*. The only way for the disadvantaged initial proposer to obtain equal earnings would be to offer nothing in the first stage, despite the fact that the theoretical prediction (assuming selfish behavior) is $2.40. The average of the observed offers, $0.63, is much closer to the egalitarian offer of zero.

The overall pattern of initial first-stage offers is shown in figure 14.3, where the horizontal axis is the pie size remaining at the start of the second and final round. Since the final round is an ultimatum game, the subgame-perfect offer amount for the proposer is the remaining pie amount that the responder can expect to earn in the final stage. This Nash prediction is the gray dashed line with a 45-degree slope. In contrast, the offer that equalizes earnings (in the presence of the asymmetric outside payments) is shown by the dotted line with a negative slope that is the opposite of the Nash prediction. The average offers for the seven treatments have a negative slope (dark thin line), bracketed by ± one standard deviation (gray lines). The actual offers are somewhat flatter than the dotted line connecting egalitarian first-stage offers, but there is an egalitarian tendency. To summarize:

Two-Stage Alternating-Offer Bargaining Results: *The first-stage offers in this experiment should be equal to the remaining pie size, as shown by the dashed Nash line in figure 14.3, but the asymmetric fixed payments were structured so that the egalitarian offers would be inversely related to the remaining pie size, with a slope of*

−1 instead of +1. This inverse relationship was generally present in the data averages (dark thin line) for the seven treatments.

The authors go on to show that the data patterns are roughly consistent with an enriched model in which people care about relative earnings as well as their own earnings. For example, a person may be willing to give up some money to avoid having the other person earn more, which is an aversion to disadvantageous inequity. Roughly speaking, think of this as an "envy effect." It is also possible that people might wish to avoid making significantly more than the other person, which would be an aversion to advantageous inequity. Think of this as a kind of "guilt effect," which is likely to be weaker than the envy effect. These two aspects of *inequity aversion* are captured by a formal model proposed by Fehr and Schmidt (1999). To get a feel for this model, let π_{self} denote a person's own payoff, and let π_{other} denote the other person's payoff. Then the Fehr-Schmidt utility function is:

(14.1)
$$U(\pi_{self}, \pi_{other}) = \pi_{self} - \alpha(\pi_{other} - \pi_{self}) \text{ if } \pi_{other} > \pi_{self} \text{ (envy)}$$
$$U(\pi_{self}, \pi_{other}) = \pi_{self} - \beta(\pi_{self} - \pi_{other}) \text{ if } \pi_{self} > \pi_{other} \text{ (guilt)}$$

where α is an envy parameter, β is a guilt parameter, and $\alpha > \beta > 0$, i.e., envy is expected to be stronger than guilt. Goeree and Holt (2000) use the two-stage bargaining game data to estimate parameters, and they conclude that both effects are present, but that the envy effect is more pronounced. To summarize:

Inequity Aversion: *Laboratory experiments often exhibit anomalies that seem to be driven by considerations of fairness and aversion to unequal earnings. The Fehr-Schmidt inequity aversion model stipulates that individuals are willing to sacrifice earnings in order to reduce inequities. The model has parameters for "envy" (aversion to having the other person earn more) and "guilt" (aversion to having the other person earn less). Empirical work confirms intuition that envy is stronger.*

It is worth emphasizing that relative payoff issues are most relevant when a single task is used for payment, as was the case here, with one of the seven decisions selected ex post for payment. Payoff differences and relative payoff issues might have been blurred if subjects had interacted repeatedly.

The attraction to equitable divisions in the two-stage bargaining setup raises the question of what might happen with more than two stages. Kloosterman and Fanning (2017) consider a setup in which the proposer makes a series of offers. A responder rejection causes the pie to shrink by a fixed proportional factor, $1 - \delta$, so the available "pie" to be split in any stage is only a fraction δ of what it was in the previous stage, with $0 < \delta < 1$. Payoff asymmetries were common knowledge, as before. Subjects bargained over splitting 100 chips, with the cash conversion values of those chips shrinking over time ($\delta = 0.95$), and with asymmetries determined by different conversion rates from chips to dollars. When conversion rates are equal for both proposers and responders, then a split of 50 chips for each will equalize the payoffs. In a second treatment with a conversion rate for proposers that is three times as high, the equal-money-payoff split is 25 chips for proposers and 75 for responders. Efficiency was measured by considering money payoffs as a fraction of the maximum that could be achieved without delay and shrinkage.

Matchings were done with a *turnpike procedure* that prevents a person from encountering the same partner again or from being matched with anyone who had encountered a previous partner, as discussed in the previous chapter. (To understand the intuition, imagine parallel lines of proposers and responders moving in different directions.) Proposers could make a series of proposed splits, which could be accepted or not by responders. Initial rejections would signal that the responder is more fair-minded, which would cause the proposer to offer a more favorable split. Not surprisingly, observed proposer offers tended to start a little low, but accepted offers in the final matchings were about equal to the equal-payoff benchmarks (50.16 in the symmetric treatment and 74.67 in the asymmetric treatment). Agreements were quick enough to yield final round efficiencies in the 92–95% range for the two treatments. To summarize:

> **Multi-stage Ultimatum Bargaining with a Shrinking Pie:** *When a proposer can make a series of offers that may be accepted or rejected, then the accepted proposals implement approximately equal money splits, with little delay and high-efficiency measures.*

14.6   Bargaining in the Shadow of Conflict

In the bargaining games considered in the previous section, a failure to reach agreement results in zero payoffs for both parties. In contrast, failure to reach agreement about a financial claim of loss could result in a judicial proceeding, with one party ending up a winner and the other a loser. Both winners and losers, however, may incur significant legal expenses in the process. Similarly, disputes between nations can result in military conflict, with serious consequences for the loser, and conflict costs for both parties. Bargaining under the threat of conflict tends to wipe away feelings of fairness or generosity, and power can be asymmetric, with one side having a higher ex ante chance of prevailing in a conflict.

Consider a stylized model in which there is $10 to be divided. The first mover ("proposer") is an aggressor who makes a demand for a share of the $10 amount.

The other player ("responder") either agrees, in which case the proposal is implemented, or refuses. A refusal results in a conflict in which each person incurs a conflict cost of $2. The treatments vary the proposer win probability from 0.2, 0.4, 0.6, to 0.8, so this probability determines the degree of power asymmetry. The winner in the conflict obtains the full amount, $10.

The conflict bargaining model can be solved with backward induction, by first considering each player's expected payoff in the event of conflict. If the proposer's win probability is 0.8, the proposer's expected payoff from conflict is $0.8(10) - 2 = 8 - 2 = 6$. Similarly, the responder's expected payoff from conflict would be $0.2(10) - 2 = 0$. Proposer offers were constrained to be integer dollar amounts, and the proposer would only have to offer the responder more than $0 to avoid conflict. So a minimal offer of $1 and an initial demand of $9 would be accepted (in equilibrium). (Note that even a cold, rational responder would be indifferent between accepting or rejecting an offer of $0, so a 50% acceptance rate in this case would induce the proposer to go ahead and cement the deal by offering $1 and demanding $9.) In this manner, the initial demands for proposer win probabilities of 0.2, 0.4, 0.6, 0.8 are $3, $5, $7, and $9. In other words, the Nash prediction (based on backward induction) is that proposers with more power will make more aggressive demands, which will be accepted with no conflict.

The experiment (Sieberg et al., 2010) began with proposers making demands in a single-round game (with no mention of what would follow). A different proposer win probability was used in each session in this between-groups design. The average proposals for this one-shot game are increasing in proposer power, as shown by the dark line in figure 14.4. These initial proposals, however, are too flat relative to the Nash predictions (dashed line), which could be due to some "pull-to-center" effect of randomness, or to inequity aversion (raising low proposals to avoid earning less, or lowering high proposals to avoid earning more). The one-shot game was followed by 6 rounds of random matching (keeping original roles), which is shown by the darker dashed line that is closer to the Nash predictions. The dotted line shows data for 6 additional rounds of random matching after a role reversal, which is quite close to the Nash predictions, even for high proposer win probabilities on the right.

The conclusion is that role reversals and repetition with random matching tend to increase conformity to Nash predictions based on backward induction. Essentially what happens is that learning occurs as subjects gain experience with the final-stage subgame. And the repetition tends to diminish the importance of fairness considerations. The sharpest deviation from Nash predictions is that conflicts occurred in about 30% of the cases for various treatments. These conflict rates are similar to those reported by Sieberg et al. (2013) in two-stage conflict bargaining games. In particular, it was not the case that conflict rates were higher when proposers had more power and made more aggressive demands, since responders made strategic concessions in such cases.

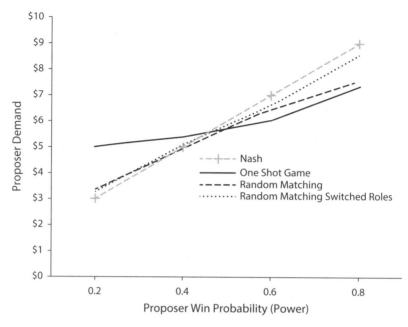

Figure 14.4. Average Proposer Demands for Conflict Bargaining Games

Conflict Bargaining Experiments: *Proposer demands are increasing in the proposer power to win in the event of a conflict, as predicted. But conflicts occur at significant rates for all proposer power treatments, which contradicts the theoretical prediction that initial demands will be set so that they are accepted immediately.*

14.7 "I will be spending years trying to figure out what this all meant."

The first ultimatum experiment was reported by Güth et al. (1982), and the results were replicated by Forsythe et al. (1988), who introduced the dictator game. There have been many experiments in other, less-structured negotiations, where the order of proposal and response is not imposed by the experimenter. For example, Hoffman and Spitzer (1982, 1985) used an open, unstructured setting to evaluate the ability of bargainers to make binding agreements on efficient outcomes, irrespective of property rights (the Coase theorem). Similarly, generalizations of the ultimatum game have been used to study bargaining in legislatures in which the proposer is selected at random (see the subsequent chapter on voting). Economists and others have been fascinated by behavior in these games where there is a high tension between notions of fairness and strategic, narrowly self-interested behavior. Ensminger (2004) reports that one of her African subjects jovially remarked: "I will be spending years trying to figure out what this all meant."

Many economists were initially skeptical of the high degree of seemly non-strategic play and costly rejections. Sefton (1992) found that allocations to the passive responder in dictator games fell by about a half when hypothetical payoffs were replaced by "normal" money payoffs. In ultimatum games, one possibility is that "irrational" rejections would diminish when the stakes of the game are increased, and that proposers would anticipate this and demand more. Hoffman et al. (1996a) increased the stakes from $10 to $100, which did not have much effect on initial proposals. Carpenter, Verhoogen, and Burks (2005) assign subjects randomly to high-stakes and low-stakes treatments, and they find no effect of going from $10 to $100 stakes with student subjects, either for proposals made in ultimatum games or demands made in dictator games. The effects of high stakes have also been studied by Slonim and Roth (1998) and List and Cherry (2000), and for a field experiment, see Carpenter, Burks, and Verhoogen (2005). Falk and Fehr (2003) present a thoughtful analysis of the value of laboratory and field experiments that focus on labor market issues. See Eckel and Grossman (1998, 2002) for a discussion of gender effects in bargaining.

Rejections are not, of course, irrational if people care about relative earnings. For example, a responder may prefer equal zero-payoff amounts to inequitable positive earnings. Bolton and Ockenfels (2000) proposed a model with preferences based on relative earnings, and Fehr and Schmidt (1999) developed a closely related model of inequity aversion that was mentioned above.

It is somewhat unusual for a seller to offer only one unit for sale to a buyer, and a multi-unit setup provides the buyer with an option for partial rejection. Suppose that there are 10 units for sale. Each costs $0 to produce, and each unit is worth $1 to the buyer. The seller posts a price for the 10 units as a group, but the buyer can decide to purchase a smaller number, which reduces each party's earnings proportionately. For example, suppose that the seller posts a price of $6, which would provide earnings of $6 for the seller and $4 for the buyer. By purchasing only half of the units, the buyer reduces these earnings to $3 for the seller and $2 for the buyer. This partial rejection option is implemented in the Veconlab software as a "squish" option (Andreoni, Castillo, and Petrie, 2003). If this option is permitted, the responder can choose a fraction that indicates the extent of acceptance: with 0 being full rejection and 1 being full acceptance.

The importance of emotions in the rejection of unfair offers is reinforced by Xiao and Houser (2005), who report that such rejections are reduced in a laboratory experiment in which responders are allowed to express their emotional reactions to offers at the same time that the accept or reject decision is communicated to the proposer. Thus, cheap talk comments provide a substitute for costly rejections of unfair offers. Here is a case where taking the social context out of the original ultimatum game experiments produced perplexing outcomes, and putting some of the context back in led to data that are closer to game-theoretic predictions!

Another angle on the process of "irrational" rejections can be obtained by monitoring brain activity as people play ultimatum games. Sanfey et al. (2003) used fMRI (functional magnetic resonance imaging) to monitor blood flows to various parts of the participants' brains. Unfair offers stimulated activity in brain areas that are associated with both cognition (prefrontal cortex) and emotion (anterior insula). The heightened activity in the later areas when unfair offers were rejected indicates the importance of emotions in this process. This anterior insula activity was higher for offers that were more unequal, and it was higher when the offers were perceived as coming from other people, as opposed to offers that were computer generated. The particular area of anterior insula activity has been identified in other studies of "disgust" involving negative physical sensations of taste and odor. Thus, the results of the fMRI ultimatum study offer evidence that unfair offers activate emotions similar to those resulting from a bad taste or smell.

Chapter 14 Problems

1.   Consider the conflict game in section 14.6, with pie size $10 and conflict costs of $2, and offers constrained to be integer dollar amounts. Explain what the proposer's initial demand should be with a win probability of 0.5?

2.   Consider a two-stage alternating offer game with a pie of size $3 in the first stage, which is reduced to $2 in the second and final stage. Find the equilibrium first-stage offer, and explain whether it will be accepted, under the assumption that people are perfectly selfish and rational.

3.   Consider a legislature composed of 5 voters, A, B, C, D, and E. There is a pot of money, $30, to be split, and if there is no agreement, most of this money is lost, since the earnings in the event of no agreement are $5 for A, $4 for B, $3 for C, $2 for D, and $1 for E. Voter A is exogenously selected to make a single proposed division, which will then be considered in a vote, where each person votes yes or no. If at least 3 yes votes are obtained, then the proposed split is enacted, and if not, then the default payments are received. No amendments or further motions are allowed. What should voter A propose on the assumption that all voters are selfish and perfectly rational? After reading about the ultimatum game, would you advise voter A to propose something different from the theoretical prediction? Explain.

4.   (non-mechanical) A "disadvantageous counter-offer" is said to happen in a two-stage bargaining game when a person rejects an initial offer of $X and the pie then shrinks so that that person's counter-offer provides themselves with

less money than the amount they just rejected. Give an example of this outcome, using specific numbers (make up your own game and the numbers for the initial offer and the counter offer). Would the Fehr-Schmidt utility function explained in section 14.5 (with $1 > \alpha > \beta > 0$) be capable of explaining disadvantageous counter-offers?

15

Trust, Reciprocity, and Principal-Agent Games

Although increases in the size of the market may promote productive specialization and trade, the accompanying increase in anonymity raises the need for trust in trading relationships. The trust game sets up a stylized situation where one person can decide how much of an initial stake to keep and how much to pass to the other person. All money passed is augmented, and the responder then decides how much of this augmented amount to keep and how much to pass back to the initial decision maker. The trust game experiment puts participants into this situation so that they can experience the tension between private motives and the potential risks and gains from trust, reciprocity, and cooperation.

The "gift-exchange" view of wage setting is that employers set wages above market-clearing levels in an effort to elicit high-effort responses, even though those responses are not explicitly rewarded ex post after wages have been set. The reciprocity game is one where each employer is matched with a worker who first sees the wage that is offered and then decides on the level of costly effort to supply. The issue is whether notions of trust and reciprocity may have noticeable effects in a market context. The reciprocity game can be generalized by allowing the employer to choose a contract with penalty and bonus provisions. These contract-based interactions are *principal-agent games*, in which the employer is the principal who seeks to provide the employee "agent" with appropriate incentives.

In contrast with earlier chapters in the book, the data discussion in section 15.2 refers to the statistical testing tools introduced in chapter 13. A reader who is unfamiliar with the earlier chapter, however, can see the overall picture in the figures provided here. In addition, some of the material from chapter 14 on bargaining (dictator game, double-blind procedures) is referenced.

> Note to the Instructor: All of the games considered can be run with the V*e*-conlab software, by selecting the desired experiment (Trust, Reciprocity, or Principal Agent) under the Bargaining/Fairness menu.

15.1 Trust Game Experiments

The trust game was studied by Berg, Dickhaut, and McCabe (1995) in a simplified setting designed to highlight trust and reciprocity under controlled conditions. A standard version endows both people in each pair with $10, which is common knowledge. The first mover must decide how much (if any) of this money to pass to the other person, and how much to keep. Money that is passed is tripled before it is given to the second mover, who decides how much (if any) to return to the first person. The first mover earns the amount kept initially plus any money that is returned. The second mover earns the amount that is kept in the second stage (from what was passed and/or endowed). The game is often explained as an "investment game," in order to avoid the suggestive "trust" terminology. If this game is only played once, as is generally the case, then the subgame-perfect Nash equilibrium is for the second person to keep all that is passed, and hence for the first mover to pass nothing. This analysis assumes selfish, rational behavior and correct first-mover expectations about the second-stage response. The action of passing money initially would signal that the first mover trusts the second mover to return a reasonable amount, perhaps due to a feeling of reciprocity generated by the initial action. Alternatively, the act of passing money that gets tripled could be motivated by altruistic preferences toward a more equal final payoff outcome. And since amounts returned are unknown, risk preferences could be a factor.

Berg, Dickhaut, and McCabe ran this experiment with 32 pairs of participants in a single-round interaction. Of these, almost all of the senders (30 of 32) passed a positive amount of money, and only about a third of the responders returned more than was sent (before being tripled). The average amount passed was $5.16 (out of $10), and after being tripled, the average amount returned was 18%. These behavior patterns are inconsistent with the prediction of a subgame-perfect Nash equilibrium for purely selfish players. The idea of a subgame-perfect equilibrium is discussed in chapter 9, but what it means here is that responders always maximize their own payoffs in the final stage, which rules out threats that are not "credible" to execute in the final stage.

Table 15.1 shows the amounts passed and returned for six pairs of individuals in a single-round demonstration experiment, which was run at the University of Virginia with full payment endowments of $10 for first movers (but not for responders), with the amount passed being tripled. Thus six first movers were given $60 in total. Based on the Berg et al. results, one would expect about half ($30) to be passed, which when tripled would become $90, and about 18% of that ($16) to be returned. For the six pairs shown in the table, the aggregate

Table 15.1. A Demonstration Trust Game Experiment

	ID1	ID2	ID3	ID4	ID5	ID6
Amount Passed	$0.00	$10.00	$2.00	$10.00	$1.00	$10.00
Amount Returned	$0.00	$0.00	$0.00	$0.00	$0.00	$10.00
First Mover's Earnings	$10.00	$0.00	$8.00	$0.00	$9.00	$10.00
Second Mover's Earnings	$0.00	$30.00	$6.00	$30.00	$3.00	$20.00

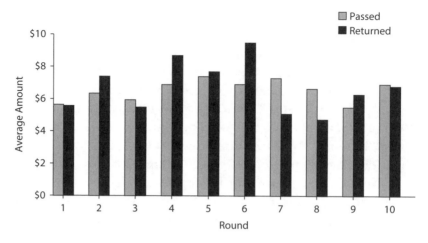

Figure 15.1. A Classroom Trust Game (University of Massachusetts)

amount passed was $33, which was tripled to $99, but only $10 was returned. The second movers in this experiment are not living up to the expectations that first movers probably had.

The results shown in table 15.1, done without responder endowments, are quite close to those of Berg et al., despite this difference. One rationale for using equal endowments for both roles is that it rules out inequity aversion as a reason for passing money.

Figure 15.1 shows the results of a classroom trust game run for multiple rounds with random matchings. This setting clearly induced a significant amount of money to be passed (light bars), on average, and average return amounts (dark) that were generally just enough to reimburse the first movers. To summarize:

Trust Game Experiment Results: *Most first movers pass positive amounts, and some pass the full endowment. Responses are variable, with some second movers returning nothing and others returning a little more than what was passed. On average, amounts returned are close to the amounts originally passed (before being tripled). So the "investment" is generally not profitable ex post.*

15.2 Motives for Passing Money in the Trust Game: The Cox Deconstruction

There has been considerable discussion of possible reasons for deviations of behavior from the Nash predictions for selfish players. Cox (2004) replicated the Berg et al. results with the tripling of amounts passed, using 32 subject pairs, with both people in each pair receiving $10 endowments in all treatments.

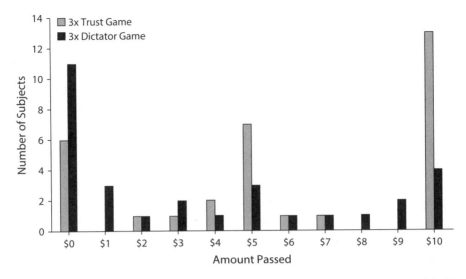

Figure 15.2. Amounts Passed in 3× Trust Game (gray) and 3× Dictator Game (dark)

The replication was done with double-blind procedures similar to those used in some of the dictator game experiments in the previous chapter. In this case, subjects used unmarked envelopes and lettered mailboxes for making decisions. With double-blind procedures, even the experimenter is unable to observe subjects' decisions, which is intended to reduce any pressure that subjects might feel to conform to norms of sharing and reciprocity.

The gray bars in figure 15.2 show the frequency counts of various amounts passed by the first movers in the trust game (gray bars). Only 6 of 32 passed nothing, and the modal amount passed was the full endowment, $10 (high gray bar on right). Some money might be passed due to noise, confusion, or distraction, but the preponderance of large amounts passed suggests that confusion is not a primary determinant of amounts passed. Another possible factor is the low "price" of transferring money to another subject, $3 is transferred for each $1 sent. Such behavior could be motivated by economic altruism (responding to the low price). Anything passed would create an earnings inequity unless more is returned, and the sizable fraction of the $10 pass amounts suggests that the motivation is not inequity aversion.

Most observers have taken the position that money is passed in anticipation of some return. One possible problem with anticipated reciprocity explanation is that the actual amounts returned are generally not larger than the amounts passed. This experiment was no exception, with an average of $5.97 passed (which got tripled) and $4.94 returned. In fact, the distribution of the 32 amounts passed is not significantly different from the 32 amounts returned (the null hypothesis of no difference cannot be rejected at standard significance levels in a two-tailed Mann-Whitney test). Of course, it could be that anticipated reciprocity was overestimated. After all, these games are played once, so there is no reason to expect that beliefs about others will not be biased.

> **Possible Motives for Passing by the First Mover in a Trust Game:** *To summarize, there are two plausible explanations for observed levels of first-stage amounts passed: (1) economic altruism that is responsive to the price, and (2) anticipated reciprocity that will provide a return on the investment.*

James Cox (2004) devised a clever procedure for deconstructing the effects of altruism and reciprocity in a trust game. A second group of paired subjects is recruited, and the first mover decides how much to pass, which is tripled as before, but the second mover cannot return anything. In other words, the second treatment is a 3× leveraged dictator game. If more is passed in the trust game than in the analogous leveraged dictator game with no possible returns, then the difference provides evidence of anticipated reciprocity. The dark bars in figure 15.2 show the frequency counts of amounts passed in the leveraged dictator game. The visual impression provided by the figure is that the dark bars are bunched on the left side (low pass amounts) and the gray bars for the trust game are bunched on the right side (high pass amounts). This impression is confirmed by a Mann-Whitney test of the null hypothesis of no difference, which is rejected at the 1% significance level. This result indicates that anticipated reciprocity is a factor in first-mover pass decisions.

Now consider the other possibility, altruism. The tendency for second movers to pass back about as much on average as what first movers initially passed is not clear evidence for altruism, since a reasonable alternative explanation is that second movers were reciprocating the favor. But the positive pass amounts in the leveraged dictator game cannot be due to anticipated reciprocity. Since responders already received an endowment of $10, anything passed in the leveraged dictator game will create inequity, so altruism seems to be a better explanation than inequity aversion.

Finally, a regression analysis of individual return decisions indicates that returns tend to be positively related to amounts sent in the trust game, also suggesting reciprocity. (You can think of reciprocity as altruism engendered by the first mover's intentional pass amounts.) This result is consistent with the return pattern light bars in the front half of figure 15.3, which shows that the return amounts were increasing in the proportion of the original endowment that was passed, when comparing small pass amounts ($0–$3), medium amounts ($4–$6), and high pass amounts ($7–$10). At this point, there is evidence for (1) anticipated reciprocity based on the higher amounts passed in the trust relative to the leveraged dictator game, and (2) altruism based on positive passed amounts in the leveraged dictator game.

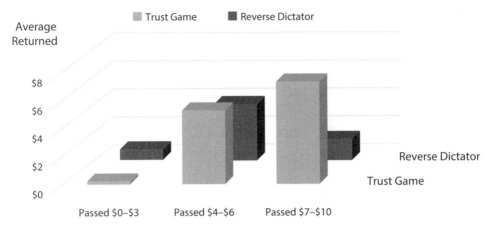

Figure 15.3. Average Amounts "Returned" in the Trust Game and Analogous Reverse Dictator Game

To further clarify these issues, there is a third treatment, in which 32 new subjects are put into the position of a second mover with a $10 base endowment that was augmented by the tripled pass amount for one of the original 32 first movers. For example, 13 of the original first movers passed $10, so there were 13 of the new subjects whose $10 endowments were augmented by 3 × $10 to $40. Each of these new subjects had to decide how much of this endowment to keep and how much to pass to the other player. The difference from second movers in the trust game is that the second movers in this third treatment did not receive their endowments as a result of the first mover's decision. The "second movers" who make a decision in this "reverse dictator" game might be motivated by altruism to pass some back to the person they are matched with, but there is no reciprocity motive since the other player did not lead off with passing money to them. If more money is passed back in the original trust game than is passed "back" in the reverse dictator treatment, this would be evidence for reciprocity. In fact, pass-backs averaged $4.97 in the trust game, which is more than twice the average of $2.06 that is passed "back" in the reverse dictator game with exogenous endowments. Cox uses a Mann-Whitney test to compare the 32 amounts passed back in the trust and reverse dictator games. The null hypothesis of no difference is rejected ($p = 0.06$, one-tailed test). Here the one-tailed test is justified by the reasonable assumption that reciprocity might matter, but even so, the result is not significant at the 5% level.

Table 15.2 lists the individual pass-back amounts for the trust and reverse dictator games. The Mann-Whitney test mentioned in the previous paragraph involved a comparison of the numbers in the middle column with those in the right column, using data ranks instead of actual pass amounts. A more sensitive test can be constructed by using actual numbers. A Wilcoxon matched-pairs test is not appropriate, since there is no way to match the choices in a given

Table 15.2. Pass-"Back" Amounts in the Trust and the Reverse Dictator Games

Second Stage Endowment	Trust Pass-Back Amounts	Dictator Pass-"Back" Amounts
$0 ($0 passed in trust game)	$0, $0, $0, $0, $0, $0	$0, $0, $0, $0, $0, $5
$6 ($2 passed in trust game)	$2	$0
$9 ($3 passed in trust game)	$0	$1
$12 ($4 passed in trust game)	$2, $5	$10, $12
$15 ($5 passed in trust game)	$0, $12, $0, $8, $10, $6, $3	$3, $0, $10, $0, $0, $1, $4
$18 ($6 passed in trust game)	$7	$0
$21 ($7 passed in trust game)	$0	$0
$30 ($10 passed in trust game)	$10, $0, $0, $0, $0, $20, $20, $9, $1, $20, $0, $17, $6	$2, $0, $0, $1, $0, $12, $0, $0, $0, $3, $0, $2, $0
	Average = $4.94	Average = $2.06

row across treatments. It is not pairs that are matched, but rather, decisions made on the basis of categories of initial amounts shown in the left column. In other words, this is a matched-clusters design instead of a matched-pairs design. Think of having each of the pass amounts as having "Trust" or "Dictator" labels. The amounts passed back should depend on the starting amounts in the left column, but under the null hypothesis of no reciprocity effect, the "Trust" or "Dictator" labels should not matter within a row. The stratified permutation test described in chapter 13 involves permuting the "Trust" and "Dictator" data labels randomly, i.e., moving the numbers *in each separate row* around randomly, while keeping half in each column. The test is based on counting the number of permutations that yield a treatment difference (in average amounts passed "back") that is larger than the difference shown in the bottom row of the table.

The number of possible permutations is quite large (there are "12 take 6" = 924 permutations in the first row alone, and permutations must be done simultaneously for all 8 rows). Therefore, the procedure used involved simulations with random permutations of the Trust and Reverse Dictator labels in each row. This was done 100,000 times, and only 1,429 out of 100,000 simulated permutations yielded a difference as great or greater than the observed treatment difference in the bottom row of the table. The ratio, 1,429/100,000, yields a ratio of $p = 0.0143$, which is the p value for a one-tailed test (32 observations per treatment). This is a lower p value than the 0.06 previously reported for the Mann-Whitney test. The takeaway is that the stratified permutation test that uses more information from the data structure provides a stronger rejection of the null hypothesis. This is evidence that reciprocity affected the pass-back decisions in the trust game. Finally, there is some pass-back even in the reverse dictator game with no possible reciprocity motive, and that provides support for the idea that there is some altruism involved. The pass "back" amounts for the reverse dictator game are shown by the dark bars in the back row of figure 15.3. Those bars do not show the increasing pattern evidenced for the trust game in the front row. To summarize, the "triadic" (three-treatment) design for the Cox deconstruction involves a standard trust game, a dictator game with the same leverage factor (tripling), and a reverse dictator game with endowments rigged to match those for second movers in the original trust game, after receiving any tripled passed amounts:

Results of Trust Game/Dictator Game Deconstruction: *First movers pass more in a 3× trust game than in a 3× dictator game, which is evidence for trust (anticipated reciprocity), since there is no role for reciprocity in the dictator game. Positive amounts passed in the leveraged dictator game seem to be due to altruism or something other than inequity aversion. Second movers pass back more in the trust game than in the analogous reverse dictator game, which is evidence for actual reciprocity, since there is nothing to reciprocate in the reverse dictator game (the "first mover" had no decision to make). The amounts passed "back" in the reverse dictator game provide evidence of altruism or some other type of other-regarding preference, since the recipients did nothing that could be a basis for reciprocation.*

Many economic interactions in market economies are repeated, and this repetition may enable trading partners to develop trust and reciprocal arrangements. For one thing, either person has the freedom to break out of a binary trading relationship in the event that the other's performance is not satisfactory. Even when a breakup is not possible, the repetition may increase levels of cooperation. For example, Cochard, Van Phu, and Willinger (2004) report results of a trust game that was run both as a single-shot interaction and as a repeated interaction for seven periods, but with payoff parameters scaled down to account for the higher number of periods. There were some other minor procedural differences that will not be discussed here. The main result is that the amounts passed and returned were higher in the repeated-interaction treatment. In the final period, which was known in advance, the amount passed stayed high, but the amount returned was very low.

Repeated interactions in market economies involve important elements of trust, and the mutual profitability of effective trust relationships suggests an evolutionary basis for this type of behavior. This evolutionary advantage of trust may be related to an apparent biological basis for trust that is reported by Kosfeld et al. (2005). They ran a standard trust game in which participants in one treatment were exposed to a nasal spray containing oxytocin, a substance that is associated with social bonding behavior. In particular, oxytocin receptors are found in brain regions that are associated with pairing, maternal care, sexual approach, and monogamous relationships in non-human mammals. The (human) subjects in the oxytocin treatment passed 17% more on average, and they passed the maximum amount at twice the rate of the control group. The return rates were not significantly different.

15.3 Pro-Sociality, Risk Aversion, and Gender

The trust game has been widely used to study differences in trusting behavior, either induced by prior experience or by gender and other demographic factors. This is an evolving literature, and some of the emerging conclusions will be discussed in this section, with references from other work deferred to the final section of this chapter.

A common practice is to use a combination of pro-sociality measures taken from one-shot games to evaluate other-regarding preferences. For example, Peysakhovich and Rand (2016) measured pro-social tendencies using a subject's decisions in a trust game, an ultimatum game, a dictator game, and a voluntary contributions game. (The voluntary contributions game will be covered in the next chapter.) Subjects made decisions in all roles of all games, with one game and role determined ex post for payment. Five of these decisions involved transferring money to another person: the gift amount in a dictator game, the proposer's offer in an ultimatum game, the amounts passed and returned in a trust game, and the contribution to the group account in the public goods game. These decisions were combined into an overall measure for each subject.

The five-component measurement of pro-sociality was motivated by the conjecture that this measure would be lower for those who previously experienced high defection rates in a prisoner's dilemma sequence. Therefore, prior to making these one-shot game decisions used to gauge pro-sociality, subjects participated in a sequence of prisoner's dilemma games with a random stopping rule, as described in chapter 11. There were two alternative environments, one designed to induce cooperation (the "C culture") with a low incentive to defect and a high continuation probability, and another designed to induce defection (the "D culture"), with a high incentive to defect and a low probability of continuation. This design difference was successful in the sense that the subjects exposed to the C culture exhibited increasing cooperation rates that leveled off above 75%, whereas subjects exposed to the D culture exhibited cooperation rates that declined and leveled off below 25%. Even though assignment to the C or D culture treatments was exogenous and random, the subsequent measures of pro-sociality were significantly higher for those exposed to the C culture. Those C culture subjects contributed more to the public good, they gave more as dictators, they offered more as ultimatum proposers, a higher proportion of them passed the endowment in the trust game, and they returned a greater proportion of what the tripled amount passed (if any). The overall pro-sociality measure was significantly related to the prior C or D culture exposure. The implication is that pro-social attitudes are, to some extent, influenced by experience.

One widely cited result of trust game experiments is the tendency for male subjects to pass more as first movers, and for female subjects to return more (as a proportion of their available cash) as second movers. This pattern was observed by Buchan, Croson, and Solnick (2008): on average, males passed $7.45 of their $10 endowments, versus $6.08 for females, and males returned 24% versus 32% for females). These results were obtained using a standard double-blind procedure. In addition, there was a treatment in which senders were given the first names of the second movers. Subsequent tests indicated that this information would enable senders to correctly identify the gender of the receiver about 95% of the time. There was no bias, in the sense that the amounts sent by either gender did not depend on the gender of the recipient. Similarly, there was no bias in terms of the proportions returned when the sender first name was provided. To summarize:

Gender Differences in Trust Games: *Male subjects tend to pass more in trust games, and female subjects return a higher percentage of funds available to them in the second round. This is generally interpreted to mean that male subjects are "more trusting," and female subjects are "more trustworthy" in this context. There does not seem to be a gender bias in the sense that the gender of the other person, if revealed by first name, has no significant effect on amounts passed or proportions returned by either gender.*

The authors provide a number of plausible reasons for the observed gender differences in trust games, but they do not stress risk. Since amounts returned cannot be known with certainty, risk aversion may matter in trust games. Evidence for the effects of risk aversion on amounts passed in trust games is mixed. Eckel and Wilson (2004) do not find a significant effect of risk aversion on amounts sent in a binary trust game (send a specified amount or not) and several measures of risk preference.

Schechter (2007) considers trust and risk with a richer structure, with five choices for how much to pass in the standard trust game. But first, subjects played an analogous risk game, with the same initial endowment as in the trust game, and with five possible amounts to bet. Amounts not bet were added to earnings. For amounts bet, a six-sided die determined the return on this bet. A roll of 1 resulted in a loss of the full amount, a roll of 2 resulted in a return of only half of the bet, 3 resulted in full recovery of the bet (but no gain), 4 resulted in earnings of 1.5 times the bet, 5 resulted in a doubling of the bet, and a roll of 6 resulted in earnings of 2.5 times the bet amount. For the trust game, done second, participants made decisions for both roles, how much to send if they ended up being a first mover, and how much to return (for each of the five sent amounts) if they ended up being a second mover.

Subjects were recruited from Paraguayan villages, with at most one person from each household. Most households surveyed sent a person to participate, and about 70% were male. Overall, 9% of the subjects bet nothing in the risk game, and 7% sent nothing in the trust game. The amounts passed in the trust game were shown to be comparable to amounts sent in a trust game in a different country, Zimbabwe (Barr, 2003).

A regression for *amount bet in the risk game* as a function of demographic measures indicated that men and people with high wealth bet more, with both of these effects being significant at the 5% level. This gender result is reasonable, since the risk game has the property that a safe earnings level can be maintained by not betting, so bets that return less can be coded as losses. As noted in chapter 3, risk elicitation tasks with a safe option tend to generate gender differences, possibly due to differential aversions to earnings that may fall below the safe amount.

A second regression was run for *amounts sent in the trust game*, again with demographic explanatory variables. Men also sent more in the trust game (significant at the 10% level), where the most salient demographic effect was that Catholics sent less (significant at the 1% level).

The third regression, also for amounts passed in the trust game, included demographics and the amount bet in the risk game. In this third regression, the two most significant effects were the Catholic and the amount bet variables (both significant at the 1% level). Once the risk aversion measure was included, both gender and wealth became insignificant as factors affecting amounts sent in the trust game. The author concluded that gender effects on amounts passed in trust games are mostly due to gender differences in risk aversion. A between-subjects design like that used by Cox would have been more convincing, due to possible sequence effects, which the author acknowledged and evaluated.

A different approach is taken by Houser, Schunk, and Winter (2010), who reach a different conclusion about the effects of risk aversion on passing in the trust game. They compare behavior in a standard trust game (their treatment T1) with that in a "risk game" in which the distribution of amounts passed back is provided to subjects (their treatment R1). The pass-back decisions were made by computer in the risk game, so this is an individual decision task. This is a between-subjects design, with each subject only making decisions in one of the treatments. The data in the trust game show more concentration of passed amounts at the lowest (pass 0) and highest (pass 10) levels, as compared with the risk game. This suggests that attitudes about trusting *actual people* are more skewed than attitudes based on "social history" risk data provided to subjects in the risk game. Risk preferences were elicited with a standard Holt-Laury menu (1× payoffs) as discussed in chapter 3. About three-fourths of the subjects made more than four safe choices (the number that would be made by a risk-neutral person), and the average was 5.86 safe choices. There was no significant gender difference in risk aversion, which is typical for this procedure, or for many procedures without a safe option. The authors divided subjects into categories: risk seeking (fewer than 4 safe choices), risk neutral (4–5 safe choices) and risk averse (6 or more safe choices). The subjects' risk aversion classifications were significantly correlated with the amount passed in the risk game, *but not in the trust game*. (Recall that the risk game is just like a trust game, but the pass-back decision is made by a computer based on an announced distribution.) The average amount passed in the risk game increased from 4.7 to 5 to 5.9 as risk classification changes from risk averse to risk neutral to risk seeking ($p < 0.05$ for the Jonckheere test that was introduced in chapter 13). This pattern is not observed in the trust game, where the measures go from 3.6 to 5.1 to 4.7 for risk averse, risk neutral, risk seeking respectively. These results suggest that amounts passed in the trust game depend on something more than risk preference (at least as measured by the Holt-Laury menu), although the measured risk categories do have some predictive power in the risk game with objective return probabilities. To summarize:

> **Gender and Risk in Trust Games:** *Even though male subjects tend to pass more money in trust games, the effects of risk aversion on pass behavior in trust games is mixed.*

15.4 A Labor Market Reciprocity Game

This game implements a setup with paired subjects, with one person setting a wage and the other choosing an effort level. The person in the employer role is free to choose a wage between specified limits. This wage is paid irrespective of the worker's subsequent effort, which must be between 0 and some upper limit. A higher level of worker effort is costly to the worker and beneficial to the employer. This is sometimes called a *reciprocity game*, since the employer may offer a high wage in the hope that the worker will reciprocate with high effort. In the experiment to be discussed, the wage was between $0 and $10, and the worker effort was required to be between 0 and 10. Each additional unit of effort reduced the worker's earnings by $0.25 and added $3.00 to the employer's earnings. Economic efficiency requires an effort of 10 in this case. Employers did not know workers' costs, and workers did not know the value of effort to the employer.

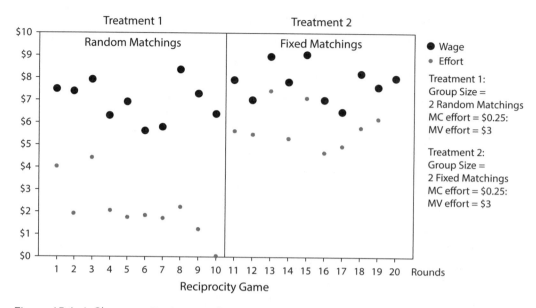

Figure 15.4. A Classroom Reciprocity Game.
Key: Average Wage (large dots) and Effort (small dots); Random Matchings (rounds 1–10) and Fixed Matchings (rounds 11–20)

The results of a classroom reciprocity game with these parameters are shown in figure 15.4. The first 10 periods were done with random matchings, which gives workers little incentive to provide acceptable efforts. The average efforts, shown by the small dots, leveled off at about 2 in rounds 3–8, followed by a sharp fall at the end. The final period efforts of 0 indicate that workers were aware that they had no incentive to be cooperative if the employers had no opportunity to reciprocate. The second treatment was done with fixed matchings, and the increased incentive to cooperate resulted in much higher effort levels (until the final period) and higher earnings. For experimental evidence of reciprocity in a related context, see Fehr, Kirchsteiger, and Riedl (1993).

15.5 Principal-Agent Games and Incentive Contracts

The reciprocity game can be given a richer structure by letting the employer propose a contract that has both a wage and some inducement for the worker to provide a specified "goal" level of effort. There are two likely directions that the employer might follow—the "stick" of stipulating a penalty if an effort goal is not achieved, and the "carrot" of stipulating an ex post bonus if the effort goal is reached. As in the reciprocity game, effort is costly for the worker and beneficial for the employer, and the worker always has the option of rejecting the contract, causing zero earnings for both. Fehr, Klein, and Schmidt (2001) ran an experiment of this type, where the ex post bonus mentioned in the contract was always purely optional, i.e., the employer could state a bonus amount if the goal effort is met and then not provide it. The penalty was a fine to be paid by the worker to the firm in the event of a low effort, but it could not always be collected, since collection required verification by a third party, which only happened with probability one-third in the experiment. The penalty was constrained to not be too large. These were one-shot games with random matching, so a selfish employer never has an incentive to keep a promise to pay a bonus, and a skeptical worker would, therefore, not expect a bonus. When employers had to choose between bonus and penalty contracts, most of the contracts proposed and accepted were bonus contracts. This structure can be implemented with the Veconlab Principal Agent Game, and it was run with 24 students in a Contract Law class at the University of Virginia (with hypothetical payments). As with the research experiment, bonus contracts were more commonly used (about three-fourths of the time). Moreover, efforts were lower for the penalty contracts.

Another option with the Veconlab program is to let the employer choose a wage, a suggested effort, and an optional ex post bonus (but no penalty). For the class experiment shown in figure 15.5, there was no bonus option in the first five rounds, and efforts were in the 2–3 range, as shown on the right axis. The introduction of a optional ex post bonus option approximately doubled efforts, even though the bonus amounts (dots between $0 and $10 in the last five rounds at the bottom on the right side) were small relative to wages (not shown).

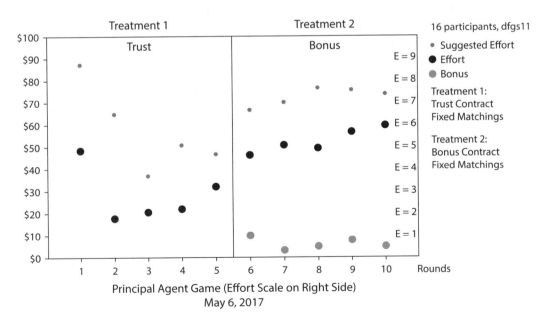

Figure 15.5. Treatment 2 Introduced the Option of an Employer Ex Post Bonus (left axis), Which Has a Clear Effect on Efforts (right axis)

A different principal-agent game is based on a model that lets the principal stipulate a contract fixed payment and a sharing rate, S, that determines what fraction of the production value is allocated to the agent. For example, S could represent a crop share rate for a farmer, and the fixed payment, if negative, is the lease payment to the landlord/principal. The agent chooses whether to accept the contract, and if so, what effort to provide. A higher share for the agent will generally result in a higher optimal effort. A sufficiently high share might mean that the agent is willing to pay the principal a fixed amount in exchange for the high share of the product value. The agent has an outside option, represented by earnings from the next best available employment. The principal needs to offer a contract that induces the agent to accept. The agent's effort decision for an accepted contract will depend on the sharing rate, the cost of effort (more effort is more costly), and the production relationship between effort and the value of the product to be shared. A low value of the outside option would make the agent more willing to accept a less favorable (lower fixed payment, lower sharing rate) contract.

Table 15.3 shows the possible effort levels from 1 to 10 on the left, the production relationship between effort and value in the center, and the effort cost on the right. If the sharing rate is $x/10$ for the agent, then the agent's optimal effort is x. For example, if the sharing rate lets the worker keep 5/10 of the value, then the worker's payoff for an effort of 5 is $100/2 - 25 = 25$, which is greater than what the worker could earn with efforts of 4 or 6 (the reader should verify this).

Another perspective on the table is in terms of marginal costs and values. The first unit of effort costs 1, the second raises the cost to 4, so the marginal cost of the second unit is 3. The marginal cost of the third unit is 5, etc. Each additional

Table 15.3. Structure of a Class Principal-Agent Effort Experiment

Agent Effort: (1=lowest; 10=highest)	Gross Return from Agent's Effort: $20×(work effort)	Costs of Effort for the Agent
1	20	1
2	40	4
3	60	9
4	80	16
5	100	25
6	120	36
7	140	49
8	160	64
9	180	81
10	200	100

unit of effort raises the total value by 20 (center column). All of the marginal costs (1, 3, 5, . . . 19) are below the marginal effort value of 20, so the optimal effort in terms of maximizing total value is 10. But the agent only gets a share S of the total value, so the agent is not necessarily interested in maximizing total value unless $S = 1$. In the absence of other factors like production or cost uncertainty, the principal specifies an optimal contract by providing the agent with the highest incentive, a sharing rate of 1 in this example, and then requiring the agent to pay a fixed fee that reduces the agent's earnings down to a level just above the value of the agent's outside option ("reservation wage"). As noted in the previous chapter, however, aggressive demands in the ultimatum game are often rejected, which suggests the dangers of pushing franchise fees down into a range where they are perceived as being unfair.

Principal-Agent Incentives: *The standard linear principal-agent model (with no uncertainty) has the property that a high sharing rate provides the agent with incentives to exert high effort and maximize the value of the product. The fixed fee in the contract is used to secure payoffs for the principal, subject to a constraint that the agent's earnings exceed the agent's best available alternative option.*

A fixed payment made to the principal is analogous to a "franchise fee" paid by a local manager in exchange for using the principal's brand and assistance with quality control, for example. Students facing this problem for the first time do not recognize the best way to incentivize the agent. Performance tends to improve with experience and role reversal, although rejections of "unfair" contract offers are common. For results of a class experiment with this type of structure, see Gächter and Königstein (2009). The purpose of a class experiment is to teach students the incentive landscape of principal-agent models and help them apply the insights to other settings, but with an awareness of behavioral and fairness considerations. For example, the principal-agent paradigm is widely used to analyze binary power relationships in political science, e.g., between a legislature and a public agency that uses funds provided by the legislature to achieve targeted outcomes.

The basic principal-agent model can be generalized in many directions that alter the nature of the optimal contract. For example, if production is uncertain, e.g., due to weather or randomness in other input costs, then an intermediate sharing rate will enable the principal and agent to share the risk. An interesting case is when the share rate is zero, $S = 0$, and the agent receives a fixed payment instead, which is *independent of effort*. The prediction of the model is that effort will be minimal unless force (or a mix of force and persuasion) is used, which is what happens on collective farms, or what happened in some of the early American colonies, as described in chapter 11. The result is an economic tragedy, e.g., mass starvation in Jamestown. A switch to private property, as happened in Jamestown after about 10 years of meager crops, generated a dramatic increase in harvests. Similarly, chapter 18 will summarize the effects of privatization in one Chinese village, where harvests increased by a *factor of seven* when the villagers signed a secret contract in 1978 to farm separate plots and keep their own harvests ($S = 1$) after making the necessary contributions to meet government quotas.

15.6 Extensions and Other Field Experiments

As the surveys in Camerer (2003) and Fehr and Schmidt (2003) indicate, experiments involving trust and reciprocity are being widely used to obtain social preference measures that are comparable across cultures and to evaluate cultural differences.

In a field setting, time and complexity issues may dictate the use of a single task to measure pro-social attitudes, and the trust game is often the one that is used. For example, Fershtman and Gneezy (2001) used the trust game to measure attitudes between subjects from two different ethnic backgrounds. The subjects were Jewish undergraduates with last names that identified them as being either from Ashkenazic or from "Eastern" backgrounds. Each person was told the last name of the person with whom they were matched, but not the person's identity. The data show a significant distrust of those from Eastern backgrounds, with less money being passed in the standard trust game. This could either be due to less concern for the earnings of those of Eastern origin, or it could be due to a belief that less will be returned. A parallel dictator game was used to distinguish these two explanations. The amounts given to the recipient in the dictator game were not affected by the recipient's ethnic origin. The authors concluded that the lower offers were driven by a fear that reciprocity would be less prevalent among those of Eastern origin.

Barr (2003) used the trust game experiment in a field study of 28 Zimbabwean villages in which transplanted farmers had to build new social relationships with strangers. Interestingly, the games themselves broke through villagers' reluctance to talk candidly and stimulated a lot of discussion, often initiated by the participants instead of the researchers. The villagers were fascinated by the metaphor of the trust game and what the results had to say about their own levels of trust and trustworthiness.

There is at least a public perception that business professionals are more selfish. Fehr and List (2004) investigated this issue by running a variant of a trust game experiment on parallel groups of students and chief executive officers (CEOs). The students were from the University of Costa Rica, and the CEOs were attending the Costa Rica Coffee Institute's annual conference in 2001. The rules of the game were the same for both subject pools, except that payoffs for the

CEOs were scaled up by a factor of 10 to equalize the importance of the incentives. The game was run as a trust game, with the proposer deciding how much to pass, and the responder deciding how much of the tripled pass amount to return. One change from the standard trust game was that the proposer had to stipulate a recommended payback amount in advance. The recommended payback was "cheap talk" in the sense that it had no effect on the way that earnings were calculated from the pass and return decisions in this treatment. The results of the experiment contradicted the authors' conjecture that CEOs would be more selfish. The CEOs passed more than students, and for a given level of transfer, they tended to return more. The authors concluded that the CEOs were more "trusting" and more "trustworthy." This result is fascinating, and it would be interesting to see it replicated with different subject pools with comparable social proximity conditions, which are difficult to assess in the Costa Rica study.

Fehr and List ran a second treatment in which the proposer had an additional option of whether to impose a fixed penalty on responders who did not return the suggested payback amount. The size of the penalty was fixed, and the penalty was paid to the experimenter out of responder earnings. The proposer in this treatment chose an amount to pass, a suggested give-back amount, and whether or not the penalty would be assessed if the responder did not meet the suggested give-back amount. Responders knew that the proposer had the option of imposing a penalty or not. Interestingly, having the option to impose a penalty and choosing *not* to impose it had a positive effect on pass-back amounts and on the proposer's earnings, both for students and for CEOs. The authors termed this a "hidden benefit" of incentives. In contrast, the imposition of the penalty option actually reduced the amount that responders tended to pass back, from 61% payback to 33% when the penalty option was imposed. This "hidden cost" of control is also documented in a parallel laboratory experiment (Falk et al., 2006).

The evidence of trusting behavior among CEOs in Costa Rica may be easier to understand in the context of a study by Karlan (2005), who used a laboratory trust game with borrowers in a non-profit village credit program in Peru. Both trusting and trustworthy behavior were generally correlated with geographic and social distance proximity. The innovative aspect of the experiment is the tie to field behavior. The experiment predicted payback behavior of the responders *more than a year later*! Responders who returned more in the trust game experiment, and hence were more "trustworthy," tended to pay back their credit union loans at a higher rate. In contrast, the savings and payback rates were lower for those who passed more in the trust game and would normally be classified as "trusting." Since payback failures are subject primarily to informal sanctions, Karlin speculates that people who pass large amounts might simply be more willing to take a risk. In Karlan's experiment, the "trustworthy" people who return more in the trust game are often involved in one-on-one loans from acquaintances, whereas those who pass more in the experiment tend to borrow bilaterally less often, which could be because they are viewed as risk takers and bad credit risks. As the author notes, the act of passing money in the trust game is itself risky, and one of the people in the proposer role remarked, "Voy a jugar." The literal translation is "I'm going to play," but on the street it means "I'm going to gamble."

Chapter 15 Problems

1. Consider a trust game in which the proposer has $10 that can be kept or passed, and what is passed is tripled. If the responder is expected to return *nothing*, then the proposer is essentially choosing a pair of money payoffs. Represent the choices by a straight "budget line" in a graph with the proposer's payoff on the horizontal axis and the responder's payoff on the vertical axis. The responder starts with $10. For consistency, put the first mover's payoff on the horizontal axis, and mark off $10 increments on each axis.

2. Suppose that the proposer for the trust game in problem 1 decides to pass $10. Now the responder gets to choose how much each person ends up earning. Represent the possible choices by a straight line in the graph for your answer to problem 1.

3. What do the indifference curves for a perfectly selfish proposer look like in the graph with the proposer's payoff on the horizontal axis?

4. An "altruistic" responder would be willing to give up some money to raise the proposer's earnings. Would the indifference curves for an altruistic responder have positive or negative slopes? Explain.

5. Think of "reciprocity" as a response to money passed by the proposer that makes the responder more altruistic, i.e., more willing to give up earnings to raise the proposer's earnings. What is the effect of reciprocity on the responder's indifference curves in a graph with proposer earnings on the horizontal axis?

6. With a sharing rate of 0.8 for the agent, show that the optimal effort for table 15.3 is 8 units of effort (by comparing agent earnings net of the fixed fee for either 7 or 9 labor units).

16

Voluntary Contributions

This chapter is based on the standard voluntary contributions game, in which the private net benefit from making a contribution is negative unless others reciprocate later or unless the person receives satisfaction from the benefit provided to others. The incentive to "free ride" on other's contributions creates a social dilemma paradigm that has a central role in social sciences. The linear voluntary contributions setup makes it possible to investigate independent variations of the private internal benefit and the public external benefit to others. These and other treatment manipulations in experiments are used to evaluate alternative explanations for observed patterns of contributions, e.g. altruism or reciprocity.

> **Note to the Instructor:** The Veconlab Voluntary Contributions game, accessed from the Public menu, can be conducted prior to class discussion. An alternative hand-run version using playing cards is easy to implement using the instructions adapted from Holt and Laury (1997), which are provided in the Instructions Appendix for this chapter at the end of the book.

16.1 Social Norms and Public Goods

The selfish caricature of *homo economicus* implies that individuals will "free ride" on the public benefits provided by others' activities. Such free riding may result in the under-provision of public goods. A pure public good, like national defense, has several key characteristics:

- It is *jointly provided* and *non-excludable* in the sense that the production required to make the good available to one person will ensure that it is available to all others in the group, i.e., access cannot be controlled.
- It is *non-rivaled* in the sense that one person's consumption of the good is not affected by another's, i.e., there is no congestion.

When a single individual provides a public good, like shoveling a sidewalk, the private provision cost may exceed the private benefit, even though the social benefit for all others' combined exceeds the provision cost incurred by that person. The resulting misallocations have been recognized since Adam Smith's (1776) discussion of the provision of street lamps.

In many cases, there is not a bright-line distinction between private and public goods. For example, parks are generally considered to be public goods, although they may become crowded and require some method of exclusion. A good that is rivaled but non-excludable is sometimes called a "common-pool resource," which is the topic of chapter 18. With common-pool resources like groundwater resources, fisheries, or public grazing grounds, the problem is typically one of how to manage the resource to prevent overuse, since individuals may not take into account the negative effects that their own usage has on others. In contrast, most public goods are not provided by nature, and the major problems often pertain to provison of the appropriate amounts of the good.

Goods are often produced by those who receive the greatest benefit, but under-provision can remain a problem as long as there are some public benefits to others that are not fully valued by the provider. Education, for example, offers clear economic advantages to the student, but the public at large also benefits from having a well-educated citizenry. A public goods problem remains when not all of the benefits are enjoyed by the provider, and this is one of the rationales for the heavy public involvement in school systems. The mere presence of public benefits does not necessarily justify public provision of such goods, given the inefficiencies and distortions due to the need to collect taxes. The political problems associated with public goods are complicated when the benefits are unequally distributed, e.g., public broadcasting of cultural materials.

In close-knit societies, public goods problems may be mitigated by the presence of social norms that dictate or reward other-regarding behavior. For example, Hawkes (1993) reports that large-game hunters in primitive societies are expected to share a kill with all households in a village, and sometimes even with those of neighboring villages. Such widespread sharing seems desirable given the difficulty of meat storage and the diminishing marginal value of excess consumption in a short period of time. These social norms transform a good that would normally be thought of as private into a good that is jointly provided and non-excludable. This transformation is desirable since the private return from large-game hunting is lower than the private return from gathering and scavenging. For example, the "!Kung" of Botswana and Namibia are a hunter-gatherer society where large prey (e.g., warthogs) are widely shared, whereas small animals and plant food are typically kept within the household. Hawkes (1993) estimated that, at one point, male large-game hunters acquired an average of 28,000 calories per day, with only about a tenth of that (2,500) going to the person's own household. In contrast, the collection of plant foods yielded an estimated return of around 5,000 calories per day, even after accounting for the extra processing required for the preparation of such food. Thus, large-game hunting by males was more productive for the village as a whole, but had a household return of about half of the level that could be obtained from unshared gathering activities. Nevertheless, many of the men continued to engage in hunting activities, perhaps due to the tendency for all to share their kills in a type of reciprocal arrangement. Hawkes questioned the reciprocity hypothesis since some men were consistently much better hunters than others, and yet all were involved in the sharing arrangements. She stressed the importance of more private, fitness-related incentives, e.g., that successful large-game hunters have more allies and better opportunities for mating.

16.2 "Economists Free Ride, Does Anyone Else?"

An early public goods experiment is reported by two sociologists, Marwell and Ames (1981), involving groups of high school students who could allocate an initial endowment between a "private exchange" and a "public exchange." Investment in the public exchange produced a net loss to the individual, even though the benefits to others were substantially above an individual's private cost. Nevertheless, the authors observed significant amounts of investment in the public exchange, with the major exception observed with a group of economics doctoral students. Their paper title began: "Economists Free Ride, Does Anyone Else?"

The Marwell and Ames paper initiated a large literature on the extent to which subjects incur private costs in activities that benefit others. A typical experiment involves giving each person an endowment of "tokens" that can be invested in a private exchange, with earnings per token that exceed the earnings per token obtained from investment in the public account. For example, each token might produce 10 cents for the investor when invested in the private account, but only 5 cents for the investor *and for each of the others* when invested in the public account. In this example, the social optimum would be to invest all tokens in the public account as long as the number of individuals in the group, N, is greater than 2, since the social benefit $5N$ would be greater than the private benefit of 10 when $N > 2$. The 10-cent private return can be thought of as the opportunity cost of investment in the public account. The ratio of the per capita benefit to the opportunity cost is sometimes called the "marginal per capita return" or MPCR, which would be $5/10 = 0.5$ in this example. A higher MPCR reduces the net cost of making a contribution to the public account. For example, if the private account returns 10 cents and the per-capita return on the public account is raised to 9 cents, there is only a 1-cent private loss associated with investment. Many of the experiments involved changes in the MPCR.

A second treatment variable of interest is the number of people involved, since a higher group size increases the total social benefit of an investment in the public exchange when the MPCR is held constant. The social benefit in the example from the previous paragraph is $5N$, which is increasing in N.

Alternatively, think about what you would do if you could give up $10 in order to return a dollar to every member of the student body at your university, including yourself. Here the MPCR is only 0.1, but the public benefit is extremely large. The motives for contribution to a public good are amplified if others are expected to reciprocate in some future period. These considerations suggest that contributions might be sensitive to factors such as group size, the MPCR, and whether or not the public goods experiment involves repetition with the same group. In multi-round experiments, individuals are given new endowments of tokens at the start of each round, and groupings can either be fixed ("partners") or randomly reconfigured ("strangers") in subsequent rounds.

The upshot is that the extent of voluntary contributions to public goods depends on a wide variety of procedural factors, although there is considerable debate about whether contributions are primarily due to kindness, to reciprocal reactions to others' kindness, or to confusion. For example, there is likely to be more confusion in a single-shot investment decision or in the first period of a repeated environment, and many people may initially divide their endowment of "tokens" equally between the two types of investment, public and private. This is analogous to the typical choice of dividing one's retirement fund contributions equally between stocks and bonds at the start of one's career. Some of the Marwell and Ames experiments involved a single decision, e.g., administered by a questionnaire that was mailed to high school students. Subsequent experiments by economists (that did *not* involve economics doctoral students) showed that repetition typically produced a declining pattern of contributions. People may stop contributing if others are observed to be free riding. One motive for making contributions may be to prevent others from behaving in this manner, in the hope that contributions will be reciprocated in subsequent rounds. This reciprocity motive is obviously weaker in the final rounds. Although contributions tend to decline, some people do contribute even in the final period.

It is instructive to begin with a summary of a one-shot experiment conducted in the field. Ensminger (2004) reports the results of a public goods experiment involving young men of the Orma, a society from Kenya that is described in chapter 14. Participants were divided into groups of four, and each person was given 50 shillings that could be kept or invested in a "group project." Investments were made by placing tokens in envelopes, which were shuffled to preserve anonymity. Then the contents were emptied and publicly counted by a member of the group before being doubled by the experimenter and divided equally among the four participants. For example, if one person contributed two shillings, these would be doubled and the resulting four shillings would be distributed, one per person. Thus a contribution of two shillings would yield a private return of only 1, so the MPCR is only 0.5.

The data for this single-round game are shown in figure 16.1. The modal contribution is four-tenths of the endowment, with a fair amount of variation. In

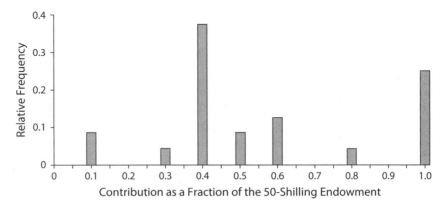

Figure 16.1. Contributions to a "Group Project" for Young Orma Males, with Group Size = 4 and an MPCR = 0.5. *Source*: Ensminger (2004).

fact, a quarter of the 24 participants contributed the entire 50 shillings. The average contribution was about 60%, which was at the high end of the 40–60% range observed in the first round of most public goods experiments done in the United States (Ledyard, 1995). The absence of complete free riding by anyone is also notable. Ensminger conjectures that contributions were enhanced by familiarity with the "Harambee" institution used to arrange for funding of public projects like schools. This practice involves the specification of income-based suggested contributions for each household, with social pressures for compliance. In fact, some of the participants commented on the similarity of the Harambee and the experimental setup. On the other hand, there is considerable free riding in this setting, since the modal decision is to contribute less than half of the endowment (and none of these participants, of course, were economics students).

MPCR Effects with Single-Round Interactions

The most salient result of voluntary contributions experiments is that contributions tend to increase with increases in the marginal per capita return. Goeree, Holt, and Laury (2002) report an experiment in which the participants had to make decisions for ten different treatments in which an endowment of 25 tokens was allocated to public and private uses. Subjects were paid $6 and were told that only one of these treatments would be selected ex post to determine additional earnings. This is a single-round experiment in the sense that there was no feedback obtained between decisions, and only one would count. A token kept was worth 5 cents in all ten cases. In one treatment, a token contributed would return 2 cents to each of the four participants. This yields an MPCR of 0.4 since each nickel foregone from the private return yields 2 cents from the public return, and 2/5 = 0.4. In another treatment, a token contributed would return 4 cents to each of the four participants, so these treatments hold the group size constant and increase the MPCR from 0.4 to 0.8. This doubling of the MPCR essentially doubled average contributions, from 4.9 to 10.6 tokens (out of the 25-token endowment). Of the 32 participants, 25 increased their contributions, 3 decreased, and 4 showed no change. The null hypothesis, that increases are just as likely as decreases, can be rejected using any standard statistical test. Since each person exhibits an increase or a decrease, the signs of the changes can be used to determine the significance level for such a test. The resulting *binomial test* is analogous to computing the chances of getting 25 or more heads from 28 flips of a fair coin. This probability is less than 0.01, so the null hypothesis of equal probabilities (fair coin) can be rejected at the 1% significance level. A more precise (Wilcoxon) test would use the signed ranks of the contribution changes instead of the signs, as noted in chapter 13.

16.3 Multi-Round Voluntary Contributions Experiments and Punishments

Economists began running multi-round public goods experiments to determine whether free-riding behavior would emerge as subjects gained experience and began to understand the incentives more clearly. The typical setup involves ten rounds with a fixed group of individuals. Figure 16.2 shows the average amounts contributed from a 10-token endowment in a classroom experiment conducted with the Veconlab default setup, which provides convenient graphs of results to motivate class discussions. Each token kept would result in $0.10 earnings, and each token contributed would increase everyone's earnings by $0.05, so the MPCR was 0.5 in this case, although this particular session was done as a class demonstration, without money payments. The 15 participants were kept in fixed groups of size 3.

The overall pattern shown here is somewhat typical of other multi-round public goods experiments; contributions tend to start in the 0.4–0.6 range, with erratic movements as individuals become frustrated and try to signal (with high contributions) or punish. Contributions generally fall in the final periods, when an attempt to signal others to contribute in the future would be useless.

A punishment procedure was implemented in round 7. This worked after each round when contributions by ID number were announced. Each person could send up to 10 punishment points to recipients designated by ID, at a cost of 25 cents to the sender per point sent. Anyone who received punishment points would lose 10% of earnings for each point received in the round. Each person makes punishment point decisions after seeing contribution results but before finding out whether any points will be received in that round. This treatment change resulted in an immediate and sustained increase in contributions. The upward trend observed in the punishment phase is typical, although the cost of the punishments tends to mitigate the resulting increase in efficiency

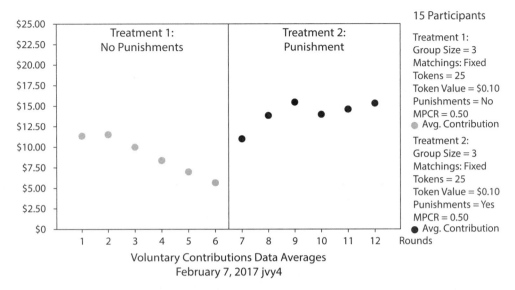

Voluntary Contributions Data Averages
February 7, 2017 jvy4

Figure 16.2. Data from a Three-Person Classroom Public Goods Game with a MPCR of 0.5

(Fehr and Gächter, 2000). It is also possible to observe counter-punishments in which a low contributor punishes others in anticipation of being punished. Punishment options are known to enhance cooperation in other contexts, e.g., management of common-pool resources like fisheries (see Ostrom, Walker, and Gardner, 1992).

MPCR and Group Size Effects in Repeated Interactions

As is the case with one-round public goods experiments, there is evidence that contributions in multi-round experiments (without punishment) respond to treatment variables that alter the benefits and costs of contributions. In a classic study, Isaac and Walker (1988b) used two different MPCR levels, 0.3 and 0.75, in a sequential design (two treatments per group), with the order being alternated in every other session. Six sessions were conducted with groups of size 4, and six sessions were conducted with groups of size 10. Group composition remained fixed for all ten rounds. The fractions of the endowment contributed are shown in table 16.1. The increase in MPCR approximately doubles contributions for both

Table 16.1. Average Fraction of Endowment Contributed for Ten Rounds

	N = 4	**N = 10**
Low MPCR (0.3)	0.18	0.26
High MPCR (0.75)	0.43	0.44

Source: Isaac and Walker (1988b).

group sizes (top to bottom comparisons in each column). The increase in group size raises contributions for the low MPCR (top row), but not for the high MPCR.

16.4   Internal- and External-Return Effects

For each of the treatments discussed in the previous section, the benefit to oneself for contributing is exactly equal to the benefit that every other person receives. In many public goods settings, however, the person making a voluntary contribution may enjoy a greater personal benefit than others. For example, benefactors give money for projects that they particularly value for some reason. It is quite common for gifts to medical research teams to be related to illnesses that are present in the donor's family. Even in Adam Smith's streetlight example, a person who erects a light over the street would pass by that spot more often, and hence would receive a greater benefit than any other randomly selected person in the town.

This difference between donor benefits and other public benefits can be examined by introducing the distinction between the "internal return" to the person making the contribution and the "external return" that is enjoyed by each of the other people in the group. One of the treatments discussed involved a return of 2 cents for oneself and for each other person when a token worth 5 cents was contributed. Thus, the internal return to oneself is $2/5 = 0.4$, and the external return to others is also $2/5 = 0.4$. Consider a treatment in which the return to oneself was raised from 2 cents to 4 cents, whereas the return to each other was held constant at 2 cents. Thus, the internal return will be increased from 0.4 to 0.8 ($=4/5$), with the external return unchanged at 0.4. This increase essentially lowers the cost of contributing regardless of the motive for contributing (generosity, confusion, etc.). This increased internal return essentially doubled the average observed contribution from 4.9 tokens to 10.7 tokens for the within-subjects single-shot games reported in Goeree, Holt, and Laury (2002). Again, the effect was highly significant, with 25 people increasing their contributions, 3 decreasing, and 4 showing no change as the internal return increased, holding the external return constant.

Next consider the effects of changing the external return (benefit to others), holding the internal return constant at 2 cents per token contributed, for an internal return of $R_I = 0.4$. An increase in the return to each of the other participants from 2 cents to 6 cents raises the external return from $R_E = 2/5 = 0.4$ to $R_E = 6/5 = 1.2$. This tripling of the external return approximately doubled the average observed contribution, from 4.9 tokens to 10.5 tokens. (Contributions increased for 23 subjects, decreased for 2, and showed no change for the other 7.) The ten treatments taken together provide a number of other opportunities to observe external-return effects. Table 16.2 shows the average number of tokens contributed with group size fixed at 4, where the external return varies from 0.4 to 0.8 to 1.2. The top row is for a low internal return of 0.4, and the bottom row is for a high internal return of 0.8. With one exception, increases in the external return (moving from left to right) result in increases in the average contribution. A similar pattern (again with one exception) is observed in table 16.3, which shows comparable averages for the treatments in which the group size was 2 instead of 4.

Finally, consider the vertical comparisons between two numbers in the same column of the same table, i.e., where the internal return is increased and the external return is held constant. The average contribution increases in all three cases. Overall, the lowest contributions are for the case of low internal and low external returns (upper-left corner of table 16.2), and the highest contributions are for the case of a high internal return and a very high external return (bottom-right corner of table 16.3).

Finally, a comparison across tables 16.2 (4 subject groups) and 16.3 (2 subject groups) affords some perspective on group size effects for the data for single-round interactions. There are four cases where group size is changed, holding the returns constant, and the increase in group size (from table 16.3 to table 16.2) raises average contributions in all cases but one.

The data in in these tables pertain to single-round public goods games. Similar effects of varying the internal and external returns are observed in multi-round public goods experiments are reported in Goeree, Holt, and Laury (2003). Subjects were paired in groups of size two, with new matchings in each round. With the internal return fixed at $R_I = 0.8$, the average number of tokens contributed (out of an endowment of 25) increased from 5 to 7.8 to 10 to 11.2 as the external return was increased from 0.4 to 0.8 to 1.2 to 2.4. Holding the external

Table 16.2. Group Size 4: Average Number of Tokens Contributed

External Return	Low External $R_E = 0.4$	Medium External $R_E = 0.8$	High External $R_E = 1.2$	Very High External $R_E = 2.4$
Low Internal Return: $R_I = 0.4$	4.9	—	10.5	—
High Internal Return: $R_I = 0.8$	10.7	10.6	14.3	—

Table 16.3. Group Size 2: Average Number of Tokens Contributed

External Return	Low External $R_E = 0.4$	Medium External $R_E = 0.8$	High External $R_E = 1.2$	Very High External $R_E = 2.4$
Low Internal Return: $R_I = 0.4$	—	—	7.7	—
High Internal Return: $R_I = 0.8$	6.7	12.4	11.7	14.5

return fixed at $R_E = 1.2$, a reduction in the internal return from 0.8 to 0.4 caused average contributions to fall from 10 to 4.4. A (random-effects) regression based on individual data was used to evaluate the significance of these and other effects. The estimated regression (with standard errors shown in parentheses) is:

$$Contribution = -1.37 - 0.37Round + 9.7R_{Internal} + 3.2R_{External} + 0.16Other_{t-1}$$
$$(3.1) \quad (0.1) \qquad (3.6) \qquad (1.2) \qquad (0.02)$$

where "$Other_{t-1}$" variable is the contribution made by the partner in the previous period. The ratios of coefficient estimates to standard errors can be used to conclude that the effects of all variables are significant at standard levels. The coefficient estimates indicate that contributions tend to decline over time, and that the internal-return effect is positive and stronger than the external-return effect. The highly significant positive effect of the previous partner's contribution suggests that attitudes toward others may be influenced by their behavior. This is consistent with the results reported by van Dijk, Sonnemans, and van Winden (2002), who measured attitudes toward others before and after a multi-round public goods experiment. Changes in attitudes were apparent and depended on the results of the public goods game in an intuitive manner.

Economic Altruism and Warm-Glow Altruism

Despite the large number of public goods experiments, there is still a lively debate about the primary motives for contribution. One approach is to model individuals' utility functions as depending on both the individual's own payoff and on payoffs received by others. Altruism, for example, can be introduced by having utility be an increasing function of both one's own payoff and the sum of others' payoffs. Anderson, Goeree, and Holt (1998) estimate that individuals in the Isaac and Walker (1988b) experiments were willing to give up at most about ten cents to give others a dollar, and similar estimates were obtained by Goeree, Holt, and Laury (2002). On the other hand, some people may not like to see others' earnings go above their own, which suggests that relative earnings matter, as discussed in chapter 14. In multi-round experiments, individuals' attitudes toward others may change in response to others' behavior, and people may be willing to contribute as long as enough others do so. Many of these alternative theories are surveyed in Ledyard (1995) and Holt and Laury (2008).

One of the motives that has been offered to explain contributions to a public good is "warm-glow altruism," which is an emotion based on the mere act of contributing. This cannot be the *only* factor at play. The increased contributions associated with the higher internal return indicate that the cost of contributions also matters to the contributor. As noted above, contributions respond positively to increases in internal return, external return, and (to some extent) group size, with the internal-return effect being strongest. Despite the individual differences and the presence of some unexplained variations in individual decisions, the overall data patterns in these experiments are roughly consistent with a model in which people tend to help others by contributing some of their tokens. To summarize:

> **Economic Altruism in Voluntary Contributions Experiments:** *Contributions are more common when (1) the private cost of contributing is reduced, (2) the benefit to each other person is increased, and (3) the number of others who benefit is increased. Even a selfish free rider may decide to contribute if it is thought that this might induce others to contribute in the future, but there is no opportunity for reciprocity in the one-round experiments in which these treatment effects are qualitatively similar. These results suggest that many individuals are not perfectly selfish free riders, and that altruistic tendencies are not exclusively a "warm-glow" feeling from the mere act of contributing, but rather, altruism is (in part) economic in the sense that it depends on both the costs to oneself and the benefits to others.* **In this sense, economic altruism depends on the price of helping others.**

16.5   Endogenous Group Formation in Social Dilemmas with Exclusion

A common type of punishment in business settings is to sever interactions and switch transactions to another firm. This section is a selective survey of a developing literature on public goods games in which the group composition can change endogenously via exit, entry, and expulsion. Some citations are included so that the interested student can read the original sources and related work, some of which was previewed in chapter 11 on social dilemmas. The summary that follows is more focused on voluntary contributions experiments.

The first paper in the series is Ehrhart and Keser (1999), who allowed people to switch to other groups in multiple-round, multiple-group voluntary contributions games. They observed "contribution chasing" by defectors, which seemed to discourage cooperators. This is an unpublished classic that stimulated much subsequent work, even though it "failed" in terms of finding a process that raises contribution levels substantially. A second failure in this sense was Coricelli, Fehr, and Fellner (2004), who implemented an auction process for accepting new group members in a voluntary contributions experiment. The trouble was that free riders tended to bid high to be included into groups of high contributors.

The next set of papers, which met with more success, implemented public goods games first, with a second game in which rewards for earlier cooperation might be expected. Barclay (2004) began with five periods of four-person voluntary contributions games, followed by a second-stage trust game (chapter 15) played with each of the others. People tended to trust and reward those who had contributed more in the public goods periods. Contributions rose when the prospect of the second game was announced in advance, which tended to generate competition to be nice, which he termed *competitive altruism*. Similarly, Sylwester and Roberts (2010) followed a repeated voluntary contributions game with a second stage in which people could indicate which of the others would be acceptable partners in a high-return two-person game. Those who tended to be more cooperative in the first stage also tended to earn more in the second stage.

The third set of papers implement various procedures for reconfiguring the groups. Gunnthorsdottir, Houser, and McCabe (2007) used observed contribution levels in the most recent round to segregate high contributors from relatively low contributors. This (unrevealed) exogenous sorting procedure increased total contributions of all groups combined, in part because the high contributors were not discouraged and were able to maintain contribution levels. Page, Putterman, and Unel (2005) achieved a similar increase in contributions using a preference rank-based sorting procedure. Each person would assign a priority rank to others, and a high rank in both directions generates a high sum of the rankings. With this "priority sum" mechanism used to reconfigure groups, contributions were approximately doubled as compared with a baseline with fixed matchings, in part because priority sum ranking tends to sort subjects by contribution tendencies.

Expulsion from the group is a severe form of punishment. Cinyabuguma, Page, and Putterman (2005) let group members view each others' histories and vote to expel particular people. Although vote-based expulsions were used sparingly, the effect was to raise contributions dramatically above a baseline with histories but no expulsion.

Much of the current work in this area is based on endogenous binary networks, e.g., pairs of people engaged in trust, prisoner's dilemma, or coordination game pairings in which some links can be broken or reconfigured after each round. Some of this work is related to evolutionary biology, e.g., Fehl, van der Post, and Semmann (2011), and is surveyed in Holt, Johnson, and Schmidtz (2017). The general finding is that more flexibility in network adjustments tends to enhance cooperation. To summarize:

> **Effects of Endogenous Groups in Social Dilemma Experiments:** *Free entry and exit does not increase contributions, due to "cooperation chasing." Exogenous sorting based on past contribution levels, and endogenous sorting based on mutual rankings, expulsions, and unilateral link breaking, can result in high contribution levels.*

16.6 Field Experiments with Charitable Giving

Selected Mail Solicitation Experiments

There is considerable interest in the study of factors that influence contributions to charities and nonprofit organizations. An analysis based only on economic theory may be insufficient, since many of the relevant issues are behavioral. For example, List and Lucking-Reiley (2002) report a controlled field experiment in which different mail solicitations for donations mentioned different amounts of "seed money" gifts, i.e., preexisting gifts that were announced in the solicitation mailing. The presence of significant seed money had a large effect on contribution levels.

In a similar vein, Falk (2007) conducted a field reciprocity experiment in which about 10,000 people in the Zurich area received a mailing from an established charitable organization, asking for donations for street children of Bangladesh. The recipients were taken from a "warm" list of possible potential donors. A third of the recipients were randomly selected to receive a small gift of a postcard with children's art, and another third received four postcards. The remaining third did not receive any gift. The letters were identical except that those with the gift stated that it could be used or given to someone else. The gifts had a surprisingly large effect on the numbers of donations, which went from 397 to 465 to 691 in the no-gift, small-gift, and large-gift treatments respectively. There was some tendency for the gifts to elicit extra small donations, but overall this effect was minor, and the size distributions of the gifts were similar. The author concluded that the inclusion of the gift was a profitable strategy for the charity in this case.

Another fund-raising technique is to offer cash matches to donors. Karlan and List (2007) sent mail solicitations to about 50,000 previous donors to a politically oriented nonprofit organization. Recipients were randomly assigned to a control (no match) and matching rates of $1:1, $2:1, and $3:1. The mere presence of a match offer (irrespective of the rate) tends to raise both the response rate and the amount donated, but there was no significant effect of the *match rate* on these measures. You can think of no match as a price of $1 per $1 donated, a 1:1 match as a price of $0.50 per $1 donated, etc. The inferred price elasticity between the no-match control and the pooled match offers was −0.30, but the price elasticity between match rate groups was essentially 0, i.e., the price reductions for higher matches had no effect, which is a perplexing result.

A cash match serves to scale up a donation, whereas a cash rebate returns a preannounced proportion of the gift to the donor, without removing any of the gift amount from what the recipient receives. Obviously, it is straightforward to design matches and rebates that are functionally equivalent. Eckel and Grossman (2008a) used a field experiment to investigate a perplexing result found in earlier lab experiments that matches have a larger effect on total donations and generate higher price elasticities than equivalent rebates. The field experiment was connected to the Minnesota Public Radio's annual fund drive, with solicitation sent to about 350,000 people, including a mix of ongoing donors, lapsed donors, and prospects. Unlike the earlier lab experiments that motivated this study, the targets of the field experiment mailing were using their own money to make donations, as compared with the "house money" (foregone earnings) used on lab experiments. Interestingly, the match price elasticity estimates for lab and field are roughly comparable. And as was the case in the lab, the match subsidies are found to be more effective than comparable rebates.

Email Solicitation Experiments

In the last decade, Internet notifications have surpassed surface ("snail") mail as a way of contacting donors. Several recent charitable giving experiments to be discussed next are based on email solicitations.

It is easy for emails to get lost in the "inbox," so it is important to stimulate potential donors to respond immediately. Castillo, Petrie, and Samek (2017) managed to do this successfully by making limited-time match offers of 80%, 100%, or 120% for donations made in *specified, narrow two-day time windows*, either two or four weeks in advance of a focal donation day known as "Giving Tuesday." Fund-raising organized around Giving Tuesday in late November of each year is a practice that originated in the United States, but has spread to more than a hundred countries. The authors partnered with nine nonprofit organizations to send out more than 39,000 emails to prospective donors. The treatments consisted of the time before Giving Tuesday (4 weeks before or 2 weeks before) and one of the match rates (80%, 100%, or 120%), with an added control group who received no match offer. All potential donors were also told that they could get a 100% match on Giving Tuesday (November 29, 2016). The early offers with time-sensitive matches resulted in a significant increase in donation rates. The rates of donations on Giving Tuesday were essentially the same for all groups and the control, so the additional donations for early offers did not displace later giving. The total dollar amounts of donations from those receiving early match offers were 80% above the amount for the control group, and the effect was largest for those receiving the email one month in advance. The match rate determines a price of giving, and the price responsiveness in this study is relatively high in absolute value (-0.86 four weeks ahead, and -1.59 for two weeks ahead). These elasticities are much larger than comparable elasticities in earlier charitable giving studies reported by Eckel and Grossman (2008a) and Karlan and List (2007).

16.7 Extensions

There have been many public goods experiments that focus on the effects of factors like culture, gender, and age on contributions in public goods games, with somewhat mixed results. For example, there is no clear effect of gender (see the survey in Ledyard, 1995). Goeree, Holt, and Laury (2002) found no gender differences on average, although men in their sample were more likely to make extremely low or high contributions.

A second strand of the literature consists of papers that examine changes in procedures, e.g., whether groups remain fixed or are randomly reconfigured each round (e.g., Croson, 1996), and whether or not the experimenter or the participants can observe individual contribution levels. Letting people talk about contributions between rounds tends to raise the level of cooperation, an effect that persists to some extent even if communication is subsequently stopped (Isaac and Walker, 1988a).

A third strand of the literature pertains to variations in the payoff structure, e.g., having the public benefits be nonlinear functions of the total contributions by group members. In the linear public goods games described in this chapter, it is optimal for a selfish person to contribute nothing, since the internal rate of return is less than one. In this case, the Nash equilibrium (with a fixed, known number of rounds) involves zero contributions by all, which is at the lower boundary of the range of possible contributions, so any noise or confusion would tend to raise contributions above the equilibrium prediction. Andreoni (1995) estimates that about half of observed contributions in linear public goods games are due to some form of "kindness" and that much of the residual may be due to confusion. This estimate was determined by running the same game, but with subject payoffs based solely on their payoff rank relative to others, which removes much of the incentive to help others, i.e., when you help some by raising their rank, you hurt others. Rank-based payoff information has a strong framing effect, which Andreoni controls for by having a treatment with rank information but regular payoffs. An alternative way to avoid the *boundary effect* is to implement a game with a Nash equilibrium somewhere in the middle of the range of possible contributions. This can be done with a non-linearity in payoff functions, so that it is actually profitable to make some contributions when contribution levels are sufficiently low. See Laury and Holt (2008) for a survey of experiments with interior Nash equilibria.

Another way to alter the payoff structure is to require that total contributions exceed a specified threshold before any benefits are derived (Bagnoli and McKee, 1991; Croson and Marks, 2000). Such *provision points* will introduce a coordination problem, since it is not optimal to contribute unless one expects that other contributions will be sufficient to reach the required threshold. Experiments with contribution thresholds will be considered in the next chapter.

All of the discussion up to this point has pertained to voluntary contributions, but many formal mechanisms have been designed to induce improved levels of provision. A critical problem is that people may not know how others value a public good, and people have an incentive to understate their values if tax shares or required efforts depend on reported values. In theory, there are some mechanisms that do provide people with an incentive to report values truthfully, and these have been tested in the laboratory. For example, see Chen and Plott (1996) or the classic survey in Ledyard (1995). In practice, most public goods decisions are either directly or indirectly made on the basis of political considerations, e.g., lobbying and "log rolling" or vote trading. In particular, provision decisions may be sensitive to the preferences of the "median voter" with preferences that split the locations of others' preferences more or less equally. Moreover, the identity of the median voter may change as people move to locations that offer the mix of public goods that suits their own personal needs. Voting experiments related to public good provision and other issues will be discussed in chapter 19.

Finally, it is worth noting that costly punishments, of the type observed in the second treatment in figure 16.2, were first used in an experiment conducted by Ostrom, Walker, and Gardner (1992). Those authors were motivated by observations of costly sanctions that were sometimes observed in small societies faced with limiting overuse of a common resource, e.g., a fishery. These types of common-pool resource problems will be considered in chapter 18.

Chapter 16 Problems

1. What was the approximate MPCR for the large-game hunters in the !Kung example described in section 16.1?

2. Suppose that Ensminger's experiment (discussed in section 16.2) had used groups of size 8 instead of size 4, with all contributions being doubled and divided equally. What would the resulting MPCR have been?

3. How would it have been possible for Ensminger to double the group size and hold the MPCR constant? How could she have increased the MPCR from 0.5 to 0.75, keeping the group size fixed at 4?

4. (non-mechanical) Ensminger wrote a small code number on the inside of each person's envelope, so that she could record which people made each contribution. The others could not see the codes, so contributions were anonymous. What if she had wanted contributions to be "double anonymous" in the sense that nobody, not even the experimenter, could observe who made which contribution? Can you think of a feasible way to implement this treatment and still be able to pay people based on the outcome of the game?

5. Suppose that Ensminger had calculated payoffs differently, by paying each of the four people in a group an amount that was one-third of the *doubled* total contributions of the *other three* participants. If contributions were 1, 2, 3, and 4 shillings for people with IDs 1, 2, 3, and 4, calculate each person's return from the group project.

6.   What would the internal and external returns be for the example in problem 5?

7.   The internal-return effects in tables 16.2 and 16.3 are indicated by the vertical comparisons in the same table. The numbers effects are indicated by the comparisons from one table to the matched cell in the other. How many numbers-effect comparisons are there, and how many are in the predicted direction (more contributions with higher group size)?

8.   The external-return comparisons in tables 16.2 and 16.3 are found by looking at averages in the same row, with higher average contributions anticipated as one moves to the right. How many external-return comparisons are there in the two tables combined, and how many are in the predicted direction? (Hint: Do not restrict consideration to averages in adjacent columns.)

9.   Which two entries in table 16.2 illustrate the MPCR effect?

10.   (non-mechanical) Explain the difference between economic and warm-glow altruism. What treatment effects might be *inconsistent* with an explanation that depends only on warm glow?

11.   (non-mechanical) Experiments with social preferences are sometimes sensitive to the way the task is framed. Give an example of a frame that you think will have an effect and explain.

17

The Volunteer's Dilemma

Sometimes it only takes a single volunteer to take an action that will benefit everyone. A dilemma arises if the per-capita value of this benefit is higher than the private cost of volunteering. Thus, each person would prefer that someone else incur this cost, but would be willing to volunteer if nobody else does. For example, each major country on the UN Security Council may prefer that a proposal by a small country be vetoed, but each would rather have another country incur the political cost of an unpopular veto. This dilemma raises interesting questions, such as whether volunteering is more or less likely in cases with large numbers of potential volunteers. A volunteer's dilemma experiment provides data that can be compared with both intuition and theoretical predictions.

Note to the Instructor: This setup is implemented by the relevant Veconlab game (on the Public menu).

17.1 Sometimes It Only Takes One Hero

After seeing the actress Teresa Saldana in the film *Raging Bull* in 1982, a crazed fan from Scotland came to Los Angeles and assaulted her as she was leaving home for an audition. Her screams attracted a group of people, but it was a man delivering bottled water who instinctively charged in and risked injury or worse by grabbing the assailant and holding him while the police and the ambulance arrived. Ms. Saldana survived her knife wounds. She has continued with her acting career and has been active with victims' support groups. This is an example of a common situation where many people may want to see a situation corrected, but all that is needed is for one person to incur a cost to correct it. Another example is a case where several politicians would like to have a pay raise, but may each prefer that someone else introduce the proposal. These are all cases where it only takes one person's costly commitment to change an outcome for all of the others. This *volunteer's dilemma* is a special type of social dilemma that is quite useful for modeling binary choices in simple games.

17.2 Initial Experimental Evidence

This dilemma was first studied by Diekmann (1985, 1986), where the focus was on the effects of increasing the number of potential volunteers. The simplest model involves N players, each facing a binary decision of whether or not to incur a cost of C and volunteer. Each receives a high payoff value of V if at least one person volunteers, and a lower payoff L if nobody in the group volunteers. Thus there are three payoff-relevant outcomes: either you volunteer and earn a certain return of $V - C$, assumed to be positive, or you do not volunteer and earn either V or L, depending on whether at least one of the $N - 1$ others volunteers.

First consider a two-person setup, with $V - C > L$, which ensures that it cannot be a Nash equilibrium for both people to refrain from volunteering. Nor can it be an equilibrium for both to volunteer, since one could save C by free riding on the other's behavior. There are also asymmetric equilibria in which one person volunteers and the other does not. With simultaneous choice, however, it would be difficult for people to coordinate on an asymmetric outcome. Finally, there is also a symmetric equilibrium in which each person volunteers with a probability p. Intuitively, one would expect that the equilibrium probability of volunteering would be lower when there are more potential volunteers, since there is less risk in relying on others' generosity in this case. This intuition is borne out in an experiment reported by Franzen (1995), with $V - L = 100$ and $C = 50$. The volunteer rates (middle column of table 17.1) decline as group size is increased from 2 to 7, and these rates level off after that point. The third column indicates that the probability of obtaining no volunteers is essentially 0 for larger groups. In other words, as individual volunteer rates level off for large groups, the probability of getting at least one volunteer goes to 1.

Table 17.1. Group Size Effects in Volunteer's Dilemma Experiment

Group Size	Individual Volunteer Rate	Rate of No-Volunteer Outcomes
2	0.65	0.12
3	0.58	0.07
5	0.43	0.06
7	0.25	0.13
9	0.35	0.02
21	0.30	0.00
51	0.20	0.00
101	0.35	0.00

Source: Franzen (1995).

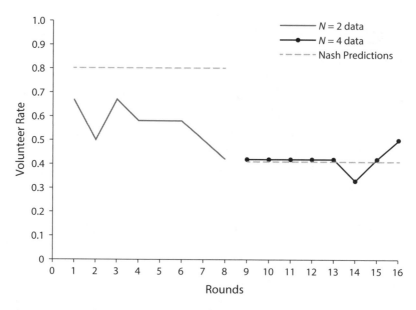

Figure 17.1. Volunteer's Dilemma with Group Size Change. *Source*: UVA, Econ 482, Fall 2001.

The Franzen experiment was done by letting subjects complete a questionnaire, and subjects received payments and results later by mail. The one-shot nature of this task raises the issue of what happens when subjects are able to learn and adjust. Figure 17.1 shows the results for a classroom experiment conducted with payoffs: $V = \$25.00$, $L = \$0.00$, and $C = \$5.00$. The setup involved 12 participants who were randomly paired in groups of size 2 for each of 8 rounds, followed by 8 rounds of random matching in groups of size 4. The round-by-round averages for the 2-person treatment are generally in the 0.5 to 0.7 range, as shown on the left side of the figure. The increase in group size reduced the volunteer rate to about 0.4. The dashed lines in figure 17.1 indicate the Nash mixed-strategy equilibrium volunteer rates that will be derived in the next section. In class, it is convenient to use a pair of treatments to make a point, e.g., that volunteer rates tend to be lower for a larger group. This figure, however, also illustrates a potential disadvantage of using a within-subjects design, since the data points track more or less continuously across the treatment change at round 9. The group size effect will be revisited in a later discussion of a research experiment with a between-subjects design.

The observation that the probability of volunteering is decreasing in group size is consistent with casual intuition and with results of social psychology experiments with "staged emergencies," e.g., Darley and Latane (1968). The observed reluctance of individuals in large groups to intervene in such emergencies has been termed "diffusion of responsibility" in this literature. Public awareness of this tendency was heightened in the 1960s by the failure of more than 30 onlookers to come to the aid of Kitty Genovese, who was raped and stabbed in the courtyard of her apartment complex in New York (although the accuracy of some newspaper accounts was later questioned). Even though each individual may have a lower chance of volunteering in a large group, the chances of getting at least one volunteer from a large group may go up, as indicated by the third column of table 17.1. The next section derives the game-theoretic predictions for these numbers effects.

17.3 The Mixed-Strategy Equilibrium

Most of the theoretical analysis of the volunteer's dilemma is focused on the symmetric Nash equilibrium with random strategies. With randomization, each person must be indifferent between volunteering and refraining, since otherwise it would be rational to choose the preferred action. In order to characterize this indifference, we need to calculate the expected payoffs for each decision and equate them. First, the expected payoff from volunteering is $V - C$ since this decision rules out the possibility of an L outcome. A decision not to volunteer has an expected payoff that depends on the others' volunteer rate, p. For simplicity, consider the case of only one other person (group size 2). In this case, the payoff for not volunteering is either V (with probability p) or L (with probability $1 - p$), so the expected payoff is:

(17.1)   Expected Payoff (not volunteer) $= pV + (1 - p)L$ (for $N = 2$).

To characterize indifference, we must equate this expected payoff with the payoff from volunteering, $V - C$, to obtain an equation:

(17.2) $V - C = pV + (1 - p)L$ (for $N = 2$),

which can be solved for the equilibrium level of p:

(17.3) $p = 1 - \left(\dfrac{C}{V-L}\right)$ (equilibrium volunteer rate for $N = 2$).

Recall that $V = \$25.00$, $L = \$0.00$, and $C = \$5.00$ for the parameters used in the first treatment in figure 17.1. Thus, the Nash prediction for this treatment is 4/5, as indicated by the horizontal dashed line with a height of 0.8. The actual volunteer rates in this experiment are not this extreme, but rather are closer to 0.6.

The formulas derived above must be modified for larger group sizes, e.g., $N = 4$ for the second treatment. The payoff for a volunteer decision is independent of what others do, so this payoff is $V - C$ regardless of group size. If one does not volunteer, however, the payoff depends on what the $N - 1$ others do. The probability that one of them does not volunteer is $1 - p$, so the probability that none of the $N - 1$ others volunteer is $(1 - p)^{N-1}$. This calculation is analogous to saying that if the probability of tails is 1/2, then the probability of observing two tails outcomes is $(1/2)^2 = 1/4$, and the probability of having N-1 flips all turn out tails is $(1/2)^{N-1}$. To summarize, when one does not volunteer, the low payoff of L is obtained when none of the others volunteer, which occurs with probability $(1-p)^{N-1}$. It follows that the high payoff V is obtained with a probability that is calculated as: $1-(1-p)^{N-1}$. These observations permit us to express the expected payoff for not volunteering as the sum of terms on the right side of the following equation:

(17.4) $$V - C = V[1 - (1-p)^{N-1}] + (1-p)^{N-1}L.$$

The left side is the payoff for volunteering, so (17.4) characterizes the indifference between the two decisions that must hold for people to be willing to randomize. It is straightforward (problem 7) to solve this equation for the probability, $1-p$, of not volunteering:

(17.5) $$1 - p = \left(\frac{C}{V-L}\right)^{\frac{1}{N-1}}$$

so the equilibrium volunteer rate is

(17.6) $$p = 1 - \left(\frac{C}{V-L}\right)^{\frac{1}{N-1}} \quad \text{(equilibrium volunteer rate)}.$$

The result in (17.6) reduces to (17.3) when $N = 2$. It is easily verified that the volunteer rate is increasing in the value V and decreasing in the value L obtained when there is no volunteer. As would be expected, the volunteer rate is also decreasing in the cost C. These claims are fairly obvious from the positions of C, V, and L in the ratio on the right side of (17.6) that is preceded by a minus sign, but they can be verified by changing the payoff parameters in the spreadsheet program for problem 8. Even though it might seem mechanical, it is useful to setting up the spreadsheet, so that changes in payoff parameters can be investigated and graphed by changing a parameter in a single cell that is referenced in the various formulas and columns of calculations.

Next, consider the effects of a change in the number of potential volunteers, N. For the setup in figure 17.1, recall that $V = \$25.00$, $L = \$0.00$, $C = \$5.00$, and therefore, the volunteer rate is 0.8 when $N = 2$. When N is increased to 4, the formula in (17.6) yields $1-(1/5)^{1/3}$, which is about 0.41 (problem 8), as shown by the dashed horizontal line on the right side of figure 17.1. This increase in N reduces the predicted volunteer rate, which is consistent with the qualitative pattern in the data (although the $N = 2$ prediction is clearly too high relative to the actual data). It is straightforward to use algebra to show that the formula for the volunteer rate in (17.6) is a decreasing function of N. This formula can also be used to calculate the Nash volunteer rates for the eight different group size treatments shown in table 17.1.

Finally, we will evaluate the chances that none of the N participants volunteer. The probability that any one person does not volunteer, $1-p$, is given in

(17.5). Since the N decisions are independent, like flips of a coin, the probability that all N participants decide not to volunteer is obtained by raising the right side of (17.5) to the power N:

$$(17.7) \quad \text{probability that nobody volunteers} = (1-p)^N = \left(\frac{C}{V-L}\right)^{\frac{N}{N-1}}.$$

The right side of (17.7) is an increasing function of N (see problem 9 for a guide to making spreadsheet calculations). As N goes to infinity, the exponent of the expression on the far right side of (17.7) goes to 1, and the probability of getting no volunteer goes to $C/(V-L)$. The resulting prediction (1/5) is inconsistent with the data from table 17.1. To summarize:

> **Deviations from Nash Equilibrium Volunteer Rates:** *The Nash equilibrium volunteer rate decreases toward 0 for larger group sizes so rapidly that the probability of finding no volunteers moves in the opposite direction, increasing with group size. In contrast, the data show that no-volunteer outcomes are almost never observed in large groups. This difference can be explained by the presence of behavioral randomness that causes occasional volunteers to be observed in any large group.*

One interesting feature of the equilibrium volunteer rate in (17.6) is that the payoffs only matter to the extent that they affect the ratio, $C/(V-L)$. Thus, a treatment change that shifts both V and L down by the same amount will not affect the difference in the denominator of this ratio. This invariance was the basis for a treatment change designed by a University of Virginia student who ran an experiment in class on this topic using the Veconlab software. The student decided to lower these parameters by \$15, from $V = \$25$ and $L = \$0$ to: $V = \$10$ and $L = -\$15$. The values of C and N were kept fixed at \$5 and 4 participants per group. Table 17.2 summarizes these payoffs for alternative setups.

All payoffs are positive in the "sure gain treatment, " but a large loss is possible in the "loss treatment." This pair of treatments was designed with the goal of detecting whether the possibility of large losses would result in a higher volunteer rate in the second treatment. Such an effect could be due to "loss aversion."

Table 17.2. Volunteer's Dilemma Payoffs for a Gain/Loss Design

Own Decision	Number of Other Volunteers	Own Payoff	Sure Gain Treatment	Loss Treatment
Volunteer	any number	$V - C$	$20.00	$5.00
Not Volunteer	None	L	$0.00	−$15.00
Not Volunteer	at least one	V	$25.00	$10.00

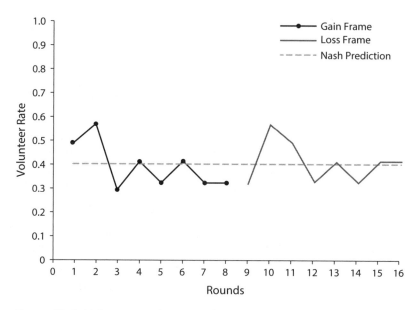

Figure 17. 2. Volunteer's Dilemma with Gain/Loss Change. *Source:* UVA, Econ 482, Fall 2001.

The results for a single classroom experiment with these treatments are summarized in figure 17.2. The observed volunteer rates closely approximate the common Nash prediction for both treatments, so loss aversion seems to have had no effect. This example was included to illustrate an invariance feature of the Nash prediction and to show how students can use theory to come up with clever experiment designs. The actual results are not definitive, however, because of the constraints imposed by the classroom setup: no replication, limited incentives, and no credible way of making a participant pay actual losses.

17.4 An Experiment on Group Size Effects

The original Franzen (1995) study was done as a single-shot game in which participants mailed in their decisions. A more appropriate test of the Nash equilibrium predictions can be done with random matchings. Goeree, Holt, and Smith (2017) ran an experiment with group sizes of 2, 3, 6, 9, and 12, with 36–48 subjects in each treatment. The subjects participated in a series of 20 volunteer's dilemma games, with random matchings and no change in group size (to avoid sequence effects). The total number of participants in each session was scaled up so that it was at least four times the group size. The parameters, $V = \$1.00$, $C = \$0.20$, and $L = \$0.20$, were such that the Nash predictions declined from 0.75 to about 0.12 as the group size increased from 2 to 12. The Nash predictions and data averages are plotted in figure 17.3. The statistical significance of this

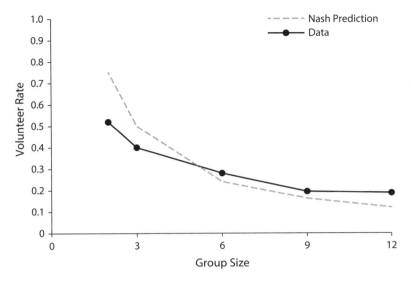

Figure 17. 3 The Effects of Group Size. *Source*: Goeree, Holt, and Smith (2017).

Table 17.3. Group Size and the Rate of No-Volunteer Outcomes

	N = 2	N = 3	N = 6	N = 9	N = 12
Predicted No-Volunteer Rate	0.06	0.125	0.19	0.21	0.22
Data Average	0.21	0.22	0.16	0.13	0.11

Source: Goeree, Holt, and Smith (2017).

numbers effect was demonstrated in chapter 13 with a Jonckheere test, using the session-by-session average volunteer rates shown in table 13.1.

The deviations from Nash predictions in figure 17.3 show a systematic pattern. The observed volunteer rate is considerably lower than the Nash prediction for the $N = 2$ treatment, as was the case with the classroom experiment in figure 17.1. In contrast, the observed rates (solid line) are higher than predicted (dashed line) for large values of N, so that the data average line is flatter than the Nash prediction line. Thus the data exhibit a "pull-to-center effect" relative to Nash predictions. An explanation for this pattern will be provided in the following section.

Next, consider the incidence of no-volunteer outcomes. When the group size is 2, the Nash volunteer rate is 0.75, and the probability of getting no volunteers is (0.25)*(0.25) = 0.06. When the group size is 3, the Nash volunteer rate is 0.5, and the probability of a no-volunteer outcome is (0.5)*(0.5)*(0.5) = 0.125, as shown in the top row of table 17.3. Note that these predicted probabilities of no-volunteer outcomes increase for larger groups, which is the *opposite* of what is observed in the data (bottom row). The higher-than-predicted volunteer rates for larger group sizes indicate that the probability of getting a no-volunteer outcome does not increase with group size. To summarize:

> **Group Size Effects in a Volunteer's Dilemma:** *With random matching and simultaneous volunteer decisions, an increase in group size reduces observed volunteer rates, as predicted in a Nash equilibrium, but the rate of no-volunteer outcomes is decreasing in group size, which contradicts Nash predictions.*

17.5 Stochastic Behavior (QRE) in the Volunteer's Dilemma

Goeree and Holt (2005b) discuss the volunteer's dilemma and other closely related games where individuals are faced with a binary decision, e.g., to enter a market or not or to vote or not. As they note, the volunteer's dilemma is a special case of a "threshold public good" in which each person has a binary decision of whether to contribute or not. The public benefit is not available unless the total number of contributors reaches a specified threshold, say M, and the volunteer's dilemma is a special case where $M = 1$. Goeree and Holt analyze the equilibria for these types of binary choice games when individual decisions are determined by probabilistic choice rules. Basically, the effect of introducing "noise" into behavior is to pull volunteer rates toward 0.5. This approach can explain why the probability of a no-volunteer outcome is *not* observed to be an increasing function of group size, as predicted in the Nash equilibrium. The intuition for why large numbers may decrease the chances of a no-volunteer outcome is that "noise" makes it more likely that at least one person in a large group will volunteer due to random causes. This intuition is inconsistent with the Nash prediction that the probability of a no-volunteer outcome is an increasing function of group size.

A simple graphical representation of the effects of stochastic choice can be used to illustrate the intuitive points made in the previous paragraph. First, recall that with 2 players, the expected payoff difference between volunteering and not is the extra earnings if nobody else volunteers, $(V - L)$ times the probability of that happening: $(1 - p)(V - L) - C$, where the final term is the cost of volunteering. Think of p as a person's belief about the other's volunteer probability, which is graphed on the vertical axis of figure 17.4. So when $p = 1$ (you think the other will volunteer for sure), the expected payoff difference is $-C$, the cost. The horizontal axis in the figure is the expected payoff difference, where negative differences are shown to the left of the vertical line in the middle. On the other hand, when $p = 0$ (the other is not expected to help at all), the expected payoff difference is $(V - L) - C$, as shown by the horizontal intercept of the dashed line in figure 17.4.

Let the expected payoff difference be denoted by $\Delta(p)$, so

(17.8) $\Delta(p) = (1 - p)(V - L) - C$ (expected payoff difference with $N = 2$)

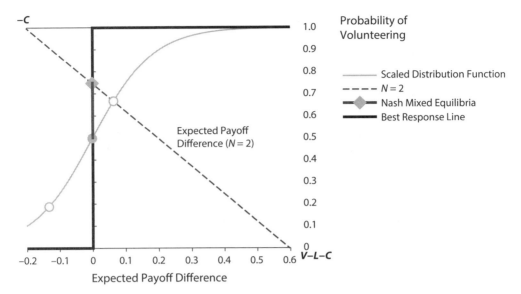

Figure 17.4. Volunteer's Dilemma Expected Payoff and Best-Response Lines when N = 2. *Notes:* The intersection of the sharp best-response line and the dashed expected payoff difference line determines the Nash equilibrium volunteer probability (large diamond at p = 0.75). The intersection of the S-shaped cumulative distribution line and the dashed expected payoff difference line determines the Quantal Response Equilibrium (circle at about p = 0.67).

This expected payoff difference is linear in p, which is why the dashed line in figure 17.4 is drawn as a straight line, with an intercept of $-C$ at the top-left side, and a horizontal intercept of $(V - L) - C$ on the right side. The Nash equilibrium in randomized strategies occurs when the player is indifferent (and therefore willing to randomize), and this equilibrium value of p is where the expected payoff difference line crosses the vertical line centered at 0.

When the expected payoff difference is positive (volunteering is better than not), which happens to the right of the vertical axis center line at 0, then the best reponse is to volunteer with probability 1 (top-right part of figure 17.4). Conversely, when the expected payoff difference is negative, the best response is to volunteer with probability 0 (lower-left part of figure). The best-response line, therefore, starts on the horizontal axis on the left, turns sharply toward the top at 0, and then continues to the right along the top with $p = 1$. This best-response line has sharp corners, since the best response is sensitive to even very small positive or negative payoff differences. With perfect rationality, it is only the *sign* of the payoff difference that matters, not the magnitude.

The S-shaped curved line, in contrast, has smoothed corners representing "better-response" tendencies that *are* affected by payoff difference magnitudes. This curved line has the typical shape of the cumulative distribution of a random variable, which starts at 0 on the left and rises to 1 on the right. The curved line represents the distribution of a random shock to the perceived

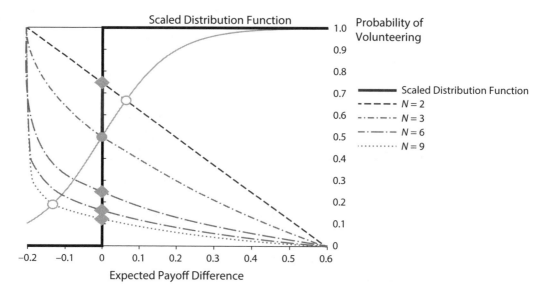

Figure 17.5. Volunteer's Dilemma with Various Group Sizes. *Notes:* The intersections of the sharp best-response line and the dashed expected payoff difference lines determine the Nash equilibria (large diamonds). The intersections of the S-shaped cumulative distribution line and the dashed expected payoff difference lines determine the Quantal Response Equilibria (circles). An increase in group size N would never push the circles below a limit of about 0.1 in the lower-left corner of the graph.

payoff difference. As the randomness goes away in the limit, the curved line would become sharp. As the randomness increases, the curved line would become flatter. The intersection of the curved line and the dashed payoff difference line represents a *quantal response equilibrium* (QRE). Notice that the quantal response equilibrium (circle in the figure) is pulled down below the Nash equilibrium at 0.75.

For the volunteer's dilemma experiment data, the observed data points are too low relative to Nash for 2 players, but are too high relative to Nash for higher group sizes. In general, a pull to center relative to the Nash prediction is observed. This observation raises the issue of how figure 17.5 would look with larger group sizes. The first step is to reconsider the expected payoff difference between the sure payoff for volunteering and the risky payoff for not volunteering. The probability that none of the $N - 1$ others volunteer is now $(1 - p)^{N-1}$, so the formula in (17.8) when $N > 2$ is modified:

(17.9) $\Delta(p) = (1 - p)^{N-1}(V - L) - C$ (expected payoff difference with N)

When $N > 2$, this equation is no longer linear in p; it shows increasing degrees of curvature as N increases to 3, 6, and 9, as shown by the curved dashed lines in figure 17.5. This increasing curvature causes the Nash equilibrium intersections with the vertical 0-expected-payoff-difference line to occur for lower values of

p as group size increases. The diamonds representing the Nash equilibria are at $p = 0.5$ for $N = 3$, and are even lower for higher values of N. This figure, therefore, illustrates the intuitive feature (noted in previous sections) that Nash equilibrium volunteer rates are decreasing in group size.

As before, the quantal response equilibria in figure 17.5 are determined by the intersections of the cumulative distribution function (curved line) and the expected payoff difference dashed lines for the various values of N. These intersections (circles) are generally pulled toward the center. There is more of a pull to center as the amount of randomness increases and the curved line becomes flatter. But as the randomness goes away and the curved line evolves toward the best-response step with sharp corners, the quantal response equilibria converge to the Nash equilibria. In other words, QRE is a generalization of the perfect-rationality Nash equilibrium.

For a fixed amount of randomness (presumably to be estimated from data), the curved line can be thought of as fixed. Then increases in group size result in equilibrium circle marks that move down and to the left. But those circle intersections will never fall below 0.1 for the randomness represented in the figure, no matter how large group size becomes. Hence, even with very large group sizes, the randomness in behavior prevents volunteer probabilities from falling all the way to 0. This has a very intuitive effect: with a large number of people, each with volunteer rates bounded away from 0, it becomes increasingly unlikely that nobody will volunteer. Thus, the chances of a no-volunteer outcome go to zero, which is consistent with the data in tables 17.1 and 17.3.

Anyone who has observed behavior in economics experiments will realize that behavior is not perfectly rational, even though people do respond to incentives in a qualitative manner. To summarize:

> **Randomness No-Volunteer Outcomes:** *With large groups, the observed deviations of increasingly rare no-volunteer outcomes from positive Nash predictions can be explained by adding randomness to individual decision making (better responses instead of best responses). This modification also explains the main qualitative feature of the observed volunteer rates, i.e., the tendency for them to be pulled toward the center relative to the Nash predictions.*

17.6   Extensions: Related Games and the Complainer's Dilemma

There have been a number of extensions of the basic model. Healy and Pate (2009) consider the effects of announced cost asymmetries. With just two players, an increase in one player's cost of volunteering should not affect that person's own volunteer rate, since otherwise the person whose cost did not change would no longer be indifferent between the two decisions. This absence of an "own-payoff effect" in a Nash equilibrium is quite unintuitive, and is not supported by data from the experiments. In particular, they report that individual volunteer rates decrease both with an increase in own cost and with a decrease in other group members' costs. These intuitive results can be explained with a quantal response equilibrium, as was the case for the intuitive violations of Nash predictions for own-payoff effects in asymmetric matching pennies games in chapter 10.

The model presented in this chapter is one of simultaneous choice, but it is sometimes more natural to introduce an ongoing time dimension. Otsubo and Rapoport (2008) examine more dynamic volunteer's dilemma games with a finite horizon and a decreasing benefit of the public good over time. Observed volunteer decisions in the experiment are generally earlier than theoretical predictions. They also report a large degree of heterogeneity across subjects in their free-riding behavior, a result that has been mentioned by others as well.

Bergstrom, Garratt, and Leo (2015) consider an interesting variation in which one person, designated as the victim, receives a benefit if at least one other person volunteers. They elicited utility payoffs from participants, using a timed mechanism to select a contributor. Subjects are motivated in part by interpersonal comparisons of volunteer decisions in this game.

The decision of whether to volunteer depends on beliefs about others' decisions. Babcock et al. (2017) consider the effects of gender sorting and find evidence that people believe that women are more likely to volunteer in mixed-gender groups. On the other hand, an ongoing analysis of the Goeree, Holt, and Smith (2017) group size experiment discussed in section 17.4 indicates that volunteer rates are clearly lower for female subjects than for male subjects for all five group size categories. The supporting statistical test was done by randomly permuting gender labels for individual volunteer rates, stratified by session. As discussed in chapter 13, this stratification essentially controls for group size effects. This result was not featured in the chapter, since gender effects in other contexts are sometimes nuanced and may not hold up in different settings (Andreoni and Vesterlund, 2001). The next step is to determine whether this surprising gender effect holds up under a wider range of parameterizations.

In some situations, it make take *more than one* volunteer to provide the public benefit. For example, if one complains about a New York taxi ride to Kennedy Airport in which the meter was not turned on, the complaint is not likely to register, unless others also complain. Greg Leo (2017b) develops the Nash equilibrium conditions for the generalized game that requires $m > 1$ volunteers, which is obtained by equating the expected payoffs of volunteering and not volunteering. The key difference from the analysis in this chapter is that volunteering is no longer a safe option if multiple volunteers are required, but otherwise the calculations are parallel. Of course, the required cutoff m might be large in some settings. For example, he notes that the Obama White House used 100 complaints as a threshold for a response.

The quantal response equilibrium used in the previous section was introduced by McKelvey and Palfrey (1995). See Goeree, Holt, and Palfrey (2016) for a unified analysis of QRE for various classes of games, with applications taken from economics and political science. There is also a chapter in that book that provides sample programs that can be used estimate noise parameters for QRE models. The particular graphical device used here is based on Goeree, Holt, and Palfrey (2017).

Chapter 17 Problems

1. Consider a volunteer's dilemma in which the payoff is 25 if at least one person volunteers, the payoff for no volunteer is 0, and the cost of volunteering is 1. Consider a symmetric situation in which the group size is 2 and each decides to volunteer with a probability p. Find the equilibrium probability.

2. How does the answer to problem 1 change if the group size is 3?

3. Suppose that the payoff is 0 if there is no volunteer, the payoff is $2 when there is at least one volunteer and the person does not volunteer, and the payoff is only $1.75 when there is at least one volunteer and the person does volunteer. The cost of volunteering has been subtracted from the value V to get the $1.75 payoff. Calculate the Nash equilibrium volunteer rate for group sizes 2 and 4.

4. For the setup in problem 3, calculate the probability of obtaining no volunteer in a Nash equilibrium, and show that this probability is 4 times as large for group size 4 as it is for group size 2.

5. In deciding whether or not to volunteer, which decision involves more risk? Do you think a group of risk-averse people would volunteer more often in a Nash (mixed-strategy) equilibrium for this game than a group of risk-neutral people? Explain your intuition.

6. The setup in problem 1 can be written as a two-by-two matrix game between a row and column player. Write the payoff table for this game, with the row payoff listed first in each cell. Explain why this volunteer's dilemma is not a prisoner's dilemma. Are there any Nash equilibria in pure (non-randomized) strategies for this game?

7. Derive equation (17.5) from (17.4), by showing all intermediate steps, and then use (17.5) to determine the Nash equilibrium predictions for each row of table 17.1.

8. To calculate group size effects for volunteer's dilemma experiments, set up a spreadsheet program. The first step is to make column headings for the variables that you will be using. Put the column headings, N, V, L, C, $C/(V-L)$, and P in cells A1, B1, C1, D1, E1, and F1 respectively. Then enter the values for these variables for the setup from the second treatment in figure 17.1: 4 in cell A2, 25 in cell B2, 0 in cell C2, and 5 in cell D2. Then enter a formula: =D2/(B2–C2) in cell E2. Finally, enter the formula for the equilibrium probability from equation (17.6) into cell F2; the Excel code for this formula is: =1 −power(E2,1/(A2–1)), which raises the ratio in cell E2 to the power $1/(N-1)$, where N is taken from cell A2. If you have done this correctly, you should obtain the equilibrium volunteer rate mentioned in the chapter, which rounds off to about 0.42. Experiment with some changes in V, L, and C to determine the effects of changes in these parameters on the volunteer rate. In order to evaluate changes in group size, enter group sizes of 6, 9, 12, and 24 in rows A3, A4, A5, and A6 respectively. The predictions for these treatments can be obtained by copying the values and formulas for cells B2–F2 downward in the spreadsheet.

9. To obtain predictions for the probability of getting no volunteers at all, add a column heading in cell G1 ("No Vol.") and enter the formula from equation (17.7) in Excel code into cell G2: = power(E2, A2/(A2–1)) and copy this formula down to the lower cells in the G column. What does this prediction converge to as the number of participants becomes large? Hint: as N goes to infinity, the exponent in equation (17.7) converges to 1.

10. Explain the volunteer's dilemma to your roommates or friends and come up with an example of such a dilemma based on some common experience from class, dorm living, a recent film, or better yet, from an economic or political setting.

11. (non-mechanical) A quantal response equilibrium for a binary choice game is a belief probability, p, for which the stochastic "better-response" probability to this belief is equal to the belief. Look at figure 17.4 for $N = 2$ and explain why $p = 0.5$ is *not* a quantal response equilibrium.

12. (non-mechanical) What feature of the "errors" represented by curved distribution function would cause this function to pass through the center point, with a height of $p = 0.5$ when the expected payoff difference is 0?

18

Externalities, Congestion, and Common-Pool Resources

Many persistent urban and environmental problems result from overuse of a shared resource. For example, an increase in fishing activity may reduce the catch per hour for all fishermen, or a commuter's decision to enter a tunnel may slow down the progress of others. Individuals tend to ignore the impact of their own activity on others' harvests or travel times. Indeed, people may not even be aware of negative effects that are small and dispersed across many other users. This externality is typically not priced in a market, and overuse can result. This chapter considers several paradigms of overuse or congestion.

The first example to be considered involves water, a Common Canal in which upstream farmers may draw irrigation away from more productive downstream users. Possible policy solutions involve use fees, auctions, and negotiations.

The second application is motivated by traffic congestion. In a Market Entry experiment, participants decide independently whether or not to enter a market (or travel on a congested road), for which the earnings per entrant are a decreasing function of the number of entrants. The earnings from the outside option of not entering are fixed. Kahneman once remarked on the tendency for payoffs to be equalized for the two decisions: "To a psychologist, it looks like magic." This impression should change for those who have participated in a classroom entry/congestion experiment in which entry equalizes travel times. Although entry rates are typically distributed around the inefficient equilibrium prediction, *entry rates are too high relative to optimal allocations since each person ignores the negative effects of their decisions on other entrants.*

The third application is a somewhat more abstract Common-Pool Resource game, in which individual efforts to secure more benefits from the resource have the effect of reducing the benefits received by others. In technical terms, the average and marginal products of each person's effort are decreasing in the total effort of all participants.

Note to the Instructors: The tragedy-of-the-common-canal experiment can be run in class using the Water Externalities program. Alternatively, the Entry/Congestion program provides for some policy options (tolls and "traffic report information") that help correct these problems. Finally, the

Common-Pool Resource program provides a more abstract framework in which each person has a range of possible extraction decisions, as compared with the binary decisions for the other two games motivated by water and traffic. All of these programs are listed on the Veconlab Public Choice menu.

18.1 The Power of a Paradigm: "Mud-pit of the Commons"

The author once attended a planning session for a major Obama-era National Science Foundation initiative tasked with redesigning the way that the broadcast spectrum frequencies are used. Participants included physicists, astronomers, electrical engineers, computer scientists, and even several economists. On the first morning, one economist from the World Bank stood up and began with the question: "Why can't the spectrum be like the sea, where people sail where they want to go, waving as they smoothly pass each other, instead of being fenced in like the land?" This magical vision resonated with several others, who one by one spoke in favor of making the spectrum like more like the sea. The conversations stopped abruptly when one person mentioned the "mud-pit of the commons" and another pointed out that local free access spectrum bands become clogged when major telecommunications companies occasionally dump overflow traffic into those bands.

The previous paragraph's mud-pit reference was to crowding in open-access grazing fields, which used to be called "commons" in the British isles. The idea of the tragedy was discussed by Hardin (1968), who attributes the example to a nineteenth-century British economist, whose insight was that each additional cow added to the increasingly muddy field tends to reduce the productivity of the rest of the herd. The next several sections describe experiments that are motivated by "common-pool" resource issues related to water, traffic, and fisheries.

18.2 Water: Tragedy of the Common Canal

Some of the most contentious allocation issues in both developing and advanced economies involve water. Consider a setup where there are upstream and downstream users, and in the absence of norms or rules, the downstream users are left with whatever is not taken upstream. This is an example of a common-pool resource, where each person may reduce the benefits obtained by others. Ostrom and Gardner (1993) describe such a situation involving farmers in Nepal. If upstream users tend to take most of the water, then their usage may be inefficiently high from a social point of view, e.g., if the downstream land is more fertile bottom land. The inefficiency is caused by upstream farmers who draw down water flows until the value of its marginal product is very low, even though water may be used more productively downstream.

For example, the Thambesi system in Nepal is one where the "headenders" have established first-priority water rights over those downstream. The farmers located at the head of each rotation unit take all of the water they need before those lower in the system. In particular, the headenders grow water-intensive rice during the pre-monsoon season, and consequently, those lower in the system cannot grow irrigated crops at this time. If all farmers were to grow a less water-intensive crop (wheat), the area under cultivation during the pre-monsoon season could be expanded dramatically, nearly tenfold. The irrigation decisions made by the headenders raise their incomes, but at a large cost in terms of lost crop value downstream. A commonly used solution to this problem is to limit usage, and such limits may be enforced by social norms or by explicit penalties. In either case, enforcement may be problematic. In areas where farmers own marketable shares of the water system, they have an incentive to sell them so that water is diverted to its highest-value uses (Yoder, 1986).

Besides overuse, there is another potential source of inefficiency if activities to increase water flow have joint benefits to both types of users. For example, irrigation canals may have to be maintained annually, and this work can be shared by all users, regardless of location. In this case, the benefit of additional work is shared by all users, so each person who contributes work will only obtain a part of the benefits, even though they incur the full cost of their own effort, unless their contribution is somehow reimbursed or otherwise rewarded. Note that the level of work on the joint production of water flow should be increased as long as the additional benefit, in terms of harvest value, is greater than the cost, in terms of lost earnings on other uses of the farmers' time. Thus, each person has an incentive to free ride on others' efforts, and this perverse incentive may be stronger for downstream users if their share of the remaining water tends to be low. To summarize:

Common-Pool Resource Inefficiencies: *There are two main sources of inefficiency, i.e., overuse and underprovision. Overuse can occur if people do not consider the negative effects that their own use decisions have on the benefit of the resource that remains for others. Underprovision can occur if the benefits of providing the resource are shared by all users, but each person incurs the full cost of their own contribution to the group effort, which may induce free riding.*

The standard economics solution to commons problems is to assign property rights that internalize the externality of appropriation. Then subsequent purchases and sales of the rights can limit appropriation to efficient levels and reallocate appropriation rights to efficient users. When property rights are difficult to enforce or politically infeasible, a set of exogenous direct regulations may improve the situation. Examples include government-enforced taxes or quotas on resource extraction. The pathbreaking work of Elinor Ostrom and her coauthors has uncovered a rich variety of home-grown institutional solutions to the commons problem—most of which avoid the need for exogenous property right assignment or heavy-handed regulation.

It is not surprising that Ostrom, a coauthor for the Nepal study and the first female to be awarded the Nobel Prize in Economics, began her career by writing a dissertation on groundwater management issues in California. She went on to study how small societies around the world solve common resource issues, with particular attention to group discussions, often supported by threats of exclusion or other punishments. Even today, however, groundwater use is not generally metered in California, and therefore is difficult to regulate, despite the fact that excessive pumping at non-sustainable levels causes the ground levels to fall at alarming rates in some locations. One solution that has been implemented, in the fine American tradition, involves a criss-crossed network of lawsuits followed by extensive negotiations, often over many years until an agreement among the landowners in a given groundwater basin is reached. Economists are especially interested in market-based alternatives to negotiation, although implementation issues can be considerable, as noted below.

The conflict between upstream and downstream water users in Nepal served as the basis for a laboratory experiment reported by Holt et al. (2012). Each session consisted of a set of 6 producers who were given roles of farmers located along a linear canal. The water supply for the canal was 12 units in each round (growing season), and each farmer had 4 fields and could use up to 4 water units. If the 3 farmers located upstream each use 4 water units, then there is no water for the 3 downstream users. Water is valuable in the sense that it triples the productivity of the field, so a field with a base value of $5 would earn $15 with irrigation. The dilemma is that the upstream users only have one high-productivity field each, with a value in the $7–$11 range randomly determined, whereas the downstream users have three high-productivity fields, as shown in table 18.1. The efficient allocation of water would be to use it only on high-productivity fields. In contrast, a first-use pattern by upstream users would reduce actual efficiency by 25%. There is an optimal use fee of $13 per unit, which would cause farmers with low-value fields to forego water use. To see this, note that the highest value draw of $6 for a low-productivity field (table 18.1) is only worth $18, when tripled, or a net gain of $12. Therefore, the $13 fee would deter low-productivity users from irrigating. Of course, regulators will not generally know what the optimal fee actually is in any given situation, and the determination of actual fees is typically a political process guided by considerations of power, historical use, etc.

Each session began with three rounds (growing seasons with renewed water stocks), with upstream users getting a chance to take the water they wish to use before any remaining amounts flow on to downstream users. As a result,

Table 18.1. Fields and Productivity Value Ranges (Tripled with Irrigation)

Field Number	Upstream Producers (IDs 1–3)	Downstream Producers (IDs 4–6)
1	$7–$11	$7–$11
2	$2–$6	$7–$11
3	$2–$6	$7–$11
4	$2–$6	$2–$6

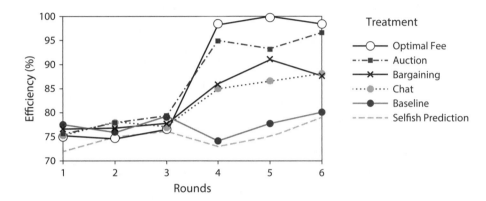

Figure 18.1. Water Use Experiment: Efficiency Results by Treatment. *Source*: Holt et al. (2012).

efficiency was low, as indicated by the lines for the left side figure 18.1. Each line in the figure is an average over three independent sessions for each treatment. The average efficiency on the left is about 75%, at about the same level as the thick gray dashed-line selfish predictions based on actual field value random draws.

The treatment effects are shown by the efficiency lines on the right side of figure 18.1. The highest efficiency line on the right is for the case in which an omniscient regulator charges the optimal $13 fee. Almost as effective is the use of an auction for water rights (dashed line with solid squares), which was implemented by letting each farmer submit 4 sealed bids, with the water going to the highest 12 bidders. Each winning bidder had to pay the highest rejected bid (of rank 13). This type of "uniform price auction" is commonly used for selling environmental permits, as will be discussed in a later chapter on multi-unit auctions. Those with high-use values tended to outbid the others in the auction, which yielded high efficiencies in the 95% range.

There were two treatments that allowed discussion in a chat room, with people identified by ID number. The "chat" treatment implemented cheap talk only, with exchanges like:

- ID1: "So how did that work out for everyone?"
- ID6: "Hahaha this game sucks."
- ID1: "I'm so sorry for you 6."
- ID6: "My plants are so thirsty."

The bargaining treatment let people make deals to sell water from upstream to downstream users, but the process was complicated by negotiation failures and the need to pay intermediate users not to take the water first. Finally, the solid gray "baseline" treatment line near the bottom of the figure shows average efficiency across sessions in which there was no treatment used, with the last three rounds being open use, just as in the first three rounds.

The experiment results illustrate the severity of the open-access dilemma, and the promise of market-based solutions, e.g., auctions, if rights can be established and use can be metered. There are, of course, many political, legal, and other implementation issues that are not present in this simplified setting. To summarize:

> **Common Canal Experiment Results:** *Overuse by upstream users tended to reduce efficiency as predicted, a situation that was largely corrected by implementing an optimal use fee or letting an auction "discover" that fee. In contrast, the chat and bilateral negotiation treatments were less effective, although they tended to reduce inefficiencies to some extent.*

18.3 Ducks and Traffic

The pressures of free entry and congestion can be illustrated with the results of a famous animal foraging experiment done with a flock of ducks at the Cambridge University Botanical Garden (Harper, 1982). The experiment was conducted by having two people go to opposite banks of the pond. Each person would simultaneously begin throwing out 5-gram bread balls at fixed intervals, every 10 seconds in one case and every 20 seconds in another. The 33 ducks quickly sorted themselves so that the expected food intake in grams per minute was equal for the groups of ducks in front of each person. A change in the time intervals resulted in a new equilibrium within about 90 seconds, in less time than it would take for most ducks at either location to obtain a single bread ball.

Even after an equilibrium had been reached in an aggregate sense, individual ducks were always in motion in a stochastic and unpredictable pattern of movements. These are the same competitive pressures that equalize urban commuting times, but here the policy issue is whether it is possible to change the system to make everyone better off. Large investments to provide fast freeways and bridges often are negated with a seemingly endless supply of eager commuters during certain peak periods. The result may be travel via the new freeways and tunnels that is no faster than the older alternative routes via a maze of surface roads.

To illustrate this problem, consider a stylized setup in which there are N commuters who must choose between a slow, reliable surface route and a potentially faster route (freeway or tunnel). However, if large numbers decide to take the faster route, the resulting congestion will cause traffic to slow to a crawl, so entry can be a risky decision. This setup was implemented in a laboratory experiment run by Anderson, Holt, and Reiley (2008). In each session, there were 12 participants who had to choose whether to enter a potentially congested activity or exit to a safe alternative. The payoffs for entrants were decreasing in the total number of entrants in a given round, as shown in table 18.2.

Think of the numbers in the table as the net monetary benefits from entry, which go down by $0.50 as each additional entrant increases the time cost associated with travel via the congested route. In contrast, the travel time associated with the alternative route (the non-entry decision) is less variable. For simplicity, the net benefit for each non-entrant is fixed at $0.50, regardless of the number who select this route. Obviously, entry is the better decision as long as the number of entrants is not higher than 8, and the equilibrium number of entrants is, therefore, 8. In this case, all 12 people earn $0.50 each, and the earnings total (net benefit) for the 12 people together is $6.00. In contrast, if only 4 enter, they each earn $2.50 and the other 8 earn $0.50, for total earnings of $14. It follows that an entry restriction will more than double the social net benefit.

It is straightforward to calculate the social benefit associated with other entry levels (problem 1), and the results are shown in table 18.3. In this setup, it is reasonable to expect that uncontrolled entry decisions would lead to approximately 8 entrants, the number that equalizes earnings from the two alternative decisions. As a result, the total benefit to all participants is inefficiently low. The reason for this is that each additional person who decides to enter lowers the net entry benefit by $0.50 for themselves *and for all other entrants*, and it is

Table 18.2. Individual Payoffs from Entry

Entrants	1	2	3	4	5	6	7	8	9	10	11	12
Payoff	$4.00	$3.50	$3.00	$2.50	$2.00	$1.50	$1.00	$0.50	$0.00	−$0.50	−$1.00	−$1.50

Table 18.3. Social Benefit for All Twelve Participants Combined

Entrants	1	2	3	4	5	6	7	8	9	10	11	12
Total Payoff	$9.50	$12	$13.50	$14	$13.50	$12	$9.50	$6	$1.50	−$4.00	−$10.50	−$18

this "external" effect that may be ignored by each entrant. For example, if there are currently 4 entrants, the addition of a fifth would reduce the earnings for the 4 existing entrants by $2 (4 × $0.50) and only increase the entrant's earnings by $1.50 (from the non-entry payoff of $0.50 to the entry payoff of $2), so total earnings decrease by $0.50. A sixth entrant would reduce total earnings by even more, $1.50 (a reduction of 5 × $0.50 that is only partly offset by the $1 increase in the entrant's earnings), since there are more people to suffer the consequences. As even more people enter, the earnings reductions caused by an additional entrant become higher, and therefore, the total payoff falls at an increasing rate. This is the intuition behind the very high costs associated with dramatic traffic delays during peak commuting periods. Any policy that pushes some of the traffic to non-peak periods will be welfare improving, since the cost of increased congestion in those periods is less than the benefit from reducing peak congestion.

In laboratory experiments, as on the open road, people are like the ducks in that they ignore the negative external effects of their actions on others' payoffs. Figure 18.2 shows the data patterns for a group of 12 participants using the payoffs from table 18.2. The average entry rate in the first 10 rounds is 0.69, which is quite close to the two-thirds (8 out of 12) rate that would occur in equilibrium. As was the case with the ducks, there is considerable variability, which in real traffic situations would be amplified by accidents and other unanticipated events.

After 10 rounds, all entrants were required to pay an entry fee of $2, which reduces all numbers in the bottom row of table 18.2 by $2. In this case, the

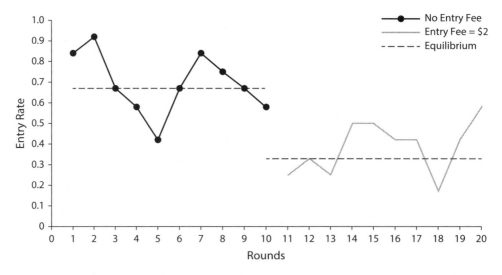

Figure 18.2. Entry Game with Imposition of Entry Fee. *Source:* Anderson, Holt, and Reiley (2008).

equilibrium number of entrants is 4, i.e., an entry rate of 0.33, as shown by the dashed line on the right side of figure 18.2. The average entry rate for these final 10 rounds was quite close to this level, at 0.38.

Even though the fee reduces entry to (nearly) socially optimal levels, the individuals do not enjoy any benefit. To see this, ignore the variability and assume that the fee reduces entry to the 0.33 rate. In equilibrium, entrants earn $2.50 minus the $2 fee, and non-entrants earn $0.50, so everyone earns exactly the same amount, $0.50, as if they did not enter at all. Where did the increase in social benefit go? The answer is in the collected fees of $8 (4 × $2). This total fee revenue of $8 is available for other uses, e.g., tax reductions and road construction. So if the commuters in this model are going to benefit from the reduction in congestion, then they must get a share of the benefits associated with the fee collections.

A second series of sessions was conducted using a redistribution of fee revenues, with each person receiving 1/12 of collections. After 10 rounds with no fee, participants were allowed to vote on the level of the fee for rounds 11–20, and then for each 10-round interval after that. The voting was carried out by majority rule after a group discussion, with ties being decided by the chair, who was selected at random from the 12 participants. In one session, the participants first voted on a fee of $1, which increased their total earnings, and hence was selected again after round 20 to hold in rounds 21–30. After round 30, they achieved a further reduction in entry and increase in total earnings by voting to raise the fee to $2 for the final 10 rounds. In a second session with voting, the group reached the optimal fee of $2 in round 20, but the chair of the meeting kept arguing for a lower fee, and these arguments resulted in reductions to $1.75, and then to $1.50, before it was raised to a near-optimal level of $1.75 for the final 10 rounds. Interestingly, the discussions in these meetings did not focus on maximizing total earnings. Instead, many of the arguments were based on making the situation better for entrants or for non-entrants (different people had different points of view, based on what they tended to do). In each of these two sessions, the imposition of the entry fees resulted in an approximate doubling of earnings by the final 10 rounds. Even though the average payoffs for entry (paying the fee) and exit are each equal to $0.50 in equilibrium, for any fee, the fee of $2 maximized the amount collected, and hence maximized the total earnings when these include a share of the collections.

Regardless of average entry rate, there is considerable variation from round to round, as shown in figure 18.2 and in the data graphs for the other sessions where fees were imposed, either by the experimenter or as the result of a vote. This variability is detrimental at any entry rate above one-third. For simplicity, suppose that there is no fee, so that entry would tend to bounce up and down around the equilibrium level of 8. It can be seen from table 18.3 that an increase in entry from 8 to 9 reduces total earnings (including the shared fee) from $6 to

$1.50, whereas a reduction in entry from 8 to 7 only raises earnings from $6 to $9.50. In general, a bounce up in entry reduces earnings by more than a bounce down by an equal amount, so symmetric variation around the equilibrium level of 8 would reduce total earnings below the amount, $6, that could be expected in equilibrium with no variation. The intuition behind this asymmetry is that additional entry causes more harm with larger numbers of entrants.

Everyone who commutes by car knows that travel times can be variable, especially on routes that can become very congested. In this case, anything that can reduce variability will tend to be an improvement in terms of average benefits. In addition, most people dislike variability, and the resulting risk aversion, which has been ignored up to now, would make a reduction in variability more desirable. Sometimes variability can be reduced by better information. In many urban areas, there is a news-oriented radio station that issues traffic reports every 10 minutes. This information can help commuters avoid entry into already overcrowded roads and freeways. A stylized type of traffic information can be implemented in an experiment by letting each person find out the number of others who have already entered at any time up to the time that they decide to enter. In this case, the order of entry is determined by the participants themselves, with some people making a quick decision, and others waiting. The effect of this kind of information is to bring about a dramatic (but not total) reduction in variability, as can be seen for the experiment shown in figure 18.3.

In a richer model, some commuters would have higher time values than others, and hence would place a higher value on the reduction in commuting time associated with reduced congestion. The fee would allow commuters to sort themselves out in the value-of-time dimension, with high-value commuters being willing to pay the fee. The resulting social benefit would be further enhanced by the effects of using price (the entry fee) to allocate the good (entry)

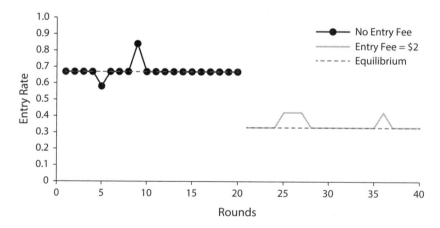

Figure 18.3. Entry Game with Information about the Number of Prior Entrants. *Source*: Anderson, Holt, and Reiley (2008).

to those who value it the most. This observation is similar to a result reported in Plott (1983), where the resale of licenses to trade, from those with low values to those with high values, raised both total earnings and market efficiency.

Traffic Experiment Results: *Excess entry into a congested route tends to cause a large reduction in total earnings, due to negative effects of entry on others' earnings, which are ignored by individual decision makers. This situation can be alleviated with an entry fee (toll), and voting on the fee level can provide dramatic improvements when the revenues are shared among voters.*

18.4 Fish

Each person in the market entry game has two decisions, enter or not. In contrast, most common-pool resource situations are characterized by a range of possible decisions, corresponding to the intensity of use. For example, a lobster fisherman not only decides whether to fish on any given day, but also (if permitted by regulations and weather) how many days a year to go out and how many hours to stay out each day. Let the fishing effort for each person be denoted by x_i, where the i subscript indicates the particular person. The total effort of all N fishermen is denoted by X, which is just the sum of the individual x_i for $i = 1 .... N$. For simplicity, assume that no person is more skilled than any other, so it is natural to assume that a person's fraction of the total harvest is equal to their fraction of the total effort, x_i/X.

In the experiment to be discussed, each person selected an effort level, and the resulting total harvest revenue, Y, was a quadratic function of the total effort:

$$(18.1) \qquad\qquad Y = AX - BX^2,$$

where $A > 0$, $B > 0$, and $X < A/B$ (to ensure that the harvest is not negative). This function has the property that the average revenue product, Y/X, is a decreasing function of the total effort. This average revenue product is calculated as the total revenue from the harvest, divided by the total effort: $(AX - BX^2)/X = A - BX$, which is decreasing in X since B is assumed to be positive. The fact that average revenue is decreasing in X means that an increase in one person's effort will reduce the average revenue product of effort for everyone else. This is the negative externality for this common-pool resource problem.

Effort is costly, and this cost is assumed to be proportional to effort, so the cost for individual i would be Cx_i, where $C > 0$ represents the opportunity cost of the fisherman's time. Assuming that the harvest in equation (18.1) is measured in dollar value, then the earnings for person i would be their share, x_i/X, of the total harvest in equation (18.1) minus the cost of Cx_i that is associated with their effort:

(18.2) $\qquad$ earnings $= (AX - BX^2)\frac{x_i}{X} - Cx_i = (A - BX)x_i - Cx_i,$

where the right side of equation (18.2) is the average product of effort times the individual's effort, minus the cost of the person's effort decision. Note that this earnings expression on the right side of equation (18.2) is analogous to that of a firm in a Cournot market, facing a demand (average revenue) curve of $A - BX$ and a constant cost of C per unit.

A rational, but selfish, person in this context would take into account the fact that an increase in their own effort reduces their own average revenue product, but they would not consider the negative effect of an increase in effort on the others' average revenue products. This is the intuitive reason that unregulated exploitation of the resource would result in too much production relative to what is best for the group in the aggregate (just as uncoordinated production choices by firms in a market result in a total output that is too high relative to the level that would maximize joint profits).

In particular, the Nash/Cournot equilibrium for this setup would involve each person choosing the activity level, x_i, that maximizes the earnings in equation (18.2) taking the others' activities as given. Suppose that there are a total of N participants, and that the $N - 1$ others choose a common level of y each. Thus the total of the others' decisions is $(N - 1)y$, and the total usage is: $X = x_i + (N - 1)y$. With this substitution, the earnings in equation (18.2) can be expressed:

(18.3) $\qquad$
$$
\begin{aligned}
\text{earnings} &= [A - B(N - 1)y - Bx_i]x_i - Cx_i \\
&= Ax_i - B(N - 1)yx_i - Bx_i^2 - Cx_i
\end{aligned}
$$

which can be maximized by equating marginal revenue with marginal cost C. The marginal revenue is just the derivative of total revenue, and the marginal cost is the derivative of total cost, Cx_i. One way to find the best level of x_i is to note that the term in square brackets at the top of (18.3) is like a demand function with slope of $-B$, so the corresponding marginal function would have a slope of $-2B$, and equating this marginal revenue with marginal cost yields (18.4). An equivalent derivation is to take the derivative of the earnings formula in (18.3) with respect to x_i and then set it equal to zero:

(18.4) $\qquad$ $A - B(N - 1)y - 2Bx_i - C = 0.$

This equation can be used to determine a common decision, x^*, by replacing both y and x_i with x^* and solving to obtain:

(18.5) $\qquad$
$$
x^* = \frac{A - C}{(N + 1)B}.
$$

The nature of this common-pool resource problem can be illustrated in terms of the specific setup used in a research experiment run by Guzik (2004), who at the time was an undergraduate at Middlebury College. He ran experiments with groups of 5 participants with $A = 25$, $B = 1/2$, and $C = 1$. The cost was implemented by giving each person 12 tokens in a round. These could either be kept, with earnings of 1 each, or used to extract the common-pool resource. Thus the opportunity cost, C, is 1 for each unit increase in x_i. With these parameters and $N = 5$, equation (18.5) yields a common equilibrium decision of $x = 8$.

The default setting for the Veconlab software presents the setup with a mild environmental interpretation to be used for teaching, i.e., in terms of a harvest from a fishery. Adjustments to the software and the use of supplemental instructions permitted Guzik to present the setup using either environmental terminology, workplace terminology, or a neutral ("magic marbles") terminology. In each of these three treatments (environmental, workplace, and marbles) there were 8 sessions, each with 5 participants. Each session was run for 10 rounds with fixed matchings. The purpose of the experiment was to see if the mode of presentation or "frame" would affect the degree to which people overused the common resource. There were no clear framing effects: the average decision was essentially the same for each treatment: 7.81 (environmental), 8.04 (marbles), and 7.8 (workplace). These small differences were not statistically significant, although there was a significant increase in variance for the non-neutral frames (workplace and environmental). The author was able to detect some small framing effects after controlling for demographic variables, but the main implication for experimental work is that if you are going to look at economic issues, then the frame may not matter much in these experiments, although one should still hold the frame constant and change the economic variables of interest.

The results reported by Guzik (2004) are fairly typical of those for common-pool resource experiments; the average decisions are usually close to the Nash prediction, although there may be considerable variation across people and from one round to the next. However, if participants are allowed to communicate freely prior to making their decisions, then the decisions are reduced toward socially optimal levels (Ostrom and Walker, 1991).

18.5 Tragedy of the Common Field (Jamestown and Xiaogang)

The auctions used for the Common Canal experiment essentially establish property rights for water, which then provide incentives for efficient use and maintenance. While the incentive effects of property rights may seem obvious, it is useful to explore what happens when property was not privately owned. The Jamestown colony in North America was organized by a joint stock company that established a farming collective in which the settlers did not own the land and were forced to turn over crop harvests to the managers. Each settler was guaranteed an equal share of the harvest, irrespective of work effort.

The disastrous consequences of this arrangement are suggested by the fact that about two-thirds of the first wave of settlers died of starvation and disease in the first year, and the following winter of 1609 was even worse for the mix of survivors and new arrivals. A new governor in 1611 found starving individuals bowling in the streets, waiting for someone else to plant the crops. Apparently, considerable energy was expended by settlers individually hunting small game at night, which supplied most of their food intake. When settlers were given 3-acre plots in 1614, the harvest reportedly increased dramatically, and all land was divided into private parcels by 1619. The effects of ownership on incentives were later described by the colony's most famous resident, Captain John Smith:

> When our people were fed out of the common store, and labored jointly together, glad was he who could slip from his labour, or slumber over his taske he care not how, nay, the most honest among them would hardly take so much true paines in a weeke, as now for themselves they woll doe in a day. (Ellickson, 1993, p. 1337)

A more recent example of the incentive failures with collective ownership is provided by the story of a secret contract drawn up in 1978 in Xiaogang Village, China. Prior to signing this agreement, each person in the village had to go work a common field area, starting at the time of the morning whistle and ending with the evening whistle. Harvests were transferred to government officials, who returned some food to the village. Food distribution was equalized instead of being based on work effort, and the amounts available were never enough. Villagers, who faced starvation, report having to go to neighboring villages to beg for food, as described in the National Public Radio Planet Money podcast entitled "The Secret document that transformed China."

The 18 signatories to the secret Xiaogang contract agreed to divide the common land into separate parcels, one for each family. They would then turn over required amounts to government officials, keeping the difference. They promised not to complain if they ended up being detained with their heads shaved. The agreement even mentioned that the non-signatories in the village had promised to take care of the children if the signatories were later jailed. The effects of the secret agreement were difficult to conceal when neighbors noticed people working after the whistle, and working hard. Moreover, the harvest increased *sevenfold* over earlier levels, a development that was impossible to hide! The parallels with Jamestown 400 years earlier are striking. Chinese government officials responded with threats, but some higher officials intervened. This experiment was extended to other farming areas, and later to manufacturing, with similarly dramatic increases in productivity.

This situation illustrates the "underprovision inefficiency" for a common-pool resource, as described in the summary in section 18.2. To an experimental economist, the "equal share" aspect of group production in a collective is akin to a very small MPCR of $1/N$ in a large-N "voluntary contributions" game. The catch is that individual differences in productivity make it harder to coordinate effort to sustain group production. Moreover, each person must decide whether to focus their own best efforts on group production or on "outside" private activities, like hunting at night and dozing during working hours. The student should consider how playing cards could be used to set up a "hand-run" class experiment that illustrates these incentive problems (problem 5 and hint). The card values could correspond to different individual productivities, and each participant would have to decide how to allocate effort between common and private production.

The lessons from this section can be summarized in terms of the effects of well-defined property rights in comparison with the absence of such rights:

> **Underprovision in Collective Production:** *In a community with forced harvest sharing, irrespective of effort, individuals have little incentive to exert much effort, the private returns of which are minimal in a large group (e.g., divided by the group size). The resulting free-riding incentives resulted in starvation and severe hardship, both in Jamestown and Xiaogang Village four centuries later.*

18.6 Extensions

The first common-pool resource experiments are reported in Gardner, Ostrom, and Walker (1990), who evaluated the effects of subject experience, and in Walker, Gardner, and Ostrom (1990), who considered the effects of changing the endowment of tokens. See Ostrom, Gardner, and Walker (1994) and the special section of the Fall 1993 *Journal of Economic Perspectives* devoted to "Management of the Local Commons," which has an interesting discussion of many applications.

For results of a common-pool resource field experiment, see Cardenas, Stranlund, and Willis (2000). The experiment was done in rural villages in Colombia. The main finding is that the application of externally imposed rules and regulations that are imperfectly monitored tends to increase the effect of individualistic motives and Nash-like results. They conclude, "modestly enforced government controls of local environmental quality and natural resource use may perform poorly, especially as compared to informal local management" (p. 1721). In other words, the imperfectly enforced rules seemed to backfire by reducing the tacit cooperation that might otherwise help solve the commons problem. Regarding the self-governed institutions, Cardenas (2003) and Cardenas et al. (2002) report on a similar set of experiments in the field aimed at exploring the potential of non-binding face-to-face communication, which was effective in previous laboratory experiments with student subjects. Their findings suggest that actual inequalities observed in the field (regarding the so-cial status and the wealth distances within and across groups) limited the ef-fectiveness of communication. Groups of wealthier and more heterogeneous villagers found it more difficult to establish cooperation via communication. Although poorer participants had more experience with actual commons di-lemmas, the wealthier villagers' incomes were more dependent on their own assets, and they seemed to have had less frequent interactions in the past in these kinds of social exchange situations.

Chapter 18 Problems

1. Calculate the social benefits associated with 4, 6, 8, 10, and 12 entrants, using the individual payoffs in table 18.2. Please show your work.

2. Use the numbers in table 18.3 to explain why unpredictability in the number of entrants is bad, e.g., why a "50–50" mix of either 6 entrants or 10 entrants is not as good as the expected value (8 entrants for sure) in terms of expected earnings.

3. Explain in your own words why each increase in the number of entrants above 4 results in successively greater decreases in the social benefit in table 18.3.

4. What would the equilibrium entry rate be for the game described in section 18.3 if the entry fee were raised to $3 per entrant? How would the total amount collected in fees change as a result of the fee *increase* from $2 to $3? Alterna-tively, explain how the total amount in collected fees would change as the result of a fee *decrease* from $2 to $1.

5. (non-mechanical) The student should think about how a deck of cards could be used in a class experiment to illustrate underprovision incentives when the collective harvest is shared equally, as discussed at the end of section 18.5. In particular, suppose that each student is dealt two playing cards, one with a high number and one with a low number, and the student has to use one of these in joint production, where the production is shared equally, and to use the other one in private production (e.g., effort spent hunting at night in Jamestown). Your task is to write a one-page set of instructions for such a class experiment, with some side discussion of what card numbers to use and how productivities of effort levels for public and private production are determined.

19

Voting and Politics Experiments

Decentralized equilibrium outcomes are efficient in many economic markets, and as a consequence, economists sometimes fall into the trap of thinking that the outcomes of a political process necessarily have desirable properties. In contrast, political scientists are well aware that the outcome of a vote may depend critically on factors like the structure of the agenda, whether straw votes are allowed, and the extent to which people vote strategically or "sincerely" (naïvely). Voting experiments allow one to evaluate alternative political structures, including newly proposed institutions. Standard game-theoretic models are often not very useful in the analysis of political interactions, and strong predictions based on median voter positions may not help in multidimensional settings. Therefore, it is important to study how people actually behave in controlled laboratory conditions that use exogenous treatment changes to infer causality. These studies are complemented by clever field experiments that create contexts with more external validity, e.g., having voters view media ads with differing formats.

> Note to the Instructor: Various voting experiments can be run with the Veconlab Voting program, which allows for simple agendas, runoffs, nonbinding polls, approval voting, and randomly determined costs of voting. Hand-run instructions are available for candidate platform competition, agenda effects, and competition between communities that offer differing public goods and associated tax levels. Class-ready adaptations of these instructions are provided in the Class Experiment Appendix for chapter 19 at the end of the book.

19.1 The Median Voter Theorem

The simplest voting model is one in which each voter has a preferred point along a line. The location of a public library on a road, for example, might involve preferred points that are a voter's home or place of work. In this case, the cost of accessing the public good is proportional to the travel distance. Suppose that voters' residences are represented by mile markers that start at 1 and end at 100 on the other end of the road. The median voter is the one with a preferred location that is central—not in location, but in terms of the numbers of voters with higher and lower addresses. If the addresses are 1, 25, 26, 35, and 99, for example, then the median location is at address 26, with two voters preferring a lower address for the public good and two preferring a higher address. In this example, candidates who propose locations for the public good would be drawn toward the preferred point of the median voter. If one candidate proposes a location of 25, then a competing candidate who proposes an address of 26 would win by 3-2. The median voter model was proposed by Hotelling (1929) and later adapted by Downs (1957) and others.

Of course, the position of the median voter will not generally be known, and the classroom experiment to be discussed shows how political competition can "discover" this point. The experiment begins with the distribution of index cards marked with integers in the range from 1 to 100, with each voter's card determining that person's preferred location. Voters' payoffs are calculated as 50 minus the absolute distance from the winning platform to their own preferred location, indicated on their index card. Two participants are initially selected to be candidates, who receive a fixed payment of 50 should their proposed location get more votes (ties decided with a coin flip). The losing candidate receives 0 and returns to being a voter, with a new "challenger" selected from the voters for the next period. It is important to note that only some of the numbers in the range from 1 to 100 were used, so nobody knew the location of voters' preferences.

Each period begins with each candidate choosing a platform on the interval [1, 100], with the incumbent (previous winner) announcing first. For the contest shown in figure 19.1, the incumbent proposed 50, which won over the challenger's proposal of 65. In the second period, a new challenger proposed 45 and won. The exact location of the median voter varied as challengers were selected (candidates were not allowed to vote), but the platforms converged to a tight range of 33–34 in periods 6–9, which corresponded closely to the median. In period 10, some new cards were exchanged for old ones, in a process analogous to redistricting, and the winning platforms quickly converged to the new median of about 57.

It is notable that the political competition tends to discover the median voter location, even though nobody knows anyone else's card numbers (discussion was not permitted). Another feature of the process is its stability. Any deviation from the median can be defeated with a proposal in the direction of the median, so that changes in the "right" direction tend to be implemented, and proposed changes in the wrong direction tend to be defeated. In fact, the median voter position in this setup is a *Condorcet winner* in the sense that it would defeat any other proposed location in a majority vote. Finally, a vote-based process like this will be sensitive to numbers of voters in each direction, but not to intensities of voter preferences, since each voter has a single vote. But if voting

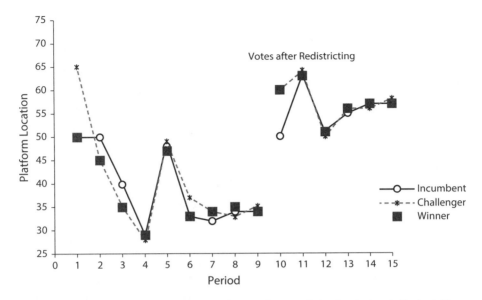

Figure 19.1. Platform Convergence to the Median, with Redistricting in Period 10. *Source*: Wilson (2005).

is costly due to the other things people can do with their time, then intensity of preference can have a role to play. In particular, preference intensity can augment turnout if voting is costly.

19.2 Voting with Your Feet (Can Be Done Outdoors!)

People who are not satisfied with median-voter or other political outcomes may have an option of relocating, or of selecting among alternative locations when they move to switch jobs. In fact, different communities may offer different levels of services and public goods that are "local" in the sense that they are not easily used by people in other communities. In a classic paper, Tiebout (1956) analyzes how migration can respond to locational differences in the provision of local public goods. In this manner, those who prefer high levels of particular public goods (and the associated higher taxes) can seek out communities that are closer to their preferences. For example, it is commonly observed that families with school-age children are willing to pay the higher taxes in districts with good schools, districts that are sometimes avoided by retirees.

Consider a simple classroom experiment in which each person has a constant marginal value for units of a public good, up to a limit determined by a numbered playing card that had been dealt to that person. Thus, a person with an 8 of hearts, for example, would earn V dollars for each unit provided, up to 8 units, with additional units having no value. Each additional unit of the good provided results in a cost of c that is divided equally among the members of the community. Hewett et al. (2005) contains instructions for a classroom experiment with this setup, which induces "single-peaked preferences" for the level of the public good. If $V = 1$, $c = 1$, and $N = 5$ community members, then the marginal cost per person is $c/N = \$0.20$, and the benefit of adding a unit up to one's maximum preferred level is $V - c/N = \$0.80$. Thus, the monetary preferences for a person with a card number X would have slope of $1 - 0.2 = 0.8$ up to X and a slope of -0.2 above X. The random allocation of playing cards would induce differing preference profiles, as shown in figure 19.2 for card draws of 2, 4, 6, 8, and 10. The person with a card of 10 has preferences that rise and peak at the right side. The "middle" person, with a card of 6, has preferences that peak at a level of 6, as shown by the gray dashed line. The preferred points for each of the 5 voters are represented on the horizontal axis by dark diamonds. Notice that the median voter in this case is determined by preferences for different levels of a public good, instead of being determined by physical location.

With the preferences in figure 19.2 and majority rule voting, the standard median voter prediction for the committee decision would be 6. This is because there is no other outcome that would beat this point in a binary contest. For example, if the voter with a preferred point of 8 were to propose an alternative level of 8, then the three voters with preferred points of 2, 4, and 6 would vote against this alternative, which would fail on a 3 to 2 vote. This reasoning can be used to show that the level 6 would beat all alternatives in two-way contests under majority voting, which is the defining characteristic of a Condorcet winner.

The example in figure 19.2 illustrates another property of the Condorcet winner, i.e., that it need not be the outcome that maximizes total earnings, assuming that side payments are not permitted. This is because raising the provision level from 6 to 7, for example, only costs a dollar, but it provides an additional dollar benefit to each of the two voters with higher preferred levels. In general, the earnings-maximizing level of the public good in this setup would be the

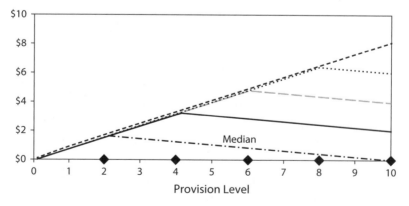

Figure 19.2. Single Peaked Preferences with a Median at 6

largest integer X such that the number of voters with card numbers at or above X is greater than the cost.

The class experiment to be discussed used playing cards to induce different voter preferences for different local public goods (determined by the suit of the card (spades, hearts, diamonds, or clubs) and the maximum preferred level (determined by the numbers on their cards). Each person received 2–3 cards. A person with a 2 of spades and an 8 of spades, for example, would earn a dollar for each added unit of the spade public good, up to a maximum of 10, and the per-person cost of provision ($2 per unit provided, divided by the group size) would be subtracted. The experiment was conducted outdoors on a nice day, with five locations indicated by manila envelopes, each with a name selected by the students who were randomly assigned to that location. A "mayor" was selected for each location, who would manage the voting and announce the results for that community.

As shown in the top row of table 19.1, the initial random allocation of voters resulted in 2 communities choosing clubs (C). For example, Metropolis in the second column selected 15 units of C, for a total cost of $30 (15 times the $2 cost per unit). With 3 residents, the tax in Metropolis in period 1 came to $10 per person. After the public goods and associated taxes were announced, people could switch communities (at most once per period). Note that Metropolis lost a resident in period 2, which resulted in an increase in the per-capita tax from $10 to $15. But the taxes there fell from $15 to $5.2 as it gained 3 residents in period 3. The influx of voters moved the median preference for the level of club good down from 15 (face cards were assigned high numbers) to 13.

A show of hands after period 2 indicated that most students were happier than before, and the total earnings (left column) more than tripled, up from $14 to $51, and subsequent switches tended to raise total earnings, which ended up at $77. This final earnings level turns out to be about 93% of the maximum

Table 19.1. Results from a Classroom Experiment with a $2 Cost per Unit

	Hoth	Metropolis	Moe's Tavern	Springfield	South Park
Round 1	3 residents	3 residents	3 residents	3 residents	3 residents
Earnings = $14	11S, tax $7.3	15C, tax $10	15D, tax $10	8H, tax $5.3	13C, tax $8.7
Round 2	3 residents	2 residents	3 residents	4 residents	3 residents
Earnings = $51	13S, tax $8.7	15C, tax $15	15D, tax $10	12H, tax $6	13C, tax $8.7
Round 3	3 residents	5 residents	3 residents	3 residents	1 resident
Earnings = $65	13S, tax $8.7	13C, tax $5.2	15D, tax $10	10H, tax $6.7	4H, tax $8
Round 4	3 residents	5 residents	3 residents	4 residents	0 residents
Earnings = $77	13S, tax $8.7	13C, tax $5.2	15D, tax $10	14H, tax $7	–

Key: C = clubs, D = diamonds, H = hearts, S = spades. Source: Hewett et al. (2005).

possible earnings that could have been obtained by a planned outcome with perfect information. Much of the earnings increase comes from individuals who migrate to communities with public goods for which they have high values (high card numbers). As people migrate, the public good provision levels tend to change as voter preference balances evolve. In total, 17 of the 19 vote outcomes in all periods were in line with the prediction of the median voter theorem. To summarize:

> **Median Voting Theorem Experiments:** *With single-peaked preferences along a single dimension, the preferred point of the middle person has strong drawing power, both in experiments and in theory. Median voter outcomes do not necessarily maximize total earnings, but the process of political competition seems to be remarkably good at discovering improvements, especially when voters also have the option of voting with their feet.*

19.3 Experimental Tests of Spatial Voting Models

When voters have to make two related decisions, it is useful to model their preferences in a two-dimensional graph. For example, you could think of the horizontal dimension as expenditure on schools and the vertical dimension as expenditure on roads. A simple spatial model is based on the assumption that voters prefer points closer to their own ideal points in the sense of Euclidian distance, so a voter's ideal point can be thought of as having concentric circular indifference curves around a utility maximum at that point.

Figure 19.3 shows the ideal points for 5 voters, represented by large black dots. Each dot could have rings of concentric circles drawn around it (not shown here), indicating indifference curves composed of points equally distant from the voter's preferred point. A straight line connecting any two voters' ideal points is a "contract curve" of points that those two might agree to if they were bargaining, since from any point off of the line, a movement in a perpendicular direction to the line takes the outcome closer to the two endpoints of the line, and hence would make both voters better off.

McKelvey and Ordeshook (1979) ran committee experiments with the setup in figure 19.3. Preferences were induced by giving each voter a contour map of the issue space, with concentric circles labeled with probabilities that increased to 1 as the ideal point was approached. The final committee decision in the two-dimensional issue space determined the voter's probability of winning a high money prize in a lottery carried out after the meeting ended; this "binary lottery" payoff procedure was intended to minimize the chances of side payments. The meetings were run by a research assistant chair, who was uninformed about the theoretical predictions. Discussion was governed by either a "closed rule"

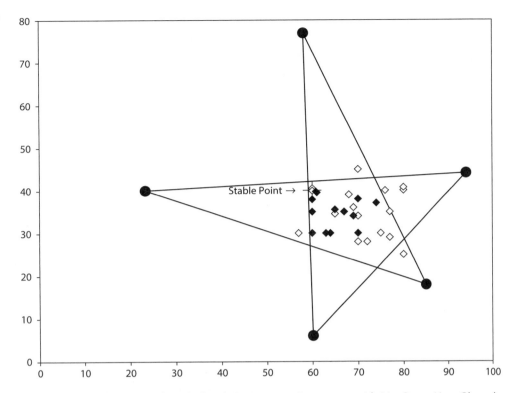

Figure 19.3. McKelvey and Ordeshook Committee Outcomes with No Core. *Key:* Closed Rule (dark diamonds), Open Rule (light diamonds)

or an "open rule." Under the closed rule, any person who wanted to speak or make a motion had to be recognized by the chair, all comments had to be directed to the chair, and only comments that pertained to a standing motion were allowed. Under the open rule, participants would speak directly with each other, without being recognized by the chair, and discussion was not limited to the motion on the floor. Under either rule, the initial position was the origin (0, 0), and any motion was required to represent a proposed movement in *only one direction*, i.e., a horizontal or a vertical movement. If a motion was seconded and passed, it would determine the new status quo. The experiment ended when a motion to adjourn was made and passed, and then final payoffs were determined by the lottery procedure.

Unfortunately, a Condorcet winner typically does not exist with multi-dimensional preferences, as the preference profile in figure 19.3 indicates. Here, the lines connecting five non-adjacent ideal points form a pentagon, which frames most of the data points from both closed rule meetings (solid diamonds) and open rule meetings (hollow diamonds). But even the points inside of this pentagon are not Condorcet winners. To see this, begin with any point in the pentagon and consider an alternative that is in a direction perpendicular to one of the faces of the pentagon. The three voters with ideal points in that direction would favor the move, but the new point could then be defeated by a different coalition that prefers a movement toward a different face of the pentagon.

The idea behind the McKelvey and Ordeshook experiments was that by limiting motions to movements in one direction only, the outcome might be driven to a "stable point" that lies at the intersection of the median of 5 voters' ideal points in the horizontal direction and the median of 5 voters' ideal points in the vertical direction. The location of this stable point is indicated by the "+" sign in figure 19.3. The results of the committee decisions are shown as solid diamonds for the closed rule meetings and as hollow diamonds for the open rule meetings. The committee outcomes do not, however, cluster tightly around this stable point under either treatment, although the closed rule does draw the outcome points a little closer to the stable point. McKelvey and Ordeshook conclude that the closed rule provides a tighter control by limiting consideration to movements in a single direction at a time, which would lead toward a median outcome in each direction. This effect of political institutions on voting outcomes parallels earlier observations of the effects of market institutions on market outcomes. To summarize:

> **Spatial Voting Experiments:** *The strong predictive power of the median voter theorem in a single direction does not carry over to two dimensions when applied separately in each dimension. The reason seems to be that, with two dimensions, there may not be a Condorcet winner that beats all other points in a binary comparison. This indeterminacy provides more "room" for behavioral effects based on factors like fairness, and for effects of changes in political institutions like voting rules or agendas to be discussed in a subsequent section.*

19.4 Legislative Bargaining and Pork: "Now *That's* the World I Live In!"

Eavey and Miller (1984) observed a strong bias toward fair outcomes in a different design where only one of the players, the agenda "setter," could propose an alternative to an exogenously designated status quo. The setter model was generalized by Baron and Ferejohn (1989), who introduced a stylized model of legislative bargaining over a fixed sum of money. The bargaining process starts with one member of the legislature being selected at random to propose a division. If a proposal passes by majority vote, it will be implemented, and if not, another randomly selected member of the legislature can make a new proposal. The theoretical prediction is that the first proposal will be structured to obtain the support of a bare majority, by offering each person in this coalition only slightly more than they can expect if the proposal is rejected and the game continues, i.e., their "continuation values." The proposer can maximize earnings by choosing the winning coalition to include those with the lowest continuation values.

A flavor of the Baron-Ferejohn model calculations can be shown with a simple symmetric example. Suppose that there are three legislators bargaining over the division of a pie of size 1. Each one is chosen with probability 1/3 to make a proposal to the others, which passes and is implemented if at least two people vote for it. If the proposal does not pass, the game continues to the next round with probability $\delta \leq 1$. Given the symmetry, consider a symmetric equilibrium in which each person has an expected payoff c if the current proposal is rejected (but before the random determination of whether to continue or not). Payoffs are 0 in the event of a random stop. Since c is what a person can expect to get if a proposal is rejected, a proposer can secure a second vote by offering one of the others a payment of c. In this case, the proposer earns $1 - c$ and the proposal passes. Then c equals the probability of continuation times an expected payoff shown in square brackets:

$$c = \delta\left[\frac{1-c}{3} + \left(\frac{2}{3}\right)\left(\frac{1}{2}\right)c\right].$$

Consider the first term in the square brackets. There is a 1/3 chance that the person will be a proposer in the next stage, and the proposer can make an accepted proposal that offers c and earns $1 - c$. The second term is the 2/3 chance of not being randomly selected to be a proposer, times the 1/2 chance of receiving a proposed payment from the proposer, times the offer of c that will be accepted with a minimal winning coalition of size 2. By collecting terms, the prediction is that $c = \delta/3$. Note that a lower continuation probability (lower δ) means that the proposer does not have to offer as much of the pie to pick up a second vote.

These predictions characterize more general versions of the model: (1) *no delay*, (2) a *minimal winning coalition*, and (3) *full rent extraction* (of the rents associated with being a proposer). As results from ultimatum bargaining game experiments (chapter 14) suggest, experiments that implement the Baron and Ferejohn legislative bargaining model do result in delays, less-than-full rent extraction, and a tendency toward larger than needed winning coalitions, all of which deviate from the sharp theoretical predictions.

The Frechette, Kagel, and Lehrer (2003) experiments implemented the Baron and Ferejohn model with groups of five voters and either an open rule or a closed rule. Under the closed rule, a proposal that is made must be voted on, whereas the open rule allows a second randomly selected person to either second or amend the proposal, with a runoff between the original and amended proposals in the event of an amendment. One difference from earlier experiments is that the authors let subjects participate in a series of committee bargaining games, which enhances learning and reduces fairness considerations if payoffs tend to equalize across games. The outcomes under the closed rule tended to approach the theoretical prediction of a minimal winning coalition in later meetings, although proposer earnings were lower than predicted. In contrast, the open rule sessions exhibited consistent super-majorities.

Christiansen, Georganas, and Kagel (2014) report an experiment in which the pie being divided can be thought of as "pork" needed to "grease the wheels" in the construction of a winning coalition. The new element is that the legislators also have different preferences for the location of a public good on a line from [0 to 100]. A randomly selected legislator's proposal consists of a location on the line and money payments to each individual that sum to the available pot of cash (pork), with the continuation probability set to 1. The preferred locations were 0, 33, and 100 for the three legislators, so in the absence of cash, the outcome should be close to the median location of 33, which is approximately what was observed (about 38) in the experimental treatment without any pork/cash. In the main treatment, with cash transfers (pork), the average location turned out to be near 50. The cash allowed proposers to extract rent and move the location in a manner to secure a second vote. *Notably, acceptance rates went up in the pork/cash treatment, as wheels were "greased."* But note that the location at 50 does not minimize travel costs (problem 6), i.e., pork transfers can have negative effects as well as any possible positive effects.

The lower acceptance rate for proposals in the no-pork treatment is reminiscent of the difficulties that the US Congress had passing spending bills after President Obama promised to veto any bill with a pork "earmark." As a result, spending bills languished and most spending was authorized with "continuing resolutions" that kept budgets at prior levels. A final point to make is that the reliance on continuing resolutions during the Obama era may not have been so bad if the effect of a veto threat on earmarks was to restrain spending increases that otherwise could result from log rolling based on pork transfers.

An interesting footnote to this story is that one of the authors was in flight to present this paper at a meeting of the Economic Science Association, and was looking over his power points. The person sitting in the next seat turned out to be a congressional lobbyist who became interested in the implications for continuing resolutions and started asking questions. At one point, she blurted out: "That's exactly what happens, now *that's* the world I live in!"

19.5 Agendas and Strategic Voting: How to Manipulate If You Must

The stability and uniqueness properties of equilibrium voting outcomes in one-dimensional environments are not general. In fact, dependence on "history" and the way that decisions are structured often makes it possible to manipulate outcomes to a surprising extent. In this section, we consider dramatic effects of agendas that control the sequence of votes to be taken over three or more options.

Status Quo Agendas with a Voting Cycle

One type of agenda is to have one person be designated to choose a proposed alternative to the status quo. A more elaborate agenda would have initial votes determine the alternatives to be considered in the final vote. Agendas essentially put structure on the voting process, and interesting issues arise when voters at an early stage might wish to think strategically about which options will fare well at a later stage of the voting. Agendas can have dramatic effects on the voting outcome in cases where voters' preferences create a natural instability, with no Condorcet winner. For example, consider the preference profiles for three voter types shown in table 19.2, with the money payoffs given in parentheses. These preferences generate a "voting cycle." To see this, suppose that A is the status quo and B is proposed as an alternative. If people vote "sincerely" according to their preferences, B would beat A by a 6 to 3 vote, but then C would beat B. To complete the cycle, note that the original starting point A would beat C.

The effect of a pre-set agenda is to provide a structure to the order in which options will be considered. The simplest agenda is one in which one option, say C, is the status quo that is run against the winner of an initial vote between A and B, as shown in figure 19.4. In this first stage, the type II and III voters might vote for B over A, since B provides higher payoffs for each of them. But then in the final stage, B would lose to C by a vote of 6 to 3. This result, which is shown by the dark arrows in the figure, would disappoint the type II voters who voted for their most preferred option B initially but ended up with their least preferred option C and a payoff of only $0.50. If the type II voters had voted strategically for the less preferred option, A, in the initial stage, then A would have won and would beat the status quo in the final stage. In this manner, *strategic voting* by the type II voters would raise their payoffs from $0.50 to $2.00. The initial non-strategic voting by type II voters for their preferred outcome in the first stage, without thinking ahead about the second stage, is called *sincere voting* in the political science literature, and it is typically referred to in less flattering terms as *myopic voting* in the economics literature.

There can be various decision rules that produce naïve voting (ignoring what might happen in later stages). The agenda in figure 19.4 first involves

Table 19.2. Preferences for a Voting Cycle

Type I (3 voters)	A ($3) > C ($2) > B ($0.50)
Type II (3 voters)	B ($3) > A ($2) > C ($0.50)
Type III (3 voters)	C ($3) > B ($2) > A ($0.50)

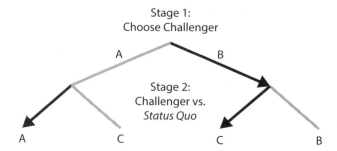

Figure 19.4. Sincere Voting for an Agenda with Option C as the Status Quo

choosing between two packages, "*A or C*" and "*B or C,*" with the final choice made in the second stage. One naïve rule would be to vote for the package with a person's best outcome. An alternative naïve rule would be to vote against the package with the voter's worst alternative, and a third rule could be based on calculating expected payoffs based on priors of 1/2 for each outcome, as suggested by Plott and Levine (1978). The Eckel and Holt (1989) agenda voting experiment was structured so that *all* three of these naïve rules predicted the same vote outcome in the first stage, which was different from what would happen with strategic voting. The authors' expectation was that voting would tend to be strategic in such a simple environment, especially in the full information treatment in which a randomly selected committee chair would go around and record people's payoffs for each option and write them on the board for all to see. This information figured in committee discussions, which were otherwise rather vague in the "no-information" treatment. The authors were surprised to observe 100% sincere voting in the initial meeting, with strategic voting only emerging after several prior meetings. But once a subset of voters figured out the advantage of voting strategically, all outcomes after that point were strategic.

When the "*C* last agenda" is implemented in class experiments, the status quo, *C*, is typically selected in the two-stage initial votes, as expected with naïve voting. With repetition using the same agenda, the type II voters eventually switch to strategic voting, causing option *A* to be selected. Agenda changes that put another option last as the status quo would result in that option being selected in the initial meeting, as predicted under sincere voting, with the strategic outcome emerging after voters gained some experience. Thus, changes in the agenda tend to change the observed outcomes, *even though voter preferences do not change*. In addition, for a given agenda, the observed outcome changes as people eventually learn to vote strategically. The main takeaway for agenda design is summarized:

Agenda Manipulation: *In many settings, especially those with voting cycles, different agendas can be constructed to yield different outcomes. Successful agenda manipulation depends on an assumption about voting, e.g., naïve or strategic. Experiments indicate that naïve voting is probably a good assumption in all but the simplest environments, unless the committee meets repeatedly, as with a university department faculty voting on hiring decisions, or unless the members have considerable experience with similar situations.*

Inefficient Outcomes and Voting Cycles with a Class Agenda Experiment

One way to add a bit of context to a class agenda experiment is to provide each person with two playing cards, where a heart indicates a preference for "Highway" construction and a spade indicates a preference for a "School." Some people received only one card, and others received both. The cards for a group of 7 voters were as shown in table 19.3. A person with a spade receives $300 if School is funded, and a person with a heart receives $300 if Highway is funded. Each of the 7 voters must pay $200 if School passes, and another $200 if Highway passes. Notice that the sum of the benefits for School (5 × $300 = $1,500) exceeds the tax cost to the 7 voters (7 × $200 = $1,400), but this is not true for Highway. The setup in the table can be replicated by adding groups of 7 new voters at a time, for group sizes of 7, 14, 21, etc.

The setup in table 19.3 was used to run a class experiment with 21 students (3 groups of 7). Various agendas provided very different outcomes. In one agenda, the first vote was Highway or Neither, and Highway passed by a vote of 13 to 8. Then School passed by a vote of 16 to 5. Finally, in a vote between Both Projects and Neither, the outcome was Neither by a vote of 12 to 9. Notice that naïve voting in this agenda provided a complete cycle, beginning with Neither being funded and ending up in the same place!

A different agenda, in contrast, resulted in both projects being funded (see Holt and Anderson, 1999 for details). A look at table 19.3 indicates how this might happen. A 4-3 majority prefer to fund Highway, even though this imposes tax costs on the three School-only people to the extent that total earnings

Table 19.3. Preferences for a Voting Cycle

Voter 1	Voter 2	Voter 3	Voter 4	Voter 5	Voter 6	Voter 7
Heart	Heart	Heart	Heart			
Spade	Spade			Spade	Spade	Spade
Highway	Highway	Highway	Highway			
School	School			School	School	School

decrease from funding this project. And a 5-2 majority prefer to fund the School, even though this imposes a tax burden on the 2 Highway-only people. This indicates how separate consideration of each project would cause both to be funded, even though total earnings would be increased by funding neither, and an agenda that considers "Both versus Neither" in the final stage would end up with neither being funded. Even greater amounts of excessive spending can result from "log rolling" in which each of two factions supports a project valued by the other, even though the net effect of funding both imposes a major tax burden on a third faction.

A Field Experiment with a Flying Club

Agenda manipulation is not relegated to the lab, as anyone with small-group voting experience can attest. Plott and Levine (1978) identifed several alternative types of non-strategic voting in somewhat more complicated agendas, and they used what they learned to manipulate the outcome of a meeting of a California flying club (Levine and Plott, 1977). The authors were members of an agenda committee for a club that was meeting to decide on a configuration for a fleet of general aviation planes. The committee surveyed club members to ascertain their preferences, which were used to structure the agenda. The authors on the agenda committee preferred to have at least one larger plane, although a significant faction of other members preferred smaller, four-seat planes. The agenda was structured to defeat this small-plane option by pitting its supporters against a coalition of smaller groups with diverse interests. The club president was forced to follow the agenda, despite attempts to deviate, which included a proposed straw vote that might have helped club members vote more strategically. The club ended up selecting a fleet with some six-seat planes, and a subsequent poll indicated that this configuration was not a Condorcet winner. The authors were later expelled from the club after an article reporting this process was published in the *Virginia Law Review*.

19.6 Coordinating Devices: Polls, Runoffs, and Approval Voting

In view of the dramatic agenda effects just discussed, it is not surprising that citizens might prefer more neutral methods of structuring the choice sequence from among multiple candidates or options. This section considers the effects of runoffs, opinion polls, and other devices that might be used by voters to coordinate their decisions. The type of coordination problem that can arise is illustrated in table 19.4, which shows the basic payoffs used in a series of experiments conducted by Forsythe et al. (1993, 1996) and Morton and Rietz (2008).

Note that voters of types I and II in this design have a strong preference for option A or B, but this majority is split in its preference for A or B. In binary contests, both A and B would defeat C. Therefore C is, in this sense, a Condorcet

Table 19.4. A Split Majority Profile

Voter Type 1 (3 voters)	Voter Type 2 (4 voters)	Voter Type 3 (5 voters)
A 1.20	B 1.20	C 1.40
B 0.90	A 0.90	A 0.40
C 0.20	C 0.20	B 0.40

loser. Under *plurality voting*, with all three options on the ballot, option C (with 5 voters) would win if the majority splits between A (3 voters) and B (4 voters), as would be the case with sincere voting. If a first-stage vote is followed by a *runoff* between the top two contenders, however, the Condorcet loser would end up being matched with one of the other options, which would win. Thus, the runoff lets the split majority figure out which of the two preferred options has the best chance of winning in the runoff. Interestingly, Kousser (1984) argued that runoff elections might have been adopted in some parts of the United States in an effort to prevent minority candidates from winning. Forsythe et al. (1993) showed that a non-binding opinion poll held before a three-way contest will also serve to help the majority coordinate on one of its preferred options in committee voting experiments. Polls, however, were not as effective as runoffs in all cases, since poll results tend to show more randomness than actual votes.

One alternative to a simple plurality procedure is to let people vote for more than one candidate, which might help coordinate voting by a split majority. The most common multiple vote procedure is called *approval voting*, since voters receive a list of candidates or options and mark each with "approve" or "not approve." The option with the most approval votes wins. For the split-majority setup in table 19.4, the use of approval voting might result in the selection of option A or B if enough type 1 and 2 voters approve of both of their top two options. In this manner, approval voting may not result in the selection of a Condorcet loser (option C in the table).

Most controlled evaluations of approval voting have used data from field experiments, with some mixed conclusions (Nagel 1984; Niemi and Bartels, 1984). One interesting field experiment is reported in Brams and Fishburn (1988), who use data from elections of officers in two professional associations. In these elections, voters indicated their preferred candidate (plurality), but they also returned a preference ranking and an experimental approval voting ballot. Approval voting is used by a number of professional associations to select board members, including the Economic Science Association mentioned in chapter 1.

Approval voting has been tried in several field experiments in which voters in the actual election are also given experimental approval voting ballots at the poll locations on election day. The experiments, which were conducted in France in 2002 and 2012, did not involve fully random assignment of separate treatments to different subjects. Instead, the procedure involved giving a separate approval ballot to those who voted in the standard single-vote first stage of the election and who were willing to participate (participation was about 80% in the selected towns used on one case). The French setting is ideal for considering approval voting and similar evaluative methods, given the relatively large numbers of candidates in these two-stage elections, with a runoff in the event that no candidate receives a majority in the first stage.

In the 2002 election, there were 16 candidates in the first stage. The two candidates selected at that stage were Jean-Marie Le Pen, a polarizing candidate from the National Front party, and Jacques Chirac, a conservative who won decisively with 84% in the runoff. Approval voting provides a person with the option of expressing support for multiple candidates. This aspect of approval voting can benefit candidates with a broader base of support. There are, however, complex strategic considerations, e.g., not approving of a strong rival candidate. The authors concluded that approval voting would have allowed the third candidate in the first stage to provide a stronger challenge to Chirac, although Chirac would have won in any case (Laslier and Van der Straeten, 2008).

A second approval field experiment was run at the time of the 2012 French election (Baujard et al., 2014). The general message that emerges from these studies is that approval voting and similar evaluative voting procedures tend to benefit inclusive candidates, as compared with exclusive, polarizing candidates who may fare better, relatively speaking, in the traditional first-stage/runoff format, especially when there is a large field of candidates. One possible issue with approval voting is that it might be easier for hackers to manipulate the outcome, since the total number of votes is not constrained by the number of registered voters. Therefore, the addition of false approval votes might not be noticed.

19.7   Participation Games and Turnout

One of the most perplexing problems for "rational-choice" models of voting behavior is to explain why people vote at all in large elections. After all, the probability of altering the election outcome is small if there are many voters. People vote anyway if the costs are low or even negative for those who enjoy voting and telling family, friends, and coworkers about it. The high turnouts in the early post-Hussain Iraqi elections provide an example where the perceived benefits of voting outweigh the costs. There is some indirect evidence that costs matter, e.g., voter registration rates in the United States are lower in localities where jury selection lists are taken from lists of registered voters. Moreover, causal observation suggests that attendance rates at faculty meetings are higher when important issues need to be decided, i.e., when the potential benefit of affecting the outcome is likely to be larger. Laboratory experiments cannot address questions about the magnitudes of voting costs and benefits in the field, but experiments can test qualitative predictions about the effects of induced costs, benefits, and group sizes on turnout rates.

The basic voter turnout model is a *participation game*, first analyzed by Palfrey and Rosenthal (1983, 1985). Voters have costs of voting, which are either deterministic or randomly drawn from a cost distribution. Think of voting costs as being explicit (time, transportation, etc.), and implicit opportunity costs as taking care of a sick child, driving a friend to the repair shop, etc. As noted above, such costs can be negative, since voting can be a social event when friends and neighbors are encountered, or when there is endless ex post discussion of intentions and motives. The formal models specify a range of voting costs, with each voter observing a random cost realization for that day. The equilibrium can consist of a threshold or "cutpoint" voting cost: those who have costs below the cutpoint tend to vote, and not otherwise. In this case, voting appears to be somewhat random to the outside observer, and the observed voting rates will be correlated with cost shocks like weather, and perhaps with the chances that one's vote is decisive.

Participation cost models generally have clear qualitative predictions, some of which have been tested by Levine and Palfrey (2007) with "large" voting experiments involving groups of 3, 9, 27, and 51 voters. The theoretical predictions are quite intuitive: that turnout will be higher with small groups of voters and with close elections (more equal numbers of the two voter types). When there are more than three voters and unequal numbers of voters of each type, turnout is predicted to be larger for the minority group. The basic features of these predictions are supported by the experiments, although precise turnout rates are less responsive to changes in treatment parameters than the theory would suggest. This indicates that there is some residual randomness in decision making due to un-modeled individual differences in beliefs or experiences. The authors use a quantal response equilibrium (discussed in chapter 11) to incorporate this residual randomness.

Participation game experiments have shown that subjects in the lab respond to parameter changes in ways that are predicted by theory, but the potentially strong effects of social interactions have been largely ignored. One exception is Großer and Schram (2006), who add a limited amount of carefully structured social interaction. As before, there are two types of voters with opposing preferences and known costs of voting. For each type, half are designated as "senders" who may decide to vote early or not. In one of the treatments, each of the senders is paired with a late voter of the same type, who is told whether or not the sender already voted. Senders who do not vote early are allowed to vote in the second stage, at the same time that the vote decisions of the late voters are made. This type of information exchange raised overall turnout rates dramatically, by about 50% as compared with a standard one-stage participation game used as a control treatment. This result is quite significant in a world where small turnout increases in targeted areas of strength for one candidate can swing an election.

19.8 Field Experiments with Context: Attack Ads and the Knock on the Door

Many key elements in the political landscape involve social context. These elements include such things as a knock on the door, a phone call, or an "attack ad," that may be difficult to replicate in a standard laboratory setting. Political scientists, therefore, have a long tradition of running field experiments, which can be roughly categorized as: (1) *enriched laboratory experiments* that may add a living room context or a randomly selected sample of voters, or (2) *pure field experiments* in which treatments are implemented in a natural setting so that the participants do not know they are in an experiment. This section contains an example of each type.

One problem with only considering non-experimental field data is that key factors are often not exogenous. For example, campaign advertising is typically targeted to areas where the outcome is close, so a simple correlation between advertising and vote shares could end up being insignificant. An ideal field experiment could involve exogenous manipulation of campaigning effort (details to follow).

Ansolabehere et al. (1994) used an enriched laboratory setting to study the effects of positive and negative campaign ads on turnout decisions. Subjects were recruited to come to a laboratory, which was furnished like a living room. They watched 15-minute tapes of a local news broadcast, which was interrupted by a commercial break with either a positive political ad, a negative political ad, or an ad for a commercial product (the control treatment). The positive and negative ads pertained to an actual upcoming election, and the substitution of positive and negative phrases was carefully done to maintain the same visuals, content, audio, and other features. In an ex post questionnaire, those exposed to a negative ad were 2.5% less likely to vote than those exposed to the neutral control, and they were 5% less likely to vote than those exposed to the positive ad. The authors concluded that the negative tone reduced the voter confidence in the responsiveness of the political institutions, thereby lowering intentions to vote. In a companion empirical study of 32 state Senate races in the 1992 elections, the authors found evidence that negative ads tended to reduce voter turnout. The controlled experiment with exogenous variations in ad negativity helps the researcher identify causality, since other factors can intervene in non-experimental settings. For example, it is quite possible for a third factor, e.g., high polarization, to trigger *both* negative ads *and* high turnout in the same election, as recent experience suggests.

The control of running this experiment in the laboratory also enabled the researchers to avoid possible biases, such as the tendency for people who remember an ad to be precisely those who were more likely to vote in the first place. On the other hand, the actual decision to vote was not directly observed, and participants' responses to survey questions about intentions to vote may be biased. This problem was cleverly addressed by Gerber and Green (2000), who used official voter records for New Haven precincts to determine whether specific participants actually voted in the election that followed the experiment. The field setting allowed them to deliver treatment stimuli in a realistic social context, without alerting recipients to the existence of a controlled experiment. The 29,380 participants were randomly assigned to treatments with messages urging them to vote. The treatments consisted of: no contact (control), direct mail, a phone call, face-to-face contact at home, or some combination. The messages were non-partisan, and urged people to vote on the basis of civic duty, neighborhood solidarity, or the closeness of the election. The main conclusion was that personal contact had a positive and substantial effect on turnout. There was a smaller direct mail effect, but phone contact did not seem to matter. This study attracted a lot of attention from political operatives, given the potentially large impacts of small turnout increases in target districts where a particular candidate is known to be more popular. Gerber and Green (2004) provide a thoughtful evaluation of the advantages of field experiments in politics.

19.9 Fake News, Polarization, and Expressive Voting

In 2016, election campaigns in Europe and the United States featured populist or exclusionist candidates, like Marine Le Pen, who bracketed more traditional candidates. The usual refrain of "no choice" seemed like a distant memory in 2016. Another surprise was the Brexit vote for the UK to withdraw from the European Community. Press accounts on the morning after the vote featured voters who had voted for Brexit, but seemed to be shocked and concerned that it had actually passed (a Google search for "Regrexit" is suggestive, but such postings can also be manipulated). This section reviews several recent experimental studies of expressive voting and polarization.

Polarization and the Anti-Median Voter Theorem

When voter preferences are single-peaked along a single dimension and candidates only care about winning, the arguments in section 19.1 indicate that candidates will locate near the center, which is a common complaint made in the United States about the similarity of Democratic and Republican candidates. Electoral contests are often asymmetric, with potentially large advantages for incumbents, film actors, relatives of popular presidents, etc. There is some theoretical work predicting that disadvantaged candidates will tend to choose more extreme locations than candidates with an advantage; see Aragones and Palfrey (2004) for an experimental evaluation of these asymmetric contests.

Another related perspective on polarization is that citizens, who have their own "ideal points," may have a stronger incentive to become candidates if their ideal points are more extreme. This effect can occur when the candidates' positions are not fully revealed, so that extreme candidates can somehow mask their views during the electoral competition. An extreme case is one in which the candidate's party, "Left" or "Right," conveys directional information about which side of the political spectrum a candidate's position lies, but not precise information. In a parallel pair of theory and experiment papers, Großer and Palfrey (2014, 2017) consider such a setting, in which each citizen can decide whether or not to incur a cost to become a candidate for the party that represents their directional preference, left or right relative to the midpoint of 50 on the range of ideal points from 0 to 100. In the event of multiple contenders, the party's candidate is selected at random from among the contenders. Voters are assumed to know the candidate's directional preference but not the candidate's ideal point, which is what the winning candidate imposes on the others. The winning candidate in the general election earned a pre-announced benefit amount. In equilibrium, only candidates with extreme views (in intervals extending inward from the boundaries of 0 or 100) compete to become candidates. The qualitative features of this polarized prediction are consistent with results of the laboratory experiment, in which entry rates are much higher for candidates with extreme preferences than for people whose ideal points are in the center. To summarize:

> **Anti-Median Voter Theorem:** *In experiments that allow general (e.g., left/right) directional party labels to obscure candidates' personal preferences, those with extreme preferences are more likely to seek office, which can yield outcomes that deviate sharply from median voter predictions.*

Expressive Voting and Partisan Information Processing ("Fake News")

In large elections in which a person's vote is extremely unlikely to be decisive (make or break a tie), voting can only be explained in terms of its psychological benefits:

> Neither the act of voting nor the direction of a vote cast can be explained as a means to achieving a particular political outcome, any more than spectators attend a game as a means of securing the victory for their team. (Brennan and Buchanan, 1984, p. 187)

One way to evaluate the importance of "expressive voting" is to manipulate the decisiveness dimension. The problem is that voters may have vague subjective ideas about the chances that their votes could be decisive, beliefs that could be difficult to measure without error.

Robbett and Matthews (2017) manipulate decisiveness by sorting subjects into groups of size 1, 5, and 25, where those in groups of size 1 will surely be decisive. Subjects were adults recruited from an online service, Amazon Mechanical Turk, and were screened with survey questions in order to identify individuals with strong partisan (Republican or Democratic) views. After the samples of partisans were determined, participants were asked to vote on factual questions with verifiable correct answers, despite a heavy overlay of political context. The questions pertained to global temperature trends, abortion rates, unauthorized immigrants, Obama's approval ratings, health insurance, and other conditions that might be perceived differently by partisan Democrats and Republicans. Each person or group was given three political questions, along with a "neutral" (non-political) question. The questions all had verifiable answers that could be displayed graphically, and each person received $1 for each question if a majority of the group voted for the correct answer from a list provided. Note that, in a group of size 1, the sole voter would automatically be decisive.

The main hypothesis is that a partisan bias due to political cheerleading would be most prevalent for voters who are less likely to be decisive. In order to measure the bias, the answers to each question were scaled on a [0, 1] interval, where 0 represents the strongest Republican perspective and 1 represents the strongest Democratic perspective. The partisan gap was large (about 13%) and statistically significant for groups of size 5 and 25, with no effect of group size. In contrast, the partisan gap was only about 5% for decisive individuals (groups of size 1). To conclude, there is a partisan gap even when the vote is sure to be decisive, but the gap is 2–3 times greater in groups of 5 or 25, which indicates that expressive voting is present even in small groups. The neutral questions provide a control for measuring the incidence of correct answers. The proportions of correct answers to neutral and political questions are not significantly different for decisive voters (group size 1), but for the groups of size 5 and 25, the proportion of correct answers to political questions is about 15 percentage points smaller than for neutral questions. To summarize the main takeaway from Robbett and Matthews (2017):

> **Expressive Voting:** *Expressive voting about political questions with factual answers generates a partisan bias that is clearly observed when a person's vote is not decisive (group sizes greater than 1), which is consistent with high turnouts in emotionally charged campaigns.*

It is interesting to consider the methodological advantages of using the online Amazon Mechanical Turk recruiting and payment platform in this particular setting. One potential problem with Mechanical Turk is that it does not provide control over what else the participant may be doing at the time. In this case, inattention was probably not a serious issue, given the 30-second voting deadlines and relatively high stakes for an online task ($1 per question). Moreover, Mechanical Turk enabled the researchers to find a balanced sample of adults with a sharp partisan divide that might not have been available with a student population at a liberal arts college. Finally, Mechanical Turk provides a simple and direct method of issuing payments, which permits large samples and high motivation.

Chapter 19 Problems

1.   Redraw the two-dimensional setup in figure 19.3 with only three voters (instead of five) who have ideal points that differ and do not lie on a straight line. Under what conditions would a Condorcet winner exist, and under what conditions would it not exist? Explain.

2.   Consider a situation with equal numbers of high-, middle-, and low-income voters. The issue to be decided is the level of expenditures on public schools, to be financed by taxes. Public education is important to the middle-income voters, who prefer high expenditure, but their second choice is low expenditure since they would send their children to private schools if the public schools are not of top quality. The high-income voters will use private schools in any case, so to minimize taxes, they prefer low expenditures to medium expenditures to high expenditures. Finally, the low-income voters prefer medium to high to low expenditure levels. Show that there is no Condorcet winner, and that these preferences induce a voting cycle.

3.   (non-mechanical) Can you think of a simple model of "log rolling" deals that would lead to majority vote outcomes with wasteful expenditures that provide lower aggregate benefits than costs?

4.   Suppose that the status quo school spending level is low, and that preferences are as given in problem 2. An agenda pits medium and high levels, with the winner in the first stage being matched against the status quo of low spending in the final stage. What outcome would you expect if voting is sincere? Would the outcome change if voting is strategic? Explain.

5.   (non-mechanical) Have you had any experience attending a meeting (club, fraternity, student organization) and thinking that the agenda was rigged to generate a specific outcome? If so, describe the agenda and why it might have had an effect.

6. (non-mechanical) Consider the setup with three voters with ideal points at locations 0, 33, and 100 on a road, with travel costs proportional to the distance from the voter's ideal point to the location of the public service. Assume that the travel costs are the same for each person. Does the median voter outcome minimize the total travel costs of the three voters? What is the extra travel cost imposed by a location at 50? How would your answers change with a travel cost that is three times as high for the person located at 100 as for the other two?

PART IV

Market Experiments

The games in this part represent many of the standard market models that are used in economics. In a *monopoly*, there is a single seller, whose production decision determines the price at which the product can be sold. This model can be generalized by letting firms choose quantities independently, where the aggregate quantity determines the market price. This *Cournot* setup is widely used, and it has the intuitive property that an increase in the number of sellers will decrease the equilibrium price. The experiment discussed in chapter 20 implements a setup with linear demand and constant cost, which permits an analysis of the effects of changing the number of competitors, ranging from one seller (monopoly) to many (competition). A second experiment involves a comparison between local monopolies and the competition that results when cross-market purchases are permitted, e.g., through elimination of tariffs or reductions in shipping costs.

   The Cournot model may be appropriate when firms pre-commit to production decisions, but it is often more realistic to model firms as choosing prices independently. Chapter 21 addresses price competition, imperfections, and the effects of alternative trading institutions when buyers are not simulated. Buyer and seller activities are essentially symmetric in the *double auction*, except that buyers tend to bid the price up, and sellers tend to undercut each others' prices. This strategic symmetry is not present in the *posted-offer auction*, where sellers post prices independently, and buyers are given a chance to purchase at the posted prices (further bargaining and discounting is not permitted). The posted-offer market resembles a retail market, with sellers producing "to order" and selling at catalogue or "list" prices. In contrast, the double auction more closely resembles a competitive "open outcry" market like that of the New York

Stock Exchange. Collusion and the exercise of market power cause larger distortions in markets with posted prices, although these effects may be diminished if sellers are able to offer secret discounts to individual buyers. Similarly, giving buyers the ability to make counter-offers and to withhold purchases at unfavorable prices can break or diminish seller market power.

Many important aspects of a market economy pertain to sales of intermediate goods along a supply chain. Chapter 22 considers experiments based on a very simple case of a monopoly manufacturer selling to a monopoly retailer, with each firm equating marginal revenue to marginal cost. As verified by experiments, this process of "double marginalization" causes the industry output to be even lower than would be the case if the two firms merged into a single monopoly. Many industries are characterized by "long" supply chains, which link manufacturers, distributors, wholesalers, and retailers. Frictions and distortions can be magnified as they echo through the supply chain (the "bullwhip effect"), which provides a rich source of topics for laboratory experiments.

The effects of asymmetric information about product quality are reviewed in chapter 23. When sellers can select a quality "grade" and a price, the outcomes may be quite efficient under full information, but qualities may fall to low levels when buyers are unable to observe the quality grade prior to purchase. The "unraveling" of quality may eventually cause a market failure in which only low-quality units are traded. Similar unraveling can cause markets for health insurance to implode as low-risk individuals decide to self-insure, which drives up the average claim amounts for the remaining insurance purchasers. The resulting premium increase may cause further erosion due to self-insurance. In this case, adverse selection (of high-risk individuals into the insurance pool) may result in high premiums and a significant proportion of uninsured individuals. Experiments are used to document such adverse selection effects and proposed remedies.

Markets for economic assets are complicated by the fact that ownership provides two potential sources of value, i.e., the benefits obtained each period (services, dividends, etc.) and the capital gain (or loss) in the value of the asset. Asset values may be affected by market fundamentals like dividends and the opportunity cost of money (the interest that could be earned in a safe account). In addition, values are sensitive to expectations about future value, and such expectations-driven values can cause trading prices to deviate from levels determined by market fundamentals. Chapter 24 summarizes some experimental work on such asset markets where price bubbles and crashes are pervasive in cash-rich (easy-credit) settings. Finally, chapter 25 considers linked markets motivated by macroeconomic issues of saving, investment, consumption, banking, and forecasting.

20

Monopoly, Cournot, and Cross-Market Price Competition

A monopolist can obtain a higher price by restricting production, and profit maximization involves finding a balance between the desire to charge a high price and to maintain sales quantity. Subjects in experiments with simulated buyers are able to adjust quantity so that prices reach monopoly levels.

The analysis of monopoly provides a bridge to the most widely used oligopoly model of the interaction of several sellers. The key behavioral assumption of this model is that each seller takes the others' quantity choices as given, and then behaves as a monopolist who maximizes profits for the resulting residual demand. The popularity of this Cournot model is, in part, due to the intuitive prediction that the market price will decrease from monopoly levels toward competitive levels as the number of sellers is increased.

The final section compares a situation with local monopolies due to cross-trade restrictions (shipping costs or tariffs) with what happens when those restrictions are lifted. In this setting, it is natural to model firms as setting prices (instead of quantities), with prices being available to buyers in other markets in the free-trade treatment. The presence of active buyers who can resist price hikes can have a restraining effect on monopoly pricing, but the effects of competition are even more powerful in terms of lowering prices and raising overall market efficiency.

Note to the Instructor: A simple monopoly experiment can be done with the Veconlab Cournot program by setting the number of sellers to be one. The setup allows price to be subject to random shocks, which adds interest and realism. Students, who may log in from remote locations, can use the price and quantity data to estimate demand and marginal revenue, and thereby to derive the monopoly price that they already discovered through trial and error. The Monopoly/Free-Trade program is also listed on the Veconlab Markets menu.

20.1 Monopoly

A monopolist is defined as being a sole seller in a market, but the general model of monopoly is central to the analysis of antitrust issues because it can be applied more widely. For example, suppose that all sellers in a market are somehow able to collude and set a price that maximizes the total profit, which is then divided among them. In this case, the monopoly model would be relevant, either for providing a prediction of price and quantity, or as a benchmark from which to measure the success of the cartel.

In antitrust analysis, the monopoly model is also applied to the case of one large firm and a number of small "fringe" firms that behave competitively (expanding output as long as the price that they can obtain is above the cost of each additional unit of output). The behavior of the firms in the competitive fringe can be represented by a supply function, which shows the quantity provided by these firms in total as a function of the price. Let this fringe supply function be represented by $S_F(P)$, which is increasing in price if marginal costs are increasing for the fringe firms. Then the *residual demand* is the market demand, $D(P)$, minus the fringe supply, so the residual demand is $R(P) = D(P) - S_F(P)$. This residual demand function indicates a relationship between price and the sales quantity that is not taken by the fringe firms; this is the quantity that can be sold by the large, dominant firm. In this situation, it may be appropriate to treat the dominant firm as a monopolist facing a residual demand function: $Q = R(P)$.

From the monopolist's point of view, the demand (or residual demand) function reveals the amount that can be sold for each possible price, with high prices generally resulting in lower sales quantities. It is useful to invert this demand relationship and think of price as a function of quantity, i.e., selling a larger quantity will reduce price. For example, a linear inverse demand function would have the form: $P = A - BQ$, where A is the vertical intercept of demand in a graph with price on the vertical axis, and $-B$ is the slope, with A and B assumed to be positive. The experiments to be discussed in this chapter all have a linear inverse demand, which for simplicity will be referred to as the demand function. The left side of figure 20.1 shows the results of a laboratory experiment in which each person was a monopolist in a market with a constant cost of $1 per unit and a linear demand curve: $P = 13 - Q$. Since the slope is minus one, each additional unit of output reduces the price by $1 and raises the cost by $1. The vertical axis in the figure is the average of the *quantity* choices made in each round. It is apparent that the participants quickly settle on a quantity of about 6, which is the profit-maximizing choice, as will be verified next.

The demand curve used in the experiment is also shown in the top two rows of table 20.1. Notice that as price in the second row decreases from 12 to 11 to 10, the sales quantity increases from 1 to 2 to 3. The total revenue, PQ, is shown in the third row, and the total cost, which equals quantity, is given in the fourth

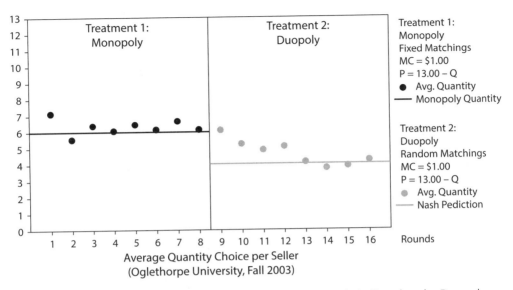

Figure 20.1. Average Quantity Choice under Monopoly (rounds 1–8) and under Duopoly (rounds 9–16)

Table 20.1. Monopoly with Linear Demand and Constant Cost

Q	1	2	3	4	5	6	7	8	9	10	11	12
P	12	11	10	9	8	7	6	5	4	3	2	1
TR	12	22	30	36					36	30	22	12
TC	1	2	3	4					9	10	11	12
Profit	11	20	27	32					27	20	11	0
MR	12	10	8	6					−6	−7	−9	−11
MC	1	1	1	1					1	1	1	1

row. Please take a minute to fill in the missing elements in these rows, and to subtract cost from revenue to obtain the profit numbers that should be entered in the fifth row. Doing this, you should be able to verify that profit is maximized with a quantity of 6 and a price of 7.

Even though the profit calculations are straightforward, it is instructive to consider the monopolist's decision as quantity is increased from 1 to 2, and then to 3, while keeping an eye on the effects of these increases on revenues and costs *at the margin*. The first unit of output produced yields a revenue of 12, so the marginal revenue of this unit is 12, as shown in the MR row at the left. An increase from $Q = 1$ to $Q = 2$ raises total revenue from 12 to 22, which is an increase of 10, as shown in the MR row. These additional revenues for the first and second units are greater than the cost increases of $1 per unit, so the increases

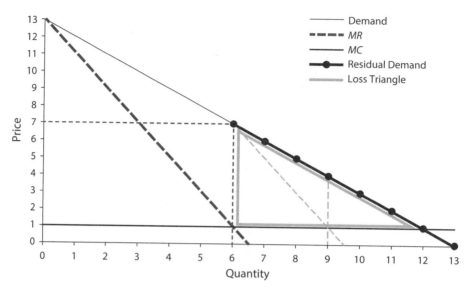

Figure 20.2. Monopoly Profit Maximization

were justified. Now consider an increase to an output of 3. This raises revenue from 22 to 30, an increase of 8, and this marginal revenue is again greater than the marginal cost of 1. Profit increases as long as the marginal revenue is greater than the marginal cost, a process that continues until the output reaches the optimal level of 6, as you can verify.

One thing to notice about the *MR* row of table 20.1 is that each unit increase in quantity, which reduces price by 1, will reduce marginal revenue by 2, since marginal revenue goes from 12 to 10 to 8, etc. This fact is illustrated in figure 20.2, where the demand line is the outer line, and the marginal revenue line is the thick dashed line. The marginal revenue line has a slope that is twice as negative as the demand line. The *MR* line intersects the horizontal marginal cost line at a quantity of 6. Thus the graph illustrates what you will see when you fill out the table, i.e., that marginal revenue is greater than marginal cost for each additional unit, the 1st, 2nd, 3rd, 4th, 5th, and 6th, but the marginal revenue of the 7th unit is below marginal cost, so that unit should not be added. (Note: the marginal revenues in the table will not match the numbers on the graph exactly, since the marginal revenues in the table are for going from one unit up to the next. For example, the marginal revenue in the sixth column of the table will be 2, which is the increase in revenue for going from 5 to 6 units. Think of this quantity as 5.5, and the marginal revenue for 5.5 in the figure is exactly 2.) In addition to the table and the graph, it is useful to redo the same derivation of the monopoly quantity using simple calculus. (A brief review of the needed calculus formulas is provided in the appendix to this chapter.) Since demand is: $P = 13 - Q$, it follows that total revenue, PQ, is $(13 - Q)Q$, which is a quadratic function of output: $13Q - Q^2$. Marginal revenue (the derivative of total revenue) is $13 - 2Q$. This marginal revenue is a line that starts at a value of 13 when $Q = 0$ and declines by 2 for each unit increase in quantity, as shown in figure 20.2. Since marginal cost is 1, it follows that marginal revenue equals marginal cost when $13 - 2Q = 1$, i.e., when $Q = 6$, the monopoly output for this market.

A case against monopoly is illustrated in figure 20.2, where the monopoly price is $7 and the marginal cost is only $1. Thus, buyers with valuations between $1 and $7 would be willing to pay amounts that cover cost, but the monopoly does not satisfy this demand, in order to keep price and earnings high. As shown in chapter 2, the loss in buyers' value can be measured as the area under the demand curve, and the net loss is obtained by looking at the area below demand, to the right of 6 and above the marginal cost line in the figure. This triangular area, bounded by light gray lines, is a measure of the welfare cost of monopoly, as compared with the competitive outcome where price is equal to marginal cost ($1) and the quantity is 12. Of course, the actual costs of monopoly can be much larger if wasteful competition (rent seeking) was involved in securing the monopoly position, as discussed in chapter 12. Another hidden cost of monopoly could be in terms of reduced incentives to innovate if the monopoly position is protected by government restrictions.

The convergence of the quantity to the monopoly prediction on the left side of figure 20.1 was observed in a context where the demand side of the market was simulated. This kind of experiment is more appropriate if one is thinking of a market with a large number of consumers, none of whom have any significant size or power to bargain for reductions from the monopoly price.

20.2 Cournot Duopoly

Next consider what would happen if a second firm were to enter the market. In particular, suppose that both firms have constant marginal costs of $1, and that they each select an output quantity, with the price then being determined by the sum of their quantities, using the top two rows of table 20.1. This duopoly structure was the basis for the second part of the experiment on the right side of figure 20.1. Notice that the quantity *per seller*, which is what is shown on the vertical axis, starts out at the monopoly level of 6 in round 9. With 2 sellers producing 6 units each, the total quantity of 12 forces price down to $1 ($= 13 - 12$). When price is $1, which equals the cost per unit, it follows that earnings were zero in that round. The average quantity is observed to fall in round 10; which is not surprising following a round with zero profit. The incentive to cut output can be seen from the graph in figure 20.2. Suppose that one seller (the entrant) knew the other would produce a quantity of 6. If the entrant were to produce 0, the price would stay at the monopoly level of $7. If the entrant were to produce 1, the price would fall to $6, etc. These price/quantity points for the entrant are shown as the dark line with dots, labeled "Residual Demand" in figure 20.2. The marginal revenue for this residual demand curve has a slope that is twice as steep, as shown by the thin dotted line that crosses MC at a quantity of 9. This crossing determines an output of 3 for the entrant, since the incumbent seller is producing 6. In summary, when one firm produces 6, the best response of the other is to choose a quantity of 3. This suggests why the quantities, which start at an average of 6 for each firm in period 9, begin to decline in subsequent periods, as shown on the right side of figure 20.1. The outputs fall to an average of 4 for each seller, which suggests that this is the equilibrium, in the sense that if one seller is choosing 4, the best response of the other is to choose 4 also.

In order to show that the Cournot equilibrium is in fact 4 units per seller, we need to consider some other best-response calculations, which are shown in table 20.2. This table shows a firm's profit for the example under consideration for each of its own output decisions (listed in rows, increasing from bottom to top) and for each output decision of the other firm (listed in columns, increasing from left to right). Consider the column labeled "0" on the left side, i.e., the column that is relevant in the monopoly case in which the other firm's output is 0. If the "Column" firm produces nothing, then the "Row" firm's profits in this column are just copied from the monopoly profit row of table 20.1: a profit of 11 for an output of 1, 20 for an output of 2, etc. The highest profit in this column is 36 in the upper-left corner, at the monopoly output of 6. This profit is indicated by a single asterisk. To test your understanding, fill in the two missing numbers in the right column. Some of the other best-response payoffs are also indicated by a *single asterisk*. Recall that the firm's best response to another firm's output of 6 is to produce 3 (as seen in figure 20.2), and this is the payoff in the third row and far right column, 9*, which is labeled with a single asterisk. A Nash equilibrium in this duopoly market is a pair of outputs, such that each

Table 20.2. A Row Seller's Own Profit Matrix

		Column Firm's Output						
		0	1	2	3	4	5	6
	6	36*	30*	24	18	12	6	
	5	35	30*	25*	20*	15	10	
Row's	4	32	28	24	20*	16**	12*	8
Own Output	3	27	24	21	18***	15	12*	9*
	2	20	18	16	14	12	10	8
	1	11	10	9	8	7	6	5

Key: * indicates Row's Best Response; ** indicates a Cournot equilibrium; and *** indicates a joint-profit maximizing outcome.

seller's output is the best response to that of the other. Even though 3 is a best response to 6, the pair (6 for one, 3 for the other) is not a Nash equilibrium (problem 1). The payoff of 16, marked with a double asterisk in the table, is the location of a Nash equilibrium, since it indicates that an output of 4 is a best response to an output of 4, so if each firm were to produce at this level, there would be no incentive for either to change unilaterally. Of course, if they could coordinate on joint output reductions to 3 each, the total output would be 6 and the industry profit would be maximized at the monopoly level of 36, or 18 each. This joint maximum is indicated by the *triple asterisk* at the payoff of 18 in the table. This joint maximum is not a Nash equilibrium, since each seller has an incentive to expand output.

The fact that the Nash equilibrium does not maximize joint profit raises an interesting behavioral question, i.e., why were subjects in the experiment unable to coordinate on quantity restrictions to raise their joint earnings? The answer is that the matchings were random for the duopoly phase. Thus, each seller was matched with a randomly selected other seller in each round, and this switching would make it difficult to coordinate quantity restrictions. In experiments with fixed matchings, such coordination is often observed, particularly with only two sellers. Holt (1985) reported patterns where sellers sometimes "walked" the quantity down in unison, e.g., both duopolists reducing quantity from 7 to 6 in one period, and then to 5 in the next, etc. This kind of "tacit collusion" occurred even though sellers could not communicate explicitly. There were also cases where one seller produced a very large quantity, driving price to 0, followed by a large quantity reduction in an effort to send a threat and then a conciliatory message, and thereby induce the other seller to cooperate. Such tacit collusion is less common with more than two sellers. Part of the problem is that when one seller cuts output, the other has a unilateral incentive to expand output, so when one shows restraint, the other has greater temptation. Also, an output expansion by one seller intended to punish another will lower the price and hurt all sellers, so these types of punishments cannot be targeted.

The Nash equilibrium just identified is also called a *Cournot equilibrium*, after the French mathematician who provided an analysis of duopoly and oligopoly models in 1838. Although the Cournot equilibrium is symmetric in this case, there can be asymmetric equilibria as well. A further analysis of table 20.2 indicates that there is at least one other Nash equilibrium, also with a total quantity of 8, and hence an average of 4, even though the two firms' quantities are not equal. Can you find this asymmetric equilibrium (problem 3)?

As was the case for monopoly, it is useful to illustrate the duopoly equilibrium with a graph. If one firm is producing an output of 4, then the other can produce an output of 1 (total quantity = 5) and obtain a price of 8 (= 13 − 5), as shown by one of the residual demand dots in figure 20.3. Think of the vertical axis as having shifted to the right at the other firm's quantity of 4, as indicated

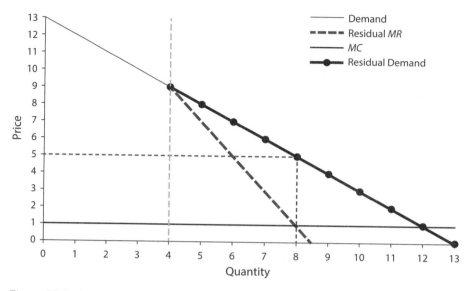

Figure 20.3. Cournot Duopoly: Maximization Given Residual Demand

by the vertical dashed line, and the residual demand dots yield a demand curve with a slope of minus 1. As before, marginal revenue will have a slope that is twice as negative, as indicated by the heavy dashed line in the figure. This line crosses marginal cost just above the quantity of 8, which represents a quantity of 4 for this firm, since the other is already producing 4. Thus the quantity of 4 is a best response to the other's quantity of 4. As seen in the figure, the price in this duopoly equilibrium is $5, which is lower than the monopoly price of $7.

20.3 Cournot Oligopoly

The classroom experiment shown in figure 20.4 involved fixed matchings, for duopoly markets (left side) and then for three-firm markets (right side). Despite the fixed nature of the matchings, participants were not able to coordinate on output reductions below the duopoly prediction of 4 per seller. In the triopoly treatment, the outputs converge to 3 on average. Notice that outputs of 3 for each of three firms translates into an industry output of 9, as compared with 8 for the duopoly case (4 each) and 6 for the monopoly case. Thus we see that increases in the number of sellers raise the total quantity and reduce price toward competitive levels.

For the triopoly (3-seller) markets, it is straightforward to redraw figure 20.3 to show that if two firms produce 3 units each, then the remaining firm would also want to produce 3 units (problem 4). Instead, we will use a simple derivation based on the fact that when demand is linear, the marginal revenue has a slope that is twice as negative. It is also convenient to derive the equilibrium for the N-firm case, and come back to the triopoly calculation afterward.

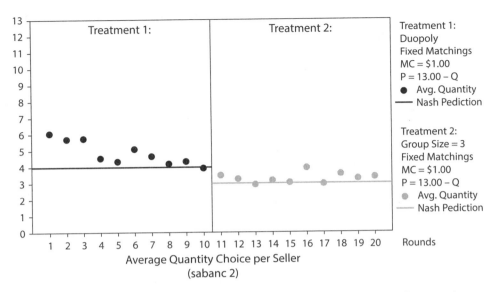

Average Quantity Choice per Seller
(sabanc 2)

Figure 20.4. Average Quantities Per Seller in a Classroom Experiment with Duopoly
(rounds 1–10) and Triopoly (rounds 11–20)

First, suppose that the $N-1$ other firms each produce an output of X units, for a total of $(N-1)X$. If the firm we are considering chooses an output of Q, then the industry output is equal to the sum of $(N-1)X$ produced by others and Q produced by the firm under consideration. The resulting price is equal to A minus B times this output, as shown in the top equation in (20.1). Then total revenue for the firm is found by multiplying this price times Q, to get the TR equation just below demand.

(20.1)
$$P = A - B(N-1)X - BQ$$
$$TR = AQ - B(N-1)XQ - BQ^2$$
$$MR = A - B(N-1)X - 2BQ$$

The marginal revenue in the bottom line of (20.1) is found by using the fact that its slope is twice as negative as that of the inverse demand, so replace the $-BQ$ term in the P equation above with $-2BQ$, as shown in the bottom equation in (20.1). (Alternatively, you could have taken the derivative of TR with respect to Q to obtain MR, as explained in the appendix to this chapter; the 2 comes from the fact that the derivative of Q^2 is $2Q$.)

As was the case for a monopolist, the firm should expand output as long as the marginal revenue is greater than the marginal cost, and the optimal output is found by equating marginal revenue with marginal cost, as shown in the top line in (20.2).

(20.2) $C = A - B(N-1)X - 2BQ$ (marginal cost = marginal revenue)
$\quad\quad\quad C = A - B(N-1)Q^* - 2BQ^*$ (using symmetry)

Just as the duopoly outputs were equal at 4 for the example in figure 20.3, there will be a symmetric equilibrium in which $Q = X = Q^*$, which denotes the common equilibrium output. This substitution produces the bottom equation in (20.2), which can be solved for the Cournot outputs:

(20.3) $$Q^* = \frac{A - C}{(N + 1)B} \quad \text{(Cournot equilibrium)}$$

Notice that the formula in (20.3) reproduces what we have shown so far for the case when $A = 13$, $B = 1$, and $C = 1$. Using $N = 1$, the formula yields the monopoly output, $Q^* = (13 - 1)/2 = 6$, and with $N = 2$, it yields the duopoly output of 4. With three firms, the equilibrium output is 3, as the average quantity series on the right side of figure 20.4 would have suggested. As N increases, the output per firm goes down, but the total industry output goes up. To see this, multiply the right side of (20.3) by N and consider what happens as N increases (problem 5).

In research experiments with quantity choices, simulated demand, and fixed matchings, it is common for tacit collusion to generate prices above Cournot predictions. But since the seller with the greatest output earns the most, rivalistic incentives can cause sellers to expand production above Cournot levels (Holt, 1985). This effect is magnified when subjects can observe others' earnings levels, so "imitation" of those with the highest earnings would tend to increase outputs (Hück, Normann, and Oechssler, 1999). To summarize:

Cournot Experiments: *In a Cournot equilibrium, each seller's quantity choice maximizes its earnings given others' quantity choices. This can be represented as each seller facing a residual demand (left over from others' quantity sales) and equating marginal revenue from this residual demand to marginal cost. With only 2 sellers in the market, prices sometimes rise above (and quantities fall below) those determined by the Cournot equilibrium. But rivalistic incentives tend to force quantities up to Cournot predictions with more than 2 or 3 sellers, especially when others' earnings can be observed.*

20.4 Monopoly versus Free Trade

Shopping these days often begins with an Internet search that displays consumer satisfaction ratings that can be used to select one or more brands. Then a second search can reveal prices and shipping costs and local availability (or the option to be notified when inventories arrive at a local outlet). Often the easiest and quickest route is to order the selected brand online, pay with a credit card, and arrange delivery to a front porch or other convenient location. This seemingly simple transaction requires a considerable level of trust, in terms of reliable consumer ratings, shipping, credit card payment, product reliability, and convenient return options. The trust built into these commercial practices has

Table 20.3. Values and Costs in Each Local Monopoly Market

	Buyer 1	Buyer 2	Buyer 3	Seller 4
1st Unit	$6.50	$5.50	$4.50	$0.50
2nd Unit	$1.50	$2.50	$3.50	$0.50
Units 3–9				$0.50 for all units

vastly expanded the scope of markets in many developed countries, especially in the United States and the European Union. Things have not always been this way, and shipping and payments are much riskier in many areas of the world. In others, tariffs and fees pose barriers to cross-market competition.

The Veconlab Monopoly and Free Trade experiment begins with several periods of trading in which sets of three buyers are located in separate local markets with a single seller who posts a price and maximum sales quantity. Buyers indicate amounts they are willing to purchase at the posted local price, and the monopolist receives the posted price and pays the unit cost for units sold, but not for unsold units offered. Buyers in the default setup have values for 2 units each, which are rotated each round, subject to the constraint that each buyer has one high-value unit (value above $4) and one low-value unit. Buyers earn the difference between their values and prices paid for units purchased. If multiple buyers indicate a willingness to purchase from the seller, then the allocation is decided by a random selection process (as if buyers arrive at the market in a random order). The values (subject to rotation) are shown in table 20.3.

Prices in the experiment are constrained to be integer amounts. So a price of $6 would result in sales of 1 unit, unless the buyer refused. Similarly, a price of $5 would result in a sale of 2 units, a price of $4 would result in a sale of 3 units, and so on. Thus, demand rises by 1 unit for each dollar reduction in price.

Total revenues can be calculated under the (tenuous) assumption that buyers do not withhold purchases strategically. At a price of $6 and quantity of 1, $TR = 6$ and $MR = 6$ for the first unit. At a price of $5 and quantity of 2, $TR = 10$, so the marginal revenue of the second unit is $10 - 6 = 4$. Similarly, it can be shown that the marginal revenues are 2 and 0 for the third and fourth units. To summarize, as quantities rise from 1 to 2 to 3 to 4, prices fall from $6 to $5 to $4 to $3, and marginal revenues fall twice as fast, from $6 to $4 to $2 to $0. Since marginal cost is $0.50, the fourth unit with a marginal revenue of $0 should not be sold, and hence, the monopolist's optimal quantity is 3 units with a price of $4. Each buyer ends up purchasing one high-value unit at this monopoly price, as indicated in the top row of table 20.3. The total surplus would be the difference between the sum of the three highest values, $6.5 + 5.5 + 4.5 = 16.50$ and the sum of the associated costs $0.5 + 0.5 + 0.5 = 1.5$. This difference is $15, which is the predicted surplus under monopoly.

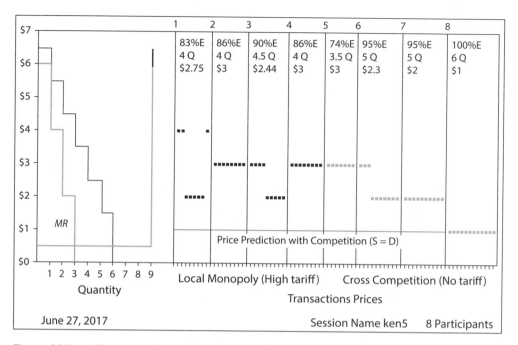

Figure 20.5. A Classroom Experiment with Local Monopoly (rounds 1–4) and Free Trade (rounds 5–8)

The second treatment involves letting buyers purchase from any seller in any market, so they see all prices posted by sellers in their own market and in markets in other localities. Another way in which this treatment change can be framed in the experiment is to have a tariff for cross-market purchases that is so high in the first treatment that it results in no cross-purchases. The tariff would be removed in the second treatment. With free trade, the price will (in theory) be driven down to $1, the lowest integer price above cost, and all 6 units in each local market will sell. The surplus would be the sum of the 6 values in table 20.3, which is $24, minus the sum of the costs for 6 units, which is $3, for a total of $21. Recall that the surplus under monopoly is $15. The tariff removal is predicted to raise efficiency from the monopoly level 15/21 to 21/21, or from 71% to 100%.

Figure 20.5 shows classroom experiment results with two local monopoly markets (3 buyers and 1 seller in each). Sellers did not know buyers' values and vice versa. In the 4 rounds of local monopoly trading, the local prices averaged about $3, with a quantity of 4. This price is lower than the monopoly prediction of $4 with 3 units traded, which was derived under the implicit assumption of full information and passive, price-taking buyers. In the 4 rounds of free cross-market trade, the price fell steadily, reaching $1 in the final round, which resulted in 100% efficiency. The figure also shows the marginal revenue line on the left (with steps at $6, $4, $2, $0), which intersects the marginal cost line at the monopoly quantity of 3. The tendency for active buyers to enhance seller competition, observed in the first 4 rounds, will be encountered again in the next chapter. To summarize:

> **Local Monopoly and Free-Trade Experiment:** *Free trade breaks up local monopoly power, and the resulting cross-market price competition drives prices down to competitive levels and raises efficiency.*

20.5 Extensions

The Cournot model is perhaps the most widely used model in theoretical work in the subfield of Industrial Organization. Its popularity is based on tractability and on the prediction that the equilibrium price will be a decreasing function of the number of sellers. This prediction is supported by evidence from laboratory experiments that implement the Cournot assumption that firms select quantities independently.

In contrast, the Cournot model is rarely used in antitrust case arguments. The obvious shortcoming of the model is the specific way in which price is determined. The implicit assumption is that firms make quantity decisions independently, and then price is cut so that all production can be sold. For example, you might think of a situation in which the quantities have been produced, so the short-run supply curve is vertical at the total quantity, and price is determined by the intersection of this vertical supply function with the market demand function. In other words, the price-competition phase is implicitly assumed to be extremely competitive. There is some game-theoretic and experimental evidence to support this view. Even so, a Cournot equilibrium would not be appropriate if it is *price*, not quantity, that firms set independently, and then produce to fill the orders that arrive. Independent price choice may be an appropriate assumption if firms mail out catalogues or post "buy now" prices on the Internet, with the ability to quickly fill orders. Some of the richer and more relevant models of price competition with discounts will be discussed in the next chapter.

Appendix: Optional Quick Calculus Review

The derivative of a linear function is just its slope. For the demand function: $P = 13 - Q$ in figure 20.2, the slope is -1 (each unit increase in quantity decreases price by \$1). To find the slope using calculus, we need a formula: the derivative of BQ with respect to Q is just B, for any value of the slope parameter B. Thus the derivative of $(-1)Q$ is -1. The slope of the demand curve is the derivative of $13 - Q$ and we know that the derivative of the second part is -1, which is the correct answer, so the derivative of 13 must be 0. In fact, the derivative of any constant is zero. To see this, note that the derivative is the slope of a function, and if you graph a function with a constant height, then the function will have a slope of 0 in the same manner that a table top has no slope. For example, consider the derivative of a more general linear demand function: $A - BQ$. The intercept, A, is a constant (it does not depend on Q, which is variable, but rather it stays the same). So the derivative of A is 0, and the derivative of $-BQ$ is $-B$, and therefore the derivative of $A - BQ$ is $0 - B = -B$.

Here we have used the fact that the derivative of the sum of two functions is the sum of the derivatives. This sum of functions rule is intuitive. Suppose Fred and Grace sell beer and sandwiches to passengers on a tourist ferry (Fred sells the beer), and their sales depend on the quantity Q of passengers. Let Fred's sales be a function $F(Q)$, and let Grace's sales revenues be $G(Q)$. Then the total sales revenues are $F(Q) + G(Q)$, and the change in the total revenue as Q increases would be the change for Fred's beer revenues plus the change for Grace's sandwich revenues, i.e., the sum of the derivatives. To summarize (ignore rules 4 and 5 for the moment):

(1) **Constant Function:** $dA/dQ = 0$.

   The derivative of a constant like A is just 0.

(2) **Linear Function:** $d(KQ)/dQ = K$.

   The derivative of a constant times a variable, which has a constant slope, is just the constant slope parameter, i.e., the derivative of KQ with respect of Q is just K.

(3) **Sum of Functions:** .

   The derivative of the sum of two functions is the sum of the derivatives.

(4) **Quadratic Function:** $d(KQ^2)/dQ = 2KQ$.

   The derivative of a quadratic function is obtained by moving the 2 in the exponent down, so the derivative of KQ^2 is just $2KQ$.

(5) **Power Function:** $d(KQ^x)/dQ = xKQ^{x-1}$.

   The derivative of a variable raised to the power x is obtained by moving the x down and reducing the power by 1, so the derivative of KQ^3 is just $3KQ^2$, the derivative of KQ^4 is $4KQ^3$, and in general, the derivative of KQ^x with respect to Q is xKQ^{x-1}.

For example, the monopolist being discussed has a constant marginal cost of \$1 per unit, so the total cost of for producing Q units is just Q dollars. Think of this total cost function as being the product of 1 and Q, so the derivative of $1Q$ is just 1 using rule 2 above. If there had been a fixed cost of F, then the total cost would be $F + Q$. Note that F is just a constant, so its derivative is 0 (rule 1), and the derivative of this total cost function is the derivative of the first part (0) plus the derivative of the second part (1), so marginal cost is again equal to 1.

Next consider a case where demand is $P = A - BQ$, which has a vertical intercept of A and a slope of $-B$. The total revenue function is obtained by multiplying by Q to get: the total revenue function: $AQ - BQ^2$, which has a linear term with slope A and a quadratic term with a coefficient of $-B$. We know that the derivative of the linear part is just A (rule 2). The fourth rule indicates how to take the derivative of $-BQ^2$, you just move the 2 in the exponent down, so the derivative of this part is $-2BQ$. We add these two derivatives together to determine that the derivative of the total revenue function is: $A - 2BQ$, so marginal revenue has the same vertical intercept as demand, but the slope is twice as negative. This is consistent with the calculations in table 20.1, where each unit increase in quantity reduces price by \$1 and reduces marginal revenue by \$2.

Chapter 20 Problems

1. Use table 20.2 to show that outputs of 6 for one firm and 3 for the other do not constitute a Nash equilibrium for the duopoly model that is the basis for that table.

2. Use table 20.2 to show that outputs of 3 for each firm do not constitute a Cournot/Nash equilibrium.

3. Find an asymmetric Cournot/Nash equilibrium in table 20.2 with the property that the total quantity is 8 but for which one seller produces more than the other. Therefore, you must specify what the two outputs are, and you must show that neither seller has a unilateral incentive to deviate. (This asymmetric equilibrium is an artifact of the discrete nature of the quantity choices, which are constrained to be integers.)

4. Redraw figure 20.3 for the triopoly case, putting the vertical dotted line at a quantity of 6 (three for each of two other firms) and show that the residual marginal revenue for the remaining firm would cross marginal cost in a manner that would make the output of 3 a best response for that firm.

5. Show that the price is a decreasing function of the number of firms, N, in the Cournot equilibrium for the linear model with the equilibrium quantity given in equation (20.3).

6. It is useful to relate the formula in (20.3) to the graph. If figure 20.3 were to be redrawn for the general linear demand function, i.e., $P = A - BQ$, then the

vertical intercept would be A and the horizontal intercept would be obtained by setting $P = 0$ and solving to get $Q = A/B$. Notice the ratio $(A - C)/B$ in equation (20.3). Now look at the horizontal line in the figure that has a height equal to marginal cost; it has length $(A - C)/B$. This line is divided into 3 equal segments in the duopoly graph in figure 20.3, just as the MR line divided it into 2 equal segments for the monopoly graph. In general, this line gets divided into $N + 1$ segments. Use geometric arguments to explain why this line has length $(A - C)/B$ and why it gets divided into $N + 1$ segments in a symmetric Cournot equilibrium.

21

Market Power, Collusion, and Price Competition

Traders in a double auction can see all transactions prices and the current bid/ask spread, as is the case with trading on the New York Stock Exchange. The double auction is an extremely competitive institution, given the temptation for traders to improve their offers over time in order to make trades at the margin. In contrast, markets with posted prices allow sellers to pre-commit to fixed, take-it-or-leave-it prices that cannot be adjusted during a trading period. This chapter begins with a consideration of the price and efficiency outcomes of such markets, in particular when sellers possess *market power*, which is, roughly speaking, the ability of a firm to raise price profitably above competitive levels. In most cases, the market efficiency is higher in the double auction, since the price flexibility built into that institution tends to bring outcomes closer to a competitive (supply equals demand) outcome.

The second part of the chapter pertains to price collusion, which is more effective when sellers are unable to offer secret discounts from posted prices to individual buyers. Such discounting tends to break collusion in laboratory experiments, and antitrust authorities have (at times) been concerned with business practices that limit discounting opportunities. This chapter is a little longer than others, but the appendix is optional, and the story in the second part of the chapter is relatively intuitive and non-technical.

Note to the Instructor: The various market institutions can be implemented with the Veconlab Double Auction, Posted Offer, or Bertrand programs.

21.1 Price Adjustment in Double Auctions and Posted Offer Markets

The common perception that laboratory markets yield efficient competitive outcomes is surprising given the emphasis on market imperfections that pervades theoretical work in industrial organization. This apparent contradiction is resolved by considering the effects of *trading institutions*. As discussed in chapter 2, competitive outcomes are typical in "double auction" markets, with rules similar to those used in many centralized financial exchanges. But most markets of interest to industrial organization economists have different institutional characteristics; sellers post prices and buyers must either buy at those prices or engage in costly search and negotiation to obtain discounts. Unlike more competitive double auctions, the performance of markets with posted prices can be degraded by the presence of market power, price-fixing conspiracies, and cyclical demand shocks.

It is common for prices to be set by the traders on the thin side of the market. Sellers, for example, typically post prices in retail markets. For this reason, theoretical price competition models are often structured around an assumption that prices are listed simultaneously at the beginning of each "period," which is known as "Bertrand competition." Using laboratory experiments, it is possible to make controlled comparisons between markets with posted prices and more symmetric institutions such as the double auction, where both buyers and sellers post bids and asks in an interactive setting that resembles a centralized stock market. Laboratory double auctions yield efficient, competitive outcomes in a surprisingly wide variety of settings, sometimes even in a monopoly when the buyers are not simulated (Smith, 1962, 1981). In contrast, prices in markets with posted prices are often above competitive levels (Plott, 1989; Davis and Holt, 1993).

Figures 21.1 and 21.2 show a matched pair of markets, one double auction and one posted-offer auction, which were run under research conditions with payments equaling 1/5 of earnings. The values and costs were arrayed in a manner such that predicted earnings per person were approximately equal. In each case, the buyer values were reduced after period 5 by shifting demand down by \$4, thereby lowering the competitive price prediction by \$2, as shown on the left sides of each figure. For example, the two buyers with the highest-value units, at \$15 in period 5, had these values reduced to \$11 in period 6, whereas costs stayed the same in all periods.

First, consider the double auction, in figure 21.1. The vertical lines to the right of the supply and demand curves indicate the breaks between trading periods, and the dots represent transactions prices in the order in which they were observed. Recall that a trader in a double auction may make a price offer (bids for buyers and asks for sellers) or may accept the best offer on the other side of the market. Thus, traders typically see a sequence of declining ask prices and/or increasing bid prices, until the bid-ask spread narrows and one person accepts another's proposal.

Despite the considerable price variation in the first period, the double auction market achieved 99% of the maximum possible surplus, and this efficiency measure averaged 98% in the first 5 rounds, with quantities of 14 that equal the competitive prediction. All traders were told that some payoff parameters may have changed in round 6, but sellers observed that their own costs had not changed.

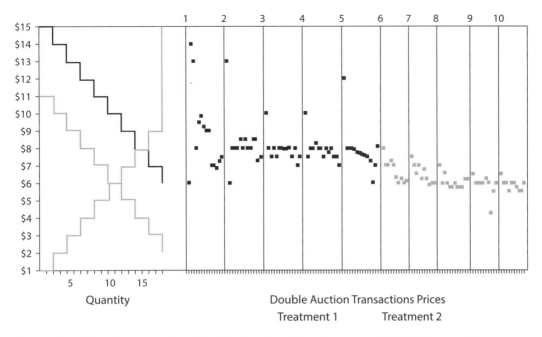

Figure 21.1. Price Sequence for a Double Auction with a Downward Shift in Demand after Round 5

However, sellers had no idea whether buyer values or others' costs had gone up or down. Transactions prices began in round 6 at the old competitive level of $8, but fell to $6 by the end of the period. This is a typical pattern, where the high-value and low-cost units trade early in a period at levels close to those in the previous periods, but late trades for units near the margin are forced to be closer to the supply-demand intersection price. The transaction quantity was at the competitive prediction of 10 in all rounds of the second treatment. As early-period prices fell toward the competitive level after round 6, the price averages were about equal to the new prediction of $6, and efficiencies were at about 96%.

Recall from the discussion in chapter 2 that buyers do not post bids in a posted-offer market. Sellers post prices independently at the start of each period, along with the maximum number of units that are offered for sale. Then buyers are selected in a random order to make desired purchases. The period ends when all buyers have finished shopping or when all sellers are out of stock. As shown in figure 21.2, prices in the posted-offer market also began in the first period with considerable variation, although all prices were far from the competitive prediction. Prices in periods 2–4 seemed to stop rising at a level of about $1 above the competitive prediction. Both the average quantity (12) and the efficiency (86%) were well below the competitive levels observed in the matched double auction. Prices fell slowly after the demand shift in period 6, but they never quite reached the new competitive prediction, and efficiencies averaged

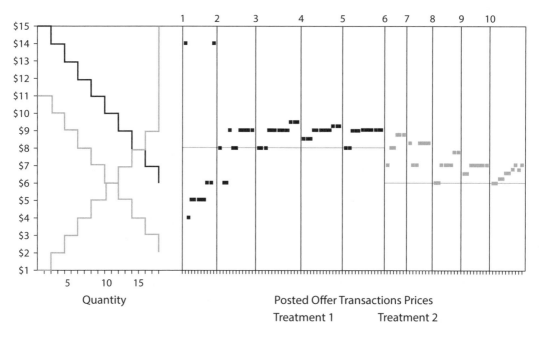

Figure 21.2. Price Sequence for a Posted-Offer Auction with a Downward Shift in Demand after Round 5

86% in the last 5 periods. The lowest efficiencies were observed in the first two periods after the demand shift, illustrating the more sluggish adjustment of the posted-offer market, which benefited sellers, as they were able to maintain prices above competitive levels. By the final period of each treatment, efficiencies had climbed above 90%, and the transactions quantity had reached the relevant competitive prediction.

The results shown in figures 21.1 and 21.2 are fairly typical. Compared with double auctions, laboratory posted-offer markets converge to competitive predictions more slowly (Ketcham, Smith, and Williams, 1984) and less completely (Plott, 1986, 1989). Even in non-monopolized designs with stationary supply and demand functions, traders in a posted-offer market generally forego about 10% of the possible gains from trade, whereas traders in a double auction routinely obtain 95–98% of the total surplus in such designs.

The sluggish posted-price responses shown in figure 21.2 are even more apparent in markets with sequences of demand shifts that create a boom and bust cycle. First consider a case where the supply curve stayed stationary as demand shifted up repeatedly in a sequence of periods, creating an upward momentum in expectations. This "boom" was followed by a sequence of unannounced downward demand shifts that reduced the competitive price prediction incrementally until it returned to the initial level. Prices determined by double auction trading tracked the predicted price increases and decreases fairly accurately, with high efficiencies. This is because the demand shifts are conveyed by the intensity of buyer bidding behavior during each period, so that sellers could learn about new market conditions as they started making sales. In contrast, prices in posted-price markets are selected before any shopping begins, so sellers cannot spot changes in market conditions, but rather, must try to make inferences from sales quantities. When posted-offer markets were subjected to the same sequence of successive demand increases and decreases mentioned above, the actual trading prices lagged behind competitive predictions in the upswing, and prices continued to rise even after demand started shifting downward. Then prices fell too slowly, relative to the declining competitive predictions (where supply equals demand). The result was that prices stayed too high on the downswing part of the cycle, and these high prices caused transactions quantities to fall dramatically, essentially drying up the market for several periods and causing severe profit reductions for sellers (Davis and Holt, 1996). To summarize:

> **Price Adjustment and Market Institutions:** *Prices in double auctions respond more sharply to shifts in supply or demand conditions, as compared with posted-price markets. If a posted-price market is subjected to a series of upward or downward demand shifts, the posted prices lag behind the supply-demand predictions. In the downward (bust) phase, prices may fall so slowly that buyers purchase very little, and the market may even "freeze up" with few transactions.*

21.2 The Exercise of Seller Market Power without Explicit Collusion

One of the major factors considered in the antitrust analysis of mergers between firms in the same market is the possibility that a merged firm may be able to raise prices, to the detriment of buyers. Of course, any seller may raise a price unilaterally, and so the real issue is the extent to which price can be raised *profitably*. Such a price increase is more likely to boost profits if others in the market are not in a position to absorb increases in sales. Therefore, the capacities of other sellers may constrain a firm's market power, and a merger that reduces others' capacities may create market power. This raises the question of whether these high prices can be explained by game-theoretic calculations. It is straightforward to specify a game-theoretic definition of *market power*, based on the incentive of one seller to raise price above a common competitive level (Holt, 1989). In other words, market power is said to exist when the competitive equilibrium is not a Nash equilibrium.

Although it is straightforward to check for the profitability of a unilateral deviation from a competitive outcome, it may be more difficult to identify the Nash equilibrium for a market with posted prices. The easiest case is where firms do not have constraints on what they can produce, a case commonly referred to as *Bertrand competition*. For example, if each firm has a common, constant marginal cost of C, then no common price above C would be a Nash equilibrium, since each firm would have an incentive to cut price slightly and capture all market sales. The Bertrand prediction for a price competition game *played once* is for a very harsh type of competition that drives price to marginal cost levels, even with only two or three firms.

Even with a repeated series of market periods, the one-shot Nash predictions may be relevant if random matchings are used to make the market interactions have a one-shot nature, where nobody can punish or reward others' pricing decisions in subsequent periods. Most market interactions are repeated, but if the number of market periods is fixed and known, then the one-shot Nash prediction applies in the final period, and a process of backward induction can be used to argue that prices in all periods would equal this one-shot Nash prediction. In most markets, however, there is no well-defined final period, and in this case, there is a possibility that a kind of tacit cooperation might develop. In particular, a seller's price restraint in one period might send a message that causes others to follow suit in subsequent periods, and sellers' price cuts might be deterred by the threat of retaliatory price cuts by others. There are many ways that such tacit cooperation might develop, as supported by punishment strategies, as indicated earlier, in chapter 11.

The Bertrand arguments in the preceding paragraphs depend on having a wide range of price choices, so that "small" price cuts are possible. This observation suggests that a tacit agreement to stick with a discrete price grid may be anti-competitive. For example, Cason (2000) uses laboratory experiments to show that the NASDAQ dealers' convention of relying on "even eighths" may have facilitated collusion. Figure 21.3 shows results for a market in which prices were constrained to be in integer dollar amounts.

The two sellers had 3 units each, with costs of $3, $4, and $4.50, as shown in the supply function on the left side of the figure. The competitive equilibrium price is $5 as determined by the intersection of supply and demand. This design was not selected for a research project, but rather, to illustrate the differences between double auction and posted-offer trading. The single session was run by hand with three periods of double auction trading, followed by three periods of posted-offer trading. Price sequences for each period are connected by a dark line, and the break in the line indicates the end of a period. The double auction prices on the left side fell to $5 at the end of all three trading periods, as sellers scrambled to sell their marginal units for a $0.50 profit. This competition resulted in 6 units being traded and an efficiency of 100% in all periods, as indicated by the percentages below the price sequences for each period. In the posted-offer periods that followed, sellers ended up raising prices to $7 and reducing sales to 4 units, which reduced efficiencies to 92%. With prices of $7, each seller would sell two units and earn $7 + $7 − $3 − $4 = $7. If one

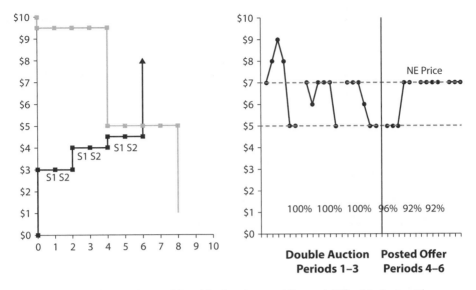

Figure 21.3. A Comparison of Double Auction and Posted-Offer Markets with Discrete Prices

seller prices at $7 and the other cuts price to $6, the one with the lower price would sell three units and earn $6 + $6 + $6 − $3 − $ 4 − $4.50 = $6.50, which indicates that there is a Nash equilibrium at $7. (You would have to check on whether a unilateral price increase is profitable; see problem 1.) Thus the prices in the posted-offer periods rose to Nash equilibrium levels.

The effects of market power in a double auction were the focus of an experiment run by Holt, Langan, and Villamil (1986), using a five-seller design that is more complex than that in figure 21.3, but with only a single unit of excess supply at prices just above the competitive level. Prices were not constrained to be in integer dollar amounts, and prices did exceed competitive levels in about half of the sessions. Even though the supply and demand design was selected to give sellers a strong incentive to exercise market power, prices in the other sessions did converge to competitive levels, which shows how competitive the double auction institution is. Davis and Williams (1991) replicated the Holt, Langan, and Villamil results for double auctions, and the resulting prices were slightly above competitive levels in most sessions. In addition, they ran a second series of sessions using posted-offer trading, which generated higher prices resulting from sellers withholding their marginal units.

Vernon Smith (1981) investigated an extreme case of market power, with a single seller in a market with a number of non-simulated human buyers. He found that even monopolists in a double auction are sometimes not able to maintain prices above competitive levels, whereas posted-offer monopolists are typically able to find and enforce monopoly prices. The difference is that a seller in a posted-offer auction sets a single, take-it-or-leave-it price, so there is no temptation to cut price late in the period in order to sell marginal units. This temptation is present in a double auction. In particular, a monopolist who cuts price late in a trading period to sell marginal units will have a harder time selling units at near-monopoly levels at the start of the next period, and this buyer resistance may cause prices in a double auction monopoly to be lower than the monopoly prediction. To summarize:

Market Power and Market Institutions: *Market power effects show up more clearly in settings where prices are posted on a take-it-or-leave-it basis. In contrast, prices conform more closely to competitive supply/demand predictions in double auctions, with continuous opportunities for within-period price reductions.*

21.3 Edgeworth Cycles and Random Prices

When all sellers offer the same homogeneous product, buyers with good price information will flock to the seller with the lowest price. This can create price instability, which may lead to randomized price choices. First, consider a specific example in which demand is inelastic at 3 units for all prices up to a limit price of $6, i.e., the demand curve is flat at a height of $6 for less than 3 units and becomes vertical at a quantity of 3 for prices below $6, as shown in figure 21.4. There are two sellers, each with a capacity to produce 2 units at zero cost. Thus, the market supply is vertical at a quantity of 4 units, and supply and demand intersect at a price of $0 and a quantity of 3. The sellers are identical, so each can only expect to sell to half of the market, 1.5 units on average, if they offer the same price. If the prices are different, the seller with the lower price sells 2 units and the other only sells 1 unit (as long as price is no greater than

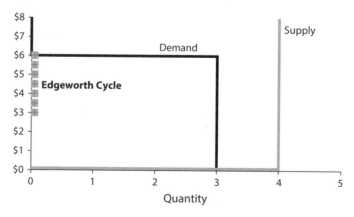

Figure 21.4. A "Box" Design with an Edgeworth Cycle

$6). If "small" price reductions are possible, there is no Nash equilibrium at any common price between $0 and $6, since each seller could increase expected sales from 1.5 to 2 units by decreasing price slightly. Nor is a common price of $0 a Nash equilibrium, since earnings are zero and either seller would have an incentive to raise price and earn a positive amount on the one-unit residual demand. Thus, there is no equilibrium in pure (non-random) strategies, and therefore, one would not expect to see stable prices.

One possibility, first considered by Edgeworth, is the idea that prices would cycle, with each firm undercutting the other's price in a downward spiral. At some point prices go so low that one seller may raise price. In the example discussed above, suppose that one seller has a price that is a penny above a level p between 0 and $6. The other seller could cut price to p and sell 2 units, thereby earning $2p$, or raise price to $6, sell the single residual demand unit, and earn $6. Thus, cutting price would be better if $2p > 6$, or if $p > 3$. Raising price would be better if $p < $3. This reasoning suggests that prices might fall in the range from $6 down to $3, at which point one seller would raise price to $6 and the cycle would begin again. The problem with this argument is that if one seller knows that the other will cut price by one cent, then the best reaction is to cut by 2 cents, but then the other would want to cut by 3 cents, etc. The declining phase of prices is therefore likely to be sporadic and somewhat unpredictable in the range of the "Edgeworth cycle." This raises the possibility that prices will be random, i.e., that there will be a Nash equilibrium in mixed strategies of the type considered in chapter 10. Since sellers must be indifferent over all prices in the relevant range to be willing to randomize, the equilibrium price distribution must be such that sellers' expected earnings are constant for all prices in this range (see the appendix for details of this calculation for the setup in figure 21.4). As we shall see, price cycles do arise over the Edgeworth cycle range, although the particular pattern predicted by a mixed-strategy Nash equilibrium is not always observed. To summarize:

> **Edgeworth Cycle:** *If sellers' capacities are constrained, each might have an incentive to "undercut" the other's price, unless the price is so low that it is better to raise price to earn more on the residual demand (that exceeds the rival's capacity).*

21.4 The Effects of Market Power

There may be several reasons for observing prices that are above competitive levels in a design like that considered in the previous section. For example, with only two sellers, a type of tacit collusion may be possible, especially if the sellers interact repeatedly. Another possible reason is that demand is inelastic and that the excess demand is only one unit at prices above the competitive level of

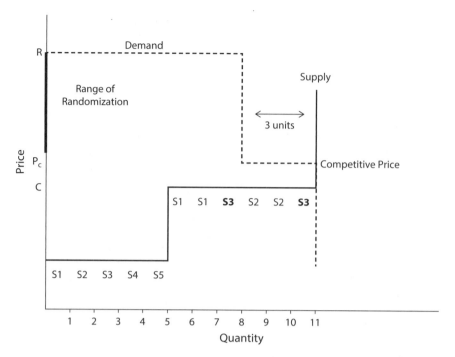

Figure 21.5. Capacity Allocations for a No-Power Design. *Note:* The Power Design is created by reallocating seller S3's two high-cost units to S1 and S2.

$0 in this example. A final reason is that earnings would be zero at the competitive outcome, which might produce erratic behavior. These types of arguments led Davis and Holt (1994a) to consider a design with two treatments, each with the same aggregate supply and demand functions, but with a reallocation of units that creates market power. In particular, a reallocation of capacity from one seller to others changed the Nash equilibrium price from the competitive price (Bertrand result) to higher (randomized) prices over the range spanned by the Edgeworth cycle.

Consider the "No-Power Design" in figure 21.5, where the market demand is shown by the dashed line. The solid line supply curve has two steps, and the units are marked with seller ID numbers. Sellers S1, S2, and S3 each have 3 units, and S4 and S5 each have a single, low-cost unit. The demand curve has a vertical intercept of R and intersects supply at a range of prices from the highest competitive price, P_c, to the level of the highest cost step, C. The demand is simulated with the high-value units being purchased first. This demand process ensures that a unilateral price increase above a common price P_c would leave the 8 high-value units to be purchased by the other sellers, whose capacity totals to 8 units. Thus, a unilateral increase from a common competitive price will result in no sales, and hence, will be unprofitable. It follows that no seller has market power in this design.

Market power is created by giving seller S3's two high-cost units (shown in bold type) to S1 and S2. With this change, each of the large sellers, S1 and S2, would have 4 units, which is more than the excess supply of 3 units for prices above the competitive level. If one of these sellers were to raise price unilaterally to the demand intercept, R, one of these 4 units would sell since the other 4 sellers only have enough capacity to sell 7 of the 8 units that are demanded at prices above the competitive level. By making the demand intercept high relative to the high-cost step, such deviations are profitable for the two large sellers, which creates market power. In this case, it is possible to calculate the price distributions in the mixed-strategy equilibrium, by equating sellers' expected payoffs to a constant (since a seller would only be willing to randomize if expected payoffs are independent of price on some range). These calculations parallel those in the appendix to this chapter, but the analysis is more tedious, given the asymmetries in sellers' cost structures (see Davis and Holt, 1994a). For the market power design, the range of randomization is shown as the darkened region on the vertical axis. Note that this design change holds constant the number of sellers and the aggregate supply and demand arrays, so that price differences can be attributed to the creation of market power and not to other factors such as a small number of sellers or a low excess supply at high prices. To summarize:

Market Power: *A concentration of seller units in the hands of a few large sellers can create market power, which is indicated by the incentive to raise price unilaterally and sell fewer units, but receive a higher price for those units.*

Figure 21.6 shows the results of an experiment in which groups of 5 sellers chose prices for 60 market periods. Demand was determined by a passive, price-taking simulated buyer, and sellers were told the number of periods and all aspects of the demand and supply structure. In three of the sessions, the first 30 periods used the No-Power design in figure 21.5, followed by 30 periods of the Power design obtained by giving S3's two high-cost units to the large sellers, S1 and S2. The price averages for these sessions are graphed as the black line in figure 21.6, with the break in the line separating the two treatments.

It is apparent that prices for the no-power treatment (dark line) on the left side of figure 21.6 start high but fall to competitive levels by the 30th round. As soon as power is created, the prices jump to the high levels (black line) that had been achieved for the other three sessions that began with the market power treatment (the gray line). In contrast, prices for the sessions that began with market power fell quickly after the power was taken away. Notice that prices did not fall all the way to competitive levels, which indicates a "sequence effect" carryover from the earlier successful tacit collusion that had been established in the initial Power design.

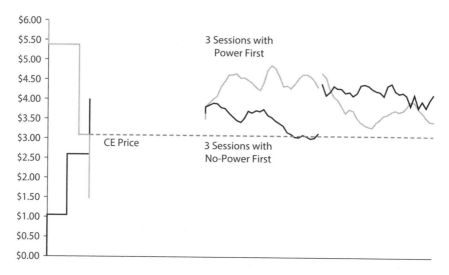

Figure 21.6. Average Prices with and without Market Power. *Key*: Power/No-Power (gray line), No-Power/Power (black line). *Source*: Davis and Holt (1994a).

Edgeworth cycles were observed in some of the sessions during periods with market power. A typical pattern was for one of the large sellers to raise price to a level near the demand intercept (*R* in figure 21.5) and thereby "carry the weight" by only selling a single unit, although at a high price. As the other sellers were then able to sell all of their units, they would raise prices. When prices reached high levels, all sellers would generally try to post prices that were high, but just below the highest of the others' prices. This attempt to avoid being the one with the highest price would sometimes drive all prices down in an orderly decline. The cycle might restart with one of the large sellers signaling again with a price near the demand intercept. Such attempts to signal and hold price up by large sellers failed in the No-Power periods, since the seller with the highest price would make no sales in this treatment.

Market power resulted in large price increases in all six sessions, holding constant the number of sellers and the aggregate supply and demand structure. Note that this is a "within-subjects" design, so each group of five sellers in a session serves as its own control. Under the null hypothesis of no effect, prices would have been just as likely to go up as down, so the chances of seeing higher prices in the power treatment are 1/2 in a single session. But getting higher prices in all six sessions has a low probability under the null hypothesis: this would be like throwing a fair coin 6 times and getting 6 heads in a row, which has probability: $(1/2)(1/2)(1/2)(1/2)(1/2)(1/2) = 1/64 = 0.016$. Thus, the null hypothesis of no effect can be rejected at the 0.05 level, or even at the 0.02 level, but not at the "1%" level. In this most extreme case, the test just described is equivalent to the Wilcoxon matched-pairs test presented in chapter 13, which is based on permuting signed ranks of differences for each pair.

Even though the creation of market power resulted in price increases in this experiment, as predicted in a Nash equilibrium, the observed cyclic autocorrelation of prices is not consistent with randomization, a point that was also made by Kruse et al. (1994). In fact, the price-increasing effect of market power was considerably greater than the difference between the Nash/Bertrand price in the No-Power treatment and the mean of the Nash equilibrium distribution of prices in the Power treatment. Since explicit communications between sellers were not permitted, it is appropriate to use the term "tacit collusion" to describe this ability of sellers to raise prices above the levels determined by a Nash equilibrium. Hence, market power has a double impact in this context: it raises the predicted mean price, and it facilitates tacit collusion that raises prices above the Nash prediction. To summarize:

> **Market Power Experiments:** *The creation of market power by reallocating seller units to a few larger sellers does increase prices in experiments, via a combination of strategic behavior and tacit collusion, although pricing is not characterized by randomization.*

The recent literature on the effects of market power can be understood by reconsidering the narrowness of the vertical gap between demand and supply to the left of the supply/demand intersection (see figure 21.5). This narrow gap implies that a seller who refuses to sell units with these relatively high costs will not forego much in the way of earnings. This seller might profit from such withholding if the price increase on low-cost units sold would more than compensate for the lost earnings on these marginal units. Suppose that the demand/supply gap is large during a high-demand "boom" period, which makes marginal units more profitable and limits the exercise of market power. In contrast, the gap might fall during a contraction, thereby enabling sellers to raise prices profitably. Thus, the effects of market power in some markets may be countercyclical (Reynolds and Wilson, 2005), i.e., more market power in low demand periods.

A dramatic example of low demand periods occurs with electricity power generation, with demand shifting dramatically between peak, shoulder, and off-peak periods. During the California energy crisis in 2000, there were sharp increases in "spot market" electricity prices (what investor-owned distributors pay for wholesale electricity). These wholesale price increases pushed local electricity distributors to the edge of bankruptcy, since they were required to acquire electricity at any cost to meet "must-serve" needs of residential and commercial users. Rassenti, Smith, and Wilson (2001) identified the problem as one in which distributors were prevented from raising prices for must-serve customers in the event of shortages. The wholesale markets had been deregulated, but the distribution markets had not. The authors suggested a thought experiment: what if airlines were forced to provide seats *to all customers on a must-serve basis, without raising prices during peak demand periods.* Their argument is that this would lead to costly, inefficient accumulation of excess service capacity.

Rassenti, Smith, and Wilson ran an experiment to study what time-of-day pricing flexibility would look like in electricity markets, even in the presence of market power. Their experiments compared pricing in "power" and "no-power" treatments, where power was created by reallocating seller units to large sellers, as was done in figure 21.5. The sellers are companies that generate electricity, and the buyers are the distributors. The authors concluded that market power, if present, would enable sellers to dramatically raise wholesale prices charged to local distributors, who would still sell to retail and industrial users at flat rates. But if the local distributors were able to raise prices to consumers in periods of peak demand or shortage, and if some fraction of those consumers were willing to reduce consumption, then the experiments indicated that price spikes would be dampened to a large extent. The title of Rassenti et al.'s paper tells the story: "Turning Off the Lights: Consumer Allowed Service Interruptions Could Control Market Power and Decrease Prices." This is another example of how the presence of active, non-simulated buyers can restrain the exercise of market power.

21.5 Price Collusion: "This *Is* Economics"

Ever since Adam Smith, economists have believed that sellers often conspire to raise price. Such collusion involves trust and coordination, and therefore, the plan may fall apart if some sellers defect. In fact, Smith's oft-quoted warning about the likelihood of price fixing is immediately qualified:

> In a free trade an effectual combination cannot be established but by the unanimous consent of every single trader, and it cannot last longer than every single trader continues of the same mind. The majority of a corporation can enact a bye-law with proper penalties, which will limit the competition more effectually and more durably than any voluntary combination whatever. (Smith, 1776, p. 144)

Price-fixing is illegal in the United States and most other developed economies, and hence it is difficult to study. Moreover, conspirators will try to keep their activities secret from those who have to pay the high prices. There can be a selection bias, since data that do surface are more likely to come from disgruntled participants in failed conspiracies. Without good data on participants and their costs, it is difficult to evaluate the nature and success of collusion, and whether the breakdowns were caused by shifts in cost or demand conditions. Laboratory experiments are not hampered by these data problems, since controlled opportunities for price fixing can be allowed, holding constant other structural and institutional elements that may facilitate supra-competitive pricing. Another antitrust issue of interest is whether the observed patterns of bids and other market conditions can be used ex post to make inferences about the presence or absence of collusion.

Given the difficulty of obtaining information about price-fixing conspiracies, it is not surprising to find differing opinions about the usefulness of prosecuting them. For example, Cohen and Scheffman (1989) argue that conspiracies are prone to failure and that enforcement costs incurred by antitrust authorities are, to a large extent, wasted. This point of view has been contested by other economists and by antitrust authorities (see Werden, 1989, and the discussion of the literature in Davis and Wilson, 2002).

The results of early laboratory experiments indicate that the effectiveness of collusion depends critically on the nature of the market trading institution. Isaac and Plott (1981) report results from double auctions with trading rules that approximate those of organized asset markets, with a continuous flow of bids, asks, and trading prices. Sellers in these double auctions were allowed to go to a corner of the room and discuss prices between trading periods. These conspiracies were not very effective in actually raising transactions prices in the fast-paced competition of a double auction, where sellers are faced with the temptation to cut prices in order to sell marginal units late in the trading period. This observation suggests that conspiracies might be more effective if mid-period price reductions are precluded. Sellers only submit a single price in a posted-offer auction, so late price discounts in response to low trading volume are not permitted.

This section will be organized around a series of market experiments reported in Davis and Holt (1998), who replicate earlier results and then relax the price rigidity in posted-offer markets in a controlled manner by introducing discount opportunities. There were three buyers and three sellers in each session. At the beginning of each period, the buyers were taken from the room under the guise of assigning different redemption values to them. Then sellers were allowed to push their chairs back from their visually isolated cubicles so that they could see each other and discuss price. They were not permitted to discuss their own production costs or to divide up earnings. Then they returned to their computers as the buyers came back into the room. At that time, sellers would enter their posted prices independently, without further discussion. The buyers were not aware of the seller price discussions. The supply and demand functions for all sessions are shown on the left side of figure 21.7. The intersection of these functions determines a range of competitive prices, the highest of which is shown as a horizontal dashed line, labeled "CE Price."

Supply and demand in figure 21.7 were structured so that if all three sellers could set a price that maximized total earnings for the three of them, then they would each sell a single unit at a price indicated by the dashed horizontal line labeled "M Price." (Buyers were required to pay a 5-cent "travel cost" prior to approaching a seller, so this reduces the M-price line to a level that is 5 cents

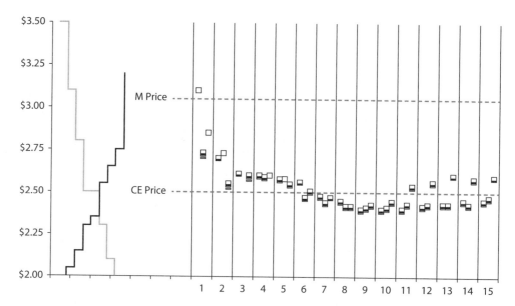

Figure 21.7. Prices for a Posted-Offer Session with No Collusion. *Key:* Posted List Prices (boxes) and Units Sold (dark dashes).

below the step on the demand curve at $3.10.) Buyers were taken from the room for value assignments, as in all sessions, but sellers were *not* permitted to fix prices. The prices for this "no-collusion" session are shown in figure 21.7 as boxes, and the units sold are shown as small black dashes in the lower part of the corresponding box. The dashes are stacked so that the thickness of these black marks is proportional to the seller's actual sales quantity at that price. The prices for each period are separated by vertical lines, and the prices for the three sellers are shown in order from left to middle to right, for sellers S1, S2, and S3 respectively. Notice that the low-price firm, S2, sells all of the units in the first round, and that the other prices fall quickly. Prices are roughly centered around the competitive price in the later periods. The competitive nature of the market (without collusion) was an intentional design feature.

In contrast, figure 21.8 shows the dramatic effects of collusion with the same market structure as before. The only procedural difference was that sellers were able to collude and make non-binding plans while buyers were out of the room. Attempts to fix a price resulted in high but variable prices in the early periods, as shown on the left side of figure 21.8. Sellers agreed on a common price in the fourth round, but the failure of seller S2 to make a sale at this price forced them to deal with the allocation issue, which was solved more efficiently by an agreement that each would limit sales to 1 unit. This agreement broke down several periods later as S2 (whose price is always displayed in the center) listed a price below the others. A high price was re-established in the final 6 periods, but prices remained slightly below the joint-profit-maximizing monopoly level.

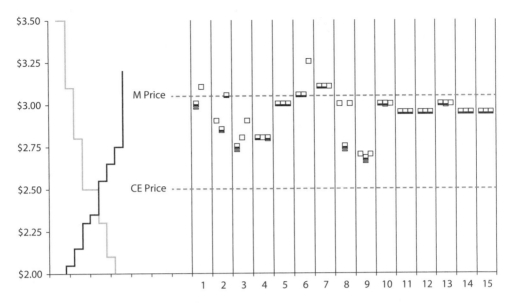

Figure 21.8. Prices for a Posted-Offer Session with Collusion. *Key:* Posted List Prices (boxes) and Units Sold (dark dashes)

In two of these periods, 10 and 13, the sellers agreed to raise price slightly and hold sales to one unit each, but on both occasions S2 sold two units, leaving S3 with nothing. These defections were covered up by S2, who did not admit to the extra sale, but claimed in the subsequent meeting that *"this is economics"* (less is sold at a higher price). The others went along with this explanation and agreed to lower price slightly.

Successful collusion was observed in all but one of the other sessions done with this treatment. In one session (not shown), sellers were not able to coordinate on a common price until round 5, but then all buyers made purchases from seller S1, which created an earnings disparity. Seller S1 then suggested that they take turns having the low price, and that *he, S1, be allowed to go first!* This agreement was adopted, and S1 made all sales in the next period. The low price position was rotated from seller to seller in subsequent rounds, much like the famous "phases of the moon" price-fixing conspiracy involving electrical equipment in the 1960s. There was some experimentation with prices above and below the joint-profit-maximizing collusive level, but prices stayed at approximately this level in most periods. Despite the high prices generated, seller profits were not as high as they could have been. The rotation scheme was quite inefficient, since each seller had a low-cost unit that would not sell when it was not that seller's turn to have the low price. To summarize:

Collusion Experiments: *Price collusion can result in near-monopoly pricing, even with buyers that are not simulated, if prices are posted on a take-it-or-leave-it basis*

*and discounts are not permitted. The "phases of the moon" rotation of sales be-
tween sellers, which is sometimes observed, can further reduce efficiency by allocating some production to inefficient high-cost units for the seller whose "turn" it is.*

21.6 Collusion with Secret Discounts

Most markets of interest to industrial organization economists cannot be classified as continuous double auctions (where all price activity is public) or as
posted-offer markets (which do not permit discounts and sales). This raises
the issue of how effective explicit collusion would be in markets with a richer
array of pricing strategies and information conditions. In particular, markets
for producer goods or major consumer purchases differ from the posted-offer
institution in that sellers can offer private discounts from the "list" prices. The
effectiveness of conspiracies in such markets is important for antitrust policy,
since many of the famous price-fixing cases, like the electrical equipment bidding conspiracy discussed above, involve producer goods markets where discounts are often negotiated bilaterally.

Davis and Holt used a third treatment to evaluate the effects of discount opportunities for the same market structure considered in figures 21.7 and 21.8.
In these sessions, sellers could collude as before, but when buyers returned and
saw the sellers' posted prices, they could request discounts. If a discount was
requested, the seller would then type in a price, which could be equal to the list
price (no discount) or lower. The seller's response was not observed by other
sellers, so the discounts were given secretly and sequentially. Sellers were free to
discount selectively to some buyers and not others, and to hold discounts until
later in a period. Prices were much lower in the collusion sessions with opportunities for discounting than in the collusion sessions with no such opportunities. The price sequence for one of the sessions with collusion and discounting
is shown in figure 21.9.

As before, the small squares indicate list prices in figure 21.9, but the black
dashes indicate actual sales that can be discounted to price levels well below
posted list prices. In period 3, for example, all sellers offered the same list price,
but S1 (on the left) sold two units at a deep discount. In periods 6, 7, and 8,
seller 2 began secret discounting, as indicated by the dashes below the middle
price square. These discounts caused S1 to have no sales in two of these periods, and S1 responded with a sharp discount in period 9 and a lower list price
in period 10. After this point, discounts were pervasive, and the outcome was
relatively competitive. *Notice that the sellers fixed a price, but the only price they
succeeded in fixing in the end was essentially the competitive price.* This competitive
outcome is similar to the results of several other sessions with discounting. This
is a between-subjects design, and treatment effects are significant using Mann-
Whitney nonparametric tests that were discussed in chapter 13.

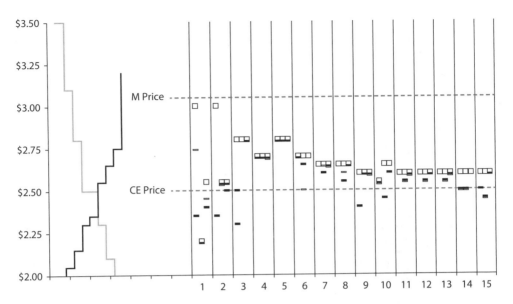

Figure 21.9. Prices for a Session with Collusion and Secret Discounts. *Key:* Posted List Prices (boxes) and Secret Discounts (dark dashes)

One factor that hampered the ability of sellers to maintain high prices in the face of secret discounts was the inability to identify cheaters. Many successful price-fixing conspiracies, especially those with large numbers of participants, have involved industries where trade associations reported reliable sales information for each seller (Hay and Kelly, 1974). This observation motivated the "trade association" treatment where sellers were given reports of all sellers' sales quantities at the end of each round. All other procedures were identical, with both collusion and the possibility of secret discounts. Even with discounting, this ex post sales information permitted sellers to raise prices about halfway between the competitive and joint-monopoly price levels. In the experiment (not shown), explicit collusion resulted in even higher *list* prices, often exceeding monopoly levels. There was not enough business to go around at these high prices, and virtually all units were sold at discount. The unequal sales quantities did expose discounters to disapproval, but all sellers were discounting, since they did not seem to trust each other. Nevertheless, the average transactions prices were closer to joint-profit-maximizing levels than to competitive levels, and earnings were much higher than in sessions with no ex post quantity reporting.

List and Price (2005) report results for a field experiment in a market with multilateral decentralized bargaining. The participants were recruited at a sports card convention. Dealers were assigned seller roles, and non-dealer attendees were assigned buyer roles. The product was homogenous; a standard, widely available sports card was used for all trades. There were three treatments that parallel those of Davis and Holt: no communication between sellers, communication with publicly observed prices, and communication without publicly observed prices. Prices were near competitive levels in the absence of opportunities for collusive discussions, and prices were much higher when collusion among sellers was allowed. For the intermediate treatment when collusion was allowed and prices were not perfectly observed by other sellers, average prices were in the middle of the averages for the other treatments. This was run like a laboratory experiment, but with more context and with participants who were experienced in their roles. In this sense, the results add a new dimension to the results of earlier laboratory experiments. To summarize:

Collusion and Secret Discounting: *The effects of market power and explicit collusion are much more severe in markets where sellers post prices on a take-it-or-leave-it basis. Opportunities to offer price reductions during the trading period make it much more difficult to coordinate price increases, and the result can be efficient competitive prices even in the presence of collusion. Ex post trade association reporting can partially mitigate the beneficial pro-competitive effects of secret discounts.*

21.7 Extensions: Cheap Talk, Mutual Forbearance, and the "V Word"

These experimental results are consistent with the antitrust hostility to industry practices that are seen as limiting sellers' options to offer selective discounts. Sales contracts and business practices that deter discounts have been the target of antitrust litigation, as in the Federal Trade Commission's *Ethyl* case, where the FTC alleged that certain best-price policies deterred sellers from offering selective discounts. The anti-competitive nature of these best-price practices is supported by results of some experiments run by Grether and Plott (1984), who used a market structure that was styled after the main characteristics of the market for lead-based anti-knock gasoline additives that was litigated in the *Ethyl* case. Holt and Scheffman (1987) provide an analysis of best-price policies and of the Grether and Plott experiments. The intuition is that a best-price policy enables a seller to resist requests for individual discounts by arguing that "if I give you a discount, my contracts with buyers will require me to do the same for the others." The courts eventually decided that such contracts were not illegal, since it is difficult to distinguish market conditions in which the effects might be pro-competitive.

There have been a number of follow-up studies on the effects of price collusion. Isaac and Walker (1985) found the effects of collusion in sealed-bid auctions to be similar to those in posted-price auctions. This is not surprising, since a sealed-bid auction is similar to a posted-offer auction, except that only a single unit or price is typically involved. Collusion in sealed-bid auctions is an important topic, since many price-fixing conspiracies have occurred in such auctions, and since many companies are relying more and more on auctions to procure supplies.

Another interesting issue is the extent to which the bid patterns might be used to infer illegal collusion after the fact. Davis and Wilson (2002) investigate this issue with a pair of treatments. In the sessions with communication, the sellers were allowed to discuss price prior to the beginning of every fourth period, which implements the idea that illegal discussions may only occur infrequently. The supply and demand setup in one treatment was such that there was a Nash equilibrium in which all sellers choose the same competitive price. Here, the "suspicious" common pricing behavior is predicted in theory even when sellers do not communicate. In the experiment, however, prices tended to be lower and more variable when communication was not allowed, and common prices were *only* observed when communication was allowed. Thus, common pricing can be an indicator of collusion in this environment. Such collusion resulted in large price increases, especially toward the end of the session after sellers had established effective collusion.

Davis and Wilson (2002) also implemented a second market structure that was intended to mimic a market for construction contracts. In this setup, each of the four firms had a limited capacity in the sense that winning a contract in one period would raise their costs of taking on an additional project in the next several periods. The motivation for this cost increase is that a firm might have to pay overtime, incur delay penalties, or subcontract some of the work when it undertakes more than one project at a time. It is natural to expect some bid rotation to develop, as companies with ongoing projects bid higher and allow others to take subsequent projects. Thus the rotation of low bids, which is sometimes considered to be a suspicious pattern, would be expected to emerge even in the absence of collusion. Such rotations were observed in both the communication and the no-communication sessions, but Davis and Wilson concluded that collusion can often be inferred from the pattern of *losing* bids. In particular, there was less correlation between losing bids and costs when the losing bidders had agreed in advance to bid high and lose. This pattern was also observed in an empirical study of bidding for school milk programs (Porter and Zona, 1993).

There is a separate series of papers on "tacit collusion" or a "meeting of the minds" which results in high prices, even in the absence of illegal direct communication. These papers identify market conditions and business practices that may facilitate such tacit collusion. In particular, there have been allegations that competitors who post prices on computer networks may be able to signal threats and cooperative intentions. Some bidders in early FCC bandwidth auctions used decimal places to attach zip codes to bids in an attempt to deter rivals by implicitly threatening to bid on licenses that rivals were trying to obtain. Similarly, some airlines attached aggressive letter combinations (e.g., FU) to ticket prices posted on the Airline Tariff Publishing (ATP) computerized price system. Cason (1995) reports that such nonbinding ("cheap talk") communications can raise prices, but only temporarily. See Holt and Davis (1990) for a similar result for a market where sellers could post non-binding "intended prices" before posting actual prices. These non-binding price announcements would correspond to the posting of intended future prices on the ATP, which are visible to competitors before such prices are actually available for consumers. Cason and Davis (1995) report that price signaling can have a larger effect in a multi-market setting, but even then the effects of purely nonbinding price announcements were limited.

For any kind of collusion (tacit or explicit) to be effective, it is important for the participants to be able to signal intentions and to identify and sanction violators. Even though explicit collusion is facilitated in sealed-bid auctions, it can be the case that more open auction formats allow people to signal and punish others. Suppose that incumbency makes licenses being auctioned off more valuable to the current provider. For example, bidder A has a value of 10 for item A and 5 for item B, and conversely, bidder B has a value of 10 for B and 5 for A. Without collusion, the items might each sell for about 5. But imagine an ascending price auction in which both bidders can signal "mutual forbearance" by not raising bids on the other's preferred item. In this case, the bidding might stop at levels below 5. If this is not the case, each person could punish the other by bidding above 5 on the other's favored item, which would be a clear signal of displeasure if the values are known. A sealed-bid auction takes away the possibilities for signaling of forbearance or punishment in incremental ways, and the result would likely be to break the tacit collusion. Plott and Li (2009) show how collusion can develop in ascending price auctions in a multi-bidder environment where each person has a preferred item, in a manner that generalizes the above example. They also show that a Dutch auction breaks this collusion. The Dutch auction, to be discussed in a later chapter, has the property that the proposed bid price is decreased until the first person stops the auction by pressing the "Buy button." The finality of the buy decision makes it harder to signal mutual forbearance.

Goeree, Offerman, and Sloof (2013) also observe mutual forbearance (demand reduction) by incumbents in ascending price laboratory auctions, which generate lower sales revenues than a parallel series of sealed-bid auctions. Their work was motivated by a widely discussed failure of the Dutch telecommunications auction to generate high revenues when incumbents stopped bidding against each other. Given the spectacular success of such auctions in neighboring countries, the Dutch auction caused considerable embarrassment. The Dutch word for auction is "veiling," and at one point, economics consultants to a Dutch government agency were advised not to use the "V word."

Appendix: Calculation of a Mixed-Strategy Equilibrium in Prices

It is useful to begin with some examples of price distributions, before calculating the equilibrium distribution for the example from section 21.3. With a continuous distribution of prices, the probability that another seller's price is less

Table 21.1. A Uniform Distribution of Prices

p	0	$1	$2	$3	$4	$5	$6	$7	$8	$9	$10
F(p)	0	0.1	0.2	0.3	0.4	0.5	0.6	0.7	0.8	0.9	1

than or equal to any given amount p will be an increasing function of p, and we will denote this probability by $F(p)$. For example, suppose that price is equally likely to be any dollar amount between 0 and $10.00. This price distribution could be generated by the throw of a ten-sided die, with sides marked 1, 2,... 10, and using the outcome to determine price. Then the probabilities would be given by the numbers in the second row of table 21.1. For each value of p in the top row, the associated number in the second row is the probability that the other seller's price is less than or equal to p. For a price of $p = \$10$, all prices that might be chosen by the other seller are less than or equal to $10 by assumption, so the value of $F(p)$ in the second row of the far-right column is 1. Since all prices are equally likely, half of them will be less than or equal to the midpoint of $5, so $F(p) = 0.5$ when $p = \$5$. A mathematical formula for this distribution of prices would be: $F(p) = p/10$, as can be verified by direct calculation (problem 2).

The function, $p/10$, is increasing in p at a uniform rate, since all prices in the range from 0 to 10 are equally likely. Other distributions may not be uniform, e.g., some central ranges of prices may be more likely than extreme high or low prices. In such cases, $F(p)$ would still be an increasing function, but the rate of increase would be faster in a range where prices are more likely to be selected.

In a mixed-strategy equilibrium, the distribution of prices, $F(p)$, must be determined in a manner to make each seller indifferent over the range of prices, since no seller would be willing to choose price randomly unless all prices in the range of randomization yield the same expected payoff. As was the case in chapter 10, we will calculate the equilibrium under the assumption that players are risk neutral, so that indifference means equal expected payoffs.

Consider the duopoly example from section 21.3, where the two sellers have zero costs and capacities of 2 units each (so costs are prohibitively high for a third unit). Thus the market supply is vertical at a quantity of 4. As before, assume that demand is inelastic at 3 units for all prices below some reservation price, which was $6 in the example. More generally, let the reservation price be denoted by V, which can be $6 or any other positive number. Since supply exceeds demand at all positive prices, the competitive price is $0, which is not a Nash equilibrium, as explained previously. Begin by considering a price, p, in the range from 0 to V. The probability that the other seller's price is less than or equal to p is $F(p)$, which is assumed to be increasing in p. When the other seller's price is lower, a firm sells a single unit, with earnings equal to p. When the other's price is higher, one's own sales are 2 units, with earnings of $2p$. Since prices are continuously distributed, we will ignore the possibility of a tie, and hence, $F(p)$ is the probability associated with having the high price and earning the payoff of p, and $1 - F(p)$ is the probability associated with having the low price and earning the payoff of $2p$. Thus the expected earnings are: $F(p)p + [1 - F(p)]2p$. This expected payoff must be constant for each price in the range over which a firm randomizes, to ensure that the firm is indifferent between those prices. To determine this constant, note that setting the highest price, V, will result in sales of only 1 unit, since the other seller will have a lower price for sure. Thus we equate expected payoffs to a constant V:

(21.1) $$F(p)p + [1 - F(p)]2p = V.$$

To summarize, equation (21.1) ensures that the expected payoff for all prices is a constant and equals the earnings level for charging the buyer price limit of V. This equation can be solved for the equilibrium price distribution $F(p)$,

(21.2) $$F(p) = \frac{2p - V}{p}$$

for $V/2 \le p \le V$. Equation (21.2) implies that $F(p) = 1$ when $p = V$. In words, the probability that the other seller's price is less than or equal to V is 1. The lower limit of the price distribution, $p = V/2$, is the value of p for which the right side of (21.2) is 0. This is also the lower bound of the Edgeworth cycle discussed above (with $V = 6$, the lower bound of the cycle is 3). See Holt and Solis-Soberon (1992) for a more general approach to the calculation of mixed strategies in markets with capacity constraints; they also consider the effects of risk aversion.

Chapter 21 Problems

1. Consider the market structure in figure 21.3. Show that a unilateral price increase above a common price of $7 is not profitable in a posted-offer market.

2. Show that the formula, $F(p) = p/10$, produces the numbers shown in the second row of table 21.1. How would the formula have to be changed if prices were uniform on the interval from $0 to $20, e.g., with half of the prices below $10, 1/4 below $5, etc.?

3. For the duopoly example in section 21.3 (figure 21.4), show that prices of $0 for both sellers do not constitute a Nash equilibrium, i.e., show that a unilateral price increase by either seller will raise earnings.

4. Consider a modification of the duopoly example in figure 21.4, where demand is now vertical at 6 units for all prices below $6, and each seller has a capacity

of 5 units at zero cost. What is the range of prices over which an Edgeworth cycle would occur?

5. Answer problem 4 for the case in which each seller's costs are constant at $1 per unit.

6. (non-mechanical) Construct an example with three sellers, in which a merger of two of these sellers makes the exercise of market power more profitable. Illustrate your answer with a graph that shows each seller's ID, and the way in which costs are reduced.

22

Supply Chains

A complex economy is characterized by considerable specialization along the supply chain; with connections between manufacturers of intermediate products, manufacturers of final products, wholesalers, and retailers. There is some theoretical and experimental evidence that frictions and market imperfections may be induced by these "vertical" supply relationships. This chapter begins by considering a very specific vertical structure: the interaction between an upstream monopoly wholesaler selling to a firm that has a local monopoly in a downstream retail market. In theory, the vertical alignment of two monopolists can generate retail prices that are even higher than those that would result with a single integrated firm, i.e., a merger of the upstream and downstream firms. The addition of more layers in the supply chain raises the question of the extent to which demand shocks at the retail level may cause even larger swings in orders and inventories at the upstream levels of the supply chain, a phenomenon that is known as the "bullwhip effect."

> **Note to the Instructor:** Vertical pricing effects can be investigated with the Veconlab Vertical Monopoly program or with the instructions for a hand-run version that can be found in Badasyan et al. (2009), which is designed to be done outdoors. The Veconlab Supply Chain program (on the Markets menu) can be used to implement a single-layer supply chain, the "Newsvendor Problem."

22.1 Double Marginalization

Since Adam Smith, economists have been known for their opposition to monopoly. Monopolization (carefully defined) is a crime in US antitrust law, and horizontal mergers are commonly challenged by the antitrust authorities if the effect is to create a monopoly. In contrast, there is much more leniency shown toward vertical mergers, i.e.. between firms that do business with each other along a supply chain. One motivation for this relative tolerance of vertical mergers is that "two monopolies are worse than one," at least when these two monopolies are arrayed vertically. The intuition is that each monopolist restricts output to the point where marginal revenue equals marginal cost, a process that reduces production below competitive levels and results in higher prices. When this marginal restriction is made by one firm in the supply chain, it raises the price to a downstream firm, which in turn raises price again at the retail level. The resulting "double marginalization" may be worse than the output restriction caused by a merger of both firms into a vertically integrated monopolist.

The effects of double marginalization can be illustrated with a simple laboratory experiment based on a linear demand function: $P = 24 - Q$, which is the demand at the retail level. There is an upstream manufacturer with a constant marginal cost of production of \$4. The retailer purchases units and is, for simplicity, assumed to incur no cost, other than the wholesale price paid for each unit. Thus there is a total production cost (manufacturing plus retail) of \$4 per unit. This would be the marginal cost of a single vertically integrated firm that manufactures and sells the good in a retail monopoly market.

First consider the monopoly problem for this integrated firm, i.e., to find the output that equates the marginal cost of 4 with the marginal revenue, as explained in the previous chapters. With a demand of $P = 24 - Q$, and an associated total revenue of $24Q - Q^2$, the marginal revenue will be $24 - 2Q$, since the marginal revenue is twice as steep as the demand curve. Equating this to the marginal cost of 4 and solving for Q yields a monopoly output of 10 and an associated price of \$14. These calculations are illustrated by the thick lines in figure 22.1. The demand curve has a vertical intercept of \$24, with a slope of -1, so it has a horizontal intercept of 24. The marginal revenue line (MR) also has a vertical intercept of 24, but its slope is twice as steep, so the horizontal intercept is 12. This MR line intersects the horizontal MC line at a quantity of 10, which leads to a price of \$14, as can be seen by moving from the intersection point vertically up to the demand curve along the light dotted line.

It is useful to organize the analysis for the vertically integrated firm:

$$\begin{aligned}
&\text{Integrated Firm Demand:} && P = 24 - Q \\
(22.1) \quad &\text{Integrated Firm Marginal Revenue:} && MR = 24 - 2Q \\
&\text{Integrated Firm Marginal Cost:} && MC = 4 \\
&\text{Integrated Firm Production:} && Q = 10, \text{ where: } 4 = 24 - 2Q
\end{aligned}$$

Next, consider the case in which the two firms are not vertically integrated. Each firm will be equating marginal revenue and marginal cost. It turns out that in these kinds of problems, it is easiest to start at the end of the supply chain (retail) and work backwards to consider the manufacturing level last. For any wholesale price, W, charged by the manufacturer, the marginal cost to the retailer will be W, since each unit sold at retail must be purchased at a wholesale price W, which is the only cost by assumption. Thus, the downstream retailer is a monopolist who will equate this marginal cost, W, to the marginal revenue, $24 - 2Q$, to obtain a single equation: $W = 24 - 2Q$. Again, it is useful to list these conditions:

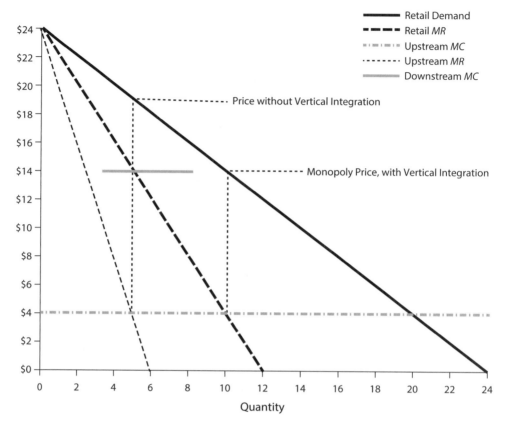

Figure 22.1. Double Marginalization

$$\begin{array}{rl}
& \text{Downstream Retailer Demand:} \quad P = 24 - Q \\
(22.2) \quad & \text{Downstream Marginal Revenue:} \quad MR = 24 - 2Q \\
& \text{Downstream Marginal Cost:} \quad MC = W \text{ (wholesale price)} \\
& \text{Downstream Firm Production:} \quad W = 24 - 2Q
\end{array}$$

The final line of (22.2), which determines the downstream firm's production quantity as a function of the wholesale price, is also the inverse demand for the upstream firm, i.e., to sell each additional unit the wholesale price must be reduced by \$2. This inverse wholesale demand function is exactly the same as the formula for marginal revenue $(24 - 2Q)$ obtained in the previous paragraph. Thus the *downstream market MR curve* (thick dashed line in figure 22.1) is the *upstream demand function*. The upstream firm's marginal revenue line, shown by the thin dashed line, has the same vertical intercept, 24, but a slope that is twice as steep: $MR_{\text{upstream}} = 24 - 4Q$. The upstream firm will want to increase output until this marginal revenue is equal to the wholesale marginal cost of 4, i.e., $24 - 4Q = 4$. This intersection occurs at a quantity of 5, which determines a wholesale price of: $W = 24 - 2Q = 24 - 10 = 14$.

$$Upstream\ Wholesale\ Demand:\quad W = 24 - 2Q$$
$$Upstream\ Marginal\ Revenue:\quad MR_{upstream} = 24 - 4Q$$
(22.3) $\qquad Upstream\ Marginal\ Cost:\quad MC = 4\ \text{(production cost)}$
$$Upstream\ Firm\ Production:\quad Q = 5,\ \text{where}\ 4 = 24 - 4Q$$
$$Upstream\ Wholesale\ Price:\quad W = 24 - 2(5) = 14$$

This wholesale price of 14 then becomes the marginal cost to the retailer, as shown by the thick flat gray line segment at $14. This marginal cost line intersects the retail marginal revenue line at a quantity of 5, which determines a retail price of $24 - 5 = \$19$. To summarize, for the market in figure 22.1 with a monopoly output of 10 and price of $14, the presence of two vertically stacked monopolists reduces the output to 5 and raises the retail price to $19. Thus the output restriction is greater for two monopolists than would be the case for one vertically integrated firm.

The market structure from figure 22.1 was used in the first 5 rounds of the classroom market shown in figure 22.2. These rounds involved pairs of students, with one retailer and one wholesaler in each pair. The wholesale and retail price predictions, $14 and $19 respectively, are indicated by the horizontal lines. The actual price averages by round are indicated by the dots converging to those lines. Notice that wholesale prices (large dots) are a little slow to converge to the theoretical prediction. This could be due to the fact that the buyer, the retailer, is a participant in the experiment, not a simulated buyer, as was the case for the retail demand in this experiment. As noted in previous chapters, a human buyer may respond to a (wholesale) price that is perceived to be unfairly high by

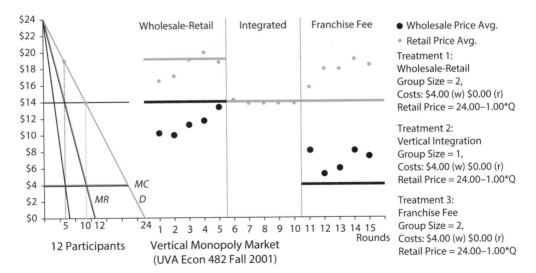

Figure 22.2. A Classroom Experiment without Vertical Integration (rounds 1–5), with Vertical Integration (rounds 6–10), and with a Franchise Fee (rounds 11–15)

cutting back on purchases, even if those purchases might result in more profit for the buyer (downstream firm in this case).

The center part of figure 22.2 shows 5 rounds in which each of the 12 participants from the first part was put into the market as a vertically integrated monopolist facing the retail demand determined by $P = 24 - Q$. The average prices converge to the monopoly level of $14 that is indicated by the horizontal line for periods 6–10. This reduction in retail price, from about $19 to $14, resulted in an increase in sales quantity, and less of a monopoly output restriction, as predicted by theory. In addition, the profits of the integrated firm are larger, since by definition, profit is maximized at the monopoly level. This increase in profitability associated with vertical integration can be verified by calculating the profits for each firm separately (problems 1 and 2).

Vertical integration may not be feasible or desirable, or even cost efficient in all cases. An alternative, which works in theory and is sometimes observed in practice, is for the upstream firm to let the downstream retain all retail revenue and require the retailer to pay a fixed franchise fee in order to be able to sell the product at all. In particular, the wholesale firm chooses a wholesale price and a franchise fee. (This is, in fact, an example of a principal agent game considered in chapter 15, where a sharing rate of $S = 1$ provides the correct incentives for the agent/retailer.) The retailer then may reject the franchise fee arrangement, in which case both earn 0, or accept and place an order for a specified number of units. The idea behind the franchise fee is to lower the wholesale price from the $14 charged when the firms are not integrated, to a level of $4 that reflects the true wholesale marginal cost, so that the retailer can maximize industry profit. When the retailer faces this marginal cost of $4, it will behave as an integrated monopolist, choosing an output of 10, charging the monopoly price of $14, and earning the monopoly profit of $(10)(14) - (10)(4) = 100$. Thus a franchise agreement that lowers the downstream firm's marginal cost to the production cost of $4 generates the same outcome that was summarized in equation (22.1), with the downstream firm behaving exactly like the vertically integrated monopolist. If this were the whole story, the wholesaler would have zero profits, since the units are sold at a price that equals marginal cost. The wholesaler can recover some of this monopoly profit by charging a franchise fee. A fee of $50 would split the monopoly profit, leaving $50 for each. In theory, the wholesaler could demand $99.99 of the profit, leaving one penny for the retailer, under the assumption that the retailer would prefer a penny to a payoff of zero that results from rejecting the franchise contract. Intuition and other experimental evidence, however, suggest that such aggressive franchise fees would be rejected (as with aggressive demands in ultimatum bargaining). In effect, retailers will be likely to reject contract offers that are viewed as being unfair, and such rejections may even have the effect of inducing the wholesaler to make a more moderate demand in a subsequent period.

The far right side of figure 22.2 shows average prices for periods 11–15 of the experiment, in which the participants were again divided into wholesaler/retailer pairs. The franchise fees, not shown, averaged about $40, nowhere near the $100 level that would capture all monopoly profit. Instead, it is apparent that the upstream sellers did not lower the wholesale price to their production cost of $4. Average prices do fall below the wholesale price levels that were observed in the first 5 rounds, but they only fall to a range of $6–$8. Thus, fairness considerations seem to be preventing the franchise fee treatment from solving the vertical monopoly problem. To summarize:

> **Double Monopoly Experiment Results:** *Double Marginalization raises observed retail prices above joint monopoly levels, which is negated if the upstream and downstream firms merge into a single vertically integrated monopolist. In laboratory experiments, sellers that remain separate and bargain over a franchise fee are not able to achieve the predicted efficient monopoly outcome.*

22.2 The Newsvendor Problem: Two Fine Student Insights and Some Calculus

The double marginalization in the vertical monopoly model shows how monopoly power at each stage can magnify distortions. In this section, we consider a model in which there is no market power to manipulate prices, which are assumed to be fixed and exogenous. Thus the firm buys units at a given wholesale price, W, and sells at a given retail price, P, with $P > W$. The catch is that final demand at the retail level is random, and units ordered lose all value if they are unsold. This is called the "newsvendor problem," since yesterday's newspaper has no commercial value. The zero value of unsold inventories adds risk (and in some cases realism), which makes for an interesting decision problem. The point of the newsvendor experiment is to study the behavioral effects of noisy feedback, i.e., responses to randomness in demand. The strategic landscape involves comparisons of the costs of unsold units (when the order quantity is too high) with the foregone earnings on lost sales (when the order quantity is too low).

In the simplest version of this model, the retail demand is assumed to be any amount on the interval from 0 to 100 units, with an equal probability for each possible demand quantity. For example, if a ten-sided die (labeled 0, 1, . . . 9) were thrown twice, with the first throw determining the "tens" digit, then the 100 possible outcomes would be the integers: 0, 1, . . . 99, and each of these would have an equal chance (1/100) of being observed. This setup can be represented as in figure 22.3, where the probability height of the dashed line is 0.01 on the vertical axis. If the order quantity is 60, for example, then there is

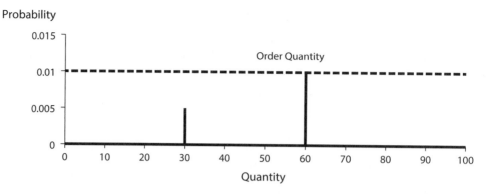

Figure 22.3. A Newsvendor Order of 60 Units and a 0.6 Probability of Having Unsold Units

a 0.6 chance that some units will remain unsold, since 60 of the 100 possible demand outcomes are to the left of 60.

For simplicity, assume that demand can be any number between 0 and 100, with fractional demands allowed, so the area under the dashed line to the left of the order quantity is the probability that demand is less then the order. This situation is illustrated in figure 22.3, where six-tenths of the area under the dashed line is to the left of 60. The flat dashed line is called a "uniform distribution" since the probability (vertical height) is uniform over the interval. When demand turns out to be less than an order quantity of 60, the amount sold will be 30 on average, since each sales amount between 0 and 60 is equally likely.

Before showing the results of an experiment, this section provides three ways to derive the optimal inventory policy. The first is the professional economist's approach of specifying an expected payoff function and setting the derivative to zero. *Most readers will want to skim over this calculus-based "instructor method" quickly to see the big picture. In contrast, the other two approaches are simple and direct.* The second method, developed by a student in the class, goes straight to the marginal comparisons and is both intuitive and insightful. The third method, devised by a second student, uses Excel in an impressive show of brute force.

Instructor Calculus Method

In general, with this uniform demand distribution from 0 to 100, the probability that the random demand is less than X is $X/100$, and the probability that it is greater than or equal to X is $1 - X/100$. Thus a firm that orders X units will sell all X units with probability:

(22.4) $$\text{Probability of Stockout} = 1 - X/100.$$

But there is a probability $X/100$ that not all units will be sold, and on average the sales in this case will be $X/2$, since each sales quantity in the interval from 0 to

X is equally likely. Multiplying probabilities times sales quantities, we get an expected sales quantity of: $X(1 - X/100) + (X/2)(X/100)$, where the first term is for the case where all X units sell and the second is for the case where (on average) only half of them sell. This expression for expected sales can be simplified to: $X - X^2/100 + X^2/200$, or equivalently, $X - X^2/200$. The firm's expected payoff is obtained by multiplying this expected sales quantity by the price, P, and then subtracting the cost of the order, XW, to obtain:

$$\text{(22.5)} \qquad \text{Expected Profit} = XP - \frac{X^2 P}{200} - XW$$

This quadratic expected payoff expression is maximized by setting its derivative with respect to X equal to 0:

$$\text{(22.6)} \qquad P - \frac{2XP}{200} - W = 0$$

Finally, (22.6) is be solved for X to obtain the optimal inventory:

$$\text{(22.7)} \qquad X = 100\left(\frac{P - W}{P}\right) \quad \text{(newsvendor optimal inventory rule).}$$

Student Intuitive Method: "Marginal Ideas from Econ 2010"

When the Veconlab newsvendor problem was first run in class in 2008, there were two students who used the optimal inventory order quantity in all rounds. One of them, Matt Cooper, when asked if he had read ahead, replied: "No, I used marginal ideas from ECON 2010." He did not elaborate, but what was clear to him only much later became clear to the author/instructor. The idea is to set marginal cost equal to *expected* marginal revenue. The marginal cost of ordering an additional inventory unit is the purchase price W. For marginal revenue, suppose that the manager is considering an order quantity X. Ordering an additional unit will result in *additional* revenue of P only if demand is greater than X, which happens with probability $1 - X/100$. This probability of selling the extra unit can be multiplied by the sale price P to obtain:

$$\text{(22.8)} \quad P(1 - X/100) \quad \text{(expected revenue from ordering an additional unit).}$$

The final step is to equate marginal cost of the additional unit, W, with the marginal revenue in (22.8): $W = P(1 - X/100)$. This $MC = MR$ condition can be solved to verify the newsvendor pricing rule shown above in equation (22.7). Matt Cooper was a Government and Foreign Affairs major, and what he did was figure out the relevant marginal condition without even taking a derivative!

Brute Force Method: "Consulting Spoils You"

A second student in the class, Vadim Elenev, was an Economics major who explained his method to the instructor as setting up an Excel table with 100 rows and 100 columns. There was one row for each possible order quantity, 1, 2, . . . 100. Each column was for one of the 100 possible demand quantities. If demand is greater than the order quantity X, then sales equal X. If demand is less than X, then sales equal demand. So you use the "min()" command in Excel. Each demand column has an associated probability of 1/100, so the probabilities are multiplied by the expected payoff calculations in each cell. Summing the products across a decision row then yields the expected payoff for that order quantity. Then a comparison of expected payoffs for the 100 order quantities reveals the optimal order. This brute force method is useful because it can be easily scaled up or adapted to situations in which demand is not uniform or there are other inventory costs besides just the order cost. After hearing about the Excel method, the instructor mentioned to Vadim that Matt had figured it out with marginal analysis, and Vadim admitted that he had suspected that there was an analytic solution, but that "consulting spoils you" in terms of going straight to computations!

To summarize:

Newsvendor Inventory Rule: *If demand is uniformly distributed on the interval from 0 to 100, the optimal inventory for a risk-neutral person is the price/cost ratio fraction ($\frac{P-W}{P}$) times the maximum demand of 100.*

22.3 Pull-to-Center Effect in Newsvendor Experiments

The newsvendor pricing rule was used in a between-subjects experiment with three treatments corresponding to different wholesale costs. If the retail price P is 4 and the wholesale cost is 2, then $(P - W)/P = 1/2$, and the optimal order quantity for a risk-neutral seller is 50. When the cost rises to 3, holding price constant, the optimal order falls to 25. Conversely, when the cost falls to 1, the optimal order quantity rises to 75. Each subject made a series of 30 order decisions in a single treatment. Information about demand and sales was provided after each round in a baseline full-information treatment, which was run with the Veconlab Newsvendor program (under the Markets menu). There were 36 subjects in the ($W = 2$) treatment, and 24 subjects in each of the other treatments ($W = 1$, $W = 3$).

Average order quantities are shown in figure 22.4, along with the theoretical predictions (dashed lines). The separation of data average lines indicates that order quantities do increase as the cost falls. There is, however, a tendency to order too much when the optimal order is below 50 and to order too little when the optimal order is above 50. When the optimal order quantity is 50, there is a slight bias upward, which seems to carry over to the other treatments as well. Thus, one might characterize two biases relative to the theoretical predictions:

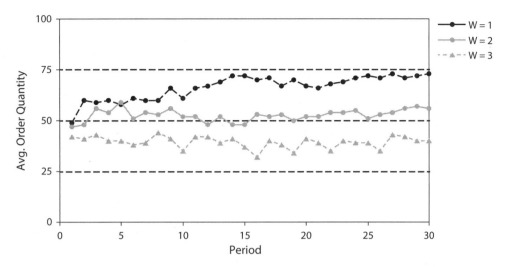

Figure 22.4. Newsvendor Average Order Quantities (dots) and Predictions (dashed lines). *Source*: Bostian, Holt, and Smith (2008).

an upward bias and a tendency for order quantities to respond sluggishly to wholesale price changes, i.e., a "pull to the center." The sluggish adjustment effect is exactly the pattern reported by Schweitzer and Cachon (2000), who consider a number of behavioral explanations *and end up rejecting them all*.

One possible approach to explaining the pattern in figure 22.4 might be to incorporate random "noise" in decision making, since randomness in decisions will tend to pull decisions up when the optimum is near the bottom and down when the optimum is near the top of the range of possible demand quantities. This is the approach taken in Bostian, Holt, and Smith (2008), who use a mix of decision error (discussed in chapter 9) and reinforcement learning presented in chapter 7.

An alternative explanation is provided by *learning direction theory* (Selten and Buchta, 1999), which postulates that adjustments will tend to be in the direction of what would have been a best response in the previous period. For example, in the $W = 3$ treatment that produces an optimal order of 25, it will be the case that demand exceeds 25 three-fourths of the time, which could tend to pull the orders up. Conversely, in the low-cost $W = 1$ treatment that produces a high optimal order of 75, demand will be less than 75 three-fourths of the time, which will tend to pull the orders down. Although this theory just predicts the directions of adjustments from period to period, not the magnitudes, it is easy to imagine how it might result in the observed "pull to the center" of the observed order decisions.

In the newsvendor literature, a tendency to base order quantities on previous period results is commonly referred to as "demand chasing." The Veconlab Newsvendor graph program has an option that lets the instructor plot current-period order quantities next to previous-period demand realizations. Some people exhibit demand chasing, but most do not. When this inventory problem is discussed in operations management courses, the advice is to "set a course" and follow it, irrespective of which way demand blows back and forth. Of course, a person might set the wrong course. For the class in which two of the students solved the newsvendor problem perfectly, the two people with the lowest earnings had order quantities that were unusually low in both treatments: with average orders of 26 and 49 when the optimal order was 75, and with average orders of 15 and 12 when the optimal order was 25, assuming risk neutrality. On the other hand, low orders protect against losses, so risk aversion or loss aversion could be having an effect for a subject whose orders are too low in both treatments. Conversely, since high order quantities are more likely to result in losses, especially in the high-cost ($W = 3$) treatment with an optimal order of 25, the upward bias that is typical for most subjects in that treatment cannot be explained solely by loss or risk aversion (problem 8). To summarize:

Newsvendor Experiment Results: *When order cost is high and optimal order quantity is low, the observed order quantities tend to be above the optimal order. Conversely, when order cost is low and the optimal order is high, observed orders tend to be too low. This pull-to-center effect, which has been observed in many experiments, could be due to a "demand chasing" tendency to raise orders after stockouts and to lower orders after ending up with unsold units. Most subjects, however, do not exhibit a clear pattern of demand chasing, and there are several alternative behavioral explanations at this point.*

22.4 The Beer Game and the Bullwhip Effect

Recall that the double marginalization problem associated with two monopolies may be alleviated to some extent by a vertical merger or by the introduction of competition downstream. Without these kinds of corrections, the analysis of section 22.1 would apply to an even greater extent with a longer supply chain, e.g., with a monopoly manufacturer selling to a distributor, who sells to a wholesaler, who sells to a retailer, where the firms at each stage are monopolists. In this context, the effects of successive "marginalizations" get compounded, unless these effects were somehow diluted by competition between sellers at each level.

The tendency for orders to be biased in a simple newsvendor problem with a single-layer supply chain suggests a second source of inefficiency. In a long supply chain, orders and inventories might not be well coordinated, especially if information is not transmitted efficiently from one level to the next. For example, Procter & Gamble found diaper orders by distributors to be too variable relative to consumer demand, and Hewlett-Packard found printer orders made by retail sellers to be much more variable than consumer demand itself. These and other examples are discussed in Lee, Padmanabhan, and Whang (1997a, 1997b).

There is a long tradition in business schools of putting MBA students into a supply chain simulation, which is known as the "Beer Game." There are four vertical levels: manufacturing, distribution, wholesale, and retail (Forrester, 1961). The participants in these classroom games must typically fill purchase orders from inventory, and then place new orders to the level above in the supply chain. There is a cost of carrying unsold inventory, and there is also a cost associated with not being able to fill an order from below, i.e., the lost profit per unit on sales. The setup sometimes involves having a stable retail demand for several rounds before it is subjected to an unexpected and unannounced increase that persists in later periods. The effect of this demand increase is to cause larger and larger fluctuations in orders placed upstream, which is known as the "bullwhip effect."

In classroom experiments with the Beer Game, the upstream sellers tend to attribute these large fluctuations to exogenous demand shifts, despite the fact that most of the fluctuation is due to the reactions of those lower in the supply chain (Sterman, 1989). Subsequent experiments, however, have shown that the bullwhip effect cannot be attributed solely to surprise demand shifts, since this effect has been observed even when retail demand shocks have an unchanging (stationary) and known distribution (Croson and Donohue, 2002, 2005, 2006).

Participants in a supply chain experiment are divided into "teams" of four participants, who are assigned to one of the four roles. Each person takes orders from below and submits orders to the person just upstream. Orders must be filled from inventory. There is a shipment delay of a couple of periods after an order is received; this delay is known and announced in advance. Subjects are given the incentive to minimize the sum of holding and backlog costs for the whole supply chain, as would be appropriate for an integrated firm.

Figure 22.5 shows the variance of order quantities as one goes up the supply chain, from Retailer (on the left) to Wholesaler, Distributor, and Manufacturer (on the right). Notice the high variability of production orders at the manufacturing level for all treatment bars on the right, as compared with wholesale purchase orders made by retailers on the left side. This is the essence of the bullwhip effect, where small variations at the retail level cause large swings in orders upstream.

The issue addressed in the Croson and Donohue experiment is whether improved information might reduce this effect. In the baseline treatment (dark bars), participants were given no information about inventory levels at the three other positions in the supply chain. The two information treatments involved providing sellers with information about inventories (or unfilled orders for the case of negative inventories) of those who are either upstream in the supply chain or downstream, depending on the treatment. The downstream

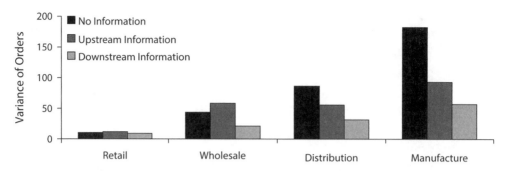

Figure 22.5. The Bullwhip Effect and the Provision of Inventory Information. *Source*: Croson and Donohue (2005).

information treatment resulted in a larger reduction in the bullwhip effect. Croson and Donohue (2005) also show that the effect is attenuated but not eliminated by providing point-of-sale information for others in the supply chain. To summarize:

> **Bullwhip Effects in the Laboratory:** *Random variations in retail demand can have magnified effects further up the supply chain, with higher variability of order quantities at the wholesale, distributor, and manufacturing levels. Even if bullwhip effects can be attenuated by better information transmission, the amplified order variations effects up and down the supply chain offer an important insight about how whole economies of interlinked markets might be sensitive to seemingly minor random shocks.*

22.5 Extensions

The newsvendor and bullwhip problems presented in this chapter are workhorse examples that are standard in any MBA operations management course. There is also a large experimental literature exploring various behavioral factors that might explain the observed biases. See, for example, the references in Bolton, Ockenfels, and Thonemann (2012), who compare the behavior of managers and students and find a pull-to-center bias for both groups of subjects. Boulou-Resheff and Holt (2017) analyze the optimal stocking decisions for a generalized inventory problem with carryover and other costs, in the presence of risk aversion. The inventory order patterns in the reported experiments continue to exhibit pull-to-center biases.

The vertical monopoly model is more closely related to economics subfields of industrial organization and antitrust. Durham (2000) uses experiments to compare price-setting behavior by an upstream monopolist who selects a wholesale price and announces it to one or more downstream sellers. There are two treatments—one with a single downstream firm and another with three downstream firms. She finds that the presence of downstream competition leads to outcomes similar to those under vertical integration. In effect, the competition at the downstream level takes out one of the sources of monopoly marginalization, so the price and quantity outcomes approach those that would result from a single monopolist. Another way of looking at this is to note that if there is enough competition downstream, then the price downstream will be driven down to marginal cost, and the total output will be a point on the retail demand curve, not a point on the retail marginal revenue curve. Then the upstream firm can essentially behave as an integrated monopolist who can select a point on the retail demand curve that maximizes total profit.

Ho and Zhang (2008) conducted a vertical monopoly experiment in which the upstream firm chooses a franchise fee along with a wholesale price, i.e., a "two-part tariff." As was the case for the classroom experiment in figure 22.2, this procedure does not solve the double marginalization problem. In fact, market efficiencies were not improved relative to the baseline case of a single wholesale price with no franchise fee. A second treatment introduced quantity discounts, which in theory provide an alternative solution to the double marginalization problem by giving the downstream firm an incentive to expand output. Efficiencies remained low, although the quantity discount procedure did provide a higher share of the surplus for the upstream firm. The authors concluded that the franchise fee fails to work because it is coded as a loss, and loss aversion makes downstream firms reluctant to accept high franchise fees. An alternative explanation, mentioned previously, is that a franchise fee might be rejected if it produces a division of the surplus that is viewed as being unfair.

It is important to note that there may be other beneficial or adverse consequences of vertical mergers or vertical price fixing. A manufacturer with a multi-market reputation to protect will want retail firms in separate local markets to provide adequate service. This could involve contracts that maintain retail prices at high levels to ensure that profit margins cover quality enhancements and in-store service. For this reason, vertical price fixing is not treated as being per se illegal in the United States. This quality-protection argument was instrumental in a landmark US Supreme Court decision on vertical price fixing. An expert witness in the case, Professor Kenneth Elzinga, argued that allowing manufacturers to enforce suggested retail prices will enhance quality and service. He noted that such price policies will "not be subject to free riding by discounting retailers who do not offer these services but free ride off the retailers who do" (quoted in Nash, 2007, who provides a brief non-technical summary of the relevant arguments in the case). The adverse effects of price competition on product quality will be considered in the next chapter.

Chapter 22 Problems

1.   Find the predicted profit for a monopoly seller in the market described in section 22.1.

2.   Find the profits for wholesaler and retailer for the market described in section 22.1 when the wholesale price is 14 and the retail price is 19, as predicted. Show that profits for these two firms are, in total, less than that of a vertically integrated monopolist, as calculated in the previous problem.

3.   Consider a market with an inverse demand function that is linear: $P = 46 - 2Q$. The cost at the wholesale level is 0 for each unit produced. At the retail level, there is a cost of 6 associated with retailing each unit purchased at wholesale. Thus the average cost for both levels combined is also 6. Find the optimal output and retail price for a vertically integrated monopolist, either using a graph or calculus. In either case, you should illustrate your answer with a graph.

4.   If you answered problem 3 correctly, your answers should imply that the firm's total revenue is 260, the total cost is 60, and the profit for the integrated monopolist is 200. Now consider the case in which the upstream and downstream firms are separate, and the upstream seller chooses a wholesale price, W, which is announced to the downstream firm on a take-it-or-leave-it basis (no further negotiation). Thus, the marginal cost downstream is $6 + W$. Equate this with marginal revenue (with a graph or with calculus) to determine the optimal quantity for the downstream firm, as a function of W. Then use this function to find the optimal level of W for the upstream seller (using a graph or calculus), and illustrate your answer with a graph in either case.

5.   Suppose that the upstream monopolist in problem 4 can charge a franchise fee. If the downstream seller is perfectly rational and prefers a small profit to none at all, what is the highest fee that the upstream seller can charge? What is the best (profit-maximizing) combination of wholesale price and franchise fee from the point of view of the upstream seller?

6.   Consider a retail firm with an exogenously given wholesale cost of $20 and an exogenously given retail price of $30. Demand is uniformly distributed from 0 to 300. What is the profit-maximizing order quantity for a risk-neutral firm?

7.   Suppose that the firm described in problem 6 orders 100 units. Find the range of realized demand quantities for which this firm will end up losing money.

8.   A firm is said to be "loss averse" if losses loom larger than gains in payoff cal-
      culations. Use intuition to guess what the effect of loss aversion would be for
      the situation in problem 6, i.e., would loss aversion tend to raise or lower order
      quantities relative to the expected profit-maximizing levels? Could loss aver-
      sion, by itself, explain the data patterns in figure 22.4?

9.   (non-mechanical) How might the standard newsvendor experiment be altered
      to eliminate possible effects of loss aversion? If losses are not possible, what
      effect might this have on average orders?

10.   (non-mechanical, advanced) How might the standard newsvendor experiment
      be altered to manipulate or control factors that produce a pull-to-center effect,
      while preserving a two-treatment design with high and low costs?

23

Adverse Selection in Lemons and Insurance Markets

When buyers observe both price and quality, there is competitive pressure on sellers to provide good qualities and reasonable prices. But when quality is not observed by buyers prior to purchase, there can be problems. For instance, a seller may be tempted to cut quality and associated costs. As a result, buyers will be hesitant to pay high prices needed to cover the costs of a high-quality seller. Then sellers may be forced to cut both price and quality.

Sellers who degrade product quality may start an unraveling process as buyers come to expect low quality. Unraveling can also occur when quality is exogenous, but sellers have to decide whether to sell given current price levels. If prices are too low, some owners of high-quality products may withdraw, which lowers the average quality in the market. If that withdrawal causes a price decline, further withdrawals may result, so that remaining units offered for sale have a low average quality. This process is known as *adverse selection*, i.e., the low-quality units are somehow selected to be offered for sale. A similar unraveling can occur in health insurance markets, as healthy people may decide to withdraw and not pay high prices for insurance, which lowers the average health of those remaining in the pool.

The "lemons market" terminology in the title of this chapter is due to George Akerlof (1970), who explained why there is no market for a new car that has just been purchased from the dealer and offered for resale, since prospective buyers will assume it is a lemon. Akerlof showed how asymmetric information and adverse selection can cause quality to deteriorate to such low levels that the market may fail to exist.

Note to the Instructor: The lemons market game can be administered using the instructions provided with the Veconlab Lemons Market program (from the Information menu). For class, it is important to use recommended default parameters and the simultaneous shopping setup, to avoid delays that result from sequential buyer shopping. It is fine to let people work in pairs or groups in order to have the numbers of buyers and sellers that are suggested in the setup process.

23.1 Endogenous Product Quality

The markets to be considered have the feature that sellers can choose the qualities of their products. A high-quality good is more costly to produce, but it is worth more to buyers. These costs and benefits raise the issue of whether or not there is an optimal quality. Most markets have the property that buyers are diverse in their willingness to pay for quality increases, so there will be a variety of different quality levels being sold at different prices. Even in this case, sellers who offer high-cost, high-quality items may face a temptation to cut quality slightly, especially if buyers cannot observe quality prior to purchase (*asymmetric information*). Quality might be maintained, even with asymmetric information about quality, if sellers can acquire and maintain reputations reported or signaled by warranties and return policies. Before considering the effects of such policies, it is useful to examine how markets might fail if buyers cannot observe quality in advance.

For simplicity, consider a case where all buyers demand at most one unit each and have the same preferences for quality, which is represented by a numerical grade, g. The maximum willingness to pay for a unit will be an increasing function of the grade, $V(g)$. Similarly, let the cost per unit be an increasing function, $C(g)$, of the grade. The net value for each grade g is the difference: $V(g) - C(g)$. The optimal grade maximizes this difference. The optimal grade may not be the highest feasible grade. For example, it is often prohibitively expensive to remove all impurities or reduce the risk of product failure to zero. The important behavioral issue in this type of market is the extent to which competition will force quality to near-optimal levels.

The features of the market can be illustrated with a simple classroom experiment with three grades. Each buyer only demands one unit of the commodity, and buyers have identical valuations that depend on grade: $4.00 for grade 1, $8.80 for grade 2, and $13.60 for grade 3. With four buyers and a grade of 2, for example, the market demand would be vertical at a quantity of 4 units for any price below $8.80, as shown by the grade 2 demand line in the middle part of figure 23.1. The other demand lines are similar, with buyer values of $4.00 and $13.60 for grades 1 and 3 respectively.

Each seller has a capacity of two units, with the cost of the second unit being $1 higher. For a grade of 1, the costs for the first and second units produced are $1.40 and $2.40 respectively, so each individual seller's supply curve would have two steps, before becoming perfectly inelastic at two units for prices above $2.40. With three identical sellers, the market supply will also have two steps, with three units on each step. The market supply for grade 1, shown in the lower part of figure 23.1, crosses the demand for grade 1 at a price of $2.40. The supply and demand curves for the other grades are shown above those for grade 1. The total surplus is the area between the supply and demand curves for a given

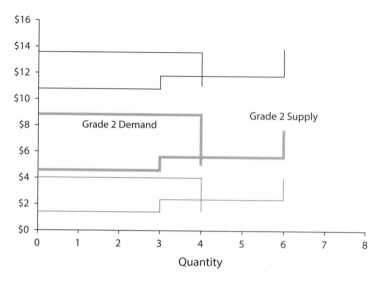

Figure 23.1. Demand and Supply Arrays by Grade

grade. It is apparent from figure 23.1 that the sum of consumer and producer surplus is maximized at a grade of 2.

The results of a classroom experiment from Holt and Sherman (1999) are shown in table 23.1. Students were divided into groups and given roles corresponding to 4 buyers and 3 sellers. In the first three rounds, sellers recorded prices and grades. When all had finished, these prices and grades were written on the blackboard, and buyers were selected in random order to make purchases. Notice that two of the three quality choices in the first round were at the maximum level. Each of these high-quality sellers sold a single unit, but the seller who offered a grade of 2 sold 2 units and earned more. Buyers preferred the grade 2 good since it provided more surplus relative to price. All three sellers had settled on the optimal grade of 2 by the third round, and the common price of $5.60 was approximately equal to the competitive level determined by the intersection of supply and demand at this grade.

The three "full-information" rounds were followed by two rounds in which sellers selected price and grade as before, but only the price was written on the board for buyers to see while shopping. Two of the sellers immediately cut grade to 1 in round 4, although they also cut the price, so buyers would be tipped off if they interpreted a low price as a signal of a low grade. In the final round, seller 3 offered a low grade of 1 at a price of $5.50, which had been the going price for a grade of 2. The buyer who purchased from this seller must have anticipated a grade of 2, and this buyer lost money in the round. This pattern is similar to the quality unraveling observed by Holt and Sherman (1990) with more quality grades. To summarize:

Table 23.1. Results from a Classroom Experiment

	Seller 1	Seller 2	Seller 3
Period 1 (full information)	$11.50 grade 3 sold 1	$6.00 grade 2 sold 2	$12.00 grade 3 sold 1
Period 2 (full information)	$5.75 grade 2 sold 2	$5.50 grade 2 sold 1	$1.90 grade 1 sold 1
Period 3 (full information)	$5.65 grade 2 sold 1	$5.60 grade 2 sold 2	$5.60 grade 2 sold 1
Period 4 (only price information)	$2.40 grade 1 sold 1	$5.60 grade 2 sold 1	$2.40 grade 1 sold 2
Period 5 (only price information)	$2.40 grade 1 sold 1	$1.65 grade 1 sold 1	$5.50 grade 1 sold 2

Source: Holt and Sherman (1999).

Unraveling in Experiments with Unobserved Product Quality: *In experiments with asymmetric information and endogenous seller quality decisions, price competition and quality-based cost savings may result in a lemons market outcome in which prices and qualities are too low relative to optimal, surplus-maximizing levels.*

There are several experimental studies of the effects of practices that alleviate the lemons market problem. Lynch et al. (1986) show that performance in double auctions with information asymmetries can be improved with certain types of warranties, requirements for truthful advertising, etc. DeJong, Forsythe, and Lundholm (1985) allowed sellers to make price and quality representations, but the quality representation did not have to be accurate, and the buyer had imperfect information about quality even after using the product. In this experiment, the sellers could acquire reputations for providing higher quality based on buyer experience. This process of reputation building prevented the markets form collapsing to the lowest quality. Miller and Plott (1985) report experiments in which sellers could make costly decisions that "signal" high quality, which might prevent quality deterioration.

23.2 Adverse Selection with Exogenous Quality

Akerlof (1970) explained how market failure can occur even if sellers are not able to choose the quality grades of the products they sell, as long as there is asymmetric information in the sense that buyers do not observe quality prior to purchase. This market failure with exogenous quality grades is caused by a process of *adverse selection* in which sellers with high-quality items drop out, since they do not expect to obtain high prices from buyers who cannot observe this quality. This exit causes a reduction in the average quality, and hence buyers' willingness to pay declines, which drives additional sellers with higher-than-average quality goods out of the market. The end result of this unraveling process may be a situation in which no trade occurs, or a complete *market failure*.

Kirstein and Kirstein (2009) conducted an experiment with a market model with asymmetric quality information. Each seller in this market was given a single unit to sell, with a randomly determined quality, g, which was equally likely to be any penny amount between 0 and 1. Imagine a random draw from a bingo cage with balls marked 0, 0.01, . . . 0.99, and 1. The quality was observed by the seller but not by buyers. In one of their designs, the money value to the seller for retaining a unit of grade g was $3g$, as shown by the seller's value line in figure 23.2. Similarly, the value to a buyer is $4g$, as shown by the buyer's value line. The value to the buyer is above the value to the seller for each quality grade on the horizontal axis of the figure, so economic efficiency would require that all trades be made.

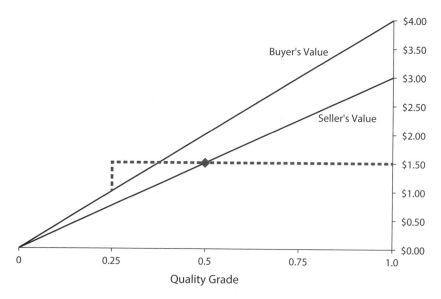

Figure 23.2. Unraveling in a Market with Grades on [0, 1]

At the start of a round, each seller was matched with a buyer, who then makes a take-it-or-leave-it price bid, p. Since follow-up communications were not permitted, the seller who knows the grade, g, should accept the bid if the value to the seller is less than price. The buyer could maximize the chances of a purchase by offering a price of $3, which the seller would accept regardless of the grade. But the average grade is only 0.5 since each grade in the range from 0 to 1 is equally likely, and the value to the buyer for a grade of 0.5 is 4*(0.5) = $2. Thus, the buyer would be paying $3 for an item that is only worth $2 on average, so losses would tend to occur. What about a lower price, say of $1.50, as shown by the horizontal dashed line in the figure? This line intersects the seller value line at a grade of 0.5, so only sellers with grades at or below 0.5 would accept this lower price. The average grade in the range from 0 to 0.5 is the midpoint, 0.25, and the value of this average grade is only $1.00 to the buyer, who is paying a price of $1.50.

As you might have guessed, these examples are easily generalized. Suppose the buyer offers a price, p, which the seller will accept if $3g < p$, or equivalently, if $g < p/3$. Thus a price offer of p will generate a sale for any seller with a grade in the range from 0 to $p/3$. Since matchings are random, all grades in this acceptance region $[0, p/3]$ are equally likely, so the average grade for an accepted bid of p is the midpoint: $(0 + p/3)/2 = p/6$. Recall that the value to the buyer is four times the grade, so four times the expected grade is $4p/6 = 2p/3$. Thus a buyer with an accepted bid will end up paying p for a unit that is, on average, only worth $2p/3$, for an expected loss of $p/3$. Higher bids have higher expected losses, so bids should decline to zero, which would result in no sales.

In the Kirstein and Kirstein experiment, observed prices tended to fall from a high initial level toward a relatively stable level in the final periods. As the prices fell, the average quality of the items traded fell as well. The market did not collapse completely to a price of 0, which the authors attribute to a limited ability of the participants to think iteratively about what the optimal bid should be. The decline in bid prices in the experiment was slow and incomplete, with prices starting at about $1.70 in period 1 and falling to about $1.00 in period 10, and leveling off in the range from $0.50 to $0.80 in periods 15–20. This pattern raises the issue of what is driving a dynamic adjustment process. In a market like this, feedback for buyers will be somewhat noisy, since there will be times when a purchase turns out to be profitable, i.e., when the seller's grade is near the top of the acceptance region. There will also be periods when the buyer's bid is too low and is rejected. In either of these cases, buyers might be tempted to *raise* their bids in the next round, which will tend to offset bid reductions that might follow rounds where the buyer loses money. These observations suggest why learning that low bids are better might be a slow process. Feltovich (2006) uses computer simulations of the learning process to explain why price declines may be quite slow in this context. The particular learning process that he considers is one in which decisions that have earned more money in the past have a higher probability of being used, which is known as "reinforcement learning," as discussed in chapter 7. To summarize:

Adverse Selection in Experiments with Exogenous Quality: *Even with exogenous (but unobserved) quality differences, adverse selection may result in high proportions of low-quality commodities and associated low prices based on low consumer expectations. The extent of market failure was less severe than the theoretical prediction of complete unraveling, perhaps due to heterogeneous individual experiences and slow learning.*

23.3   Unraveling in Health Insurance Markets

A similar analysis can provide some perspective to unraveling in insurance markets, which seemed to be occurring with the Affordable Care Act Insurance Exchanges in some US states in 2018. Suppose that each person in a particular healthcare pool has an overall health grade, g, where high values correspond to better health. For simplicity, think of g as the probability of not having a claim of a standard amount, e.g., a hospital stay. Therefore, a grade of 1.0 on the right is for a healthy person with no chance of a claim, and a g value of 0.9 is for a person with a 10% chance of a claim. This grade determines the expected cost of insurance claims for that person. This expected cost will be a decreasing function of g, as shown by the downward-sloping expected daily cost line in figure 23.3. Those with the lowest health grade of 0 have expected insurance claims of $80 on the left side. If the price of insurance exactly covers the *expected* claim cost, most people would still be quite willing to purchase insurance to reduce the *risks*, which can cause actual expenses to be higher than expected values in a given month. Therefore, the upper dotted willingness-to-pay line is higher than the expected healthcare cost line.

For simplicity, suppose that people do not know their health grades (this will be relaxed) and that grades are uniformly distributed along the line. With a uniform distribution of grades between 0.6 and 1, the average grade would be 0.8, which has an expected daily claim cost of $40, calculated as a 0.2 chance of a $200 claim. In this case, an insurance premium of $40 per day would cover expected claims, and people would be willing to purchase the insurance to reduce the risk if they did not know their health grades.

As a person's health status evolves and more is known, people with very high grades may consider not purchasing the insurance at the $40 premium price. In figure 23.3, those would be people with grades above about 0.85, whose maximum willingness to pay is below the $40 premium. If the healthiest people were to withdraw, the average grade in the insured pool would fall, which would

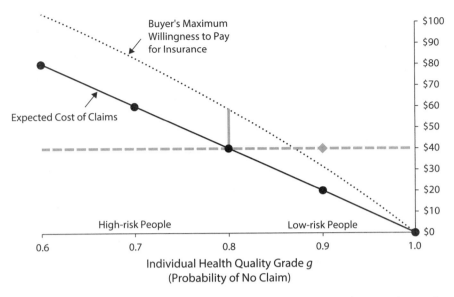

Figure 23.3. Unraveling in a Market with Health Insurance. *Note:* Low-risk people on the right have low expected claim costs (solid line). Risk aversion causes the willingness to pay for insurance (dotted line) to be higher than expected claim costs. The insurance premium (dashed line) is above the dotted willingness-to-pay line for low-risk people on the right side, so they may opt to self-insure, which lowers the average health grade of those in the insurance pool, and raises the premium, which can cause even more low-risk people to self-insure. In theory, this unraveling will cause the market to collapse as fewer and fewer people buy insurance.

require a rise in premiums. A rise in premiums would cause additional withdrawals and premium increases, which could lead to a complete collapse.

Riahia, Levy-Garboua, and Montmarquette (2013) ran an experiment that was based on a discrete version of the model in figure 23.3. There were two types of people, those with low risk ($g = 0.9$) and those with high risk ($g = 0.7$). The low-risk people on the right had expected claim costs of 20, and the high-risk people had expected claim costs of 60. With equal numbers of each type, the expected claim cost for the combined pool was the midpoint of 40. There were four insurance sellers who would submit premium prices, and all purchases would be made from the seller with the lowest premium price. This severe (Bertrand) price competition would be expected to drive the premium price down to the expected claim cost, which it did. The insurance buyers were informed of their own risk class (high or low), and knowing this, would consider the premium and decide whether or not to purchase insurance.

By paying subjects in lottery tickets for a final prize, the authors induced a utility function that was curved and exhibited risk aversion (this method is mentioned briefly in chapter 6). For our purposes here, it is sufficient to understand that this risk aversion is what caused the dotted willingness-to-pay line in figure 23.3 to be above the expected claim cost line. The risk aversion measure induced in the experiment was such that low-risk people would be willing to pay at most $32 to buy insurance and reduce the risk, even though the expected value of their claim was only $20. This $32 willingness to pay is below the diamond point at $40 on the dashed premium amount in the figure, so the prediction is that the low-risk individuals will not purchase the insurance and self-insure. If everyone who does purchase is a high-risk person, the premium would have to rise to $60 to cover the expected claims. This is not what happened in the experiment. Only about half of the low-risk people did not purchase insurance, and the competition among sellers drove the premium down to about $40, well below the predicted level of $60. Sellers did not lose money because a significant fraction of low-risk people did purchase insurance, which reduced expected claim costs and allowed sellers to maintain lower premiums. There was some unraveling in the sense that insurance was purchased more frequently by high-risk individuals, even though the pool of purchasers contained a significant fraction of low-risk individuals. As in the previous section, the experiment shows an intermediate result, with less than complete market collapse. To summarize:

Experiment Results for Insurance Markets: *When buyers know their own risk types, those with low risk tend to self-insure more frequently than those with high risk. The resulting adverse selection, however, is only partial and does not cause a collapse of the insurance market, as sellers manage to avoid losses on average and continue to offer policies.*

The authors also ran a second treatment in which insurance firms were permitted to offer a contract with a high deductible, so instead of paying the full loss of $200, the policy would only pay a fraction of the loss. In this case, the relevant theory (Rothschild and Stiglitz, 1976) predicts that low-risk individuals will select into the plan with the high deductible and that only the high-risk individuals will purchase the more expensive plan that offers full coverage. Thus, the theoretical prediction was *complete separation*, so that the seller could infer a buyer's risk type from the insurance contract (deductible or not) selected. But the most dramatic prediction is that the market would not collapse, and that all individuals would purchase some type of insurance. The separation observed in this treatment was less than complete, although high-risk individuals were more likely to choose the full coverage. Moreover, the ability to offer a deductible did not solve the adverse selection problem, since significant fractions of people did not purchase insurance at all (about 50% uninsured for low-risk and about 25% uninsured for high-risk individuals). In fact, the proportion of uninsured individuals was not even lower in the treatment with deductible options than it was in the baseline treatment with no deductibles. The authors noted that there seemed to be some heterogeneity in the way that different people perceive risks, which caused individual purchase decisions to be less predictable than would be expected based on pure theory. To summarize:

Deductibles and "Screening" Based on the Risk of Loss: *The use of insurance policies with and without deductibles did not solve the adverse selection problem in a laboratory experiment. The results indicate less-than-perfect separation based on whether or not the deductible was selected. Moreover, the incidence of uninsured individuals was not reduced relative to a baseline treatment with no deductibles.*

23.4 Extensions

The discussion of health insurance market failures in the previous section was somewhat abstract. Hodgson (2014) presents an easy-to-implement class experiment with a lot of interesting detail taken from US healthcare markets. Each person is assigned a "character" with a health status, e.g., "35 years old and in good health" or "57 years old with uncontrolled diabetes." Each character is presented with probabilities of major and minor health expenditures and is given a choice of three different health plans, with different deductibles. As students choose plans, the instructor enters them into a spreadsheet that calculates the expected claim costs of those enrolled in each plan. As healthy people opt out, the expected claim costs and premiums rise, so that students can watch as the market collapses. This class experiment can be used to guide discussions current on policy issues, e.g., the effects of forcing insurers to accept applicants with preexisting conditions.

Capra, Comeig, and Banco (2014) use a laboratory experiment to study a screening mechanism that is often implemented in credit markets. They show how a lender can screen high- and low-risk borrowers to prevent quality deterioration and credit market failure by offering a pair of incentive-compatible contracts. Those contracts include two features each: interest rate (price) and collateral (warranty). High-quality borrowers (with low-default risk) self-select by choosing the contract with high collateral and low interest rate (better price), while low-quality borrowers (with high-default risk) prefer the contract without collateral and a higher interest rate (worse price). The low-risk borrower is more willing to put up collateral that would be lost in the unlikely event of a default. The experiment, which is well suited for class use, is motivated by a study of contracts used in actual loan markets (Comeig, Del Brio, and Blanco, 2014).

Chapter 23 Problems

1. Use the unit costs and redemption values for the supply-demand configuration in figure 23.1 to calculate the total surplus for grade 1.

2. In a competitive equilibrium for a given grade, the price is determined by the intersection of supply and demand for that grade. Compare (or estimate) the equilibrium total seller profits for each grade for the market in figure 23.1.

3. (non-mechanical) For the market structure shown in figure 23.1, consider a small increase in the cost for each seller's first unit *for grade 2 only*. An increase of $0.25 in this cost would mean that the competitive equilibrium profits for each seller are lower for grade 2 than for the other grades. What effect do you think this cost increase would have on the tendency for grade choices by sellers to converge to the optimal grade (2)?

4. Groucho Marx once remarked: "Please accept my resignation, I don't want to belong to any club that will accept me as a member." How is this situation different from (or similar to) the exogenous quality model discussed in section 23.2?

5. Consider the exogenous quality model in section 23.2, with the value of a grade g item to the seller being $3g$ (as before) but the value to a buyer is changed to $1 + 3g$. Discuss what might be learned from an experiment with this modified structure.

24

Asset Markets and Price Bubbles

The featured experiment in this chapter involves trading of shares of an asset that pays randomly determined dividends each period. Traders are endowed with both shares and cash balances. The latter can be used to either purchase shares or earn interest in a safe account. The interest rate is an opportunity cost that determines the value of the asset on the basis of fundamentals: dividends and associated probabilities, and the redemption value of each share after a pre-announced final round of trade. This fundamental value serves as a benchmark from which price bubbles can be measured. Bubbles and crashes are typical, especially for inexperienced traders, and the resulting discussion can be used to teach ideas about present value, asset returns, price forecasting, etc.

Note to the Instructor: The asset market setup is easy to implement using the Veconlab Limit Order Market program (on the Finance/Macro menu). The Leveraged Asset Market offers more options, including price forecast elicitation and buying with limited down payments. The STOP button on the admin results page of both programs lets you terminate a period if anyone is too slow.

24.1 Tulip Fever

Despite the widespread belief that equity values rise steadily over the long term, these markets exhibit strong price swings that do not seem to be justified by changes in the underlying conditions. Keynes's explanation was that many (or even most) investors are less concerned with fundamentals that determine long-term future profitability of a company than with the price the stock might sell for in several weeks or months. Such investors will try to identify stocks that they think other investors will flock to. This herding may create upward pressure on price in a self-fulfilling prophecy. The psychology of the process is described by Charles Mackay's (1841) account of the Dutch "tulipmania" in the seventeenth century:

> Nobles, citizens, farmers, mechanics, seamen, footmen, maid-servants, even chimney-sweeps and old clotheswomen, dabbled in tulips. Houses

and lands were offered for sale at ruinously low prices, or assigned in payment of bargains made at the tulip-mart. Foreigners became smitten with the same frenzy, and money poured into Holland from all directions.

Of course, tulips are not scarce like diamonds. They can be produced, and it was only a matter of time until the correction, which in Mackay's words happened:

> At last, however, the more prudent began to see that this folly could not last forever. Rich people no longer bought the flowers to keep them in their gardens, but to sell them again at cent per cent profit. It was seen that somebody must lose fearfully in the end. As this conviction spread, prices fell, and never rose again. Confidence was destroyed, and a universal panic seized upon the dealers.

Even if traders realize that prices are out of line with production costs and profit opportunities, a feeling of overconfidence may lead them to believe they will be able to "sell high," or at least that they will be able to sell quickly enough in the case of a crash. The problem is that, should a crash happen, there are no buyers at any price levels. Often all it takes is slight decline, perhaps from an exogenous shock, to spook buyers and stimulate the subsequent free fall in prices.

Asset price bubbles generate roller-coaster mood swings that go along with strong price surges and corrections. In housing markets, for example, young couples may feel envious of acquaintances who have already purchased a house and "enjoyed" a rise in assessed value. A steady price increase may trigger impatience, a desire to purchase before prices are too high to be affordable. Homes may only be listed for short periods before being sold, and regret or anticipated regret are strong emotions that arise naturally. Risks may seem low during periods of steady price increases, which engenders overconfidence. Emotions triggered by economic ownership can be mixed with other passions. This is the theme for the 2017 film *Tulip Fever*, in which a passionate seventeenth-century couple resorts to investing what they have in the high-stakes tulip market.

Over the years, some economists argued that the tulip mania and similar episodes were not necessarily price bubbles, but rather, were rooted to a large extent in "fundamentals." A careful account is provided by Garber (1989, p. 535):

> Most of the "tulipmania" was not obvious madness. High but rapidly depreciating prices for rare bulbs is a typical pattern in the flower bulb industry. Only the last month of the speculation, during which common bulb prices increased rapidly and crashed, remains as a potential bubble.

This skeptical view of bubbles is not rare. A new doctoral student once arrived at the University of Virginia, intent on writing a dissertation on price bubbles, having just witnessed the "dot-com" bubble and the Russian bond market collapse while working as a financial trader. Despite her enthusiasm, her faculty advisor at the time discouraged her from studying patterns that "do not exist."

As the careful analysis provided by Garber and others suggests, it is difficult to distinguish sharp price movements driven by unfounded speculation from movements triggered by economic fundamentals. Sudden price declines may be caused by a mix of economic events (like bank failures) and other events (like the 1907 San Francisco earthquake). Sometimes the origin of a downturn is essentially random. The author once encountered a professional trader who remembered spending a weekend in 1987 making massive sell orders, as requested by his company. The orders were executed on the following Monday ("Black Monday") and triggered program trading that then led to a massive stock crash. In this particular case, the stocks were probably not overpriced, since after falling about 30%, the market surged back and ended the year where it started. The trouble with assessing whether assets are overpriced is that the world is always changing. In any case, the phrase "this time is different" is not a rationale for dismissing a bubble, as indicated by the book title: *This Time Is Different: Eight Centuries of Financial Folly* by a pair of economic historians (Reinhart and Rogoff, 2011).

Eugene Fama, who later won a Nobel Prize for his work in financial economics, once suggested that the housing market was efficient because buyers think carefully about comparable prices for something that is likely to be the largest purchase they make in their lifetimes. Just before the US housing bubble broke in 2008, Fama quipped: "The word 'bubble' drives me nuts." He went on to suggest high valuations of technology stocks during the "Internet bubble" were justified by 1.4 firms of the same size as Microsoft (quoted in Clement, 2007). In retrospect, the 2008 housing price collapse shown in figure 24.1 was a bubble, fueled by unsound lending practices that provided easy credit to high-risk borrowers. Between 1991 and 2007, housing prices in most cities had increased by factors of 2 or 3 above base values, with rapid increases at the end of this period. Real estate comparable price lists ("comps") are important, but they are more useful for assessing relative prices than for identifying whether *all* prices are too high.

The obvious advantage of a laboratory experiment is that exogenous factors like technology changes or artificial shortages can be controlled, while economic conditions (e.g., easy credit or "excess cash") that have been associated with bubbles in stock and real estate prices can be varied. Alternatively, price bubbles and crashes can be re-created in computer simulations by introducing a mix of trend-based and fundamentals-based traders who execute strategies that are determined by their types. In simulations, price surges are stimulated by positive exogenous shocks that trigger programmed trading by types that tend to extrapolate price changes. Such simulations can even produce negative bubbles that follow negative shocks (Steiglitz and Shapiro, 1998). In a "negative bubble," the simulated trend traders are selling because prior price decreases

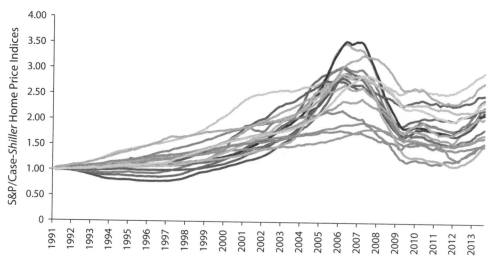

Figure 24.1. Case-Shiller Housing Prices for 16 US Cities Relative to 1991 Levels. *Source*: Coppock et al. (2015).

are expected to continue and this sell pressure draws prices down if the fundamentals-based traders do not have the resources to correct the situation.

Computer simulations are suggestive, especially if they mimic the price and trading volume patterns that are observed in stock market booms and crashes. However, the obvious question is how long human traders stick to mechanical trading rules as conditions change. On the other hand, a problem with running laboratory experiments with human subjects is to decide how owners of unsold shares are compensated when the experiment ends. The next section describes experiments in which the asset has no value at all after a pre-announced final period.

24.2 Declining Fundamental Value Experiments: The Literature

Smith, Suchanek, and Williams (1988) addressed the endpoint problem by using a redemption value for the asset after a pre-announced final period. Traders were endowed with asset shares and cash, and they could buy and sell assets at the start of each period (via double auction trading). Assets owned at the end of a period paid a dividend that was randomly determined from a known distribution. Cash did not earn any interest, and all cash held at the end was converted to earnings at a pre-announced rate.

In most markets, the final redemption value was set to zero. To a risk-neutral person, each share would be worth the sum of the expected dividends for the periods remaining. For example, consider an asset that pays $0.50 or $1.50, each with probability 1/2, so the expected dividend is $1.00. This asset would only be worth $1 in the final period. In the second to last period, with two periods remaining, the asset would be worth $2, etc. Thus the fundamental value

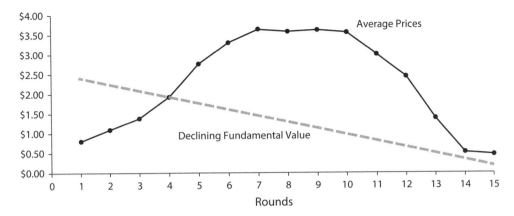

Figure 24.2. Average Transactions Prices in a 15-Round Double Auction with a Declining Fundamental Value

of the asset is proportional to the number of remaining periods, which declines linearly over time. With this declining value setup, price bubbles were observed in many (but not all) sessions.

Figure 24.2 shows the results of a strong price bubble for a session where the final redemption value was $0.00 and the expected dividend was $0.16. Since there were 15 periods ("rounds"), we can calculate that the initial fundamental value is equal to 15 × $0.16 = $2.4. Hence, the line starts at $2.40 and declines by $0.16 per period, reaching $0.16 in the final round. The figure also shows the average transactions prices for each round, as determined by double auction trading. The prices were below the fundamental value line in early rounds, before rising steadily above the falling value line. People who bought in an early round and then saw the price rise would often try to buy again, and others wanting to join in on the gains would begin to buy as well. As the final round approached, it became obvious that nobody would pay much more than the expected dividend in the final round, which causes the bubble to collapse. Such bubbles have been observed with diverse groups of participants: undergraduates, graduate students, business students, and even a group of commodity traders. The speculative bubbles could form, *even if* people knew that there was a fixed endpoint.

Smith, Suchanek, and Williams did run some sessions with a non-zero final redemption value, which was set to be equal to the sum of the realized dividend payments for all rounds. Thus, the final redemption value would not be known in advance, but rather, would be determined by the random dividend realizations, and would generally be decreasing from period to period (problem 1).

The declining value structure that results from a zero redemption value has been used extensively in tests of various aspects of behavior in asset markets. Some of the main results can be listed briefly (more references will be provided in the final extensions section of this chapter):

- *Excess Cash:* Bubbles tend to be enhanced by treatments that increase the cash in the system through injections such as incomes, dividends, or a longer series of market periods. Bubbles are reduced if cash injections are dampened, e.g., by delaying dividend payments until the end of the final period, or by paying positive and negative dividends that tend to cancel out.
- *Experience:* Bubble magnitudes diminish if the subjects with prior experience in a one-market sequence are brought back to participate in a second-market sequence with the same underlying structure. However, bubbles can be rekindled with experienced subjects after changes in dividend structure and injections of liquidity.
- *Biological Markers:* Bubble amplitudes are larger with groups of male subjects who have been administered doses of testosterone.
- *Gender Effects:* Bubble amplitudes are larger with groups of women in some studies. Gender effects, however, are not always observed and may be sensitive to the procedural elements, culture, and whether the fundamental value is declining or flat, as discussed later in this chapter.
- *Cognitive Skills:* Subjects who score higher on tests of cognitive ability tend to earn more, and groups with high average cognitive abilities tend to exhibit lower bubble magnitudes.
- *Futures Markets and Short Sale Opportunities:* These factors tend to reduce bubble intensity, as would be expected, although bubbles are not eliminated with these options.
- *Risk Aversion:* Correlations between individual risk aversion measures and shares held at asset price peaks are weak. This could be because bubbles are driven more by anticipated capital gains and extrapolated price trends than by risk tolerance.

The general impression is that bubbles can be enhanced or diminished, but they are pervasive in asset markets with significant liquidity. To summarize:

Experiment Asset Market Results: *Price bubbles above a declining fundamental value are enhanced by factors that increase excess cash, testosterone, and low scores on cognitive response tests. Bubble amplitudes are reduced by prior trading experience in an asset market, delayed dividend payments, and short sales.*

24.3 Discounting and Present Values When Interest Is Paid on a Safe Asset

Since most financial assets do not have predictable, declining fundamental values, it is instructive to set up experiments with constant or increasing values. One approach is to use a fixed probability of breakage to induce a preference for the present (when the asset is not broken) over the future. Even though discounting and present value considerations can be induced by a probabilistic stopping rule, it is natural to consider a setup that is closer to the primary economic motivation for discounting future payments, which is that a dollar in the future is worth less than a dollar today because the money today can be invested and increased via compound interest. Suppose that a person is considering purchasing a share at a price of V today that will pay a random dividend with an expected value of D. Cash not used to purchase the share earns interest at a rate r. If no capital gains are anticipated, then the expected benefit of the purchase is D, and the opportunity cost of tying up the cash for a period is rV. The expected returns from the safe and risky assets are the same if $rV = D$, or equivalently,

(24.1a) $\qquad r = \dfrac{D}{V}$ (interest on \$1 = dividends per dollar invested)

(24.1b) $\qquad V = \dfrac{D}{r}$ (share value = present value of future payments).

Another perspective on this equality follows from a final period analysis. Suppose that an asset will sell for D/r in the final period and will also be redeemed for D/r, so there are no capital gains. Then the interest on the D/r used to purchase a share would be $r(D/r)$, which equals the dividend D received at the end of the period. Finally, the D/r ratio on the right side of (21.1b) can be shown to equal the present value of future dividends, which is the next task.

The shares actually used in experiments will pay dividends for a finite number of periods, followed by a redemption value payment, so the value of the share (in the absence of capital gains) is the present value of these future payments. Present value considerations are induced by paying interest on accumulated cash balances, at a rate of r per period. Suppose that an asset only pays a dividend of D dollars once, at the end of the first period, and then loses all value. Valuing the asset means deciding what to pay now for a dividend of D that arrives one period later. This asset is worth less than D now, since you could take D dollars, invest at an interest rate r, and earn $(1+r)D$, by the end of the period. The result would be greater than D. Thus, it is better to have D dollars now than to have an asset that pays D dollars one period later. This is the essence of the preference for present over future payments, which causes one to "discount" such future payments.

To determine how much to discount the future dividend payment, consider an amount that is less than D. In particular, consider $D/(1+r)$ dollars at the beginning of the current time period. If this amount is invested at rate r, the amount obtained at the end of the period is the amount: $(1+r)D/(1+r)$, which equals D. Therefore, the *present value* of getting D dollars one period from now is $D/(1+r)$. Similarly, the present value of any amount F to be received one period from now is $F/(1+r)$. Conversely, an amount V invested today yields an amount

$V + Vr$ after one period, where the V is the principal and the Vr is the interest. These observations can be summarized:

Present and Future Values for a 1-Period Horizon

Future value of a present payment V: $F = (1 + r)V$

Present value of a future payment F: $V = \dfrac{F}{1+r}$

Similarly, to find the future value of an amount V that is invested for one period and then reinvested at the same interest rate, we multiply the initial investment amount by $(1+r)$ and then by $(1+r)$ again. Thus, the future value of V dollars invested for two periods is $V(1+r)^2$, which represents compounded interest. Conversely, the present value of getting F dollars two periods from now is $F/(1+r)^2$.

Present and Future Values for a 2-Period Horizon

Future value (2 periods later) of a present payment V: $F = (1 + r)^2 V$

Present value of a future (2 periods later) payment F: $V = \dfrac{F}{(1+r)^2}$

In general, the present value of an amount F received t periods into the future is $F/(1+r)^t$. With this formula, one can compute the value of a series of dividend payments for any finite value of t. For example, the present value of an asset that pays a dividend D for two periods and then is redeemed for R at the end of period 2 would be: $D/(1+r) + D/(1+r)^2 + R/(1+r)^2$, where the final term is the present value of the redemption value.

Recall the result in equation (24.1b) that the purchase of a share selling for V dollars has the same expected return as that of holding cash and earning interest if $V = D/r$. This value also turns out to be the present value of an infinite sum of dividends in the future:

(24.2) $$V = \frac{D}{1+r} + \frac{D}{(1+r)^2} + \cdots \frac{D}{(1+r)^t} + \cdots$$

One way to derive the "perpetuity" present value is to factor out a common term on the right side of (24.2) and then use the definition of V in (24.2) to rewrite it as an expression of V on the right side of (24.3):

(24.3) $$V = \frac{1}{1+r}\left[D + \frac{D}{1+r} + \frac{D}{(1+r)^2} + \cdots\right] = \frac{1}{1+r}[D + V].$$

Therefore, $V = \frac{1}{1+r}[D + V]$, which can be solved for $V = D/r$. For example, if the interest rate is 0.05 $(=1/20)$ and the dividend is \$1, then the present value of the infinite series of dividend payments is $\frac{1}{1/20} = \$20$. To summarize:

> **Discounting and the Present Value of a Perpetuity:** *If subjects have access to a safe asset that pays an interest rate r, then the incentives induce a preference for present over future payments, which must be discounted. Future payments after t periods can be discounted by dividing by $(1+r)^t$, so later payments are discounted more heavily. Even though the sum of a constant dividend amount D over an infinite number of periods is infinite, the present value of the infinite sum of dividends is a finite amount, D/r, that equates to the expected returns on the safe and risky assets.*

24.4 A Flat Fundamental Value: "Looking Beyond the Lampost"

The declining value setup discussed is a marvelously simple way of setting up an asset with a well-defined fundamental value, and it is remarkable that bubbles are commonly observed even though subjects know that the asset will be worthless in the end. On the other hand, most assets do not decline in value like a gold mine or oil field with a known amount of reserves. For example, instead of stating that the asset will pay dividends of D for each of 15 periods, the asset could be described as having 15 units of gold in a mine, each worth D, with the mine having no value after the last gold unit is mined. This description tends to mitigate or even eliminate observed bubbles (Kirchler, Huber, and Stöckl, 2012). While acknowledging the value of having multiple parallel studies using the same declining value format, Jörg Oechssler (2010) made a strong case for considering more realistic settings with flat values and multiple assets in a paper entitled "Searching beyond the Lamppost: Let's Focus on Economically Relevant Questions."

In this section, we consider a design with a flat fundamental value. Subjects can earn interest on cash at a rate of r per period (e.g., an insured savings account). In contrast, shares of the risky asset pay random dividends that are either high, H, or low, L, each with probability 1/2. The expected value of the dividend will be denoted by $D = (H+L)/2$. As shown in the previous section, the value of the asset would be D/r if the number of periods were infinite. In the first experiment to be discussed, $H = \$1.00$, $L = \$0.40$, so the expected value of the dividend is: $D = \$0.70$. With an interest rate of 10%, $r = 0.1$, the present value of an infinite-period asset would be $\$0.70/0.1 = \$7.00$. This is the present value of future dividends, and it does not change as time passes, since the future is always the same. The trick in setting up a *finite-horizon* experiment with an asset value that is constant over time is to have the final-period redemption value be equal to D/r, which would be the present value of the dividends if they were to continue forever. In this manner, the infinite future is incorporated into the redemption value per share in the final period, and the resulting asset value will be flat over time, even as the final round is approaching. Thus the redemption value was set to be $7.00. Similarly, a redemption value above (or below) D/r would induce an increasing (decreasing) fundamental value (Holt, Porzio, and Song, 2017).

In the experiment, anyone with shares could submit a sell order, consisting of a minimum sale price and a number of shares offered. Anyone with cash could submit a buy order, consisting of a maximum purchase price and a desired number of shares. Trades were arranged by using these orders to determine a single market-clearing price at which the number of shares offered for sale equals the number of shares demanded. All trades are executed when the market is closed or "called." This setup is typically referred to as a call market, since the process ends and trades are determined at a pre-specified time, e.g., at the end of the business day, or before the financial markets open in the morning. The market "call" in the experiment occurs when all traders have submitted orders, or when the experimenter presses the "Stop" button on the Admin Results page (whichever comes first).

Once the market is called, buy orders are ranked from high to low in a demand array, and the sell orders are ranked from low to high in a supply array. These supply and demand arrays are crossed to determine the market-clearing price and quantity. For example, if one buyer bids $5 for 2 shares and if another bids $4 for 2 shares, the demand array would have steps at $5 and $4. Suppose in this case the only sell order involved 3 shares at a limit price of $3. Then 3 shares would trade at a price of $4, since there are 4 shares demanded at any price below $4 and only 2 shares demanded at any higher price.

Figure 24.3 shows the Veconlab "admin results" graph of a typical market bubble with the flat fundamental value of $7.00. The participants were students in a finance class at the University of Valencia, who had been asked to bring laptops or sit next to someone with a laptop. Prices in initial periods typically begin a little above or below the fundamental value. In this case, these finance students started trading right at the predicted value. That did not last long, as prices inched up and speculative instincts were aroused. The rising phase of a bubble can be sharp or slow and steady as in this case, but the end of the boom usually catches many traders by surprise.

Trading volume is indicated by the width of the horizontal segment for each round. Sometimes there is a decline in volume before the bubble breaks, which was definitely not the case for the market shown here. There can be gaps in trading after prices start to fall as people are reluctant to sell at a loss. The students in Valencia applauded when the final round closed, which reminded the author of the applause at the close of the New York Stock Exchange. The experiment generated a lively discussion, and the instructor reported that an unusually large crowd showed up at her office hours the next day.

Even though the sharp price declines are likely to trigger negative emotions (regret, anxiety, disappointment), a more subtle question is whether emotional

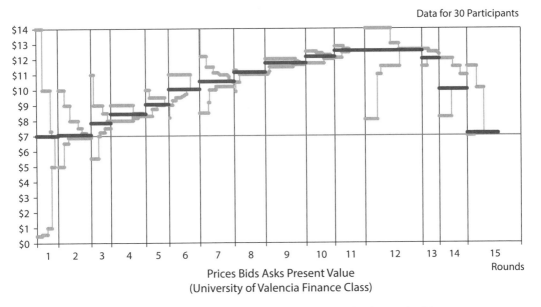

Figure 24.3. A Classroom Asset Market with a Flat Fundamental Value at $7.00

states have predictive value. Breaban and Noussair (2017) use facial recognition software to gauge and classify emotional states. They report that positive emotional states *measured before a market opens* are correlated with larger bubbles, and that negative pre-market emotional states are associated with lower prices.

Another factor that enhances bubbles is the presence of high cash reserves. For example, Caginalp, Porter, and Smith (2001) endowed subjects with higher amounts of cash in a declining value experiment, which resulted in more price speculation and bubble activity. A similar effect can be induced in flat value markets by scaling up both interest rate and dividend payments while holding cash and share endowments constant. For example, the fundamental value, D/r, would stay the same if both the interest rate and expected dividends are doubled. The baseline treatment, with a $0.70 dividend and a 0.1 interest rate, has the same fundamental value as the high-cash treatment with a $1.40 dividend and a 0.2 interest rate. Note that the final period redemption value remains unchanged at $7. The compounded interest and dividends in the high-cash treatment results in very large cash reserves that can then lead to larger bubbles. This "high-wealth" treatment generated strong bubbles in all three sessions, as shown by the thick gray lines in figure 24.4. In all three of these cases, the bubbles were followed by crashes back to the fundamental value of $7.00.

The tendency for prices to drop sharply in the final rounds of some markets provided the motivation for running long markets with 40-round horizons, but with the baseline parameters (10% interest and an expected dividend of $0.70). These were, in a sense, high-wealth sessions since the accumulated

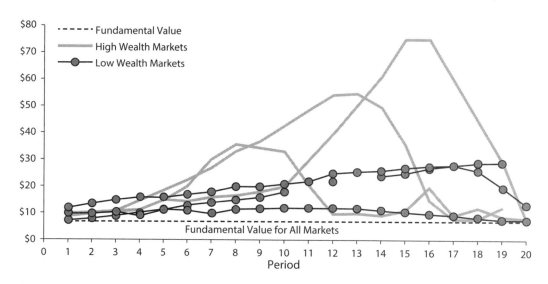

Figure 24.4. Six Limit Order Markets with a Constant Fundamental Value of $7.00.
Source: Bostian, Goeree, and Holt (2005).

effect of interest and dividends over the 40-period horizon was much larger than it is over the 20-period horizon. One of these 40-period markets resulted in a dramatic surge after period 25, with prices peaking at $257 in round 31, which is more than 30 times the fundamental value! The crash, when it came, started slowly and then accelerated with a $100 drop in round 34. This was the most extreme bubble observed, but other long sessions also generated strong bubbles.

Letting investors borrow to make asset purchases can magnify the effects of excess cash in the market. Coppock and Holt (2014) compare the outcomes of asset markets in which borrowing is easy with other markets in which it is much too costly. In the easy credit treatment, investors only had to come up with cash to cover 20% of the purchase price. The tight credit treatment was intended to offer the same degree of task complexity, but with an impossibly high collateral requirement set at ten times the amount borrowed. In either case, an investor who borrowed was forced to sell if the share price fell to a level that equaled the amount originally borrowed to finance the purchase. Figure 24.5 shows the average asset price trajectories across markets in each treatment, as compared with the flat fundamental value of $28 determined by the ratio of the $1.40 expected value of the dividend for each risky share to the 5% interest paid on cash.

Not surprisingly, bubbles were higher with low (20%) collateral requirements. Note that being able to borrow generated much higher bubbles initially. In some low-collateral markets, the crash was sharpened when subjects were forced to sell after the downturn put their loans "under water." To summarize:

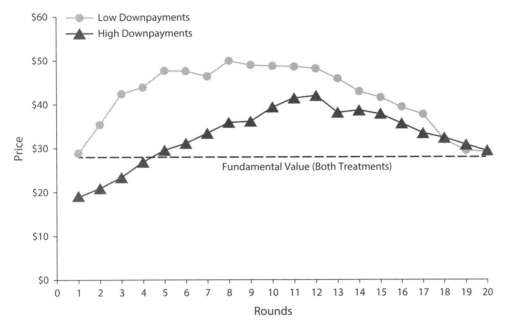

Figure 24.5. Average Asset Prices with Loose and Tight Collateral Requirements. *Source*: Coppock and Holt (2014).

Experiment Results with a Flat Fundamental Value: *In an asset market for shares that pay random dividends, with an alternative safe asset that pays a known interest rate, observed share prices generally exhibit bubbles and crashes. A doubling of dividends and interest, which leaves the D/r fundamental value constant, results in bubbles with higher amplitudes, as does letting investors borrow with low collateral requirements.*

24.5 Forecasting, Gender, and Cognitive Response Differences

John Maynard Keynes, who was an avid speculator himself, once asserted that investors were motivated by "animal spirits." In particular, he noted that most investors fixated on short-term price movements instead of worrying about fundamental values:

> An investor can legitimately encourage himself with the idea that the only risk he runs is that of a genuine change in the news over the near future, the likelihood of which he can attempt to form his own judgment, and which is unlikely to be very large. . . . he need not lose his sleep merely because he has not any notion what his investment will be worth ten years hence. (Keynes, 1965, p. 153)

Such short-term gains can have strong psychological effects, even if the investor is unsure about fundamental value and does not plan to liquidate shares in the immediate future. Consider, for example, a market in which shares pay an expected dividend of $1.40 and the interest rate is 5%. In this case, the dividend/interest ratio would be $1.4/0.05 = $28, which would be the flat fundamental value if the redemption value were set at $28. In this case, an anticipated capital gain of only $1.40 per period would match dividend payouts and thus, in some sense, justify a share price of $56, about twice as high as the fundamental value.

Price forecasting experiments have generally been performed in stationary environments, with exogenous, random price series. Early experiments identified a simple type of "adaptive" forecasting behavior: anchor on the previous forecast and adjust up or down depending on whether the forecast was too low or too high. When prices are not stationary, a natural procedure involves "extrapolation" of previous trends, as observed by Haruvy, Lahav, and Noussair (2007). Asset market experiments with booms and busts provide a fine setting for studying actual forecasting behavior when prices are nonstationary and endogenous.

Figure 24.6 shows the market price trajectories (dark lines) for a pair of 25-period markets run with a flat $28 value (expected dividends of $1.40, interest of 5%, and a redemption after period 25 of $28). Prior to trading in each period, the nine subjects in each market were asked to predict the market price for the current period, one period ahead, and two periods ahead, with small payments to be received for each forecast with a range of $2.50 of what the actual price subsequently turns out to be. The short gray segments in the figure show the forecast averages. Each segment connects three dots, which are the forecast averages for the current period, one period ahead, and two periods ahead. For the market in the right panel, the first grey segment on the left indicates that average forecasts before any trading in the first period were about $19, $24, and $27 for the first three periods.

Forecasts in figure 24.6 tend to be too low when prices are rising, and too high when prices are falling. Moreover, short-term forecasts are generally too flat relative to observed price movements. Finally, the overshooting at the peak indicates that the timing of the downturn was not anticipated. These patterns were confirmed by an analysis of all the other sessions and can be summarized:

Price Predictions in Asset Markets: *Short-term forecasts tend to be too low and too flat during the boom phases, and too high and too flat during the bust phases. The downturn at the peak price level is not generally anticipated.*

Forecasting Models

The simplest adaptive forecast F_t of the time t price is calculated by anchoring on the previous forecast, F_{t-1} and adding a correction based on the difference between the prior price P_{t-1} and the prior forecast F_{t-1} of that price:

(24.4) $F_t = F_{t-1} + \beta(P_{t-1} - F_{t-1})$ with $0 < \beta < 1$. (*adaptive expectations*)

If $0 < \beta < 1$, the adaptive forecast is a weighted average of the previous forecast and the most recently observed price.

   When the price series is not stationary, a well-known problem with adaptive forecasting is that it may produce *systematic, correctable forecast errors*. Suppose, for example, that the price sequence is an increasing sequence: 1, 2, 3, . . . and that the initial forecast was 1 in period 1. Then the period 2 forecast would also be 1, for any weighting of the initial forecast and the observed price of 1. For the third period, the adaptive forecast would be a weighted average of the prior forecast (1) and the prior price (2), which would again be less than the observed price of 3. In this manner, all forecasts would be too low. A model that produces persistent forecast errors is generally dismissed by theorists, although *persistent forecast errors are the norm* for the two markets shown in figure 24.6.

   As noted above, when there is a trend, an alternative approach is to anchor on the *prior price* and partially extrapolate the previously observed trend in prices: $F_t = P_{t-1} + \beta(P_{t-1} - P_{t-2})$, with $0 < \beta \leq 1$. Trend extrapolation rules are known to add instability into particular macroeconomic systems, as is confirmed by experiments in which subjects are paid for accurately forecasting variables that are endogenously determined by their own forecasts.

   The idea that traders extrapolate recent price trends has considerable appeal, and it is natural to specify a *double-adaptive forecasting model*, with adaptive adjustments in both price *levels and trends*. Let F_{t-1}^+ denote the one-period-ahead forecast made in the previous period. Then the current forecast F_t can be

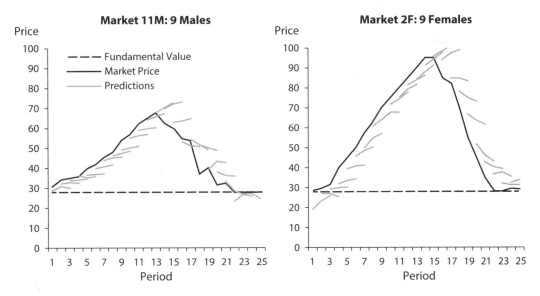

Figure 24.6. Average Asset Prices (dark lines) and Predictions (gray lines) for Gender-Segregated Markets. *Source*: Holt, Porzio, and Song (2017).

anchored on this previous forecast, with parameters that correct for errors in price levels (β) and price trends (γ):

$$(24.5) \qquad F_t = F^+_{t-1} + \beta(P_{t-1} - F_{t-1}) + \gamma[(P_{t-1} - P_{t-2}) - (F^+_{t-1} - F_{t-1})]$$

with $0 < \beta, \gamma < 1$. Notice that the period t price forecast, F_t (left side), is specified to be F^+_{t-1} (first term on the right representing the prior one-period-ahead forecast for period t made in period $t-1$), plus two adaptive adjustment terms. The β parameter is used to adjust for the most recently observed forecast error in price levels ($P_{t-1} - F_{t-1}$), whereas the γ term adjusts for the most recently oberved error in price trends, i.e., the bracketed difference between the most recently observed price change, $P_{t-1} - P_{t-2}$, and the predicted change in $t-1$: $F^+_{t-1} - F_{t-1}$. Price levels are fully corrected if $\beta = 1$, and trends are fully extrapolated if $\gamma = 1$.

Equation (24.5) was estimated using the change in forecasts $F_t - F^+_{t-1}$ as the dependent variable and the most recent forecast error $P_{t-1} - F_{t-1}$ and change error $(P_{t-1} - P_{t-2}) - (F^+_{t-1} - F_{t-1})$ as independent variables constructed from observed prices and forecasts for all individuals in all markets, using ordinary least squares. The estimates for β and γ (not shown) are significantly different from 0 and 1, indicating the importance of both adaptive terms. The parameter estimates are similar for males and females, although the γ estimates reveal a slight tendency for females to extrapolate price trends more than males. The approximate similarity of forecasting behavior for males and females is consistent with the absence of clear gender effects in bubble formation to be discussed next.

Gender Effects

The two markets in figure 24.6 were run in a cash-rich setting, with initial endowments of 6 shares and $70, and per-round exogenous incomes of $30 per period. The relatively large number of periods, 25, resulted in high cash accumulations, which were expected to provide a fertile environment for speculation. Prices exceeded the $28 fundamental value by a large margin in most markets, including the two that are shown. The market in the left panel was done with a group of 9 males, and the market on the right was done with 9 females. There were 16 sessions in total, 8 with each gender, and there was no significant gender difference in terms of bubble magnitudes or other standard measures of bubble amplitude. This absence of a gender effect was also observed in a second, robustness check treatment with shorter (15-period) sequences and 5 markets for each gender.

Holt, Porzio, and Song (2017) also report a parallel treatment *with a declining value setup* that did generate larger price bubbles with men. This gender effect is consistent with the original Eckel and Füllbrunn (2015) experiment that had previously reported higher bubbles with groups of men in a 15-round declining value setup, with explicit gender sorting. However, in a follow-up study (also with declining values), the same authors find no gender difference, but

Table 24.1. Do Groups of Men Generate Higher Asset Price Bubbles?

	Declining Fundamental Value	Flat Fundamental Value
Obscured Sorting	Holt et al. (2017); males higher Eckel/Füllbrunn (2017): no difference	Holt et al. (2017): no difference
Explicit Sorting	Eckel/Füllbrunn (2015): males higher Wang et al. (2017): males higher Wang et al. (2017): China, no difference	

this time with hidden gender sorting using the same subject pool (Eckel and Füllbrunn, 2017). More mixed results are reported by Wang, Houser, and Xu (2017), who replicate the Eckel and Füllbrunn (2015) gender difference (declining values, explicit sorting) with groups of men and women in the United States, but find no gender difference with Chinese subjects under the same conditions. Table 24.1 summarizes the overall pattern of these results.

It is worth considering the rationale for revealing or hiding gender sorting. Explicit gender division may trigger experimenter demand effects, as people respond to or react against stereotypes. Moreover, gender is not apparent in most actual financial markets where trading is not done on the floor, and actual markets are not segregated by gender. On the other hand, there may be situations in which gender can be inferred to some extent, e.g., in terms of names used on trading terminals, and in such cases explicit gender sorting can be used to study important elements of group dynamics and common expectations. To summarize:

Gender Effects in Asset Market Experiments: *The current evidence for gender effects in asset markets is mixed, and may depend on the value structure (declining versus flat values), the nature of gender sorting (hidden or explicit), and culture. At this point, there is no evidence of gender differences in markets with flat fundamental values and with hidden gender sorting, both of which are more characteristic of actual financial markets that do not typically have declining values or explicitly segregated gender sorting.*

CRT Scores

Since an asset market bubble creates opportunities to earn and lose money relative to what others earn, it is natural to consider any effects of cognitive differences. A standard test used by psychologists is the *cognitive response test*, which consists of five questions. Generally speaking, each question has an obvious but incorrect answer that results from impulsive thinking. Holt, Porzio, and Song (2017) only used two of the five standard questions (Q1 and Q2), along with a speed question (Q3):

Q1: A hockey stick and puck cost 110 Canadian dollars in total. The stick costs 100 more than the puck. How much does the puck cost?

Q2: In a lake, there is a patch of lily pads. Every day, the patch doubles in size. If it takes 48 days for the patch to cover the entire lake, how long would it take for the patch to cover half of the lake?

Q3: A person drives 60 kilometers at 60 kilometers per hour, and then turns around and drives back to the starting point at 20 kilometers per hour. What was the average speed?

The standard wording of the first question was changed from baseball terminology to hockey. The second question is the standard; the impulsive answer is that it takes 24 (half of the 48) days to cover half of the lake, which misses the fact that the area being covered doubles each day. The third question was included to be more mathematical, given that asset valuation tasks tend to be mathematical. The third question (adapted from a GRE study book) can be answered correctly by taking the ratio of total distance (120) to total time (4 hours), instead of just averaging the two speeds. The score is the number of correct answers, which ranges from 0 to 3.

The data indicate that people who score low on this test earn less than those with high scores. This result is consistent with Corgnet et al. (2014), who also find that people with low CRT scores tend to be net purchasers when the share price exceeds the (declining) fundamental value. In contrast, correlations between average CRT scores for groups of traders and bubble amplitudes are rather weak, which suggests the importance of other factors like a tendency to extrapolate gains.

24.6 Extensions: "Thar She Blows"

All of the asset price bubbles shown thus far have been for sessions with inexperienced participants, i.e., those who have not previously encountered a bubble in an asset trading experiment. In contrast, sessions with experienced subjects did not produce bubbles reliably (Smith et al., 1988; Van Boening, Williams, and LaMaster, 1993). Peterson (1993) ran some asset market sessions with people who had been in two previous markets, and prices did not rise much above the declining fundamental value in this "super-experienced" treatment. When discussing the implications of this difference, Vernon Smith once commented that crashes are relatively infrequent, sometimes decades apart, and many new investors enter the market in the meantime. Dufwenberg, Lindqvist, and Moore (2005) report that bubbles are eliminated or largely abated when as few as a third of the traders are experienced. This suggests that bubbles are uncommon in mixed-experience settings, which may explain why severe surges and crashes are relatively few and far between in major financial markets. However, Hussam,

Porter, and Smith (2008) show that bubbles can even be *rekindled* with experienced subjects, as long as there are exogenous structural changes, e.g., increases in liquidity and dividend uncertainty in the second or third market sequence. This result provides an affirmative answer to the question in the title of their paper: "Thar She Blows: Can Bubbles Be Rekindled with Experienced Subjects?"

Asset market experiments are highly simplified relative to the complex types of investment instruments that are available to sophisticated traders. Futures markets and short sale options provide investors with strategies that might diminish price bubbles, an intuition that is confirmed by experiments (King et al., 1993; Haruvy and Noussair, 2006; Noussair and Tucker, 2006; Porter and Smith, 1995).

Most asset market experiments cited here use double auction trading, in which price evolves during a trading period. This process is analogous to the evolution of ticker tape prices on centralized exchanges. On the other hand, most individuals who trade in designated stocks end up conveying orders to their brokers that specify limits in amounts to be paid or spent, either in total or on a per-share basis. One advantage of call market experiments is that limit orders can be executed quickly, which permits large numbers of trading periods. Lahav (2011) used call markets to run sessions with 200 trading periods, which generated a wider range of patterns than those single-peak patterns shown in the figures in this chapter.

Giusti, Jiang, and Xu (2016) report that bubbles are diminished with fundamental values that are increasing due to high redemption values, as compared with flat fundamental value settings. Finally, Ball and Holt (1998) use a random stopping rule to induce present value considerations and a flat fundamental value.

There is a large subliterature on forecasting in asset markets and the effects of such forecasts in macroeconomic models. Williams (1987) first showed that forecasting behavior was well explained by a simple adaptive rule in *stationary* settings. Haruvy et al. (2007) first studied forecasting in asset market experiments with human subjects, and they specified a model that extrapolates trends. The double adaptive model discussed in the chapter is a special case of an "autoregressive" linear function of the two most recent prices, which has been shown to provide a good fit for individual forecasting data (Hommes et al., 2005). For a summary of more recent work, see Assenza et al. (2014).

The cognitive response tests discussed in the previous section are generally not incentivized and rely on internal motivations instead. There are clear gender differences in CRT scores, presumably due to differences in majors or secondary school coursework. These and other effects are surveyed in Brañas-Garza et al. (2016). Also, a broader perspective on CRT scores and economic behavior is provided in a special issue of the *Journal of Economic Behavior and Organization* (Brañas-Garza and Smith, 2016).

See Palan (2013) for a recent survey of the literature on asset market experiments. Different ways to measure bubbles in terms of average deviations from fundamental value are proposed and evaluated in Stöckl, Kirchler, and Jürgen (2015). The effects of testosterone administered to men in asset market experiments are documented in Nadler et al. (2016). Cipriani, Fostel, and Houser (2012) were the first evaluate the price-boosting effects of leverage (buying on margin) in financial markets.

Finally, there is an unrelated literature on using asset markets to aggregate information to provide predictions about elections or other events. For example, see the survey in Wolfers and Zitzewitz (2004), and the election applications in Forsythe et al. (1992) and Davis and Holt (1994b).

Chapter 24 Problems

1. Consider an asset market with no interest payments. Suppose that the experiment lasts for five periods, with dividends that are either $1 or $3, each with probability 1/2, and with a final redemption value that equals the *sum* of the dividends realized in the five periods. Thus, the actual final redemption value will depend on the random dividend realizations. Calculate the expected value of the asset at the start of the first period, before any dividends have been determined. On average, how fast will the expected value of the asset decline in each round?

2. A project costs $2,000 to build in the present period, and yields $5,000 in earnings ten periods later. What is the present value of the future payout using a 5% interest rate? Is this a worthwhile investment?

3. (non-mechanical) Price forecasts are generally incentivized, with payments made for accurate forecasts within a permitted range of error. Do you think the payments for these forecasts should be given to subjects as cash during the experiments, affecting their available cash positions? Is there a danger that payments for accuracy might affect trading behavior? How might experimenters deal with these possible problems?

4. (non-mechanical) Do you think unannounced gender sorting is deceptive?

5. In an asset market with an interest rate of 10% on cash and expected dividends of $5 per period, what redemption value would have to be used to ensure a flat fundamental value? What redemption value range would lead to increasing fundamental values?

25

Bank Runs and Macro Experiments

Just as the Great Depression in the early twentieth century caused Keynes and other economists to reconsider the role of macroeconomic policies, the Great Recession of 2008 shook the prevailing confidence in our understanding of forces that drive the macro-economy. Consequently, there has been a resurgence of experimental work motivated by macroeconomic issues. Despite the obvious scale limitations of such experimentation, it is easy to imagine how this work may yield useful insights. For example, biases that affect individual behavior can be magnified by interactions. Even in a simple social dilemma game like the traveler's dilemma (chapter 11), the amplification of behavioral biases can have a dramatic effect on average decisions, moving the data away from unique theoretical predictions to the opposite side of the set of feasible decisions. Such dramatic effects are driven by connections between players' expectations of others' actions that are affected by others' expectations, etc. In a more macro context, markets for a single financial asset can be volatile when forecasts and forecast-driven price movements interact, as in the experiments considered in the previous chapter.

There are, however, several dimensions in which macroeconomic interactions are far more complex than single games or markets. First, the accumulation of savings and capital stocks transforms simple problems into multi-period dynamic decisions that raise difficult-to-manage tradeoffs between the present and the future. Moreover, there are interconnections across markets for labor, capital, and financial assets even within a single production period. Another type of interconnection involves linkages up and down a supply chain, from retailers to wholesalers, distributors, and manufacturers. The bullwhip effect encountered in chapter 22 shows how shocks in one sector can be magnified by lags in orders, inventories, etc. Such lags are important features of macroeconomic models. Finally, there is the threat of an economy-wide slowdown, fueled by coordination failures to deliver supplies, which may be rational if other deliveries do not materialize. A smoothly functioning macro-economy requires coordination and trust, and when trust is shaken, it is hard to re-establish, as can happen with bank runs in a financial crisis. This chapter reviews laboratory experiments that are motivated by macro issues of consumption, banking, and multi-market production.

Note to the Instructor: The relevant experiments are listed on the Vecon-lab Finance/Macro menu. The Bank Run program is easy to run in class, and it often generates some excitement. The Discount Window Stigma game discussed in the second section is listed on the same Finance/Macro menu. The Macro Markets program takes more time, since labor markets clear before production, and monetary injections can have delayed effects. The end-of-period reports provide participants with aggregate information on employment, prices, and wages. A third option is based on the Lever-aged Asset Market that features price forecasts and asset purchases funded by loans and margin purchases. This program also requires more class time, and may not be appropriate for people who have encountered price bubbles in the asset market trading experiment in the previous chapter. In addition, this program has a consumption option that lets participants allocate as-sets between consumption and saving for future consumption, all in the shadow of a market for a risky asset. Consumption is subject to diminishing marginal utility, so optimal consumption involves a smooth consumption pattern, while using the power of compound interest to save for later peri-ods, e.g., retirement.

25.1 Bank Runs and Coordination: "Confidence Is Everything"

Bank failures are often indicators of a full-scale financial crisis. A failure occurs when a bank cannot meet the demands of depositors and creditors who wish to withdraw funds. To understand how such liquidity issues might arise, it is important to recognize that banks typically use deposits from individuals and businesses to fund loans or other investments that have high rates of return. These investments reduce the liquidity available for short-term needs, but this is not normally an issue when individual depositor withdrawals are more or less random, independent, and predictable in the aggregate. The bank's comfort zone narrows sharply in a crisis when worrisome financial news might trigger waves of withdrawals with cascading effects. It is sometimes said that vener-able old bank buildings featured grand lobbies that prevented street traffic from seeing a line of customers waiting to withdraw funds. Bank runs in the twenty-first century have a new dimension in the sense that major depositors are often institutional, e.g., other banks. The first major bank failure in the "great reces-sion" of 2008, for example, was Northern Rock in the UK. Even though the mortgages held by this bank were performing adequately at the time, this bank had borrowed funds from other banks to fund some of its investments, so those banks were essentially large depositors. When some of the institutional deposi-tors withdrew their credit in 2007, individual depositors queued up to withdraw funds, and the bank failed.

In the United States, the Federal Deposit Insurance Corporation quickly scaled up the limits that would be covered by deposit insurance. But the collapse was so severe in Iceland that the government struggled for years to pay back a portion (up to £15,000) of depositors' losses. The point is that deposit insurance typically has limits, risks, and possible delays. And there are many investment funds and investment banks that are not covered by regulatory safeguards designed for commercial banks. In 2008, for example, some investment funds imposed limits on clients' cash withdrawals from money market funds. Although commercial banks in the United States can borrow from the Federal Reserve Bank through its "discount window," that option was not widely used in 2008 due to an associated stigma that might cause other depositors to withdraw funds. Bernanke (2009) for example, noted "In August 2007, . . . banks were reluctant to rely on discount window credit to address their funding needs. The banks' concern was that their recourse to the discount window, if it became known, might lead market participants to infer weakness—the so-called stigma problem." This fear is well founded. In 2007, a BBC leak that Northern Rock had borrowed from the Bank of England was instrumental in that bank's demise.

A banking crisis is exacerbated when withdrawals spread fear that leads to further withdrawals. Depositors in such situations face a dilemma: If others are going to withdraw, early withdrawal is a good option for getting one's funds back safely without delay. However, if others will not withdraw funds at a rate that stresses the bank, it may be better to leave one's cash in the bank for the safety and returns that it may provide. As noted by Diamond and Dybvig (1983) in their classic paper on bank runs, the situation is analogous to a coordination game in which one equilibrium involves withdrawals by all depositors, and another equilibrium involves only minimal "normal" withdrawals.

A classroom bank run experiment entails assigning all students to be depositors who are linked to banks. In each period, students must decide whether to withdraw funds or not. The simulated banks fail if withdrawals exceed a specified threshold. If the bank does not fail, then those who withdraw do get their cash back, but once it fails, then the depositors who have not withdrawn at that point receive a lower return, or even nothing at all if there is no deposit insurance.

The classroom experiments to be described were structured so that each bank had eight depositors (students), and each student was assigned to the same bank for all periods. Depositors had cash balances of $10 in their bank at the start of each period, which would increase to $15 if the money was not withdrawn and the bank did not fail in that period. The bank liquidity threshold was set at four, so the first four depositors to withdraw could recover their $10 stake. But as soon as four depositors withdrew funds, others would be left with $0 in the no deposit insurance treatment, and with $9 in the 90% deposit insurance treatment. The "show withdrawals" setup option allows people to see

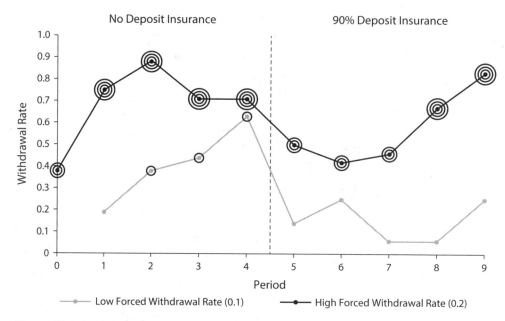

Figure 25.1. Bank Run Depositor Withdrawal Rates, Before and After the Advent of 90% Deposit Insurance after Dashed Line. *Notes:* The dark line shows the overall withdrawal rate in a session with a high (0.2) forced withdrawal rate. The advent of 90% deposit insurance in the second half did not prevent continued withdrawals and bank failures, which are indicated by multiple concentric circles. Bank failures are less common with a low forced withdrawal rate (0.1), represented by the gray line, which shows minimal withdrawals, mostly forced, in the second half with deposit insurance.

any withdrawals from their own bank or from other banks. The instructor can decide whether to avoid framing effects with the neutral bank names provided as defaults, although the setup does permit the specification of more vivid bank name labels like "Germany" or "Spain."

Normal transactional needs were implemented by specifying a probability that any depositor would be exogenously forced to withdraw due to normal or even unexpected liquidity needs. The gray line in figure 25.1 tracks the sequence of withdrawal rates for a session, with the rings indicating bank failures (explained below). This line shows withdrawal rates that immediately rise above the 0.1 level at the exogenous forced withdrawal rate, a level that would be expected with purely exogenous withdrawals. The single dark ring in rounds 2, 3, and 4 indicates that one of the banks in that session failed in those rounds. Since pessimism and the absence of trust are persistent emotions, it is not surprising that it was the same bank that failed each time. Those failures ended after the imposition of 90% deposit insurance (rounds to the right of the vertical dashed line) dampened withdrawal rates to about the levels determined by the exogenous withdrawals.

In contrast, the dark line shows high withdrawal rates (endogenous and exogenous) in a separate session with an exogenous random withdrawal rate of 0.2. In fact, the two rings in the very first round indicate that two banks failed, as total withdrawals occurred at about twice the exogenous rate in that round. The three concentric rings indicate that all three banks failed in the next four rounds. The imposition of 90% deposit insurance reduced withdrawals, but two banks still failed, and all three banks failed in the final rounds of this session. Sometimes the imposition of partial deposit insurance eliminates bank runs, and sometimes it does not. It is useful to save data from prior classes to show the variety of possibilities. If the imposition of deposit insurance fails to reduce withdrawals, it is useful to begin the discussion by saying "confidence is everything, it is very difficult to recover when trust is gone."

It is crucial to remember that behavior in a bank run experiment may be difficult to predict, given the exogenous randomness in forced withdrawals, individual differences in risk tolerance and trust, and endogenous expectations that may push behavior in either direction. The continuous classroom game offers students flexibility in terms of timing withdrawals in response to observed trends if other withdrawals can be observed. This continuous game can be modeled more simply by letting each person make simultaneous decisions about whether or not to withdraw, if withdrawal is not forced. If the exogenous withdrawal rate is not excessive, one equilibrium is for everyone to refrain from withdrawing, which results in payoffs of $15 for each person with high probability. But if everyone decides to withdraw, each person is better off trying to withdraw in the hopes of being one of the four depositors who would be selected at random for payment in this case. (See problem 1 for a simplified setting with two depositors.) To summarize:

> **Bank Run Experiment Results:** *The generic bank run setup can be modeled as a coordination game with multiple equilibria, one involving depositor withdrawals and another involving no withdrawals unless forced. High withdrawal rates and associated bank failures are common in the absence of deposit insurance. But the implementation of insurance for a high fraction of deposits does not necessarily remedy the situation if confidence has been destroyed by previous bank failures.*

25.2 Coordination and Discount Window Stigma

Most central banks have an obligation to loan funds to banks in need of liquidity. Such loans can support monetary policy by establishing an upper limit on interest rates which, together with the lower rate paid on overnight bank deposits, determines an interest rate corridor. The upper boundary of this corridor can become "leaky" if banks avoid borrowing from the central bank due to a fear that investors might view such loans as a sign of weakness, as indicated by Ben Bernanke's comment that was quoted above.

Central bank loans are intended to help a bank that is temporarily *illiquid* due to an imbalance between long-term assets and short-term liabilities, even though the bank may be solvent in the sense that assets exceed liabilities. The problem is that private investors may not be able to distinguish solvent banks from insolvent ones, due to asymmetric information. As a consequence, banks with liquidity problems face the prospect of having to pay high interest rates or to engage in "fire sales" of assets. Alternatively, the bank can turn to the lender of last resort, the central bank. Borrowing from a central bank can be risky, given the possibility of leaks such as the one mentioned earlier in the chapter that triggered the demise of Northern Rock in 2007. One policy solution designed to eliminate the stigma is to force randomly selected banks to borrow from the central bank, so that borrowing conveys less information about a bank's financial condition. In fact, a forced-borrowing policy was recently implemented by the Bank of England.

Armantier and Holt (2017a) evaluate a simplified forced-borrowing policy in a laboratory setting that is otherwise prone to an undesired outcome with no discount window borrowing. The setup differed from the previously discussed bank run experiment in that subjects were given both investor and bank roles (with randomly matched pairs of subjects, one investor per bank). Investors in the experiment earn 100 if "their" bank turns out to be solvent, but they would want to withdraw funds from an insolvent bank, given the likelihood that it might later fail and yield 0 earnings, as compared with earnings of 50 from the investor's best alternative opportunity. Half of the banks in each round were randomly determined to need liquidity to continue short-term operations, so those banks had to borrow from the discount window at a cost of 20 or incur a higher cost of 40 from selling assets to maintain confidentiality. Investors were told that two-thirds of those illiquid banks would end up being insolvent and yielding a return of 0 for an investor who had not withdrawn funds.

There is, however, a chance that borrowing from the central bank will be detected by investors. In practice, this discovery can result from leaks, lawsuits, or inferences made by looking at Federal Reserve regional aggregate summaries and indirect signals provided by market activity. It is generally believed that there is "safety in numbers" in the sense that the chances of borrowing *by any individual bank* being undetected are increased as more banks take this route. This setup generates a Nash equilibrium in which all liquidity-constrained banks borrow from the discount window, even though investors who detect such borrowing will withdraw funds, causing the bank to earn 0. This will be referred to as the "no-stigma" equilibrium. When all illiquid banks do borrow from the discount window, the safety-in-numbers aspect as implemented in the experiment has the effect of reducing the detection probability to 0.30, which helps overcome

Table 25.1. Illiquid Bank Expected Earnings Conditional on Other Illiquid Bank Behavior

	Others Borrow at DW	Others Sell Assets
Sell Assets	earn **60** (= 50 + 50 – 40)	earn **60** (= 50 + 50 – 40)
Borrow at DW	earn **65** (= 80, or 30 if detected)	earn **42.5** (= 80, or 30 if detected)

the stigma effect. In contrast, if all banks needing liquidity avoid the discount window, then the chances of detection of a single borrower are raised to 0.75, which turns out to be high enough to deter this borrowing. This situation results in a no-borrowing "stigma" (Nash) equilibrium.

The two rows in table 25.1 correspond to the two decisions that can be made by an illiquid bank, i.e., borrow at the discount window (DW) or sell assets. Each bank, illiquid or not, earns 50 from ongoing operations. In addition, each bank earns another 50 from an investment, as long as the investor who provided the capital does not withdraw the credit after detecting DW borrowing. An illiquid bank that sells assets has no risk of detection, so its earnings are 50 (ongoing operations) + 50 (from the investment) – 40 (cost of asset sales), for a total of 60 as shown in the top row of the table. Note that this safe earnings level does not depend on whether other banks borrow from the DW (left column) or not (right column).

In contrast, an illiquid bank that borrows from the DW has earnings of 50 (ongoing operations) + 50 times the probability of being undetected, – 20 (cost of DW borrowing). Thus, borrowing at the DW is risky, resulting in total earnings of either 80 or 30, depending on whether it is detected by investors who withdraw funds and cause investment earnings to be 0. The probability of being undetected is endogenously determined by the number of banks that are illiquid and borrow from the discount window, and the expected payoffs turn out to be 65 from such borrowing in the no-stigma equilibrium when other illiquid banks are also borrowing, as compared with the lower expected payoffs of 42.5 associated with DW borrowing when no other banks are doing so and the probability of being undetected is only 0.25. These expected payoffs, which are determined by the detection probabilities and illiquid bank proportions used in the experiment, are shown in the bottom entries of table 25.1 (details not provided). Expected payoff comparisons indicate that it is best to borrow when others are doing so (no-stigma equilibrium in the lower left box), but not to borrow when others do not either (stigma equilibrium, upper right).

The *expected* values of an illiquid bank's payoffs shown in bold in the bottom row for DW borrowing mask the *high risk* that results in realized payoffs of either 30 or 80. This added risk should be particularly focal among actual banks in an economy that is beginning to slide into a financial crisis. Observed borrowing in this treatment converged to the stigma equilibrium, with fewer than 20% of illiquid banks borrowing in the final rounds. At the individual subject level, borrowing from DW was negatively correlated with risk aversion measured with a bomb task discussed in chapter 3 (with 100 boxes).

A follow-up treatment implemented a situation in which one of the six subjects with bank roles in each session was selected at random each period and forced to borrow from the central bank. Forced borrowing makes it harder for investors to make inferences about bank solvency from discount window borrowing, and it helps nudge behavior toward a no-stigma equilibrium in which all banks borrow from the discount window when they are illiquid. Observed behavior in this forced-borrowing treatment converged toward this no-stigma equilibrium, as predicted by theory and intuition: About 80% of illiquid banks decided to borrow from the discount window, and investors who detected this did not always cut off funding. The funding rate contingent on detection was about four times higher in this forced-borrowing treatment (at 40%) than in the control treatment. An analysis of individual data indicates that the random forced borrowing, which blurs the signal provided by detected borrowing, has a confidence-building effect, both when detection does not occur and when it does but investment continues. To summarize:

> **Discount Window Borrowing Stigma:** *Borrowing from the lender of last resort can carry a stigma that causes investors who discover such borrowing to withdraw funds. If the probability of detection depends on the numbers of other banks who are also borrowing, then the resulting expected payoffs can create a coordination game with two equilibrium configurations: (1) a no-stigma equilibrium in which illiquid banks do borrow from the discount window, and (2) a stigma equilibrium in which illiquid banks pursue a safer but more costly strategy, e.g., selling assets. Discount window stigma is observed in a baseline treatment, but the no-stigma borrowing outcome is predominant if a randomly selected bank is forced to borrow from the discount window, whether or not it is illiquid.*

25.3 Monetary Shocks in a Macro Economy with Labor and Goods Markets

This section presents results of a classroom experiment that implements a closed economy with markets for both labor and goods. Students are either given the role of a worker/consumer or of a producer. Both types are endowed with amounts of fiat money. Consumers use fiat money to purchase goods produced by firms, and firms use fiat money to purchase labor from consumers. The quantity of money is fixed exogenously, and can be augmented during the experiment in order to alleviate cash imbalances and liquidity constraints that

Table 25.2. Workers' Leisure Consumption and Firm's Production Technology

Workers' Leisure Values				Firm's Production Values						
Leisure Units	1st	2nd	3rd	Labor Input	1st	2nd	3rd	4th	5th	6th
Leisure Value	3	1	0	Marginal Product	12	8	6	4	2	0

may arise in the trading process, even without bank failures of the type considered in the previous section.

The production technology shown on the right side of table 25.2 exhibits a diminishing marginal product. The first labor input unit yields 12 product units. Similarly, the marginal products of the second and third labor units are 8 and 6.

The firm's first marginal product of 12 is represented by the step at 12 on the left side of figure 25.2. It would make sense to produce this unit if the marginal product of a unit of labor (12) times the product price P is greater than the wage cost W for that labor unit. In other words, if MP_L is the marginal product of a labor unit, then the unit should be demanded if $P(MP_L) > W$, or equivalently, if $MP_L > W/P$, which is the real wage. Therefore, the vertical axis of the figure can be thought of as the real wage that determines a firm's demand for labor. The second labor unit used adds 8 product units to the total production, as represented by the step at 8. The third step is at 6, followed by steps of 4, 2, and 0 moving down and to the right. These steps trace out the firm's demand for labor as a function of the real wage, which is the array of decreasing marginal products.

Consumers are each endowed with three units of time that can be consumed as leisure or sold to firms as labor. The first unit of time consumed as leisure is worth 3 units of goods consumption, as shown in the first unit column on the left side of table 25.2. Think of it this way: if the consumer gives up a unit of leisure and receives a money wage W, then this wage can be used to buy W/P units of the product. It would be better for the consumer to take leisure, worth 3, if the real wage W/P is less than 3. Hence the consumer would require a real wage of at least $3 to supply all 3 leisure units, so the right-side step in the labor supply function has a step at $3. The second unit of leisure consumed is only worth 1, so the worker would supply 2 units of labor at any real wage above $1. The third unit of leisure consumed is worth 0, so any positive real wage will induce the worker to supply a unit of labor, and the labor supply function starts with a step at 0 on the left side, followed by steps at 1 and 3.

The labor market clears in odd-numbered rounds (1, 3, . . .), after firms post wages and workers decide what is the least amount of money they would work for. Then firms use hired labor units to produce a product. In even-numbered rounds (2, 4, . . .), firms post prices for units produced and workers decide what is the most they would be willing to pay using money from wages or savings.

Figure 25.2. Money Wages (circles), Money Prices (small diamonds), and Real Wages (large dots). *Notes*: The money supply doubled in period 6 and increased by another 25% in periods 9, 10, 11, and 12, which raised prices and wages.

Each person begins with an endowment of cash that could be used to make purchases or payments. Experiment earnings did not depend on final cash holdings, as indicated in the instructions summary with the phrase: "you can't take it with you." Final earnings were determined by firms' consumption of units not sold and workers' consumption of leisure and of units purchased, as explained:

- *Worker Shopping Procedure:* After prices are posted, each worker selects an upper limit on what they are willing to pay and a maximum number of product units to request at that limit price. Workers must already have sufficient cash to make the requested purchases.
- *Goods Allocation:* If workers request more product units from a firm than the firm's specified limit, then available units will be allocated randomly among those workers.
- *Firm Consumption:* A firm retains and consumes all units that are not sold or offered for sale. Each product unit consumed adds $0.30 to the firm's final experiment earnings.
- *Worker Consumption:* A worker consumes all units of the product that are purchased, plus any that result from home production with retained leisure. Each product unit consumed adds $0.30 to the worker's final experiment earnings.

Therefore, all production by firms is either consumed by consumers or the firms themselves, so there is no inventory carryover. Moreover, consumers enjoy the benefits of leisure consumption (measured in product units) which can be thought of as "home production." In other words, the consumers have utility functions that are linear in product units and nonlinear in leisure, with diminishing values shown on the left side of table 25.2.

The market labor supply, which is the horizontal sum of individual supplies, has steps at $0, $1, and $3. Similarly, the market demand for labor is aggregated across firms, with steps at real wages of $12, $8, etc. With twice as many workers as firms, these functions intersect at a real wage in a range between $2 and $3, and a total production based on each worker supplying 2 units of labor, and retaining a third unit for consumption. When each worker sells 2 units of labor, each firm hires 4 units since there are twice as many workers as firms in this session. As shown on the right side of figure 25.2, money wages rose after the money supply doubled in period 7, but with prices in the $2–$4 range, the resulting real wages were generally close to the $2–$3 prediction band.

An alternative perspective is provided by the "home production" interpretation of the leisure values on the left side of table 25.2. Suppose that each worker retains a single unit of leisure, with home production of 3 product units. In this case, each worker sells 2 labor units, and each firm would hire 4 labor units, with marginal products of 12, 8, 6, and 4. The fifth labor unit with a marginal product of 2 would be less productive than using that unit in home production, where it produces 3. In other words, the intersection of labor supply and demand on the left side of figure 25.2 determines employment that maximizes total production (including home production) for the given amount of available labor.

The predicted employment (2 labor units from each worker, 4 units hired by each firm) can be combined with the firm's production functions to generate a predicted total production. The first four marginal products, 12, 8, 6, and 4, sum to 30 per firm, so the predicted production is 30 times the number of firms, plus 3 times the number of workers (home production of the single leisure unit). Actual production in each even-numbered round is shown as a proportion of optimal production in figure 25.3. Production ratios for this session start at about 0.7 (70% of optimal) and rise to about 0.9 by the final rounds. The low production ratios in initial rounds can be attributed, in part, to the cash-in-advance requirement, i.e., purchases on credit were not permitted. The money supply was doubled in period 6 by providing additional endowments to each worker and firm. Production optimality ratios tended to rise after this "helicopter" monetary injection. This adjustment indicates that an expansion of the nominal money supply can have a real effect in a cash-constrained economy. The labor market effect of the monetary expansion, shown on the right side of

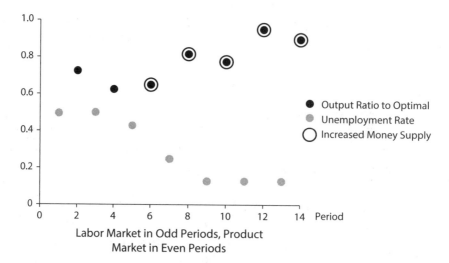

Figure 25.3. Unemployment (gray dots) and Output as Proportion of Optimal (dark dots). *Notes*: The dark circles indicate periods after the fiat money supply doubled in period 6. The money supply rose by another 25% in periods 10–12. These monetary injections eased the effects of the cash-in-advance constraint, as indicated by higher output ratios and reduced unemployment.

figure 25.2, was to raise nominal wages, with little obvious impact on real wages in that session.

Labor market participation can be measured in terms of the number of leisure units offered by workers after they have reviewed the wage offers from firms. Then labor units actually employed can be used to calculate an unemployment rate (actual employment as a ratio of units offered by workers). In particular, if six workers offer 10 time units for sale in the aggregate and only 9 are purchased by firms, then the unemployment rate would be measured as 10%, even if all workers sell at least one time unit. The resulting unemployment rates are plotted as gray dots in figure 25.3.

The story thus far is that a crisis of confidence can produce bank runs and credit contractions that limit firms' abilities to hire and operate productively. The takeaway from figure 25.3 is that a proactive monetary policy can have important real effects on employment and production in a financial crisis. To summarize:

Classroom Macro Markets Experiment: *A large monetary expansion in a cash-constrained economy can generate a real effect that shows up as a higher ratio of actual production to the optimal level, and as a lower unemployment rate. A monetary expansion also tends to have more of an effect on nominal wages and prices than on real wages.*

25.4 Anatomy of a Financial Crisis: Looking Inside the Box

A major advantage of experiments is that it is possible to examine individual decisions in the context of the incentives that people face. In other words, it is often possible to "look inside the box" at the process of adjustment in an attempt to understand causal factors. In most experiments, the investigator is looking at aggregate results as they come in, and the subjects are kept in the dark about the big picture in order to interpret interactions as independent decisions in separate games. In contrast, firms, consumers, and investors are eager to stay informed about what is happening in the macro-economy. An interesting procedural aspect of the macro experiment discussed in the previous section is that individuals received regular reports providing wage and price indices and data on total output, employment, and unemployment rates. These reports are patterned after the types of aggregate information that are freely available from news and media sources in advanced economies. Table 25.3 shows what that report looked like after the final period of another session, with five firms and ten workers.

With twice as many workers as firms, each worker is predicted to sell two labor units (20 in total) and each firm is predicted to hire 4 labor units and produce 30 product units, calculated by summing up the marginal products (12+8+6+4). As before, with 5 firms, the predicted and optimal production is 5 × 30 = 150 units. It can be seen in the left column of table 25.3 that actual production was 142 units, well over 90% of the maximum, with only one in ten workers not hired.

This experimental economy, however, quickly became depressed as employment and production deteriorated sharply in periods 3 and 4 (second column

Table 25.3. Macroeconomic Report for Workers and Firms

Period	1, 2	3, 4	5, 6	7, 8	9, 10	11, 12	Notes
Production	142	102	107	90	94	127	(excludes home production)
Price Index	$1.25	$1.17	$0.95	$0.96	$0.88	$1.21	(prices weighted by quantities)
Employment	18	16	15	14	13	19	(measured in labor units)
Wage Index	$4.67	$3.63	$3.30	$3.57	$3.08	$5.47	(wages weighted by labor units)
Unemployment	10%	30%	30%	50%	40%	11%	(sought work and not employed)
Money Supply	150	150	150	150	150	465	(held by workers and firms)

of the table), a trend that continued through period 10. Notice that each seller initially received cash endowments of $20. The predicted real wage of $2–$3 would require real wage payments totaling $40–$60 ($8–$12 per seller), and money wage payments would need to be higher if prices went above a dollar, as was the case in early periods. In any case, liquidity was not an issue in the first period, and employment was relatively high.

The labor and goods market trading process in this session, however, created cash imbalances that seemed to interact with the cash-in-advance constraints to curtail hiring and production. By period 9, the 5 sellers only had combined cash of about $75 (as compared with their initial endowment holdings of $100). Even worse, more than half of the seller cash reserves at this point were held by a single seller who had high product sales in the prior period. This imbalance prevented several sellers from hiring, even though workers offered to provide labor, which caused unemployment to spike. Then in period 11, each worker and seller received an additional $21 in fiat money. This dramatic increase in the money supply, which was not anticipated, reduced unemployment to 10% and raised production to 127 (relative to a theoretical maximum of 150). In addition to these real effects, the cash injection had a "nominal" effect, with wages rising by almost 80% and prices by about 40%.

The economy represented in table 25.3 began and ended in a reasonably good situation, with a severe decline and recovery in between. This decline and recovery illustrate how normal market activities can generate a decline in economic activity when credit is constrained, perhaps as a result of a prior banking crisis. To summarize:

A Laboratory Business Cycle: *Normal trading activity can create cash imbalances that result in curtailed hiring and production in the absence of credit availability. This adverse situation can be corrected with monetary injections that end up having both real effects on production and nominal effects on wages and prices.*

25.5 Retirement and Consumption

Common sense and economic theory suggest the importance of saving money in high-income periods to be used later when income reductions are anticipated, e.g., in retirement. The intuition is that the utility reduction caused by savings during high-income periods may be much less than the utility gain during low-income retirement periods based on the notion of diminishing marginal utility. This incentive to save is enhanced if savings balances grow by interest compounding, i.e., if there is interest paid on interest. Compounding causes savings balances to increase at ever-increasing rates. Laboratory experiments indicate that people do not fully appreciate the power of compound interest (Levy and

Tasoff, 2016), which is one possible explanation for low savings amounts that are observed for many individuals. A recent study (Merrill Edge, 2016) reported that about 40% of US workers doubted they would be able to reach the necessary savings for a comfortable retirement, and 17% indicated that they would have to win the lottery in order to reach their retirement goals! Another explanation for low retirement savings is a type of "present bias," although it is difficult to imagine how important present bias would be in a laboratory experiment that evolves over an hour or so. Some people, however, may suffer from a typical tendency to procrastinate and delay savings for a later period, which is more of a mistake in the presence of interest compounding. Finally, there is the issue of perceptions. Groneck, Ludwig, and Zimper (2017) conclude that young people tend to undersave because they underestimate how long they will live.

Regardless of the cause of undersaving, the laboratory offers an attractive environment that controls for many confounds, like misperceived chances of survival. The experiment to be described in this section used interest on savings to provide an incentive to save early. In addition, the experiment implements a nonlinear transformation of lab cash into "take-home pay" to induce diminishing utility associated with increases in consumption.

The consumers in the previous section's Macro Markets experiment received earnings of $0.30 for each unit of product consumed. In other words, the utility of consumption, c_t, in period t was linear: $u(c_t) = 0.3c_t$. In this case, a consumer would be indifferent between consuming one's income today, tomorrow, or waiting until one's final year of life and then consuming all accumulated savings. However, a pattern of "binge" consumption is not what consumers strive for when there is a diminishing marginal utility of consumption. For example, consider a "power" utility function with an exponent of 2/3: $u(c_t) = (c_t)^{2/3}$. A graph of this function would have an "uphill" shape, increasing sharply for small amounts of consumption, and more slowly for larger amounts of consumption. This diminishing marginal utility is reflected in the derivative of utility:

$$(25.1) \qquad\qquad u'(c_t) = \frac{2}{3}(c_t)^{-1/3}.$$

The derivative is decreasing in the amount consumed, as indicated by the negative sign on the exponent of the derivative. (The power-function derivative rule is explained in the appendix to chapter 20.) It follows that equal increases in consumption amounts have smaller and smaller utility increments as consumption increases, as shown in table 25.4. Notice that the first 100 lab dollars converted to take-home cash yield $1.11, whereas the second time that 100 lab dollars are converted only adds $1.88 − $1.11 = $0.77. With this utility function, it is not optimal to stagger consumption by starving and binging.

Subjects in the experiment were endowed with shares of an asset that paid pre-determined $1 dividends for each share. Shares were traded in a call-market

Table 25.4. Selected Transformations of Lab Dollars into Take-Home Earnings

Lab Cash Converted:	$100	$200	$300	$400	$500	$600	$700	$800	$900	$1000
Take Home Earnings:	$1.11	$1.88	$2.53	$3.11	$3.64	$4.14	$4.61	$5.06	$5.49	$5.91

process, as described in the previous chapter, with a final-period redemption value that induced a flat fundamental value for the asset. In addition, subjects received a pre-announced exogenous money income at the start of each period. Any cash held at the end of the period, but before dividend payments, would earn 5% interest. There were 18 decision periods with incomes, followed by a nineteenth period in which any remaining cash balances were automatically converted into take-home pay, using the same nonlinear utility function. Additional procedural details are provided in Bohr, Holt, and Schubert (2017).

At the start of each period, the subject would decide how much of the available cash (from incomes or cash carryover from the previous period) to convert to take-home earnings. Since the nonlinear conversion to take-home earnings has not been discussed previously, it is useful to see how this process is explained in the experiment instructions, and what hints are provided.

- *Cash Conversion Decision:* At the beginning of each period, you will be asked to indicate the amount of cash (lab dollars) that you wish to convert into final earnings (in dollars). Once lab cash is converted to final earnings, there is no interest paid, and it cannot be converted back to lab dollars.
- *Conversion Strategy:* As noted previously, the rate of return on conversions is diminishing as the amount converted becomes large. Therefore, converting large amounts in the final periods will not do much to increase your final earnings. On the other hand, converting all of your lab cash early on will not give you the liquidity you need to earn lab dollars via interest, dividends, asset purchases and sales, etc. The conversion formula to be used is shown in the rows of the drop-down menu that follows:

- *Note:* For very small lab cash conversions of several dollars, the drop-down menu shows that the value received is negative. Therefore, your take-home earnings are reduced if you convert nothing or only make very small lab cash conversions in a round.

All take-home earnings for each conversion were reduced by a constant (20 cents) to make it plain that converting $0 and losing $0.20 is not an attractive option, an observation that was mentioned in the preceding instructions. The final summary also stressed that any remaining cash at the end of the pre-announced period would be automatically converted into earnings using the same conversion table. The hints about consumption and savings were intended to reduce confusion and highlight the strategic landscape in a way that parallels the general advice provided by friends, media, and employer advisors. These hints were held constant across the two treatments, which will be discussed next.

The motivation was to evaluate defined-benefit retirement systems that are prevalent in many European countries. The high tax rates in those countries are used to build up "assisted savings" amounts that provide a high, pre-announced fraction of pre-retirement earnings. In our Private Savings treatment, no such assistance was provided and subjects had to save for their own retirement years (periods 15–18) in which no exogenous income was received. In particular, subjects in this treatment received $100 per period for the first 14 periods, and nothing thereafter. In contrast, subjects received a constant $80 amount in all 18 periods for the Assisted Savings treatment. These amounts were selected to generate approximately equal present values of income for each treatment, subject to a desire to use focal multiples of $10 amounts for income payments.

The optimal consumption path can be calculated with a simple intuitive observation. With an interest rate of 5%, a dollar saved in one period will enhance savings by $1.05 in the next. Thus, the optimal savings decision must equate the marginal utility of consuming $1 less in period t with consuming $1.05 more in period $t+1$:

(25.2) $-u'(c_t) + (1.05)u'(c_{t+1}) = 0$ or equivalently, $u'(c_t) = (1.05)u'(c_{t+1})$.

The marginal utility for the function in equation (25.1) is used to express this equality condition as:

(25.3)
$$\frac{2}{3}(c_t)^{-1/3} = (1.05)\frac{2}{3}(c_{t+1})^{-1/3}.$$

The (2/3) terms can be cancelled and the consumption terms cross-multiplied to obtain:

(25.4)
$$(c_{t+1})^{1/3} = (1.05)(c_t)^{1/3}.$$

Finally, both sides of (25.4) can be cubed to obtain an equation that determines the rate of growth in optimal consumption from one period to the next:

(25.5)
$$c_{t+1} = (1.05)^3 c_t.$$

The final step is to experiment with different initial consumption levels, c_1, and let it grow by a factor of $(1.05)^3$ each period, with unspent income being saved. The right starting consumption level is the one that leaves just enough final cash in period 19 so that the final forced consumption is exactly $(1.05)^3$ times the consumption in period 18.

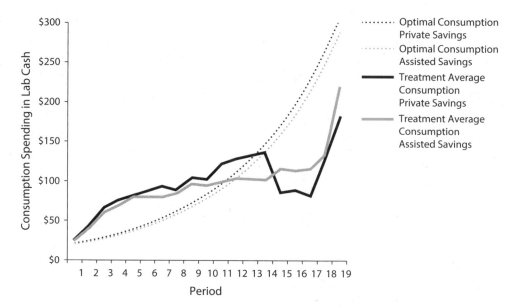

Figure 25.4. Optimal and Actual Consumption by Period, with Private Savings (dark) and Assisted Savings (gray). *Source*: Bohr, Holt, and Schubert (2017).

This optimal initial consumption, obviously, will depend on the income stream. The two income streams (80 in all periods with assisted savings and 100 for 14 periods with private savings) were selected so that the optimal consumptions for the two treatments were approximately equal, as shown by the closely aligned dotted lines in figure 25.4. An exact equivalence could have been obtained with non-integer income streams, but approximate equivalence was achieved with focal, integer-valued incomes.

The dotted line optimal consumption trajectories for each treatment are upward sloping due to the power of interest compounding. If the utility function had exhibited less curvature (marginal utility diminishing at a slower rate), the optimal consumption would have started lower and increased at a faster rate. But the most important feature of the optimal dotted-line consumption paths is that the income dip in the Private Savings treatment gets smoothed out so that consumption is *not* predicted to dip when income plummets from $100 to $0 in period 15. The motivation for adding the asset market was to provide a riskier savings strategy and to permit visceral emotions due to speculation and overconfidence to interfere with careful savings decisions.

The thick lines in figure 25.4 show the actual consumption paths, averaged over six sessions in each treatment. Notice that observed consumption levels in each treatment exceed optimal levels in early periods and are far too low in later periods. At the retirement point (period 15), consumption dips sharply for the Private Savings treatment, a dip that is not observed with Assisted Savings. These treatment differences generate economically significant differences. In terms of statistical significance, consumption with the Private Savings is higher

Table 25.5. Asset Price Peaks by Session for Each Treatment

Session	1st	2nd	3rd	4th	5th	6th	Avg.	FV
Assisted Savings	42.5	21.25	30	26	43	38	**33.5**	20
Private Savings	42	36	53	61.5	38.5	70	**50.2**	20

in pre-retirement periods 10–15 and lower in the retirement periods (15–18), with p values below 0.01 in both cases using Mann-Whitney tests with session-level data (six observations per treatment).

Such consumption dips at retirement are commonly observed in the United States, and this observation is used to infer that savings are sub-optimal, even though optimal levels are not directly observed in naturally occurring markets. See Ackert and Deaves (2010, p. 303) for further discussion of the retirement consumption dip and associated references.

A final treatment difference is in terms of asset price bubbles. Recall from the previous chapter that the fundamental value of an asset can be flattened out by setting a redemption value per share to be the ratio of the dividend to the interest rate (in this case $1 to 0.05), which is $20. The intuition is that a $1 dividend on a risky asset costing $20 provides a 5% return, which matches the interest rate on the safe asset. Asset prices exhibited bubbles with peaks significantly above this fundamental value in all sessions, and the bubbles were significantly higher in the Private Savings sessions, with peak prices shown in table 25.5. Notice that average peak prices of about $50 for the Private Savings treatment were more than double the asset fundamental value of $20. The statistical significance of this difference can be evaluated with a Mann-Whitney test (problem 5). To summarize:

Consumption Experiment with Asset Markets, Interest, and Diminishing Marginal Utility: *Subjects tend to over-consume relative to optimal levels in early periods, and under-consume in later periods of a known finite lifetime. These directional biases are worsened in a Private Savings treatment in which income drops to zero in a retirement phase. The consumption dip at retirement observed for many subjects is clear evidence of non-optimal saving in this treatment. No such consumption dip is observed in the Assisted Savings treatment. Moreover, asset price bubbles are more pronounced in the Private Savings treatment, which provides more cash for investors to speculate with in the formative middle periods of the experiment.*

The flip side of saving is debt. When income patterns are anticipated to increase sharply over time, it can be optimal to borrow instead of save. Meissner (2016) recently ran an experiment with an increasing income pattern in one of the treatments and a decreasing pattern in another. Subjects exhibited a type of "debt aversion" by borrowing too little in the increasing-income treatment. On the other hand, most economists do not worry about debt aversion, given the massive amounts of student-loan and housing debt observed in many countries. Eckel et al. (2007), for example, use a combination of laboratory and field observations to conclude that there is not strong evidence of debt aversion in terms of postsecondary educational loans. A related perspective is provided by Brown, Camerer, and Chua (2009), who use beverage rewards to study impatience and consumption in a controlled experiment with thirsty subjects. The implication is that visceral emotions, e.g., physical desires, conspicuous consumption, or real estate speculation, may play a role in impulsive decisions related to undersaving and debt. Such emotions may not be present in standard laboratory environments, but are clearly present in the asset markets that were incorporated into the consumption/savings experiment summarized in this section.

25.6   Extensions: Asset Pricing and Present Values

Duffy (2016) provides an up-to-date survey of the expanding literature on macroeconomics experiments. There are, for example, sections in that survey devoted to summarizing research on optimal consumption and bank runs. Balkenborg, Kaplan, and Miller (2011) describe a classroom bank run experiment that is similar to the one presented in the first section of this chapter. That paper also provides a nice summary of modern bank runs, illustrated with examples from the 2008 recession. Moreover, their version of the bank run game is also available on the web (for free) on the "FEELE" site at the University of Exeter (https://projects.exeter.ac.uk/feele/LecturerStart.shtml).

The Veconlab Macro Markets experiment is based on an earlier hand-run version (using playing cards for money and goods). In that setup, fiat money was represented by red cards, and product units were represented by black cards. Trades were enacted by trading red cards for black cards, and a monetary injection was accomplished by dealing more red cards to participants. For details, see Goeree and Holt (1999a), which also provides a "four-quadrant" graph that represents the real and nominal aspects of the equilibrium. See Duffy and Puzzello (2017) for a more recent experiment involving monetary policies, inflation, and production.

There have been a number of interesting papers investigating savings and consumption behavior in dynamic, multi-period experiments. One issue that arises is how to implement present value incentives. In a seminal laboratory study of consumption, Hey and Dardanoni (1988) induced present value considerations by using *both* interest on cash and a fixed probability of ending the decision sequence after each period. They concluded that subjects tend to misperceive the independence of termination probabilities, which generated unpredicted time dependencies and insensitivity to discounting incentives provided by the random stopping procedure. Noussair and Matheny (2000) also had different treatments with either random stopping or interest-based present value considerations. They observed non-optimal consumption spikes in both treatments, but such "consumption binges" were a little more common in the random stopping treatment. It is worth noting that both random stopping (survival risk) and interest rate considerations can affect a person's preference for present values outside of the laboratory. Nevertheless, when running an experiment, it may be advisable to use prior experiment results to select only one of these methods in order to reduce confusion.

There is a recent literature with experiments that are motivated by the Lucas asset pricing model of interactions between asset markets and consumption behavior. Two of these papers did not observe price bubbles. First, Crocket and Duffy (2015) conjecture that the presence of concave utility with the incentive to smooth consumption may dampen or eliminate price bubbles that are common in simple single-market environments. In particular, speculation might tend to arise in simple settings in which subjects have nothing else to do. Similarly, Asparouhova et al. (2016) do not observe asset price bubbles, which they attribute to "incessant incentives to trade" induced by heterogeneous endowments, state-dependent incomes, etc. An alternative (or additional) explanation for the absence of price bubbles is that excess cash is not allowed to accumulate in either of these experiments. In the Asparouhova et al. (2016) experiment, for example, all cash that is not invested in assets is consumed immediately at the end of each period, which prevents cash accumulation. An important takeaway from earlier asset market bubble experiments is that excess cash is a key causal factor, and the higher excess cash in the Private Savings treatment did seem to generate higher price bubbles in the experiment discussed in the previous section. These observations are consistent with a recent experiment run by Fenig, Mileva, and Petersen (2017), who implement a rich macroeconomic setting with labor and asset markets and endogenous policy rules. One of the macroeconomic policies considered (the prospect of a binding leverage constraint) tends to increase labor supply in their experiment, resulting in the accumulation of cash balances and associated price bubbles.

Chapter 25 Problems

1. Consider a bank with only two depositors and a threshold of 1 in the sense that the bank fails if at least one of the depositors withdraws. Write out the payoffs for a 2-person, 2-decision game when each person begins with a deposit of $10 and must decide whether to withdraw the $10 (or attempt a withdrawal) or leave the deposit with the bank, in the hopes of earning $12 if the other depositor does the same. Unlike the classroom experiment discussed in the first section of the chapter, decisions are simultaneous, and there are no forced withdrawals. If both try to withdraw at the same time, then the tie is decided by a coin flip, i.e., each has a 1/2 chance of recovering their deposit, and a 1/2 chance of recovering nothing. If only one withdraws, then that person

retrieves their $10 deposit, and the other earns $0. Both players are assumed to be risk neutral, so that the payoffs in each cell of the payoff matrix can be written as expected values if there is randomness. Verify that this game is a coordination game with two Nash equilibria in pure strategies.

2. For the game in problem 1, how would the payoff table change with the introduction of deposit insurance that pays a depositor a minimum of $9 in the event of failure and an inability to recover the original deposit? To be clear, if both withdraw, then each has a 1/2 chance of earning $10 and a 1/2 chance of not getting the deposit back and only getting paid the $9 insurance payment. If only one withdraws, then that person earns $10 and the other earns the $9 insurance payment. How is this different from or similar to a coordination game?

3. For the game described in problem 1, determine the decision of a level 1 player who makes a best response to a belief that the other player is a level 0 (perfectly random) player who withdraws with probability 1/2? (Assume that player 1 is risk neutral.)

4. Using the leisure and production values from table 25.2, determine the predicted real wage and number of labor units employed when there are 5 firms and 15 workers, i.e., 3 workers per firm instead of 2.

5. For each session peak price in the Assisted Savings treatment (top row of table 25.5), count the number of binary wins over session-level peak prices in the other treatment. Use the minimum binary win count and the number of observations in each treatment (6) to determine the p values for one-tailed and two-tailed Mann-Whitney tests. The relevant p values are provided in table 13.2 of chapter 13.

6. (non-mechanical) The discussion of asset market experiments in chapter 24 identified a number of factors that are associated with larger price bubbles. Can one or more of these factors provide an explanation for the higher price bubbles observed in the Private Savings treatment in section 25.5? Explain.

7. (non-mechanical) What kind of behavior would you expect to see if depositors in the bank run experiment in section 25.1 could decide which bank to use? Do you think bank failures would be more frequent or less frequent than with exogenous assignments?

8. Consider the effect of lowering the cost of discount window borrowing from 20 to 10. How would this affect the illiquid bank's earnings for each entry in the bottom row of table 25.1?

PART V

Auctions and Mechanism Design

Auctions can be very useful in the sale of perishable commodities like fish and flowers. The public nature of most auctions is also a desirable feature when equal treatment and "above-board" negotiations are important, as in the public procurement of milk, highway construction, to note just a few examples. Internet auctions can be particularly useful for a seller of a specialty item who needs to connect with geographically dispersed buyers. Auctions are being increasingly used to sell bandwidth licenses, as an alternative to administrative or "beauty contest" allocations, which can generate wasteful lobbying efforts.

Most participants find auctions to be exciting, given the win/lose nature of the competition. The two main classes of models are those where bidders know their own private values, and those where the underlying common value of the prize is not known. Private values may differ from person to person, even when the characteristics of the prize are perfectly well known. For example, two prospective house buyers may have different family sizes or numbers of vehicles, and therefore, the same square footage may be much less useful to one than to another, depending on how it is configured into common areas, bedrooms, and parking. An example of a common value auction is the bidding for an oil lease, where each bidder makes an independent geological study of the likely recovery rates for the tract of land being leased. Winning in such an auction can be stressful if it turns out that the bidder overestimated the prize value, a situation known as the "winner's curse."

Private value auctions are introduced in chapter 26, where the focus is on comparisons of bidding strategies with Nash equilibrium predictions for different numbers of bidders. There is also a discussion of several interesting variations, e.g., a "second-price" rule that only requires the high bidder to pay the second highest bid price.

Common value auctions are considered in chapter 27. One issue is the identification of factors like the numbers of bidders that affect the severity of winner's curse issues. The buyer's curse is a similar phenomenon that arises with a bidder seeking to purchase a firm of unknown profitability from the current owner, so a takeover bid could be successful because the value of the firm was overestimated. In each case, the curse effect arises because bidders may not realize that having a successful bid is an event that conveys useful information about others' valuations or estimates of value.

The Internet has opened up many exciting possibilities for new auction designs, and laboratory experiments can be used to "testbed" possible procedures prior to the final selection of the auction rules. For example, the Georgia Irrigation Reduction Auction, which was designed and run by experimental economists, was structured without a pre-announced final period and with some other features that were intended to defeat possible attempts at collusion by bidders (farmers). This was a multi-unit auction: the state government purchased irrigation reduction commitments from a large number of farmers, based on a selection of those who were willing to accept the lowest payments per acre for such reductions.

Another type of multi-unit auction involves blocks of emissions permits that are purchased by firms, with one permit required for each ton of greenhouse gas emissions emitted by a firm. Auctions for greenhouse gas emissions designed by experimental economists are now standard regulatory tools in the United States (California and the northeast states) and in the European Union. The Federal Reserve Bank of New York uses auctions to sell portfolios of various financial assets, e.g., mortgage-backed securities, which raises issues of how to use a single auction to buy or sell diverse securities, with bids on different securities being normalized by estimates of relative values known as "reference prices." Auction design and policy issues that arise from these multi-unit auctions are discussed in chapter 28.

Chapter 29 pertains to auctions in which bidders can purchase *packages* of commodities or licenses. For example, a telecommunications company may place a high value on acquiring a combination or "network" of geographically adjacent broadcast licenses. In such cases, large companies may prefer to bid on packages of licenses, to avoid the "exposure" problem that arises when some of the desired licenses in the network are not obtained, which degrades the values of licenses that are obtained. The FCC typically runs simultaneous auctions for individual broadcast licenses, but in 2008 it used a form of package bidding that solves the exposure problem by letting firms submit all-or-nothing bids on combinations of licenses. The hierarchical package structure used in that auction was first tested with a laboratory experiment. This project and related developments in combinatorial auctions will be discussed.

Price-based auctions may not be desirable in settings in which it is not considered appropriate to give an advantage to high-income families, e.g., the allocation of dorm rooms, slots in urban schools, medical residency positions, sorority membership, etc. In such cases, allocations can be based on preference rankings submitted by those on one or both sides of the market. A good matching mechanism provides incentives for eliciting truthful, non-strategic rankings and for arranging matches that are stable in the sense that matched pairs would not prefer to break away from assigned matchings. Experiments to evaluate alternative matching algorithms are reviewed in chapter 30.

26

Private Value Auctions

Internet newsgroups and online trading sites offer a fascinating glimpse into the various ways that collectibles can be sold at auction. Some people just post a price, and others announce a time period in which bids will be entertained, with the sale going to the person with the highest bid at the time of the close. These bids can be collected by email and held as "sealed bids" until the close, or the highest standing bid at any given moment can be announced in an ascending bid auction. Trade is motivated by differences in individual values, e.g., as some people wish to complete a collection and others wish to get rid of duplicates. This chapter pertains to the case where individual valuations differ, with each person knowing only their own "private" value. The simplest model of a private value auction is one where bidders' values are independent draws from a distribution that is uniform in the sense that each money amount in a specified interval is equally likely. For example, one could throw a ten-sided die twice, with the first throw determining the "tens" digit and the second throw determining the "ones" digit. The chapter begins with the simplest case, where the bidder receives a value and must bid against a simulated opponent, whose bids are in turn determined by a draw from a uniform distribution. Next, we consider the case where the other bidder is another participant. In each case, the slope of the bid/value relationship is compared with the Nash prediction.

Note to the Instructors: These auctions can be done by hand with dice, or with the Veconlab Private Value Auction program. This online auction has setup options that include a "second price auction." The Shanghai license auction discussed at the end of chapter 28 is a private value auction that can be run from the Veconlab Multi-Unit Auctions program, using the "single round discriminatory" or the "continuous discriminatory" setup options.

26.1 Introduction

The rapid development of e-commerce has opened up opportunities for creating new markets that coordinate buyers and sellers at diverse locations. The vast increase in the numbers of potential traders online also makes it possible for relatively thick markets to develop, even for highly specialized commodities and collectibles. These markets are structured as auctions that permit bids and asks to be collected 24 hours a day over an extended period. Gains from trade arise because different individuals have different values for a particular commodity. There are, of course, many ways to run an auction, and alternative sets of trading rules may have different performance properties. Sellers, naturally, would be concerned with selecting the type of auction that will enhance sales revenue. From an economist's perspective, there is interest in finding auctions that promote the efficient allocation of items to those who value them the most. If an auction fails to find the highest-value buyer, this may be corrected by trading in an "after market," but such trading itself entails transactions costs, which may be 5% of the value of the item, and even more for low-value items where shipping and handling costs are significant.

Economists have typically relied on theoretical analysis to evaluate efficiency properties of alternative sets of auction rules. The seminal work on auction theory can be found in a 1961 *Journal of Finance* paper by William Vickrey, who later received a Nobel Prize in economics. Prior to Vickrey, an analysis of auctions would likely be based on a Bertrand-type model in which prices are driven up to be essentially equal to the resale value of the item. It was Vickrey's insight that different people are likely to have different values for the same item, and he devised a mathematical model of competition in this context. The model is one where there is a probability distribution of values in the population, and the bidders are drawn at random from this population. For example, suppose that a buyer's value for a car with a large passenger area and low gas mileage is determined by the person's daily commuting time. There is a distribution of commuting distances in the population, so there will be a distribution of individual valuations for the car. Each person knows their own needs, and hence their own valuation, but they do not know for sure what other bidders' values are. Before discussing the Vickrey model, it is useful to begin with an overview of different types of auctions.

26.2   Auctions: Up, Down, and the "Little Magical Elf"

Suppose that you are a collector of cards from *Magic: The Gathering*. These cards are associated with a game in which the contestants play the role of wizards who duel with spells from cards in a deck. The cards are sold in random assortments, so it is natural for collectors to find themselves with some redundant cards. In addition, rare cards are more valuable, and some out-of-print cards may sell for hundreds of dollars. Consider the thought process that occurs as you contemplate a selling strategy. Suppose that you log into the newsgroup site and offer a bundle of cards in exchange for a bundle that may be proposed by someone else. You do receive some responses, but the bundles offered in exchange contain cards that you do not desire, so you decide to post a price for each of your cards. Several people seek to buy at your posted price, but suppose that someone makes a price offer that is a little above your initial price in anticipation of excess demand. This causes you to suspect that others are willing to pay more as well, so you post another note inviting bids on the cards, with a one-week deadline. The first bid you receive on a particularly nice card is $40, and then a second bid comes in at $45. You wonder whether the first bidder would be willing to go up a bit, say to $55. If you go ahead and post the highest current bid on the card each time a new high bid is received, then you are essentially conducting an *English auction* with ascending bids. On the other hand, if you do not announce the bids and sell to the high bidder at closing time, then you will have conducted a *first-price sealed-bid auction*. David Lucking-Reiley (2000) reports that the most commonly used auction method for *Magic* cards is the English auction, although some first-price sealed-bid auctions are observed.

Reiley (a.k.a. Lucking-Reiley) also found a case where these cards were sold in a descending-bid *Dutch auction*. Here the price begins high and is lowered until someone agrees to the current proposed price, which stops the auction immediately. This type of descending-bid auction is used in Holland to sell flowers. Each auction room contains several clocks, marked with prices instead of hours. Carts of flowers are rolled into the auction room on tracks in rapid succession. When a particular cart is "on deck," the quality grade and grower information are flashed on an electronic screen. Then the hand of the clock falls over the price scale, from higher to lower prices, until the first bidder presses a button to indicate acceptance, which stops the auction suddenly. This auction sequence proceeds quickly, with a high sales volume that is largely complete by late morning. The auction house is located next to the Amsterdam airport, so that flowers can be shipped by air to distant locations like New York and Tokyo.

If you know your own private value for the commodity being auctioned, then the Dutch auction is like the sealed-bid auction. This argument is based on the fact that you learn nothing of relevance as the clock hand falls in a Dutch auction (until it's too late). Consequently, you might as well just choose your stopping point in advance, just as you would select a bid to submit in a sealed-bid auction. These two auction methods also share the property that the winning bidders pay a price that equals their own bid. In each case, a higher bid will raise the chances of winning, but the higher bid lowers the value of winning. In both auctions, it is never optimal to bid an amount that equals the full amount that you are willing to pay, since in this case you would be indifferent between winning and not winning. To summarize, the descending-bid Dutch auction is strategically equivalent to the first-price, sealed-bid auction in the case of known private prize values.

This equivalence raises the issue of whether there is a type of sealed-bid auction that is equivalent to the ascending-bid English auction. With a known prize value, the best strategy in an English auction is to stay in the bidding until the price just reaches your own value. For example, suppose that one person's value is $50, another's is $40, and a third person's value is $30. At a price of $20, all three are interested. When the auctioneer raises the price to $31, the third bidder drops out, but the first two continue to nod as the auctioneer raises the bid amount. At $41, however, the second bidder declines to nod. The first bidder, who is willing to pay more, will agree to $41 but should feel no pressure to express an interest at a higher price since nobody else will speak up. After the usual "going once, going twice" warning, the prize will be sold to the first bidder at a price of $41. Notice that the person with the highest value purchases the item, but only pays an amount that is approximately equal to the second highest value.

The observation that the bidding in an English auction stops at the second-highest value led Vickrey to devise a *second-price sealed-bid auction*. As in any sealed-bid auction, the seller collects sealed bids and sells the item to the person with the highest bid. The winning bidder, however, only has to pay the second-highest bid. Vickrey noted that the optimal strategy in this auction is to bid an amount that just equals one's own value. To see why this is optimal, suppose that your value is $10. If you bid $10 in a second-price auction, then you will only win when all other bids are lower than your own. If you decide instead to raise your bid to $12, you increase your chances of winning, but the increase is *only* in those cases where the highest of the other bids is above $10, causing you to lose money on the "win." For example, if you bid $12 and the next highest bid is $11, you pay $11 for an item that is only worth $10 to you. Thus, it is never optimal to bid above value in this type of auction. Next consider a reduction to a bid, say to $8. If the highest of the other bids were below $8, then you would win anyway and pay the second bid with or without the bid reduction. But if the highest of the other bids turned out to be above $8, say at $9, then your bid reduction would cause you to lose the auction in a case where you would have won profitably. In summary, the best bid is your own value in a second-price auction. If everybody does, in fact, bid at value, then the high-value person will win, and will pay an amount that equals the second highest value. But this is exactly what happens in an English auction where the bidding stops at the second-highest value. Thus, the second-price "Vickrey" auction is, in theory, equivalent to the English auction.

Reiley (2000) points out that stamp collectors have long used Vickrey-like auctions as a way of including bidders who cannot attend an auction. For example, if two distant bidders mail in bids of $30 and $40, then the bidding would start at $31. If nobody present at the auction entered a higher bid, then the person with the higher bid would purchase at $31. If somebody else agreed to that price, the auctioneer would raise the price to $32. The auctioneer would continue to "go one up" on any bidder present until the higher mail-in bid of $40 is reached. This mixture of an English auction and a sealed-bid second price auction was achieved by allowing "proxy bidding," since it is the auctioneer who is entering bids based on the limit prices submitted by mail. It is a natural extension to entertain only mail-in bids and to simulate the English auction by awarding the prize to the high bidder at the second bid. Lucking-Reiley found records of a pure second-price stamp auction held in Northampton, Massachusetts in 1893. He also notes that the most popular online auction, eBay, allows proxy bidding, which is explained:

> Instead of having everyone sit at their computers for days on end waiting for an auction to end, we do it a little differently. Everyone has a little magical elf (a.k.a. proxy) to bid for them and all you need to do is tell your elf the most that you want to spend, and he'll sit there and outbid the others for you, until his limit is reached.

Notice that a bidder may not recognize the incentive to bid at value in a second-price sealed-bid auction, whereas a bidder in an ascending-price English auction is very likely to see the advantage of raising one's bid as long as the competition is below one's own private value. To summarize:

First and Second Price Auctions with Known Private Values: *A declining-bid "Dutch" auction is strategically analogous to a sealed-bid first-price auction in which the high bidder wins and pays their own winning bid. An increasing-bid "English" auction is strategically equivalent to a sealed-bid second-price auction in which the high bidder only has to pay the second highest bid. Strategic equivalence does not necessarily imply behavioral equivalence if the continuous Dutch and English auctions help bidders make more careful decisions.*

26.3 Bidding against a Uniform Distribution

This section describes an experiment in which bidders received private values, and others' bids were simulated by throws of ten-sided dice. This experiment lets one begin to study the tradeoffs involved in optimal bidding without having to do a full analysis of how others' bids are actually determined in a market with real (non-simulated) bidders. At the start of each round, the experimenter would go to each person's desk and throw a ten-sided die three times to determine a random value between $0.00 and $9.99. This would be the person's private value for the prize. Since each penny amount in this interval is equally likely, the population distribution of values is uniform. After finding out their own values, each person would select a bid knowing that the "other person's" bid would be randomly determined by three subsequent throws of the ten-sided die. If the simulated other person's bid turns out to be lower than the person's own bid, then the person bidding would earn the difference between their private value and their own bid. If the simulated other bid were higher, then the person bidding would earn nothing.

Suppose that the first three throws of the die determine a value that will be denoted by v, where v is now some known dollar amount between $0.00 and $9.99. The only way to win money is to bid below this value, but how much lower? The strategic dilemma in an auction of this type is that a higher bid will increase the chances of winning, but the value of winning with a higher bid is diminished because of the higher price that must be paid. Optimal bidding involves finding the "sweet spot" in this tradeoff, given one's willingness to tolerate risk, which can be considerable since the low bidder in the auction earns nothing.

This strategic tradeoff can be better understood by considering the bidder's expected payoff under a simplifying assumption of risk neutrality. This expected payoff consists of two parts: the probability of winning and the payoff conditional on winning. A person with a value of v who wins with a bid of b will have to pay that bid amount, and hence will earn $v - b$. Thus the expected payoff is the product of the winner's earnings and the probability of winning:

(26.1)   Expected Payoff $= (v - b)\Pr(\text{winning with bid } b).$

The probability of winning with a bid of b is just the probability that this bid is above the simulated other bid. The other bid is equally likely to be any penny amount: $0.00, $0.01, \ldots $9.99. For simplicity, we will ignore ties and assume that the other person will win in the event of a tie. Then a bid of 0 would win with probability 0, a bid of $10.00 would win with probability 1. This suggests that the probability of winning is $b/10$, which is 0 for a bid of 0 and 1 for a bid of 10. For a bid of $5, the probability of winning is exactly 1/2 according to this formula, which is correct since there are 500 ways that the other bid is below $5 ($0.00, $0.01, \ldots $4.99), and there are 500 ways that the bid is greater than or equal to $5 ($5.00, $5.01, \ldots $9.99). Using this formula ($b/10$) for the probability of winning, the expression for the bidder's expected payoff in (26.1) can be expressed:

(26.2)   Expected Payoff $= (v - b)\dfrac{b}{10} = \dfrac{vb}{10} - \dfrac{b^2}{10}.$

This expected payoff exhibits the strategic dilemma discussed earlier. The payoff conditional on winning, $v - b$, is decreasing in the bid amount, but the probability of winning, $b/10$, is increasing in b. The optimal bid involves finding the right balance between these factors, a high payoff and a high probability of winning.

The bidder knows the value, v, at the time of bidding, so the function on the right side of (26.2) can be graphed to find the highest point, as in figure 26.1 for the case of $v = $8. The bid, b, is on the horizontal axis, and the function starts with a height of 0 when $b = 0$ since a 0 bid has no chance of winning. Thus at

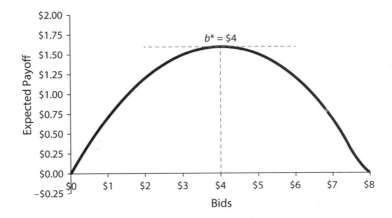

Figure 26.1. Expected Payoffs with a Private Value of $8.00

the other end, a bid that equals the value v will also yield a 0 expected payoff, since the payoff for bidding the full value of the prize is 0 regardless of whether one wins or loses. In between these two points, the expected payoff function shows a hill-shaped graph, which rises and then falls as one moves to higher bids (from left to right).

One way to find the best bid is to use equation (26.2) to set up a spreadsheet to calculate expected payoffs for each possible bid and find the best one (problem 1). For example, suppose that $v = \$4$ (instead of the $8 value used in the figure). With a value of $4, the expected payoffs are: $0 for a bid of $0, $0.30 = 0.1(\$4-\$1)$ for a bid of $1, $0.40 for a bid of $2, $0.30 for a bid of $3, and $0.00 for a bid of $4. Filling in payoffs for all possible bids in penny increments would confirm that the best bid is $2 when one's value is $4. Similarly, it is straightforward to show that the optimal bid is $2.50 when one's value is $5. These calculations suggest that the best strategy (for a risk-neutral person) is to bid one-half of one's value.

The graphical intuition behind bidding half of value is shown in figure 26.1 for a value of $8. The expected payoff function starts at the origin of the graph, rises, and falls back to $0 when one's bid is equal to the value of $8. The expected payoff function is quadratic, and it forms a hill that is symmetric around the highest point. The symmetry is consistent with the fact that the maximum is located at $4, halfway between $0 and the value of $8.

At the point where the function is flat, the slope of a dashed tangent line is zero, and the tangency point is directly above a bid of $4 on the horizontal axis. This point could be found graphically for any specific private value. Alternatively, we can use calculus to derive a formula that applies to all possible values of v. (The rest of this paragraph can be skipped by those who are already familiar with calculus, and those who need even more review should see the appendix to chapter 20.) For those who do not skip to the next paragraph, the only thing you will need besides a little intuition is a couple of rules for calculating derivatives (finding the slopes of tangent lines). First consider a linear function of b, say $4b$. This is a straight line with a slope of 4, so the derivative is 4. This rule generalizes to any linear function with a constant slope of k: the derivative of kb with respect to b is just k. The second rule that will be used is that the derivative of a quadratic function like b^2 is linear: $2b$. The intuition is that the slopes of tangent lines to a graph of the function b^2 become steeper as b increases, so the slope, $2b$, is an increasing function of b (the 2 is due to the number 2 in the exponent of b^2). This formula is easily modified to allow for multiplicative constants, e.g., the derivative of $3b^2$ is $3(2b)$, or the derivative of $-b^2$ is $-2b$.

The expected payoff in equation (26.2) consists of two terms. The first one can be written as $(v/10)b$, which is a linear function of b with a slope of $v/10$. Therefore, the derivative of the expected payoff will have a $v/10$ term in it, as can be seen on the right side of (26.3):

(26.3)   Derivative of Expected Payoff $= \dfrac{v}{10} - \dfrac{2b}{10}.$

The second term in the expected payoff expression (26.2) is $-b^2/10$, and the derivative of this term is $-2b/10$ because the derivative of b^2 is $2b$. To summarize, the derivative on the right side of (26.3) is the sum of two terms, each of which is the derivative of the corresponding term in the expected payoff in (26.2).

The optimal bid is the value of b for which the slope of a tangent line is 0, so the next step is to equate the derivative in (26.3) to 0:

(26.4) $\dfrac{v}{10} - \dfrac{2b}{10} = 0.$

This equation is linear in b, and can be solved to obtain the optimal bidding strategy:

(26.5) $b^* = \dfrac{v}{2}$ (optimal bid for a risk-neutral person).

The calculus method is general in the sense that it yields the optimal bid for all possible values of v, whereas the graphical and numerical methods had to be done separately for each value of v being considered.

To summarize, the predicted bid in equation (26.5) is a linear function of value, with a slope of 0.5. The actual bid data in the Holt and Sherman (2014) experiment formed a scatter plot with an approximately linear shape, but most bids were *above* the half-value prediction. A linear regression yielded the estimate:

(26.6) $b = 0.14 + 0.667v$ ($R^2 = 0.91$),

where the intercept of 14 cents was not significantly different from 0. The slope, with a standard error of 0.017, was significantly different from 1/2.

By bidding above one-half of value, bidders are obtaining a higher chance of winning, but a lower payoff if they win. A willingness to take less money in order to reduce the risk of losing and getting a zero payoff may be due to risk aversion, as discussed in the next section. There could, of course, be other explanations for the overbidding, but the setup with a simulated other bidder does permit us to rule out some possibilities. Since the other bidder was just a roll of the dice, the overbidding cannot be due to issues of equity, fairness, or rivalistic desires to win or reduce the other's earnings.

26.4 Bidding Behavior in a Two-Person, First-Price Auction

The experiment described in the previous section with simulated "other bids" is essentially an individual decision problem. An analogous game can be set up by providing each of two bidders with randomly determined private values drawn from a distribution that is uniform from $0 to $10. As before, a high bid results in earnings of the difference between the person's private value and the person's own bid. A low bid results in earnings of 0. This is known as a *first-price auction* since the high bidder has to pay the highest (first) price.

Under risk neutrality, the Nash equilibrium is to bid one-half of one's private value. The proof is essentially an application of the analysis given above, where the distribution of the other's bids is uniform on a range from 0 to $10. Now suppose that one person is bidding half of value. Since the values are uniformly distributed, the bid of one-half of value will be uniformly distributed from 0 to a level of $5, which is one-half of the maximum value. If this person's bids are uniformly distributed from 0 to $5, then a bid of 0 will not win, a bid of $5 will win with a probability that is essentially 1, and a bid of $2.50 will win with probability 1/2. So the probability of winning with a bid of b is $b/5$. Thus, the expected-payoff function is given in equation (26.2) if the 10 in each denominator is replaced by a 5. Equations (26.3) and (26.4) are changed similarly. Then, multiplying both sides of the revised (26.4) by 5 yields the $v/2$ bidding rule in (26.5).

Recall that the predicted bidding strategy is linear, with a slope of 1/2 when bidders are risk neutral. Bids exceeded the $v/2$ line in an experiment with simulated other bids, and the same pattern emerges when the other bidder is a subject in the experiment. Figure 26.2 shows the bid/value combinations for ten rounds of a web-based classroom experiment. There were ten bidding teams, each composed of one or two students on a networked PC. The teams were randomly matched in a series of rounds. The bidding pattern is approximately linear, except for a leveling off with high values, which is a pattern that has also been noted by Dorsey and Razzolini (2003). Only a small fraction of the bids are at or below the Nash prediction of $v/2$, which is graphed as the solid line. In fact, the majority of bids are above the dashed line with slope of 0.67 obtained from a linear regression for the case discussed in the previous section; the bid-to-value

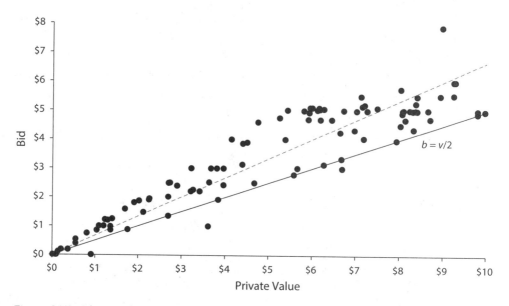

Figure 26.2. Observed Bidding Strategies for Ten Subjects in a Classroom Experiment

ratio averaged over all bids in all rounds is about 0.7. This pattern of overbidding relative to the Nash prediction is typical, and the most commonly mentioned explanation is risk aversion. An auction model with risk-averse bidders is presented in the appendix to this chapter. It is shown that bidding two-thirds of value is consistent with bidders having a "square root utility function" $u(x) = x^{1/2}$, which implies a coefficient of relative risk aversion of 1/2. This amount of risk aversion is in the range of estimates previously discussed in chapter 3.

> **Private Value Auction Experiment Results:** *Bids in first-price auctions with private values are approximately linearly related to values, as predicted, although some second-order departures from linearity have been reported. Bids tend to be higher than Nash predictions.*

26.5 Numbers of Bidders, Risk Aversion, and Regret

Up to now, we have only considered two-person auctions with risk-neutral bidders. Auctions with more bidders are more competitive, which will cause bids to be closer to values. The formula in (26.5) can be generalized to the case of N bidders drawing from the same uniform private value distribution:

$$(26.7) \quad b^* = \frac{(N-1)v}{N-r} \quad \text{(equilibrium bids with uniform value distributions)}$$

where r is a coefficient of constant relative risk aversion for a utility function, $u(x) = (x)^{1-r}$ for $0 \le r < 1$, which is assumed to be the same for all bidders.

Notice that (26.7) specifies bidding half of value when $N = 2$ and bidders are risk neutral ($r = 0$), and increases in risk aversion result in higher bids. As N increases, bids rise as a proportion of value for larger group sizes. As the number of bidders increases, the ratio, $(N-1)/(N-r)$, becomes closer and closer to 1, and bids converge to values. This increase in competition causes expected payoffs to converge to zero as bid/value differences shrink.

Some of the earliest experimental tests of the Vickrey model are reported by Coppinger, Smith, and Titus (1980). In particular, they observed overbidding (bidding too high) relative to Nash predictions in a private value, first-price auction, assuming risk neutrality. The effect of risk aversion on individual behavior in private value first-price auctions was explored in a series of papers by Cox, Roberson, and Smith (1982), and Cox, Smith, and Walker (1985, 1988), who conclude that risk aversion is the main reason that subjects bid above Nash predictions based on risk neutrality. All of the papers that compare bidding across different group sizes find more competitive outcomes (bid value ratios closer to 1) for larger groups, as implied by the bidding function in (26.7). To summarize:

First-Price Sealed-Bid Auction Experiments: *For experiments with randomly determined private values, a robust finding is that bids exceed Nash equilibrium predictions implied by risk neutrality, which is often attributed to risk aversion. But experiments do support the qualitative Nash prediction (with or without risk aversion) that bids will converge to values for large group sizes.*

The risk aversion explanation of overbidding has been one of the most hotly debated topics in experimental economics. Harrison (1989) was one of the first to question this explanation, with a provocative paper in the leading professional journal, the *American Economic Review*, which was followed by a lively exchange of replies and comments. Harrison made a valid point that expected payoffs in these settings are very flat near the maximum, as is the case for the flat hilltop for the quadratic function graphed in figure 26.1. With a flat maximum, the loss in expected payoffs from slight overbidding will be relatively small. In figure 26.1, for example, the loss would be less than a quarter for overbidding by one or two dollars. These expected payoff losses would be even less if others are overbidding.

The flat-maximum critique suggests the plausibility of deviations from risk neutrality predictions, but it does not explain the *direction* toward overbidding instead of underbidding. The most commonly mentioned alternative to risk aversion for explaining a directional bias is the notion of *regret aversion*. Remember that these auction experiments typically involve sequences of auctions, so bidders may respond to disappointing results by adjusting bids relative to values. It is useful to consider *loser's regret*, which may be experienced by an unsuccessful bidder when another person's winning bid is lower than the bidder's own private value. In other words, loser's regret occurs when a losing bidder realizes that a higher bid might have won and earned a profit. Conversely, *winner's regret* can occur if the winning bidder notices that a lower bid could have won. A winning bidder will experience more regret when the winning bid is considerably above the second highest bid, which is referred to in the trade as "leaving money on the table." The US Department of the Interior has, at times, declined to release losing bids in timber lease auctions, presumably in an effort to prevent winner's regret and induce higher bids and auction revenues.

If all bids are released, then there is a possibility of both winner's and loser's regret, which tend to pull in different directions. Englebrecht-Wiggans and Katok (2008) make a strong argument that loser's regret is a stronger emotion, and hence, that the regret model explains overbidding in first-price sealed-bid auctions.

There are several interesting variations of the regret explanation. Filiz-Ozbay and Ozbay (2007) point out that bidders do not have to actually experience and react to regret, but rather, may *anticipate* regret even in a single auction setting. By manipulating the information about others' bids that will be released after the auction, they provide evidence that an over-responsiveness to anticipated loser's regret can explain overbidding in auctions. A closely related explanation is provided by Neugebauer and Selten (2006), who use *learning direction theory* to argue that bidders tend to adjust bids in the direction of what would have been a best response in the most recent auction. This directional response would be to raise bids after losing and lower bids after winning. When there are more losers than winners and when the psychological effect of losing is stronger, this directional learning can explain overbidding. To summarize:

Other Explanations of Overbidding in Experiments: *Losers may regret having bid too low, and winners may regret having bid too high. The net effect could be an upward bias in bids if loser's regret is stronger, or if losing bids are not released after the auction, which diminishes winner's regret. This is a dynamic explanation, based on bid adjustments in the direction of what would have been a best response in the most recent auction. Support for learning-direction and regret-based explanations of overbidding in first-price auctions is provided by experiments that manipulate information released to bidders.*

One useful procedure for isolating causality in a complicated interactive settings is to pre-sort subjects into groups based on measures of the primary factor of interest. Millner and Pratt (1991) are the first economists to adopt this approach. They used a price-list measure of risk aversion to divide subjects into groups. They found that groups of more risk-averse subjects tended to exert less effort in rent-seeking tournaments, as discussed in chapter 12. More recently,

Füllbrunn, Janssen, and Weitzel (2016) used the "bomb" risk aversion measure to sort people into auctions on the basis of the numbers of boxes checked in this task. The version of the bomb task that they used involved 100 boxes, with a bomb hidden in one of them that would eliminate all earnings if encountered. As noted in chapter 3, a risk-neutral person will check half of the boxes (50 in this case), and a risk-averse person will reduce the bomb risk by checking fewer than half of the boxes.

The 20–24 subjects recruited in each session were given the bomb task and then ranked according to measured risk aversion. Then market groups of 4 were constructed by putting the 4 most risk-averse subjects into one group, the next most risk-averse subjects into a second group, etc. Of the 46 market groups, the 9 with the highest average measures of risk aversion were put into a High Risk Aversion (HRA) category; those subjects only selected about 29 out of 100 boxes on average in the bomb task. Similarly, the 9 groups with the lowest average measure were put into a Low Risk Aversion (LRA) category; they selected about 58 out of 100 boxes. The remaining groups were classified as being in the Moderate Risk Aversion group (with people who checked 43 boxes on average).

In addition to the three risk aversion treatments, there was a no-regret treatment in which subjects were only informed about whether or not they won the auction, and a regret treatment in which the losing bidders were told the value of the winning bid and the amount of the "missed opportunity." This missed opportunity was the difference between the losing bidder's value and the winning bid (if that difference was positive). This treatment was designed to stimulate loser's regret and overbidding.

Each group participated in a sequence of 50 private value auctions with fixed matchings, with the first 25 being a no-regret procedure, followed by 25 auctions with the regret (information revelation) procedure. In all auctions, the Nash equilibrium bid prediction under risk neutrality would be determined by equation (26.7) with $N = 4$ and $r = 0$. In this case, the prediction is that each bid will equal $(N-1)/N = 3/4$ of the person's randomly determined private value. Values were draws from an interval from 0 to 10,000 currency units. People with low values tend to submit erratic "throw-away" bids, so the focus at the market level was on the ratio of the winning bid (market price) to the predicted winning bid (three-fourths of the highest private value in the group). As shown in figure 26.3, the overpricing percentage relative to Nash risk neutrality predictions is increasing as the group risk aversion category increases from left to right. These risk aversion effects are highly significant at the 1% level, both for binary Mann-Whitney tests (HRA vs. LRA) and for an ordered test (HRA > MRA > LRA).

There is also some evidence of a regret effect, as can be seen by comparing matched pairs of light (no-regret) and dark (regret) bars. However, the regret treatment was done in the second set of 50 rounds, and there was enough of

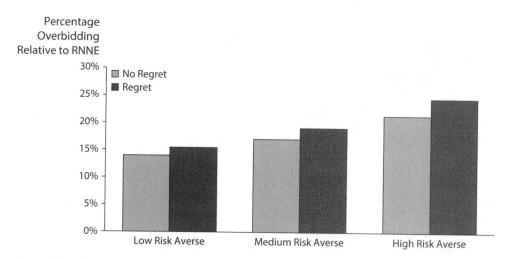

Figure 26.3. Percentage Overpricing Relative to Risk-Neutral Nash Equilibrium Predictions with Bret Risk Aversion Pre-sorting. *Source:* Füllbrunn, Janssen, and Weitzel (2016).

an increase in overbidding over time to cause the regret dummy variable to become insignificant in a regression done with individual overbidding measures.

In contrast, the experiment design does enable the authors to report a "robust, causal, and positive effect" of bidders' risk aversion on overbidding in first-price auctions, irrespective of the information release (regret) treatment. This overbidding was costly, as measured by the increase in profits that a bidder could have earned by submitting the risk-neutral Nash equilibrium (RNNE) bid (three-fourths of value) against others' actual bids, as compared with actual earnings. Bidders in the high risk aversion group could have increased their earnings by about 65% by using the RNNE bid in all rounds. Even the low risk aversion bidders could have increased their earnings by about 20% by a *unilateral* switch to the RNNE bid. To summarize:

Experiment Results with Pre-sorting for Risk Aversion: *Overpricing relative to risk-neutral Nash predictions is significantly higher for more risk-averse groups of bidders. Bidders sacrifice significant earnings relative to what they would have obtained from a unilateral switch to the risk-neutral Nash bid. There is some evidence of a regret effect as well, but the treatment sequence used does not permit a clear identification of this effect.*

Deviations from sharp game-theoretic predictions in private value auctions can also be due to bounded rationality that injects behavioral noise into the process. Goeree, Holt, and Palfrey (2002) considered a first-price auction with randomly determined private values from a discrete distribution spaced so that bidding above the Nash level (half of value) was not very costly in one treatment, but bidding below was very costly in terms of foregone profits. This asymmetry could generate an upward bias in bids. Think of an expected payoff function that rises sharply as bids are increased in the direction of the Nash prediction, but that falls very slowly with further increases. This asymmetry was reversed in a second treatment that was structured so that bidding above the Nash prediction (half of value) was very costly, which might generate a downward bias. The motivation was to run an experiment that would explain overbidding with bounded rationality in one treatment and underbidding in another. The experiment, however, produced overbidding relative to the Nash prediction in *both* treatments, although overbidding was even more pronounced in the treatment that was intended to produce an upward bias. The authors concluded that risk aversion is a factor, even in models with noisy behavior. Their evidence against other explanations like joy of winning was, however, indirect. In contrast, the Füllbrunn et al. (2016) presorting procedure provides sharper evidence of risk aversion effects, since the risk preference presorting is exogenous in their treatment setup.

Next, consider the effects of gender. An informal survey of RAs and experimental economics students suggests that people generally expect males to bid higher in first-price auctions, based on conjectures about aggressive behavior. This is not the case. For example, in the rent-seeking contests considered in chapter 12, women tend to bid higher by exerting more effort. There is also evidence that women bid higher in first-price private value auctions. Füllbrunn and Holt (2017) use explicit gender pre-sorting to divide subjects into groups that are all male, all female, or a 50-50 mix. Each person participated in 50 private value sealed-bid auctions, in a group of 4, with fixed matchings as in the previously discussed experiment. People in mixed groups were told that the group consisted of two men and two women. As before, the effects of throwaway bids by people with low values were suppressed by using an overpricing measure that is the percentage difference between the winning bid and the risk neutrality prediction (three-fourths of the highest value among bidders in the auction).

As shown in figure 26.4, there is overpricing relative to Nash predominant in both treatments. The overpricing percentage relative to Nash risk-neutrality predictions is not correlated with the gender mix in the first half, rounds 1–25 (gray bars), but is increasing in the proportion of females in the second half (dark bars), as the proportion of females increases from 0 to 50% to 100% moving from left to right. These gender effects are significant for the second half when the all-female sessions are compared with the all-male sessions, as is also the case with an ordered test (Female > Equal Mix > All Male). Gender effects with pre-sorting emerge and become pronounced after more periods, as it may overcome stereotypes about males bidding aggressively that may arise in the presence of explicit gender sorting.

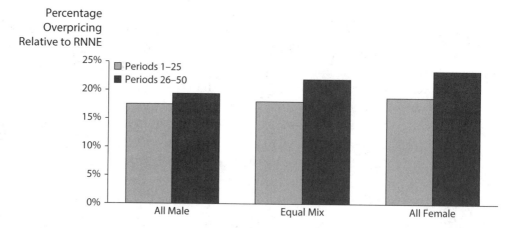

Figure 26.4. Percentage Overpricing Relative to Risk-Neutral Nash Equilibrium Predictions with Gender Pre-sorting. *Source*: Füllbrunn and Holt (2017).

One interesting point about the gender effects just discussed is made by Chen, Katuscak, and Ozdenoren (2013) in a paper entitled "Why Can't a Woman Bid Like a Man." Women subjects do bid higher in their first-price auction experiments, and they earn less as a result, but there is no gender difference in second-price auctions. The authors show that overbidding by women is systematically related to the menstrual cycles of women taking oral contraceptives. To summarize:

> **Gender Effects in First-Price Auction Experiments:** *There is some evidence that women bid higher in first-price auctions than men, especially in later periods. This overbidding by female subjects is correlated with biological factors.*

26.6 Extensions: Online Ad Auctions

One promising extension is the analysis of auctions for positions on lists of online advertisers. With only a single slot, a second-price auction would give incentives to bid at one's own private value, since the amount a winner pays is someone else's bid. A *generalized second-price auction*, adopted by Google, extends this to the case of multiple, ranked positions by giving the top position to the high bidder, who is required to pay a price that equals the second highest bid. The second rank on the display that consumers see goes to the second highest bidder, who pays a price that equals the third highest bid, etc. One problem is that bidding at value is not necessarily a Nash equilibrium in this process. For example, suppose that there are two positions that yield 100 "clicks" in the top position, and 70 "clicks" in the second position. The private values *per click* are $10, $8, and $5 for bidders I, II, and III respectively. If bidders II and III bid their values of $8 and $5 respectively, the earnings for bidder I with a bid at the $10 value would be ($10 − $8)(100) = $200, but a deviation to a second-ranked bid of $6 would earn ($10 − $5)(70) = $350, which is higher.

An alternative to the generalized second-price auction is a *Vickrey-Clarke-Groves (VCG) auction*, used by Facebook, in which each winning bidder pays an amount that equals the "externalities" that are the lost earnings of lower ranked bidders when they are moved down one notch in the ranking. In the previous example with two positions, a bid of $10 for bidder I would require a payment of ($8)(100 − 70) = $240, for the 30 clicks that the second-ranked bidder lost by being moved from the top position (100 clicks) to the second position (70 clicks). In addition, bidder I would have to pay the $350 that bidder III would have earned in the second position (with a value of $5 for 70 clicks) relative to the 0 clicks for ending up at the bottom. In this case, the total amount that bidder 1 would pay for the top position is the sum of the externalities: $240 + $350 = $590. Payments, of course, are made to the platform, not to the other bidders. The resulting earnings for this high-value bidder are $10(100 clicks) − $590 = $410, which is higher than what could be earned by deviating to the second position and paying the $350 externality for pushing bidder III down to the third position. McLaughlin and Friedman (2016) compared these two formats in a situation in which bidders could submit bids and change them in continuous time, with prices being computed each time bids were updated. They report that efficiency is significantly higher in the Vickrey VCG auction for most parameterizations used in their experiment. The two methods, however, did yield approximately equivalent sales revenues. Given the increasing economic importance of the online ad auctions, this work should stimulate additional research on position auctions. The next chapter considers a wide range of other multi-unit auctions.

Appendix: Risk Aversion

The analysis in this section shows that a simple model of risk aversion can explain the general pattern of overbidding discussed above. The analysis uses simple calculus, i.e., the derivative of a "power function" like kx^p, where k is a constant and the variable x is raised to the power p. The derivative with respect to x of this "power function" is a new function where the power has been reduced by 1 and the original power enters multiplicatively. Thus the derivative of kx^p is kpx^{p-1}, which is called the "power-function rule" of differentiation. A second rule that will be used is that the derivative of a product of two functions is the first function times the derivative of the second, plus the derivative of the first function times the second. This is analogous to calculating the change in room size (the product of length and width) as the length times the change in width plus the change in length times the width. This "product rule" is accurate for "small" changes.

Before using the power-function and the product rules, we must obtain an expression for a bidder's expected utility. The most convenient way to model risk aversion in an auction is to assume that utility is a nonlinear function, i.e., that the utility of a money amount $v - b$ is a power function $(v - b)^{1-r}$ for $0 \leq r < 1$. When $r = 0$, this function is linear, which corresponds to the case of risk neutrality. If $r = 1/2$, then the power $1 - r$ is also $1/2$, so the utility function is the square root function. A higher value of r corresponds to more risk aversion.

First consider the simple case of bidding against a uniform distribution of other bids on the interval from 0 to 10. Thus the probability of winning is $b/10$. With risk aversion, the expected payoff function in (26.2) is replaced by expected utility, which is the utility of the payoff times the probability of winning:

$$(26.8) \qquad \text{Expected Utility} = (v - b)^{1-r} \frac{b}{10}.$$

The optimal bid is found by equating the derivative of this function to zero. The expected utility on the right side of (26.8) is a product of two functions of b, so we use the power-function rule to obtain the derivative (first function times the derivative of the second plus the derivative of the first times the second function). The derivative of $b/10$ is $1/10$, so the first function times the derivative of the second is the first term in (26.9) below. Next, the power-function rule implies that the derivative of $(v - b)^{1-r}$ is $-(1 - r)(v - b)^{-r}$, which yields the second term in (26.9):

$$(26.9) \qquad (v - b)^{1-r} \frac{1}{10} - (1 - r)(v - b)^{-r} \frac{b}{10}.$$

This derivative can be rewritten by putting parts common to each term in the parentheses, as shown on the left side of (26.10):

$$(26.10) \qquad (v - b)\left(\frac{(v - b)^{-r}}{10}\right) - (1 - r)b\left(\frac{(v - b)^{-r}}{10}\right) = 0.$$

Multiplying both sides by $10/(v - b)^{-r}$, one obtains:

$$(26.11) \qquad (v - b) - (1 - r)b = 0.$$

This equation is linear in b, and can be solved to obtain the optimal bidding strategy:

$$(26.12) \qquad b^* = \frac{v}{2 - r} \quad \text{(optimal bid risk aversion)}.$$

This bidding rule reduces to the optimal bidding rule for risk neutrality (bidding half of value) when $r = 0$. Increases in r will raise the bids. When $r = 1/2$, the bids will be two-thirds of value, which is consistent with the results of the regression equation (26.6).

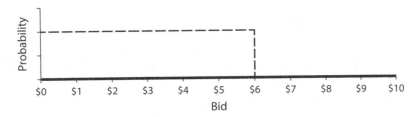

Figure 26.5. A Uniform Distribution Bid on [0, 6]

Up to this point, we only have a risk-averse person's bid against a simulated bid that is uniform on the interval from 0 to $10. Now consider an auction with two bidders, and suppose that the equilibrium bidding strategy with risk aversion is linear: $b = \beta v$, where $0 < \beta < 1$. Then the lowest bid will be 0, corresponding to a value of 0, and the highest bid will be 10β, corresponding to a value of 10. The distribution of bids is represented in figure 26.5. For any particular value of β, the bids will be uniformly distributed from 0 to 10β, as indicated by the dashed line with a constant height representing a constant probability for each possible bid. The figure is drawn for the case of $\beta = 0.6$, so bids are uniform from $0 to $6. A bid of 0 will never win, a bid of $10\beta = 6$ will win with probability 1, and an intermediate bid will win with probability of $b/6$. In the general case, a bid of b will win with probability $b/10\beta$.

In a two-person auction when the other's bid is uniform from 0 to 10β, the probability of winning in the expected payoff function will be $b/10\beta$ instead of the ratio $b/10$ that was used earlier for the expected utility in (26.8). The rest of the above analysis of optimal bidding in that section is unchanged, with the occurrences of 10 replaced by 10β, which cancels out of the denominator of equation (26.10) just as the 10 did. The resulting equilibrium bid is given in equation (26.12) as before. This bidding strategy is again linear, with a slope that is greater than one-half when there is risk aversion ($r > 0$). Recall that a risk aversion coefficient of $r = 1/2$ will yield a bid line with slope 2/3, so the bids in figure 26.2 are roughly consistent with a risk aversion coefficient of at least 1/2.

Chapter 26 Problems

1. This problem lets you set up a simple spreadsheet to calculate the optimal bid. Begin by putting the text "$V =$" in cell A1 and any numerical value, e.g., 8, in cell B1. Then put the possible bids in the A column in fifty-cent increments: 0 in A3, 0.5 in A4, 1 in A5, etc. (You can use penny amounts if you prefer.) The expected payoff formula in column B, beginning in cell B3, should contain a formula with terms involving $B$1 (the prize value) and A3 (the bid). Use the expected payoff formula in equation (26.2) to complete this formula, which

is then copied down to the lower cells in column B. Then change the value to 4 in cell B1 and verify the expected payoff numbers for that value that were reported in section 26.3. By looking at the expected payoffs in column B, one can determine the bid with the highest expected payoff. If the value of 8 is in cell B1, then the optimal bid should be 4. Then change the value in cell B1 to 5 and show that the optimal bid is $2.50.

2. Recall that equations (26.1)–(26.5) pertain to the case of a bidder who is bidding against a simulated other bidder. Write out the revised versions of equations (26.2), (26.3), (26.4), and (26.5) for the two-person auction, assuming risk neutrality.

3. (non-mechanical) How could the risk aversion sorting experiment in section 26.5 be redesigned in a way that permits a clear test of whether or not overpricing in auctions is affected by regret?

4. (non-mechanical) In what sense is the bomb task used in section 26.5 similar to a first-price auction? Is there a problem with using this measure, in your opinion?

27

The Winner's Curse

This chapter begins with a consideration of a situation in which a buyer cannot directly observe the underlying value of some object, e.g., the fundamental value of a takeover target firm. A trade is possible if the buyer's value is higher than that of the seller, perhaps because the buyer is a better manager or owns synergistic production facilities. Even if a trade is possible, a buyer's bid will only be accepted if it is higher than the value to the current owner. The danger is that a purchase is more likely to be made precisely when the buyer overestimates the owner's value and bids high, which may lead to a loss for the buyer.

The private value auctions considered in the previous chapter have the property that each buyer knows the exact value of an object that might be purchased, but different buyers may have different needs and values. The situation to be considered here is the opposite: the prize value is the same for all bidders, although none of them can assess exactly what this "common value" will turn out to be. Each bidder may obtain an estimate of the prize value, e.g., the amount of oil available in a designated tract of land. Bidders with higher estimates are more likely to make higher bids. As a result, the winning bidder may overestimate the value of the prize and end up paying more than it is worth. This *winner's curse* is exacerbated in the presence of a large number of competing bidders, in which case the winner is more likely to have overestimated the value of the object at auction.

Note to the Instructor: The buyer's curse game with a single buyer and a single seller can be done with ten-sided dice or with the Veconlab Takeover Game on the Auctions menu, which scales up easily for a large class. Similarly, a common value auction with multiple bidders can either be run with dice or with the Veconlab Common Value Auction, which has default settings that permit an evaluation of the tendency for increases in the number of bidders to produce a more severe winner's curse. Finally, the marshmallow guessing game and associated auction, discussed in section 27.3, can be done in a large class with a classroom response system ("clickers") or with bids submitted through cell phone data collectors (Bostian and Holt, 2013). To avoid an annoying recount, it is best to count marshmallows or M&Ms in groups of 50 at a time, using a cup and making hash marks on a sheet of paper to record the number of cups of 50 that are emptied into the container.

27.1 *Wall Street* (the Film)

In the 1987 Hollywood film *Wall Street*, Michael Douglas plays the role of a corporate raider who acquires TELDAR Enterprises, with the intention of increasing its profitability by firing the union employees. After the acquisition, the new owners become aware of some previously hidden business problems (a defect in an aircraft model under development) that drastically reduce the firm's profit potential. As the stock is falling, Douglas turns and gives the order "Dump it." This event is representative of a wave of aggressive acquisitions that swept through Wall Street in the mid-1980s. Many of these takeovers were motivated by the belief that companies could be transformed by infusions of new capital and better management techniques. The mood later turned less optimistic as acquired companies did not achieve profit goals.

With the advantage of hindsight, some economists have attributed these failed mergers to a selection bias: it is the *less profitable* companies that are more likely to be sold by owners with inside information about problem areas that raiders may not detect. In a bidding process with a number of competitors, the bidder with overly optimistic expectations is likely to end up making the highest tender offer, and the result may be an acquisition price that is not justified by subsequent profit potential. Even with only a single potential buyer, a bid is more likely to be accepted if it is too high, and the resulting potential for losses is sometimes called the "buyer's curse."

In a sense, winning in a bidding war can be an informative event. A bid that is accepted, at a minimum, indicates that the bid exceeds the current owner's valuation. Thus there would be no incentive for trade if the value of the company were the same for the owner and the bidder. But even when the bidder has access to superior capital and management services, the bidder may end up paying too much if the intrinsic valuation cannot be determined in advance.

27.2 A Takeover Game Experiment

Some elements that affect a firm's intrinsic profitability are revealed by accounting data and can easily be observed by both current and prospective owners. Other aspects of a firm's operations are private, and internal problems are not likely to be revealed to outsiders. The model presented in this section is highly stylized in the sense that all profitability information is private and known only by the current owner. The "raider" or prospective buyer is unsure about the exact profitability and has only probabilistic information on which to make a bidding decision. The prospective buyer, however, has a productivity advantage that will be explained below.

The first scenario to be considered is one where the value of the firm to the current owner is equally likely to be any amount between 0 and 100. This current owner knows the exact value, V, and the prospective buyer only knows the

Table 27.1. Round 1 Results from a Classroom Experiment

	Proposer 1	Proposer 2	Proposer 3	Proposer 4	Proposer 5	Proposer 6
Buyer Bid	60	49	50	36	50	0
Owner Value	21	23	31	6	43	57
Owner Response	Accept	Accept	Accept	Accept	Accept	Reject
Buyer Value	32	35	46	10	65	86
Buyer Earnings	−28	−14	−4	−26	15	0

Source: University of Virginia, Spring 2002.

range of possible equally likely values. In an experiment, one could throw a ten-sided die twice at the owner's desk to determine the value, with the throw being unobserved by the buyer. To provide a motivation for trade, the buyer has better management skills. In particular, the value to the buyer is 1.5 times the value to the current owner. For example, if a bidder offers 60 for a company that is only worth 50 to the current owner, then the sale will go though and the bidder will earn 1.5*50 minus the bid of 60, for a total of 15.

Table 27.1 shows the first-round results from a classroom experiment with this structure. The proposer names (and their curses!) are not provided. Notice that proposers in this round were particularly unfortunate. Except for proposer 6 (who knew "too much" as we will see below), the proposer bids averaged 49. All of these bids were accepted, and the owner values (21, 23, 31, 6, and 43 for proposers 1 to 5 respectively) averaged 25. The first four proposers lost money, as shown by the bottom row of the table. The fifth proposer earned 1.5 times 43 on an accepted bid of 50, for earnings of 15 cents. This tendency for buyers to make losses is not surprising ex post, since a bid of about 50 will only be accepted if the seller value is lower than 50. The five seller values averaged 25, and 1.5 times this amount is 37.5, which is still below 50.

The tendency for buyers to lose money is illustrated in figure 27.1, where payoffs are in pennies. Owner values are uniformly distributed from 0 to 99, with a probability of 1/100 of each value, as indicated by the dashed line with a height of 0.01. A bid of 60 will only be accepted if the owner's value is lower, and since all lower bids are equally likely, the expected owner value is 30 for an accepted bid of 60. This average owner value of 30 translates into an average value of 1.5 times as high, as indicated by the vertical line at 45. So an accepted bid of 60 will only yield expected earnings of 45. This analysis is easily generalized. A bid of b will be accepted if the owner value is below b. Since owner values for accepted bids are equally likely to be anywhere between 0 and b, the average owner value for an accepted bid is $b/2$. This translates into a value of $1.5(b/2)$, or $(3/4)b$, which is less than the accepted bid, b. Thus, any positive bid of b will generate losses on average in this setup, and the optimal bid is zero!

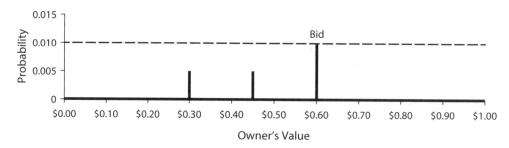

Figure 27.1. Expected Outcomes for a Bid of 60

The feedback received by bidders is somewhat variable, so it is difficult to learn this through experience. In five rounds of bidding in the classroom experiment, only three of the 13 accepted bids resulted in positive earnings, and losses resulted in lower bids in the subsequent round in all cases. But accepted bids with positive earnings were followed by bid increases, and rejections were followed by bid increases in about half of the cases (excluding the proposer, who bid 0 and was rejected every time). The net effect of these reactions caused average bids to level off at about 30 cents for rounds 2–5. The only person who did not finish round 5 with a cumulative loss was the person who bid 0 in all rounds. This person, a second-year physics and economics major, had figured out the optimal bidding strategy on his own. The results of this experiment are typical of what is observed in research experiments summarized next.

The first buyer's curse experiment was reported in Bazerman and Samuelson (1983), who used a discrete set of equally likely owner values (0, 20, 40, 60, 80, and 100). Ball, Bazerman, and Caroll (1990) used a grid from 0 to 100 with MBA subjects, and the bids did not decline to zero. The average bid stayed above 30, and the modal bid was around 50. To summarize:

Takeover Game Experiment Results: *When the seller value is uniformly distributed from a lower bound of 0 up to an upper limit, a buyer acquisition will increase efficiency if the value to the buyer is some multiple (above 1) times the value to the seller. As long as this value multiplier is below 2, any bid by the buyer will tend to generate a loss in an expected value sense, and hence the optimal bid for the buyer is 0. In experiments with a multiplier of 1.5, bids of 0 are the exception, and losses are commonly observed for most subjects, even MBA students.*

27.3   The Wisdom of the Crowds and the Wisdom of Avoiding a Crowd in a Common Value Auction

Once the author asked several flooring companies to make bids on replacing floor tile in a kitchen. Each bidder would estimate the amounts of materials needed and the time required to remove the old tile and install the floor around the various odd-shaped doorways and pantry areas. One of the bids was somewhat lower than the others. Instead of expressing happiness over getting the job, he exhibited considerable anxiety about whether he had miscalculated the cost, although in the end he did not withdraw the bid. This story illustrates the fact that winning an auction can be an informative event, or equivalently, the fact that you win produces new information about the unknown value of the prize. A rational bidder should anticipate this information in making the bid originally. This is a subtle strategic consideration that is almost surely learned by (unhappy) experience.

The possibility of paying too much for an object of unknown value is particularly dangerous for one-time auctions in which bidders are not able to learn from experience. Suppose that two partners in an insurance business work out of separate offices and sell separate types of insurance, e.g., business insurance from one office and life insurance from the other. Each partner can observe the other's earnings, but cannot determine whether the other person is really working. One of the partners is a single mother who works long hours, with good results despite a low earnings-per-hour ratio. The other partner, who enjoys somewhat of a lucky market niche, is able to obtain good earnings levels while spending a large fraction of each day socializing on the Internet. After several years of happy partnership, each decides to try to buy the whole business from the stockholders, who are current and former partners. When they make their bids, the one with the profitable niche market is likely to think the other office is as profitable as the person's own office, and hence to overestimate the value of the other office. As a result, this partner could end up acquiring the business for a price that exceeds its value.

The tendency for winners' value estimates to be biased has long been known in the oil industry, where drilling companies must make rather imperfect estimates of the likely amounts of oil that can be recovered on a given tract of land being leased. For example, Robert Wilson, a professor in the Stanford Business School, was once consulting with an oil company that was about to submit a bid for less than half of the estimated lease value. When he inquired about the possibility of a higher bid, he was told that firms that bid this aggressively on such a lease are no longer in this business (source: personal communication at a conference on auction theory in the early 1980s).

The intuition underlying the unprofitability of aggressive bidding is that the firm with the highest bid in an auction is likely to have overestimated the lease value, a phenomenon known as the *winner's curse*. The resulting possibility of winning "at a loss" is more extreme when there are many bidders, since the highest estimate out of a large number of estimates is more likely to be biased upward, even though each individual estimate is ex ante unbiased. For example, suppose that a single-value estimate is unbiased. Then the higher of two unbiased estimates will be biased upward. Analogously, the highest of three unbiased estimates will be even more biased, and the highest of 100 estimates may show an extreme bias toward the largest possible estimation error in the upward direction. Knowing this, bidders in an auction with many bidders should treat their own estimates as being inflated, which will likely turn out to be true if they win. This *numbers effect* can be particularly sinister, since the normal strategic reaction to increased numbers of bidders is to bid higher to stay competitive, as in a private value auction.

The claim that the winner's curse is more severe with large groups may seem perplexing at first. After all, if each individual's estimate is unbiased, then a group discussion of the object's value might aggregate this information and produce a good group estimate. If group members decide to vote on a group forecast, then the median voter theorem (chapter 19) would suggest that the median estimate might prevail. With a large group, the median could provide an accurate estimate. The paradox is due to the fact that the winner in an auction is the highest bidder, whose estimate may be far from the median of the other estimates.

These observations can be illustrated with the results of a class experiment reported in Bostian and Holt (2013). The authors purchased several bags of pastel colored mini-marshmallows, counted them, and wrote the resulting number on a scrap of paper that was placed in a translucent rectangular plastic freezer container with the marshmallows. The container was shown to about 200 students in a behavioral finance class, and it was shaken and rotated for all to see. Then students were asked to use clickers to submit guesses about the number, with the understanding that the person with the best guess would receive a prize of $10.

Before announcing the number of marshmallows and the guessing game winner, the students were asked to submit a bid to purchase a prize worth 1 penny for each marshmallow in the container, again using classroom response clickers. The rules provided that the high bidder would win a penny for each marshmallow in the container, but would have to pay the amount of their own winning bid. Students were permitted to submit any penny bid amount above 0. They were told that if someone ended up paying more than the prize value, they would have to settle up with the professor afterward, without any specifics provided.

Figure 27.2 shows the scatter plot of the guesses (horizontal axis) and the penny bids (vertical axis). Notice that most of the dots are to the left of the vertical line at 1250, which was the actual number of marshmallows in the container. Thus there was a clear bias, which is probably associated with misperceiving the three-dimensional nature of a container with depth. This downward bias in guesses is commonly observed with class experiments (see Thaler, 1988), which is ironic since the Internet is loaded with similar game results in which groups of people are surprisingly accurate. In those Internet clips, the average guess is often closer to the true number than any single individual guess. This notion was popularized by James Surowiecki's (2004) book with the memorable title:

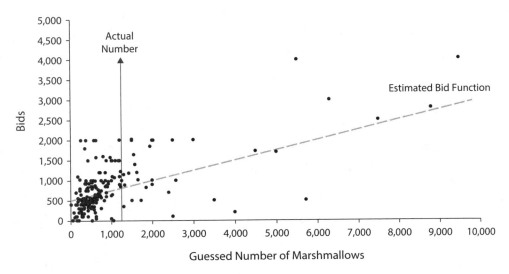

Figure 27.2. Bids and Guesses for the Box of Marshmallows

Table 27.2. Guesses and Bids for the Classroom Marshmallow Game

	Average	Median	Maximum
Guesses	970	550	9475
Bids	$8.12	$6.00	$40.24

Source: University of Virginia, Spring 2002.

The Wisdom of Crowds: Why the Many Are Smarter Than the Few and How Collective Wisdom Shapes Business, Economies, Societies and Nations.

The means and medians for the marshmallow guess and bids are shown in table 27.2. Obviously, the actual number of marshmallows, 1250, is not well predicted by either the mean or the median, both of which are biased downward. This result (and similar ones for other class experiments) suggests that general claims about the "wisdom of the crowds" should be taken with a grain of salt.

What is going on with the flurry of correct crowd guesses for games with M&Ms and jelly beans, etc., that can be found on the Internet? The author's best guess is that this phenomenon is probably just a *selection bias*, with only those games that provide accurate forecasts being posted on the web. More disturbing is the possibility of a confirmation bias that would induce the experimenter to stop asking people for guesses at a point when the average is accurate.

The final column in table 27.2 illustrates the winner's curse. The person with the highest guess (over 9,000 marshmallows) was far off the mark, and that person turned out to be the person who submitted the highest bid of over $40, for a prize worth $12.50! The dashed line in the figure is a regression line, with a positive slope of about 0.3, which illustrates the fact that bids are correlated with guesses, a correlation that will almost invariably cause the high bid to exceed the prize value when there is a large number of bidders.

27.4 Common Value Auctions with Private Signals

The auction consultant mentioned in the previous section later specified a model with this common value structure and showed that the Nash equilibrium with fully rational bidders involved some downward adjustment of bids in anticipation of a winner's-curse effect (Wilson, 1969). Each bidder should realize that their bid is only relevant for payoffs if it is the highest bid, which means that they have the highest value estimate. The bidder should consider what they would want to bid, knowing that other estimates are likely to be lower than theirs in the event that they have the high bid, which is the only relevant event for estimating the payoff. The a priori correction of this over-estimation produces bids that will earn positive payoffs on average.

Consider a simple situation where each of two bidders can essentially observe half of the value of the prize, as would be the case for two bidders who can drill test holes on different halves of a tract of land being leased for an oil well. Another example would be the case for the two insurance partners discussed in the previous section. Each person might then double their observed value to come up with a total value estimate. In particular, let the resulting value estimates for bidders 1 and 2 be denoted by v_1 and v_2 respectively. The actual prize value is the average:

$$(27.1) \qquad \text{Prize Value} = \frac{v_1 + v_2}{2}.$$

Both bidders know their own estimates, but they only know that the other person's estimate is the realization of a random variable. In the experiment to be discussed, each person's value estimate is drawn from a distribution that is uniform on the interval from $0 to $10. Bidder 1, for example, knows v_1 and that v_2 is equally likely to be any amount between $0 and $10.

A classroom auction was run with these parameters, using five pairs of bidders in each round. One bidder had a relatively high value signal of $8.69 in the first round. This person submitted a bid of $5.03 (presumably the three-cent increase above $5 was intended to outguess anyone who might bid an even $5). The other person's signal was $0.60, so the prize was only worth the average of $8.69 and $0.60, which is $4.64. The high bidder won this prize, but paid a price of $5.03, which resulted in a loss. Three of the five winning bidders in that round ended up losing money, and about one out of five winners lost money in each of the remaining rounds.

The prize value function in equation (27.1) can be generalized for the case of a larger number of bidders with independent signals, by taking an average, i.e., dividing the sum of all signals by the number of bidders. In a separate classroom auction: conducted with 12 bidders, the sole winning bidder ended up losing money in three out of five rounds. As a result, aggregate earnings were zero or negative for most bidders. A typical case was that of a bidder who submitted the high bid of $6.10 on a signal of $9.64 in the fifth and final round. The average of all 12 signals was $5.45, so this person lost 65 cents for the round.

Despite some losses, earnings in the two-bidder classroom experiment discussed above averaged $3.20 cents per person for the first five rounds of bidding. In contrast, earnings per-person in the 12-bidder auctions averaged *minus* 30 cents for five rounds of bidding. These results show a tendency for the winner's-curse effect to be more severe with large numbers of bidders. Think of it this way: the person with the highest signal often submits one of the highest bids. Since the maximum of 12 signals from a uniform distribution is likely to be *much* higher than the maximum of 2 signals, the person with the highest signal in the 12-person auction may be tempted to bid too high. Another way to think about this situation is that each signal is unbiased, since the value of the prize is the average of the two signals, but the maximum of a number of signals will be a biased estimate of the common value, and this bias is larger if there are more bidders. To summarize:

Winner's Curse: *In common value auctions with an unknown prize value, the bidder with the winning bid is likely to be someone who overestimated the prize value, even if individual value estimates are unbiased or biased downward. This selection effect can cause the winning bidder to pay more than the prize is actually worth. Such winner's curse effects are even more likely to occur in settings with large numbers of bidders.*

The author was once asked to consult for a large telecommunications company that was planning to bid in the first major US auction for radio wave bandwidth to be used for personal communications services. This company was (and still is) a major player, but at the last minute, it decided not to bid at all. The company representative mentioned the danger of overpayment for the licenses at auction. This story does illustrate a point: that players who have an option of earning zero with non-participation will never bid in a manner that yields negative expected earnings. The technical implication is that expected earnings in a Nash equilibrium for a game with an exit option cannot be negative. This raises the issue of how bidders rationally adjust their behavior to avoid losses in a Nash equilibrium, which is the next topic.

27.5 The Nash Equilibrium in a Common Value Auction

As was the case in the chapter on private value auctions, the equilibrium bids will end up being linear functions of the signals, of the form:

(27.2) $\qquad b_i = \beta v_i \quad$ where $0 < \beta < 1$ and $i = 1,2$.

Suppose that bidder 2 is using a special case of this linear bid function by bidding exactly one-half of the signal v_2. In the next several pages, we will use this assumed behavior to find the expected value of bidder 1's earnings for any given bid, and then we will show that this expected payoff is maximized when bidder 1 also bids one-half of the signal v_1, i.e., $\beta = 0.5$. Similarly, when bidder 1 is bidding one-half of their signal, the best response for the other bidder is to bid one-half of their signal too. Thus the Nash equilibrium for two risk-neutral bidders is to bid half of one's signal.

Since the arguments that follow are a little more mathematical than most parts of the book, it is useful to break them down into a series of steps. We will assume for simplicity that bidders are risk neutral, so we will need to find bidder 1's expected payoff function before it can be maximized. (It turns out that risk aversion has no effect on the Nash equilibrium bidding strategy in this 2-bidder case, as noted later in the chapter.) In an auction where the payoff is 0 in the event of a loss, the expected payoff is the probability of winning times the expected payoff conditional on winning. So the first step is finding the probability of winning given that the other bidder is bidding one-half of their signal value. The second step is to find the expected payoff, *conditional on winning with a particular bid*. The third step is to multiply the probability of winning times the expected payoff conditional on winning, to obtain an expected payoff function, which will be maximized using simple calculus in the final step. The result will show that a bidder's best response is to bid half of the signal when the other one is bidding in the same manner, so that this strategy is a Nash equilibrium.

Before going through these steps, it may be useful to review how we will go about maximizing a function. Think of the graph of a function as a hill, which is increasing on the left and decreasing on the right, as for figure 26.1 in the previous chapter. At the top of the hill, a tangent line will be horizontal (think of a graduation cap balanced on the top of someone's head). So, to maximize the function, we need to find the point where the slope of a tangent line is 0. The slope of a tangent line can be found by measuring its slope, but this method is not general since specific numbers are needed to draw the lines. In general, the slope of the tangent line to a function can be found by taking the derivative of the function. Both the probability of winning and the conditional expected payoff will turn out to be linear functions of the person's bid b, so the expected-payoff product to be maximized will be quadratic, with terms involving b and

b^2. A linear term, like $b/2$, is a straight line through the origin with a slope of $1/2$, so the derivative is the slope, $1/2$. The quadratic term, b^2, also starts at the origin, and it increases to 1 when $b = 1$, to 4 when $b = 2$, to 9 when $b = 3$, and to 16 when $b = 4$. Notice that this type of function is increasing more rapidly as b increases, i.e., the slope is increasing in b. Here all you need to know is that the derivative of a quadratic expression like b^2 is $2b$, which is the slope of a straight line that is tangent to the curved function at any point. This slope is, naturally, increasing in b. Armed with this information, we are ready to determine the expected payoff function and maximize it.

Step 1. Finding Bidder 1's Probability of Winning for a Given Bid

Suppose that the other person (bidder 2) is known to be bidding half of their signal. Since the signal is equally likely to be any value from $0 to $10, the second bidder's signal is equally likely to be any of the 1,000 penny amounts between $0 to $10, as shown by the dashed line with height of 0.001 in figure 27.3. When $b_2 = v_2/2$, then bidder 1 will win if $b_1 > v_2/2$, or equivalently, when the other's value is sufficiently low: $v_2 < 2b_1$. The probability of winning with a bid of b_1 is the probability that $v_2 < 2b_1$.

The probability of winning can be assessed with the help of figure 27.3. Suppose that bidder 1 makes a bid of $2, as shown by the short vertical bar. We have just shown that this bid will win if the other's signal is less than $2b_1$, i.e., less than $4 in this example. Notice that four-tenths of the area under the dashed line is to the left of $4. Thus, the probability that the other's value is less than $4 is 0.4, calculated as $4/10 = 2b_1/10$. A bid of 0 will never win, and a bid of $5 will always win, and in general, we have the result:

$$(27.3) \qquad \text{Probability of Winning (with a bid of } b_1) = \frac{2b_1}{10}.$$

Step 2. Finding the Expected Payoff Conditional on Winning

Suppose that bidder 1 bids b_1 and wins. This happens when $v_2 < 2b_1$. For example, when the bid is $2 as in figure 27.3, winning would indicate that $v_2 < \$4$, i.e.,

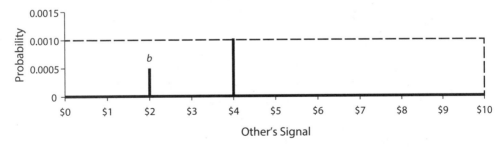

Figure 27.3. A Uniform Distribution on the Interval [0, 10]

to the left of the higher vertical bar in the figure. Since v_2 is uniformly distributed, it is equally likely to be any penny amount less than \$4, so the expected value of v_2 would be \$2 once we find out that the bid of \$2 won. This generalizes easily; the expected value of v_2 conditional on winning with a bid of b_1 is just b_1. Bidder 1 knows the signal v_1 and expects the other signal to be equal to the bid b_1 if it wins, so the expected value of the prize is the average of v_1 and b_1:

(27.4)   Conditional Expected Prize Value (winning with a bid of b_i) $= \dfrac{v_1 + b_1}{2}$,

where v_1 is the bidder's own signal and b_1 is the bidder's expectation of what the other's signal is, conditional on winning.

Step 3. Finding the Expected Payoff Function

The expected payoff for a bid of b_1 is the product of the probability of winning in equation (27.3) and the difference between the conditional expected prize value in equation (27.4) and the bid:

(27.5) $$\text{Expected Payoff} = \frac{2b_1}{10}\left(\frac{v_1 + b_1}{2} - b_1\right) = \frac{b_1 v_1}{10} - \frac{b_1^2}{10}.$$

Step 4. Maximizing the Expected Payoff Function

In order to maximize this expected payoff, we will set its derivative equal to 0. The expression on the far right side of equation (27.5) contains two terms, one that is quadratic in b_1 and one that is linear. Recall that the derivative of $(b_1)^2$ is $2b_1$ and the derivative of a linear function is its slope coefficient, so:

(27.6) $$\text{Expected Payoff Derivative} = \frac{v_1}{10} - \frac{2b_1}{10}.$$

Setting this derivative equal to 0 and multiplying by 10 yields: $v_1 - 2b_1 = 0$, or equivalently, $b_1 = v_1/2$. To summarize, if bidder 2 is bidding half of value as assumed originally in equation (27.2), then bidder 1's best response is to bid half of value as well, so the Nash equilibrium bidding strategy is for each person to behave in this manner.

(27.7) $$b_i = \frac{v_i}{2} \quad i = 1,2 \quad \text{(equilibrium bids)}.$$

Normally in a first-price auction, one should "bid below value," and it can be seen that the bid in (27.7) is less than the conditional expected value in (27.4). Finally, recall that this analysis began with an assumption that bidder 2 was using the strategy in (27.2) with $\beta = 1/2$. This may seem like an arbitrary assumption, but it can be shown that the only linear bidding strategy for this auction must have a slope of 1/2

27.6 The Winner's Curse

When both bidders are bidding half of the value estimate, as in (27.7), then the one who wins will be the one with the higher estimate, i.e., the one who over-estimates the value of the prize. Another way to see this is to think about what a naïve calculation of the prize value would entail. One might reason that since the other's value estimate is equally likely to be any amount between $0 and $10, the expected value of the other's estimate is $5. Then, knowing one's own estimate, say v_1, the expected prize value is $(v_1 + 5)/2$. This expected prize value, however, is not conditioned on winning the auction. Notice that this uncon-ditional expected value is greater than the conditional expected value in (27.4) whenever the person's bid is less than $5, which will be the case when bids are half of the value estimate. Except for this boundary case where the bid is exactly $5, the naïve value calculation will result in an overestimate of value, which can lead to an excessively high bid and negative earnings.

Holt and Sherman (2014) conducted a number of common value auctions with a prize value that was the average of the signals. Subjects began with a cash balance of $15 to cover any losses. Figure 27.4 shows bids and signals for the final five rounds of a single session. One of the bidders can be distinguished

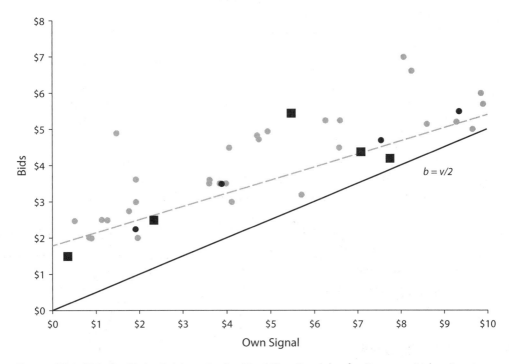

Figure 27.4. Bids for Eight Subjects in the Final Five Rounds of a Common Value Auction. *Key*: Solid line: Nash equilibrium; dashed line: regression for all sessions. *Source*: Holt and Sherman (2014).

from the others by the large square marks. Notice that this person is bidding in an approximately linear manner, but that this person's bids are above the $v/2$ line, which represents the Nash equilibrium. In fact, almost all bids are above the $v/2$ line, which indicates the earnings-reducing effects of the winner's curse in this session. The dashed line shows the regression line that was fit to all bids over all sessions, so we see that this session was a little high but not atypical in the nature of the bid/value relationship.

27.7   Extensions

The winner's curse was first discussed in Wilson (1969), and applications to oil lease drilling were described in Capen, Clapp, and Campbell (1971). Kagel and Levine (1986) provided experimental evidence that it could be reproduced in the laboratory, even after subjects obtain some experience. The literature on common value auctions is surveyed in Kagel (1995). A complete treatment of this and related topics can be found in Kagel and Levine (2002).

The takeover game described in section 27.2 has the property that the optimal bid is zero. This is no longer the case when the lowest possible owner value is positive. When seller values range from 50 to 100 and buyer values are 1.5 times the seller value, for example, the optimal bid is at the upper end of the range, i.e., at 100 (see problems 2–4). Holt and Sherman (1994) ran experiments with this setup and found bids to be well below the optimal level. Similarly, this high-owner-value setup was used in periods 6–10 of the classroom experiment described above, and the average bids were 75, 76, 78, 83, and 83 in these rounds, well below the optimal level of 100. Holt and Sherman attributed the failure to raise bids to another type of error, i.e., the failure to realize that an increase in the bid will pick up relatively high-value objects at the margin. For example, a bid of 70 will pick up objects with seller values from 50 to 70, with an average of 60. But raising the bid from 70 to 71 will pick up a purchase if the seller value is at the upper end of this range, at 70. Failure to recognize this factor may lead to bids that are too low, which they termed the *loser's curse*. To summarize:

> **Loser's Curse:** *Bidders in a common value auction may be subject to a loser's curse that is due to the failure to realize that raising one's bid tends to pick up high-value objects at the margin, which can lead to underbidding when there is a single bidder, as in some parameterizations of the takeover game.*

Chapter 27 Problems

1.   How would the analysis of optimal bidding in section 27.2 change if the value to the bidder were a constant K times the value to the owner, where $0 < K < 2$

and the owner's value is equally likely to be any amount between 0 and 100? Thus, an accepted bid yields a payoff for the bidder that is K times the owner's value, minus the amount of an accepted bid. In particular, does the optimal bid depend on K?

2.   Suppose that the owner values are equally likely to be any amount from 50 to 99, so that a bid of 100 will always be accepted. Assume that owners will not sell when they are indifferent, so a bid of 50 will be rejected and will produce zero earnings. The value to the bidder is 1.5 times the value to the owner. Show that the bid of 50 is not optimal. (Hint: Compare the expected earnings for a bid of 50 with expected earnings for a bid of 100.)

3.   For the setup in problem 2, a bid of 90 will be accepted about four-fifths of the time. If 90 is accepted, the expected value to the owner is between 50 and 90, with an average of 70. Use this information to calculate the expected payoff for a bid of 90, and compare your answer with the expected payoff for a bid of 100 obtained earlier.

4.   The analysis for your answers to problems 2 and 3 (with owner values between 50 and 100) suggests that the best bid in this setup is 100. Let's model the probability of a bid b being accepted as $(b - 50)/50$, which is 0 for a bid of 50 and 1 for a bid of 100. From the bidder's point of view, the expected value to the owner for an accepted bid of b is 50 plus half of the distance from 50 to b. The bidder value is 1.5 times the owner value, but an accepted bid must be paid. Use this information in a spreadsheet to calculate the expected payoff for all bids from 50 to 100, and thereby to find the optimal bid. The five columns of the spreadsheet should be labeled: (1) Bid b, (2) Acceptance Probability for a Bid b, (3) Expected Value to Owner (conditional on b being accepted), (4) Expected Value to Bidder (remember to subtract the bid), (5) Expected Payoff for Bidder of Making a Bid b (which is the product of columns 2 and 4). Alternatively, you may use this information to write the bidder's expected payoff as a quadratic function of b, and then use calculus by setting the derivative of this function to zero and solving for b. (Note: A classroom experiment for the setup in problems 2–4 resulted in average bids of about 80. This information will *not* help you find the optimal bid.)

5.   For the model in problem 4, discuss the optimal bid for a risk-neutral bidder. (Hint: Use the midpoint of the range for your answer to the previous question to obtain the probability that a bid will be accepted. Then one approach is to multiply the probability that a bid is accepted times the difference between the expected value of an accepted item and the bid price paid for that item. The resulting expected payoff will be a quadratic function of the bid price, which can be maximized.)

28

Multi-Unit Auctions

Emissions, Water, License Plates, Securities, . . .

The chapter begins with a case study of auction design for multiple, identical prize units, in this case, irrigation permits. The auction enabled farmers to essentially sell their irrigation usage, i.e., it determined which tracts of land in drought-stricken Southwest Georgia would *not* be irrigated during the 2001 growing season. Laboratory experiments and a field test (run onsite in South Georgia) were used to refine the auction design, and the actual auction was conducted with a web-based network of computers at eight locations.

Laboratory experiments also provided a basis for designing auctions for regulating greenhouse gas emissions. A good example of a consulting report that incorporates laboratory experiments is: *Auction Design for Selling CO2 Emissions Allowances Under the Regional Greenhouse Gas Initiative* (RGGI), which can be found on the RGGI website. The auction proceeds by ranking bids from high to low (like a demand function), and accepting the bids to the left of the auction quantity (vertical supply function), with winning bidders paying the highest rejected bid, which is the market-clearing price. RGGI implemented the recommended auction procedures, and many interesting variations of this basic auction design have been subsequently adopted for cap-and-trade programs in California and the European Union Emissions Trading System (EUETS).

Auctions can also be used to let participants simultaneously buy and sell. In 2014, California launched an ambitious greenhouse gas emissions program with a major auction component that allows purchases and sales. Regulated utilities that receive free allocations of permits are required to "consign" them for sale and then buy back what they need in the same auctions. Consignment auctions have promising applications in water markets, where the problem is often that those with free allocations may not be aware of sales opportunities. These and other topics will be covered in the final sections of the chapter, which includes examples of both *reverse auctions* (government purchases) and *forward auctions* (sales). The wide range of applications of multi-unit auctions that are considered will help the reader understand the ways that experiment design and context decisions can be tailored to fit the needs of policy makers in a particular context.

Note to the Instructor: This chapter is several pages longer than others, although the material is not particularly technical. For class experiments, there are three options: (1) The Georgia water auction format can be run using the Veconlab Water Auction program listed under the Auctions menu. For this program, set the Government Purchase Budget at $25,000 times the number of bidders and set the Target Number of Acres at 300 times the number of bidders. (2) The Veconlab Opportunity Costs experiment, listed under the Micro Principles menu, is simple to run (use the default settings) and can be used to simulate opportunity cost considerations that motivated the auctions-versus-grandfathering experiment discussed in section 28.2. The graph button lets you use the SHOW ID option to identify the people who earned the most and the least, to show that ignoring opportunity cost can be costly. (3) A uniform price or a simple clock auction can be set up with the Emissions Permits experiment, also on the Auctions menu. On the first admin setup page for this program, set ALL of the following options to "no": Banking, Spot Market, California AB32, and Consignment. The list at the bottom of that page lets you switch from the uniform price default to a clock auction if desired. Then set equal numbers of high and low users on the page that follows to a level appropriate for the class size. The auction quantity should be about three times the total number of participants which, with an exogenous product price of $21, will yield a predicted auction-clearing permit price of $8, as illustrated in figure 28.2.

28.1 Dry 2K

In early 2000, just after the publicity of the Y2K computer bugs, a severe drought plagued much of the Southeastern United States. Some of the hardest hit localities were in South Georgia, and an Atlanta newspaper ran a regular "Dry 2K" update on conditions and conservation measures. Of particular concern were the record low levels for the Flint River, which threatened wildlife and fish in the river. In April 2000, the Georgia legislature passed the Flint River Drought Protection Act, mandating an "auction-like process" to restrict agricultural irrigation in certain areas if the director of the Georgia Environmental Protection Department called a drought emergency. The unspecified nature of the auction created an ideal situation for laboratory testing of alternatives.

The Flint River, which begins from a drainage pipe near the Atlanta airport, grows to a size that supports barge traffic by the time it reaches the Florida state line. Most of the water usage in the river basin is agricultural. Farmers have permits, which were obtained without charge, for particular circular irrigation systems that typically cover areas from 50 to 300 acres. Water is not metered, and therefore, it is liberally dumped into the fields during dry periods, creating the green circles visible from the air, which makes restrictions on irrigation easy to monitor. The idea behind the Drought Protection Act was to use the economic incentives of a bidding process to select relatively low-use-value land to retire from irrigation. Farmers would be compensated to reduce any negative political impacts. The state legislature set aside $10 million, taken from a multi-state tobacco industry settlement, to pay farmers not to use irrigation permits. Besides being non-coercive and sensitive to economic use value, an auction format has the advantage of being fair and easy to implement relative to administrative processes. Speed was also an important factor, given the limited time between the March 1 deadline for the declaration of a drought emergency and the optimal time for planting crops several weeks later. Finally, the scattered geographic locations of the farmers required that the auction collect bids from diverse locations, which suggested the use of web-based communications between officials at a number of bidding sites.

Any auction would involve a single buyer, the state Environmental Protection Department (EPD), and many sellers, the farmers with permits. Permits varied in size in terms of the numbers of acres covered. This is a multi-unit auction, since the state could "purchase" many permits, i.e., compensate the permit holders for not irrigating the covered areas for the specified growing season. The goals of the people running the auction were that the auction not be viewed as arbitrary, and that the auction take out as much irrigation as possible (measured in acres covered by repurchased permits) given the available budget. Of course, economists would also be concerned with economic efficiency, i.e., that the auction would take less productive land out of irrigation.

A number of different types of auctions were considered and tested. All of these formats involved bids being made on a per-acre basis so that bids for different sized tracts could be compared and ranked, with the low bids being accepted. One method, a *discriminative auction*, would have subjects submitting sealed bids, with the winning low bidders each receiving their bid amounts. For example, if the bids were $100, $200, $300, and $400 per acre, and if the two lowest were accepted, then the low bidder would receive $100 per acre and the second low bidder would receive $200 per acre for not irrigating. This auction is "discriminative" since different people receive different amounts for approximately the same amount of irrigation reduction per acre. In contrast, a *uniform price auction* would establish a cutoff price and pay all bidders at or below this level an amount that equals the cutoff price. If the cutoff price were $200 in the above example, then the two lowest bidders would each receive $200, despite the fact that one bidder was willing to accept a compensation of only $100 per acre. This uniform price auction would have been a multi-unit, low-bidder-wins version of the "second-price auction" discussed in chapter 26, whereas the discriminative auction is analogous to a first-price auction. Just as bidding behavior will differ between first- and second-price auctions, bidding behavior will differ between discriminative and uniform price multi-unit auctions, so it is not evident without experimentation which one will provide the greatest irrigation reduction for a given expenditure.

The initial experiments were run in May 2000 at Georgia State University, almost a year before the actual auction (details are provided in Cummings, Holt, and Laury, 2004). Early discussions with state officials indicated that discriminative auctions were preferred, to avoid the apparent "waste" of paying someone more than they bid, which would happen in a uniform price auction in which all bidders would be paid the same amount per acre. After some initial experiments, it became clear that a multi-round auction would remove a lot of the uncertainty that bidders face in such a new situation. In a multi-round auction, bids would be collected, ranked, and provisional winners would be posted, but the results would not be implemented if the officials running the auction decided to accept revised bids in a subsequent round. Bids that were unchanged between rounds would be carried over, but bidders would have the option of lowering or raising their bids, based on the provisional results. This process was perceived as allowing farmers to find out approximately what the going price would be and then to compete at the margin to be included in the irrigation reduction. A decision was made not to announce the number of rounds in advance, to make it more difficult to collude and to maintain flexibility in terms of the required time that involved coordination of bid collection at multiple bid sites in the Flint river basin.

State EPD officials observed some of the early experiments, and a conscious decision was made to provide an amount of context and realism that is somewhat unusual for laboratory experiments. This reverse auction in which low bids win was thought to be sufficiently unfamiliar that it would be worthwhile to explain the setup in terms of the actual irrigation reduction situation. The money amounts used were in the range of a dollar per acre, so the experimenters did not worry that participants would bring in "homegrown values" for what an acre would be worth. Atlanta students and local participants from South Georgia were recruited in groups ranging in size from 8 to 42, and were told that they would have the role of "farmers." There was a field test with more than 50 local participants and students bidding simultaneously at three different locations near Albany, to check the software and communication with officials in Atlanta. The final auction, conducted in March 2001, involved about 200 farmers in eight different locations.

Since most farmers had multiple tracts of differing sizes and productivities, each subject was given three tracts, with a specified number of acres and use value, the amount of money that would be earned if that tract would be irrigated. Participants were told that the land would not be farmed if the owners sold their irrigation permits. Consider a person with three permits, with acreage and use values per acre shown in the first two columns of table 28.1. The

Table 28.1. Sample Earnings Calculations for the Irrigation Reduction Auction

	Total Acres	Use Value (per acre)	Bid (per acre)	Auction Outcome	Earnings
Permit 1	100	100	120	Accepted	12,000
Permit 2	50	200	250	Accepted	_____
Permit 3	100	300	400	Rejected	_____

first permit covers 100 acres, and has a use value of 100 per acre, so the bidder would require an amount above 100 per acre as compensation for not irrigating. In this example, the bid was 120 per acre on permit 1, which was accepted, so the earnings are 120 (per acre) times 100 (acres), which equals 12,000, listed in the right-hand column. If this bid had been rejected, the irrigation would have resulted in earnings equaling the use value of 100 per acre times the number of acres, for a total of 10,000.

After experimentation with several sets of procedures and consultation with EPD officials, the focus was narrowed to a multi-round discriminatory auction in which all bids would be collected and ranked from low to high at the end of each round. Then, starting with the low bids, the total expenditures would be calculated for adding permits with higher bids. A cutoff for inclusion was determined when the total expenditures reached the amount allotted by the auctioneer. For example, if the amount to be spent had been set at 50,000, and if the lowest 20 bids yielded a total expenditure of 49,500, and a twenty-first bid would take the expenditure above this limit, then only 20 bids would be accepted. After the cutoff is calculated, the permits with bids below the cutoff would be announced as being "provisionally accepted." Then new bids would be accepted, which would be ranked, with a new announcement of which tracts had provisionally accepted bids. This process would continue until the experimenter decided to stop the auction, at which time the accepted bids would be used to determine earnings. Earnings on tracts with rejected bids would be determined by use values that were distributed to subjects at the beginning of the auction.

A typical session began with a "trainer" auction for colored writing pens. This practice auction was intended to convey the main features of submitting different bids for different tracts, which would be ranked, with some low bids being accepted and others not, without providing practice with the actual payoff parameters to be used in the auction. The trainer lasted about 45 minutes, and it was followed with a *single* auction for real cash earnings. The decision not to run a series of repeated auctions was made to mimic the field setting in which farmers would be participating in the auction for the first time. There was concern that farmers who knew each other well would try to collude, so participants in the experiment were allowed to collude on bids in any manner, except blocking access to the bid submission area (which was discussed in one

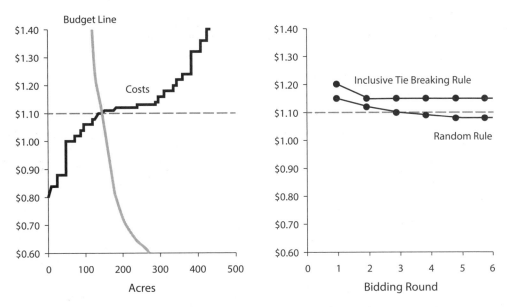

Figure 28.1. Results of Two Sessions with Different Tie-Breaking Rules. *Source*: Cummings, Holt, and Laury (2004).

of the pilot sessions). Some people discussed price in small groups while others made public speeches.

Figure 28.1 shows the results of two sessions with identical cost structures and a procedural (tie-breaking) difference to be explained subsequently. The line labeled "cost" shows the permit use values per acre, with each step having a width equal to the number of acres for the permit with a use value at that step. These opportunity costs, arrayed from low to high, constitute a supply function.

In each auction, there was a fixed total budget to be used to purchase permits. A low total acreage can be purchased at a high price per acre, and a high total acreage can be purchased at a low price per acre. This fixed budget then generates a curve that is analogous to a demand function. If B is the total budget available to purchase Q total acres at a price per acre of P, then all money is spent if $PQ = B$, or if $P = B/Q$, which generates the negative relationship between P and Q that is graphed on the left side of figure 28.1. The right side of the figure shows the average prices per acre of the provisionally accepted bids, by round. Notice that these price averages are close to competitive predictions, despite the fact that participants could freely collude.

One advantage of laboratory experiments is that procedures can be tested, and unanticipated problems can be discovered and fixed before the real auction. In one of the sessions shown in the figure, a participant asked what would happen if a number of bidders tied at the cutoff bid that exhausted the announced budget, and if there was not enough money in the budget to cover all bids at the tie level. This possibility was not spelled out clearly in the instructions, and the experimenter in charge of the session announced that all of those tied would be included as provisional winners, or as final winners if this were the final round. In the second round of this session, a tie arose at a price about 5% above the competitive level, and all tied bids were provisionally included. In the subsequent round, more bids came in at the tie level from the previous round, and this accumulation of bids at the focal tie point continued. The resulting payments to subjects, which were needed to include all tied bids, ended up being twice the budgeted amount. This would have been analogous to spending 20 million in the actual auction instead of the 10 million budgeted by the legislature!

The price sequence for the session with the inclusive tie-breaking rule is the upper line on the right side of figure 28.1. A second session was run later with different people and a random tie-breaking rule. The prices converged to the competitive level, as shown by the lower price trajectory in the figure.

A number of other procedural changes were tested and implemented. For example, it became apparent that low bids tended to increase after the announcement of the cutoff bid in each round. This pattern of increasing low bids is intuitive, since the low bidders faced less risk with bid increases if they knew about how high they could have gone in the previous round. In subsequent sessions, the list of provisional winners posted at the end of each round included the permit numbers, but not the actual bids. This procedure resulted in less of an upward creep of low bids in successive rounds, although high bids tended to fall as bidders scrambled to become included. These modifications in tie-breaking rules and the post-round announcement procedures were incorporated into the final auction rules.

The actual auction was conducted in April 2001, under the direction of Susan Laury and Ron Cummings, with the assistance of other experimental economists from several universities (Mark Van Boening, Lisa Anderson, Regan Petrie, and the author). The results of each round were collected and displayed to top EPD officials, who met in the Experimental Economics Laboratory at Georgia State, where most of the laboratory tests had been done. Almost 200 farmers turned up at 8 locations at 8 am on the designated Saturday morning, along with numerous television reporters and spectators. The procedures were quite close to those that had been implemented in the experiments, except that the EPD did not reveal what they intended to spend. After bids were collected, transmitted, and ranked, officials in Atlanta would discuss whether to stop the auction, and if not, how much money to release to determine provisional winners for that round. This non-fixity of the budget differed from the procedures used in the experiments, but it did not contradict any of the published auction rules. The increases in the provisionally released budget prevented significant drops in the cutoff bid, which was $125 per acre in round 4. The director of the EPD then decided to release more money to buy back more irrigation. After the bids were received for the fifth round, a cutoff bid to $200 was used, after which the auction was terminated.

The auction was a success. Bids were received on about 60% of the acres eligible for retirement from irrigation. In total, about 33,000 acres were taken out of irrigation, at an average price of about $135 per acre. A second auction was run the following year, using a sealed-bid mail-in form that also performed well (Petrie, Laury, and Hill, 2004). To summarize:

Georgia Irrigation Auction Results: *Experiments implemented a multi-round reverse auction with discriminatory price payments to winning bidders. Pilot tests were used to justify a random tie-breaking procedure and a practice of not announcing provisionally winning bid amounts to keep prices from rising above competitive predictions. These procedures were used in the actual auction.*

28.2 Cap and Trade of Emission Permits: Auctions Versus Grandfathering

There are two major ways to regulate greenhouse gas emissions: (1) the imposition of a "carbon tax" to be paid by emitting firms, and (2) a "cap and trade" program that distributes a fixed number of permits or "allowances," each good for one ton of carbon dioxide emissions. In principle, economists tend to prefer the price-based tax approach, but the determination of an appropriate social cost of carbon is an obstacle, and taxes get eroded over time due to inflation and political resistance. A carbon tax program in Australia, for example, was dismantled when the winning candidate for prime minister ran on an "axe the tax" platform. Environmentalists tend to favor the cap-and-trade approach, with a clear emissions quantity cap being determined as a rollback of total carbon emissions to a focal prior year, e.g., 1990. A fixed cap is more resilient to political or economic erosion. Regardless of how the permits are allocated, the "and trade" aspect of these programs helps correct misallocations and gives a boost to producers who are using cleaner technologies.

In 2007, the regulators in ten northeastern US states signed a memo of understanding that they would establish parallel cap-and-trade programs to cover emissions from electric power generation, known as the Regional Greenhouse Gas Initiative (RGGI). The immediate issue was how permits should be distributed. Most firms preferred "grandfathering," namely receiving permits for free with allocations based on historical emissions. This procedure would provide high allocations to firms that used coal-fired facilities, which emit about twice as much carbon as facilities that use natural gas. As permit prices rise, trading in a spot market would redirect permits as natural gas generation would expand in line with the added costs of emissions. An alternative to free allocations would be to auction off the limited supply of permits to the highest bidders, who would tend to be firms with the lowest costs (including costs associated with permit acquisition).

Regardless of the initial allocation, permits would be bought and sold in secondary "spot" markets. Firms would be required to acquire permits to cover their own emissions or face steep penalties. Regulators were concerned by warnings that making firms pay for permits might result in higher final product (electricity) prices. A broad-brush summary of the relevant economic theory is based on the idea that tradeable permits have *opportunity costs*, irrespective of whether they are purchased or acquired at no cost. Firms should consider opportunity costs when setting prices, and therefore, the initial allocation method would not affect product prices. This theoretical argument was greeted with skepticism (eyes rolled) in a 2007 open meeting of stakeholders and regulatory officials in New York City.

The opportunity cost argument can be illustrated with a simple example taken from the Veconlab Opportunity Cost experiment, with instructions:

- You have the capacity to produce and sell up to 3 units of a product.
- Each unit actually sold results in an additional fuel input cost: **$1.00** for unit 1, **$3.00** for unit 2, **$5.00** for unit 3.
- Each product unit that you sell also requires a single permit. You have been allocated **2 permits** at no cost in this round, and if you use additional permits, they must be purchased for **$3.00** each. If you sell fewer than 2 units of product and hence do not use all of your permits, you can sell them and earn **$3.00** for such sale.
- **Note:** The going product market price in this round is **$5.50** and all units that you sell will be sold at this price.
- Now choose a production quantity by deciding whether to produce or not for each of your 3 units of capacity.

At this point, the reader should decide whether to produce with 1, 2, or 3 capacity units, knowing that the production from each unit will generate a revenue of $5.50.

A possible reaction is to conclude that all three units should be produced since price exceeds marginal fuel costs for all units. But the third unit requires purchase of a permit for $3.00, so that unit's total cost of $5 + $3 is greater than the $5.50 price. What about the second unit? It can be produced at a cost of $3 without purchasing a permit, since there are two free permits provided. But producing the second unit will lower a person's earnings (problem 1). This is because using the second free permit entails an opportunity cost of $3, the price at which it can be sold. In fact, the optimal production decision is independent of whether or not any permits are obtained for free. To summarize:

Opportunity Costs: *In a cap-and-trade program, free allocations have an opportunity cost that is determined by the market price of those permits, and optimal production decisions are not affected by whether permits are distributed for free. This is because optimal production is based on comparing marginal revenues and all marginal costs, explicit and implied opportunity costs. In particular, requiring regulated entities to purchase emissions permits should not result in higher product prices due to mistaken "pass through" of permit purchase prices.*

The intuitive appeal of the incorrect pass-through argument motivated the Goeree et al. (2010) study. There were two treatments, with initial allocations of "permits" to subjects being either free or sold at auction. In both cases, initial allocations were followed by a spot market in which subjects could buy or sell permits, prior to submitting bids to sell units produced (electricity) in a product market with a downward sloping demand curve. Production costs were randomly drawn from uniform distributions that resulted in approximately linear supply functions, subject to random variation. Since product price could be affected by the magnitudes of cost draws, it was important to use the same cost draws for each treatment. Because a particular set of random draws might be biased in favor of one particular allocation method, the researchers used different sets of random cost draws for each pair of grandfathering and auction sessions. Each session consisted of six subjects who made trading and production decisions in nine market periods, with an initial allocation (auction or free), a spot market for permits, and a product market that allocated demand to sellers with the lowest price offers.

The equilibrium calculation involves simultaneously determining the price of permits and the price of the final product that is produced (analogous to electricity). Figure 28.2 illustrates the solution in the product market. If permits cost nothing and are freely available, then the market supply for the product depends on the costs of high and low users. The three high users (high emitters) in the experiment have unit costs that range from $4 to $8, and the uniform distribution of these costs results in a supply function that rises from $4 to $8, as shown by the thick gray line. These "high users" are analogous to facilities that burn coal and require two permits per production unit, but when permits are free, there are no permit costs and the supply is completely determined by production costs, which are lower for coal. The experiment design featured an equal number of three low users (who would emit less and only require one permit per production unit). The costs of low users are higher, and range uniformly from $8 to $12 in the experiment, which results in the dark line supply segment on the lower supply function in figure 28.2. To summarize, think of the lower supply line in the figure as being composed of steps, with high

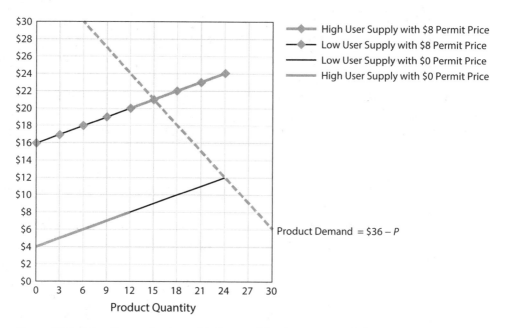

Figure 28.2. Price Determination with Cap and Trade. *Note*: With free permit allocations, the high user supply (thick gray line) is on the left side of the lower supply function, and the low user supply (thin black line) is on the right side of the lower supply function. The $8 equilibrium price of permits raises the marginal cost lines for "low users" (who only require a single permit) by $8, and it raises the marginal costs of "high users" by $16 since they require 2 permits to cover emissions for each product unit produced. These cost increases move the high user supply (thick gray line) from the left side of the lower supply line to the right side of the upper supply line. The upper supply function intersects market demand at a product price of $21, with 15 product units produced, mostly by low-emitting users.

emitters operating on the lower cost steps and with low emitters operating on the higher, upper-right side cost steps. When permits are free, the intersection of this lower supply function with the market demand determines a product price of $12 and a production quantity of 24, with 12 product units produced by low users and 12 by high users. The total emissions would be 36 (12 from low users and 24 from high users who emit twice as much).

Emissions are restricted in the experiment by limiting the number of permits to 18, which is half of the 36 permits that would be demanded at a $0 price. This cap will determine the permit value. It can be shown algebraically that the equilibrium permit price is $8. This outcome can be illustrated graphically from figure 28.2. With an $8 permit price, the cost for operating each production unit rises by $8 for low users, so the black line supply segment shifts up by $8. But the cost of operation rises by $16 for each high-user production unit, since emissions are 2 units and permits cost $8. Therefore, the gray line segment of the lower supply function shifts up by $16, which pushes it to the right side of the upper supply function with an overlay of diamond symbols. The new equilibrium product price is \$21, with 15 units produced (12 by low users who use 12 permits, and 3 by high users, who use 6 permits). The total permit usage in this equilibrium matches the cap of 18 permits.

Recall that each session of the experiment involved six subjects—three low users and three high users. Each producer was given 4 production units, so production would be 24 units if permits were freely available. In the grandfathered treatment, each high user was endowed with four permits (enough to operate 2 of the 4 production units) and similarly, each low user was given two permits, which requires a total usage of 18 permits. In the auction treatment, the 18 permits were sold in an initial auction at a uniform price determined by the highest rejected bid. After the initial allocation by auction or endowment, permits could be bought or sold in a spot market, which was run as a two-sided call market with a single clearing price determined by the intersection of bid and offer arrays. After the call market cleared, each of the six producers could make an offer to sell product at a specified price. The intersection of the offer price array and the exogenous product demand ($Q = 36 - P$) determined a production quantity and product price. Sellers with rejected offers to sell in the product market could "bank" their permits for future periods.

The motivation for the experiment was the argument that making firms buy permits in an auction could result in higher product prices if firms incorrectly "pass through" permit costs. The counter-argument was that if permits could be bought and sold, then the permit market value determines an opportunity cost that firms would build into their product price offers. Hence, the implication of economic theory is that product prices should reach equilibrium levels of \$21 *in both treatments*, and in particular, should not be higher in the auction treatment.

The main result, shown in figure 28.3, is that forcing producers to pay for permits in the auction treatment did not raise product market prices, shown as dark lines at the top of the figure. In fact, the product prices for the grandfathered treatment (dots) are a little above the product prices for the auction treatment, although this difference is not statistically significant. The null hypothesis of no treatment difference was done by comparing average product prices in the nine sessions for the auction allocation treatment with nine average product prices for the nine sessions in the grandfathered allocation treatment. The actual comparison was done on matched pairs, since the first session run in each treatment used one set of random cost draws, the second session in each treatment used a second set of random cost draws, etc. Therefore, a Wilcoxon matched-pairs test was used for the nine matched pairs of sessions (see problem 2).

In the auction treatment, the auctioned permit prices track the \$8 prediction in figure 28.3, which was derived from an analysis of the permit supply and demand. The spot market permit prices, as shown by the gray lines, are somewhat above the predicted \$8 level, especially in the grandfathered treatment (dots).

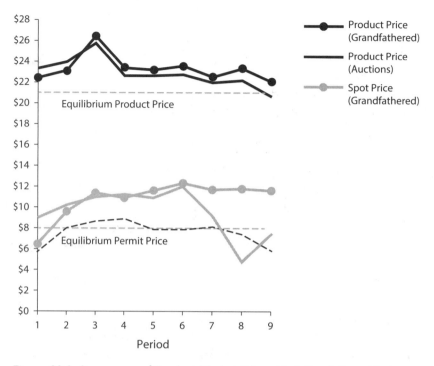

Figure 28.3. Sequences of Product Market Prices (dark lines), Spot Market Prices (gray lines), and Permit Auction Prices (dark dashed line). *Key:* Prices in the grandfathered allocation treatment are indicated by dots.

The advantage of a laboratory experiment is that it is possible to look closely at the trading process to determine causal mechanisms. In this case, offers to buy and sell in the spot market were fairly close to permit values, with one exception. The offers made to sell permits in spot markets following grandfathered allocations were well above permit values. This is because the grandfathered allocations target two-thirds of the free allocations to high users (based on high historical emissions), and the equilibrium involves those high users only using 6 of the 18 permits, as indicated by the earlier discussion of figure 28.2. The prediction follows that high users who receive free permits will sell about half of those permits to low users in the spot market. The actual offers to sell made by high users correspond to a supply function that is shifted upward and to the left (above use values), as if they were withholding supply in an attempt to raise price.

This exercise of market power by high users explained the higher spot prices, especially in the final periods of this grandfathered treatment. Those high spot prices, in turn, probably kept product market prices high, although product prices tended to drift down toward the supply and demand prediction of $21 in the final periods in both treatments. To summarize:

Auctions versus Grandfathering Experiment: *Product prices do not increase when firms are required to purchase permits instead of receiving them for free.*

In terms of efficiency, the consumer surplus was about the same in both treatments, which is not surprising given the similarity of product price averages. The main difference is that seller profit (producer surplus) was only about a quarter as high in the auction treatment, with the difference going to the government as auction revenue. Since 2008, the RGGI system has generated large amounts of auction revenue, about $2.7 billion in the first decade of auctions, even though permit prices were unexpectedly low for many years. Those low prices were caused by several years of favorable weather conditions (warm winters and cool summers) and by the 2008 recession that depressed electricity demand. In addition, the discovery of large natural gas reserves lowered fuel costs and helped reduce the impact of the emissions cap. Nevertheless, auction revenues have been substantial and have been used for strategic energy initiatives, among other things.

28.3 Consignment Auctions and the California AB32 Program

The California cap-and-trade program for greenhouse gas emissions implemented the same uniform price auctions that RGGI had been using on the East Coast. In order to soften the impact on firms' revenues, however, free allocations were provided to many classes of firms, with the provision that the free allocations would be gradually reduced over time. The auctions differed from the RGGI setup in a number of ways, but one especially interesting feature was the ability of firms to "consign" permits for sale in the auction. Consigned permits that were sold in the auction would return revenue to the seller calculated as the uniform clearing price times the number of permits sold.

Even more interesting is the case of forced consignments. In particular, regulated utilities that received free allocations were *required* to consign their free allocations to the auction and then bid to buy back what permits they needed for operations. For example, a firm that consigned 1,000 permits to the auction and then bid $10 per permit for 1,000 permits would end up selling the permits and buying nothing if the auction closed above their $10 bid, which would be rejected. Alternatively, the firm would end up with the original number of permits (1,000) if the auction closed below their bid, which would be accepted. If the auction closed at $9, then the firm would receive $9,000 for permits consigned and would pay $9,000 for the permits purchased, and the firm's final position would be identical to the initial position, 1,000 permits and no net payment. In other words, the firm's bid served as a reserve price with the message: "sell 1,000 permits if you can get a price at $10 or higher, otherwise we'll keep the permits." Consignment makes it easy for firms with high free allocations, as determined by historical emissions, to sell some or all of them to more efficient or lower-emissions producers.

One concern with consignment is that those with high endowments who must consign will be net sellers, and may try to manipulate the auction-clearing price to achieve higher consignment revenues. Then a high auction price signal might elevate prices in the relatively thin spot market that would follow. The counter-argument is that consignment enhances auction quantities (higher market "liquidity") and provides better price discovery as the buyback bids of consigners convey value information and as more bidders will participate in the auction, instead of relying on post-auction spot markets. This section reports on an ongoing experiment by Holt and Shobe (2017) that is motivated by these issues.

The experiment involves scaling up the parameters used in the auctions-versus-grandfathering experiment, with twice as many bidders (6 low users and 6 high users) and twice as many permits (36 instead of 18). There was no product market, so the product price was not endogenous, but rather, was set at the $21 level that was an equilibrium in the prior setup. The cost distributions matched those in figure 28.2, but the actual random cost draws generated random permit values that changed from period to period. For a low user with a cost draw of $10, for example, the permit value would be the earnings from using this permit, i.e., $21 − $10 = $11. For a high user, the permit value would be the difference between price and cost, divided by 2, which is the required number of permits. So, a high user with a cost draw of $7 would have a permit value of (21 − 7)/2 = 14/2 = $7. As before, each firm had four production units with randomly determined costs for each. The left side of figure 28.4 shows the permit values (steps composed of thick gray line segments) for the final period 12. This value array intersects the vertical auction supply at 36 permits to determine a "Walrasian" prediction of $8, as indicated by the horizontal dashed line. The lightly shaded rectangle is referred to as the "Walrasian revenue," which provides a benchmark for comparisons with actual auction revenues.

As before, the initial endowments were skewed toward the high users, who received 24 permits in total, as compared with 6 for low users, with 6 set aside for the auction in the no-consignment treatment. In contrast, all free allocations were consigned in the forced consignment treatment, so the auction quantity was 36. The motivation for the experiment was to determine whether consignment would result in lower prices (due to the beneficial liquidity effects) or higher permit prices (due to the exercise of market power by net sellers).

The triangles on the right side of figure 28.4 track the average auction prices across all sessions, and the dots ("spots") track the spot prices. The dark triangles and dots for the forced consignment treatment are significantly lower than the gray triangles and dots for no-consignment. There is no evidence that net sellers were able to push auction prices above the Walrasian (dashed-line) prediction. This result is roughly consistent with the earlier Shobe, Holt, and Huetteman (2014) experiment, in which consignment worked smoothly and prices were not skewed away from Walrasian predictions. In contrast, that earlier paper did document adverse effects of "holding limits" on the permit banks for individual bidders.

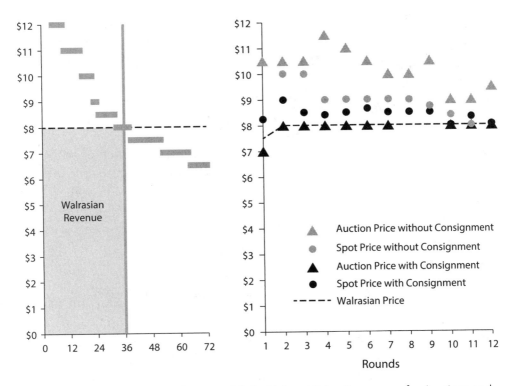

Figure 28.4. Permit Values and Demand (left side) and Price Sequences for Auctions and Spot Markets: Consignment (black markings) and with No Consignment (gray markings). *Key*: Triangles indicate auction prices, and dots indicate spot market prices.

Consignment auctions may have important applications in markets for commodities like water rights in which trading is often quite limited. For example, surface water rights typically have a "use it or lose it" aspect that encourages farmers to keep irrigating, even if the crop production is of low value relative to other operations downstream. If the use-it-or-lose-it feature were changed to use-or-sell-it (or lose it), then the added liquidity offered by consignment auctions would be quite attractive. Zelland (2013) discusses the performance of some initial irrigation consignment auction experiments and the implications for water resource management. Similarly, Ledyard and Szakaly-Moore (1994) discuss the application of consignment auctions to the sale of pollution permits. Both of these papers compared consignment auctions (forcing sales through a uniform price auction) with a standard double auction in which buyers and sellers submit bids and asks in real time, with good price information as in a stock market. Not surprisingly, the double auction does as well or better, since double auctions are notoriously efficient. On the other hand, regulators would be very reluctant to set up continuous-time, centralized markets for trading permits. In the absence of centralized water markets, it is easier for farmers to consign sales into an auction instead of arranging trades with individual buyers. In any case, auctions are powerful mechanisms for providing liquidity and price discovery on a large scale, and consignment is convenient in that it transforms a standard sales auction into a two-sided auction.

28.4 Alternative Auction Methods: Collusion, Loose Caps, Price Discovery

There are many different ways to run a multi-unit auction with N permits. Three of these will be considered in this section: discriminatory, uniform price, and clock auctions. As noted above, in a discriminatory auction, sealed bids are collected and the N highest bids are accepted, with each winning bidder paying their own bid price. This "pay as bid" auction is discriminatory in the sense that different bidders end up paying different amounts, as was the case in the Georgia Water Auction (although that was a "reverse auction" in which the low bids were winners). A uniform price auction is also a sealed-bid procedure with sales made to the N highest bidders, but the winning bidder only has to pay the highest rejected bid. (Alternatively, the cutoff could be set equal to the lowest accepted bid.) So, if three permits are to be sold and the bids are 55, 44, 33, and 22, then the top three bidders only have to pay 22. The same rules apply if a single bidder can submit multiple bids, as is typically the case.

Unlike the other two auction methods, a *clock auction* evolves in a series of stages, with an announced provisional price at each stage. At each stage, bidders indicate how many permits they desire to purchase at the current provisional price, and if this revealed "demand" exceeds the auction quantity N, then the provisional price is raised by a predetermined increment, as if by the hands of a clock. The clock price stops rising as soon as the demand is less than or equal to N. On the left side of figure 28.4, for example, if the clock started at $6.50, then bidders would be willing to purchase all 72 units, and the clock price would rise, say to $7. The actual sum of quantity bids at $6.50, of course, might fall short of 72 if some bidders are withholding demand. If bidders fully revealed demand at $8, then the thick dark step at that point indicated that there would be a small excess demand, so the clock price would go up one notch and then stop.

In any uniform price auction, bidders could attempt to manipulate the market-clearing price. Suppose that two units are being sold, and the bids are 5 and 3 from one bidder, and 4 and 2 from the other. The ranked bids for the two units are, therefore, 5, 4, 3, 2, and the market-clearing price determined by the highest rejected bid would be 3, with each bidder purchasing one unit at that price. By lowering the marginal rejected bid of 3, a bidder could reduce the amount paid on the winning bids. Of course, a bidder would not know ex ante whether a particular bid will end up being the highest rejected bid, but strategic bidding in this case often involves bidding lower for a second unit that has a lower marginal value to the bidder. Such strategic bidding has been observed in both laboratory and field settings. For example, List and Lucking-Reiley (2000) sold thousands of dollars of tradable sports cards in two-person, two-unit uniform price auctions. These auctions were conducted at a display table at a sports card collectors' convention, with cards that had identical book values. By comparing bids, the authors concluded that bidding is typically below value on the second unit. In a follow-up field experiment with sports cards, this strategic demand reduction was less pronounced with larger numbers of bidders (Engelbrecht-Wiggans, List, and Reiley, 2006). This type of profitable demand reduction would be less likely to occur in a single auction, or with lots of bidders and randomly changing permit values.

Demand reduction can also occur in a clock auction if a person reduces their quantity bid in order to stop the clock. Suppose that two units are being auctioned, and that private values are 5 and 3 for one bidder, and 4 and 2 for another. If the initial starting price is 2 and each bidder indicates a bid quantity of two units, the clock price would rise. But suppose that bidders guess from experience that they will only end up purchasing a single unit, and therefore, they reduce their demands to 1 right away. In this case, the clock will stop at 2 and each bidder would purchase a permit for which they have a high value (5 for one bidder and 4 for another) at this low reserve price.

A clock auction was first used for tradable emissions permits in the 2004 Virginia NOx auction. The prizes were thousands of one-ton permits, and bidders indicated their demand quantities as the clock price was raised, until demand fell to the available number of permits. A decision was made not to reveal excess demand to bidders when the clock advanced to the next stage, and the auction administrator (Bill Shobe) believed the procedure prevented the auction from stopping prematurely. The clearing prices were 3–6% above the current spot prices for those permits. It is possible that the success of this auction was due to the values bidders placed on being able to bid for large blocks of permits instead of having to buy "bits and pieces" in the spot market. This auction, which was considered a success in terms of transparency, speed, and revenue generation, was implemented after a series of laboratory experiments that demonstrated the feasibility of the clock format (Porter et al., 2009). Interestingly, the state government RFP (request for proposals) for administering the auction did not mention a clock format, since this format had never been used for selling emissions permits. The successful brokerage firm was the only one to propose a clock auction, which it had learned about through an unpublished auction theory paper that turned up in a routine Internet search!

Several years later, a team of experimental and environmental economists at the University of Virginia and Resources for the Future was formed to design the original RGGI auctions. The initial tests of these mechanisms involved groups of 12 bidders, in a simple setup with no banking of permits and new random cost draws (that determine random permit values) in each round. The setup was fairly competitive in the sense that the auction quantity was only about 66% of the number of permits that bidders could use at full capacity. All three auction methods performed well in this competitive environment, with revenues at about the Walrasian benchmark level and efficiencies close to 100%. (An efficiency of 100% would mean that the *N* bidders with the highest values obtain the permits.)

The competitive conditions in the initial tests were relaxed in a second wave, with fewer bidders, some opportunities for collusion, a loose cap (90% of bidder needs), and an unanticipated shift in the demand for permits. Each session consisted of six bidders with fixed matchings and random cost draws that stayed the same. Moreover, auctions were followed by spot markets to give bidders a chance to acquire permits if collusion failed. These sessions consisted of a longer series of auctions (12) to give collusion a chance to develop. Half of the sessions were done with no communication opportunities, and half were done with a short period in which bidders could post comments in a chat room.

The most common plea in the chat sessions was for people to bid the reserve price, which was $2 per permit. There was some frustration, and one person even remarked, "Economics is evil." Another quipped, "every man for himself, this is war." One apologized: "sorry guys, I cheated every time," and another responded, "we all did." Group dynamics were important in some cases, especially when good humor and trust could be established. The most effective collusion tended to develop in the clock auctions. One sequence of comments contained the following:

> "Again, bid for fewer permits earlier on so we can get permits cheaper."
>
> *"So why doesn't everyone bid exactly the same amount that we ended last round with, since we keep getting the same clearing price?"*
>
> "This will go 5× faster and we will all make LOTS of money if everyone just cooperates the first time."
>
> "Well that was thrilling."

The perceptive subject's suggestion, shown in italics, was to *start* with bids that equal the permit quantities bidders actually purchased in a previous clock auction. Since the final award quantities exactly match the auction quantity, this suggestion would stop the clock at the low starting price, i.e., at the reserve price.

Figure 28.5 shows the round-by-round series of auction price averages, reported in Burtraw et al. (2009). The prices in the communication treatments on the right side are generally lower, as expected, with the exception of the discriminatory auction. Overall, the uniform and discriminatory auctions perform about the same in terms of average prices, although the discriminatory auction is a little more resistant to collusion. On the other hand, the clock auction yields significantly lower prices in both treatments, communication or no

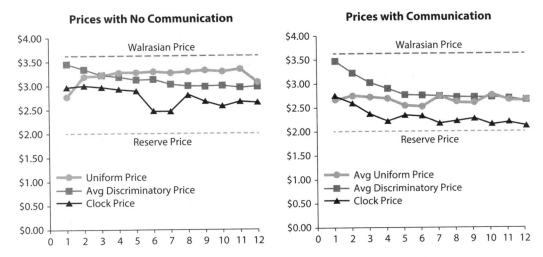

Figure 28.5. Average Auction Prices for Uniform Price, Discriminatory, and Clock Auctions.

communication, than is the case for the two sealed-bid auction formats. In fact, more than half of the clock auctions stopped at the reserve price or one bid increment (15 cents) above.

In addition, the clock auction did not do any better than either of the sealed-bid auctions in other tests involving a "loose cap" or an unanticipated demand shock about halfway through the auction sequence. The discriminatory auction raised a little more revenue than the uniform price auction in the loose cap treatment (Shobe et al., 2010), but it did worse in terms of tracking an unanticipated demand shift in a "price discovery" test (Burtraw et al., 2010).

Factors other than price and seller revenue may matter. In particular, participants in a stakeholder meeting complained about the day-long series of bid rounds in the earlier Virginia clock auction. Several other stakeholders indicated that they liked the idea of a uniform price, which prevents their competitors from acquiring permits at a lower price than they ended up paying. The final recommendation was to adopt a uniform price auction, since it was more resilient to collusion than the clock auction and simpler to implement. To summarize:

Emissions Permit Auction Experiments: *All three auction formats produced high revenue and efficiency measures in competitive conditions with large numbers of bidders and tight caps. With less competitive conditions, the discriminatory and uniform price auctions generated roughly equivalent auction permit prices, with the discriminatory auction being a little more resilient to the effects of collusion. Both of these sealed-bid auctions generated higher prices and revenues than the clock auction, both with and without communication opportunities. More than half of the clock auctions stopped at the reserve price or at one bid increment above it.*

28.5 Extensions: Supply Buffers, License Plates, and Securities

This section will briefly summarize studies that indicate the broad applications of multi-unit auction designs.

Supply Buffers

Policy makers and regulators who set emissions caps always struggle with the issue of how tight to set an emissions cap. If it is too loose, permit prices will fall and firms will have little incentive to invest in emission abatement. If the cap is too tight, the high permit prices might stress firms' earnings or result in higher electricity prices that harm consumers. The problem is complicated by unanticipated changes in climate and economic conditions that change permit demand. Some of these demand shocks are temporary, but some, like the discovery of new natural gas reserves, have long-lasting effects.

Figure 28.6 represents an extreme case in which permit demand is either high or low, as shown by the two downward-sloping lines. If these two conditions occur with equal probability, then the average demand would be the dotted line in between, which intersects the vertical supply at a permit price of $10. When demand is high, the outer demand is relevant, and the auction clearing price would be high. But when demand is low, the clearing price would be $2, as indicated by the "□" mark at the reserve price in the figure. Notice that at this low price, the low demand is less than the auction quantity, but the producers are likely to buy all permits offered if demand is expected to recover. These extra purchases may end up creating an excessive permit bank that will depress permit bids in future periods.

The vertical segment of the auction supply in figure 28.6 has a horizontal segment on the left, which corresponds to a *reserve price* that is supported by reducing the auction quantity if the clearing price was below $2. Conversely, the horizontal segment at the high price of $22 on the right side of the figure represents the effects of a *price cap* that is maintained by releasing permits into the auction to prevent the clearing price from rising above the cap. But the vertical segment in between the reserve price and the cap will cause demand shifts to have sharp price effects, although these could be offset by banking unneeded permits for later use (problem 4).

Now suppose that the regulator adopts a rule that up to five permits will be withheld from the auction if the clearing price is below $6. This rule essentially adds a step at $6 to the auction supply, as shown by the dashed line in the figure. The intersection of the low demand with this dashed line step is at $6 and a quantity of 30. The intuition behind the policy is that, in low demand conditions, a quantity restriction will soften the price impact, thereby reducing price volatility, and prevent a glut of permits from accumulating in the market. The effect of the step is to make the auction supply look more like an upward sloping supply.

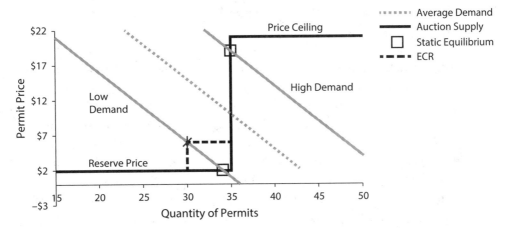

Figure 28.6. A Market for Permits with High and Low Demand Shocks, with a Reserve Price of $2 and a Supply Buffer (ECR) shown by the Dashed Step. *Note:* When demand is low for several periods in a row, firms will not buy to bank for the future, and the price will fall. The dashed line lets the regulator remove permits from the auction in order to prevent the clearing price from falling below $6. Thus, the benefits of low permit demand are shared by issuing fewer permits (lower emissions over time) but holding the price of permits up, in order to preserve the benefits of offering an incentive to continue abatement efforts.

In September 2017, the Regional Greenhouse Gas Initiative instituted this single-step supply buffer, referred to as an Emission Containment Reserve (ECR). Permits that remain unsold to support the price at the step will be retired permanently, to prevent the accumulation of excessive permit banks in the hands of electricity producers. The policy is intended to share the benefit of reduced permit demand between environmental interests (reduced emissions) and economic interests. The policy was adopted after the RGGI administrators considered a memo with supporting arguments that were based, in part, on preliminary laboratory experiments done by the members of the original RGGI design team. A natural extension would be to introduce multiple steps.

Shanghai Auction for License Plates

Several cities in China have recently adopted policies to control auto traffic by limiting license plates. In Beijing, these plates are distributed by lottery, which generates no revenue and creates some ancillary costs. In contrast, Shanghai sells licenses in regular auctions that raise about a billion dollars a year. The revenues have been used to improve the subway system, among other things. This program can be thought of as "cap and *no* trade" since it is not legal to resell licenses, although that does happen sometimes. The prohibition against resale, intended to prevent speculation, also enhances the importance of using an auction procedure that is efficient in the sense that high-value users obtain the licenses.

The auctions began in 2003, using a discriminative "pay as bid" procedure. If 10,000 plates were offered, for example, then the 10,000 highest bidders would win and pay their own bid amounts. The auctions worked well, but by 2008, the price of a license had approximately tripled, reaching a level that matched the cost of a new economy car! The high prices generated some concern, and in 2008 officials decided to switch to a different procedure that was designed to lower prices. The new Shanghai auction begins with an initial bid window of several hours in which anyone can submit bids online. The initial bids are ranked and the lowest accepted bid is announced. Then there is a bid revision phase in which each bidder has the option to revise their bid or not, subject to a constraint that the revised bid must be within a very narrow price band around the current lowest accepted bid. Bidders can submit at most two bid revisions prior to the final closing time, so there tends to be "sniping" at the end of the bid revision phase. As the permissible bid range moves over time, previously accepted bids may get pushed into the rejection region. With lots of bidders and limited numbers of licenses and bid revisions, many bidders are unsuccessful.

The new auction mechanism resulted in a dramatic reduction in prices, but the effect was clouded by city officials offering about twice as many licenses for sale after the new auction rules were applied. As the Chinese economy continued to grow quickly, license prices climbed back up to former levels, and some price caps were imposed.

Holt and Liao (2016) report an experiment that controls for external factors like the number of licenses and changes in valuations due to income growth. There were three treatments in the basic design: a discriminatory pay-as-bid auction matching the pre-2008 setup, a uniform price auction in which all winners pay the same price that is set to equal the highest rejected bid, and the Shanghai auction with the two-stage bid revision process. Each treatment consisted of three independent sessions (12 subjects and 10 auctions, with randomly generated private values). Table 28.2 lists the average auction revenues and efficiency measures for each session, where efficiency is calculated as the ratio of the sum of winning bidders' private values to the maximum value sum that would have resulted from awarding licenses to those with the highest values.

The revenues for sealed-bid auction formats, uniform and discriminatory, are equivalent, with no significant difference. Those auction revenues approach the Walrasian benchmark, calculated in a manner analogous to the area of the rectangle on the left side of figure 28.4. The main result is that revenues are *much* lower in the Shanghai bid revision auction, a difference that is both economically and statistically significant (problem 5). This intended revenue reduction comes at a cost in terms of efficiency, which is reduced by about 10 percentage points below the near-perfect levels for the sealed-bid auctions. An

Table 28.2. Revenues and Efficiencies by Session and by Treatment

	Session Revenues: Treatment Average	Session Efficiencies: Treatment Average
Uniform Price (sealed bid)	74.6, 76.6, 82.2: **78.5**	96.3, 99.2, 96.6: **97.4**
Discriminatory (sealed bid)	74.9, 73.6, 80.5: **76.3**	97.1, 97.6, 98.3: **97.6**
Shanghai Auction (limited bid revisions)	46.6, 51.5, 57.3: **51.8**	85.7, 84.6, 88.1: **86.1**

Source: Holt and Liao (2016).

analysis of individual bid patterns provides an explanation. When a bidder decides to revise a bid to a level in the narrow price band around the current cutoff, it is typically a bid reduction that reduces revenues. But limits on bid revisions and last-second sniping have the effect that bidders with the highest values are not necessarily those who end up winning licenses, which reduces efficiency. Analysis indicated a low correlation between bidder values and final bids in the narrow revision band, which is much lower than the bid-value correlations for the other auction formats. To summarize:

> **Shanghai Auction Experiment Results:** *The two-phase Shanghai auction with limited bid revision opportunities yields lower revenues and efficiencies than the sealed-bid Uniform Price and Discriminatory auctions that have no bid revision opportunities. Bidder competition and constraints imposed in the bid revision phase of the Shanghai auction lower revenues (as intended), but efficiencies are also lowered as a byproduct.*

Reference Price Auctions for Mortgage-Backed Securities

In late September 2008, as the US economy was in freefall, Congress passed the Troubled Asset Relief Program (TARP). TARP authorized a massive Treasury purchase of "toxic" mortgage-backed securities, with a budget that exceeded $500 billion at one point. The US Treasury immediately began making plans to set up a reverse auction for those purchases. There were more than 8,000 different securities (identified by CUSIP numbers), most of which were not being actively traded and were of questionable value. With the upcoming November election and rapidly deteriorating conditions, there was no time to run separate reverse auctions for each CUSIP, but a combined auction that made purchases from the lowest sale offers would have resulted in mostly "junk" purchases. The lead design team, working with Olivier Armantier at the New York Federal Reserve Bank, proposed using relative value estimates to normalize bids from various securities in order to create a level playing field and enhance competition between sellers of different securities. For example, if security A was deemed to be half as valuable as security B, the reference prices would be 1 for A and 2 for B. If an offer to sell A was 5 and an offer to sell B was 8, for example, the normalized bids would be 5/1 and 8/2, or 5 and 4. Thus the higher sell offer on B (normalized to 4) would be accepted before the offer of 5 on A, even though the unnormalized offer on B is higher. However, there were issues with rapid software procurement, and significant risk associated with a massive auction just weeks before a presidential election. The TARP auction was suspended in late October, although this was not announced publicly due to the volatility of the stock market at that time. (The design team received a phone call with a suggestion to "take a vacation.") The team efforts switched to setting up an experiment that would evaluate the effectiveness of reference price auctions, should such an auction ever be used in the future (Armantier, Holt, and Plott, 2013).

The situation was reversed a decade later when the federal government owned about $4 *trillion* in mortgage-backed securities that were acquired after 2008, mostly through direct negotiations with banks. In late 2017, the government started the process of unwinding this portfolio, primarily by not replacing assets that mature. One alternative, of course, is to sell securities. Armantier and Holt (2017b) report an experiment that evaluates reference price *sales* in a *forward auction* in which high bids win. As before, the proposal envisions combining multiple securities in each auction, with bids to purchase being normalized by dividing by reference price relative value estimates. This normalization process is used to rank bids for different securities for inclusion, up to a point at which desired sales revenues are reached. The experiments also evaluate a procedure in which reference prices are *endogenously* adjusted on the basis of bids received in the auction, before bids are normalized. The endogenous reference price procedure improved auction outcomes, even in a treatment in which bidders were informed about it. This general approach could be implemented even if normalizations are done in the background after bids are received, without following explicit pre-announced procedures (Klemperer, 2010).

Chapter 28 Problems

1. Consider the example from the Opportunity Cost experiment summarized by the instructions insert in section 28.2, and calculate seller earnings when the second unit is produced and when it is not.

2. The table below shows matched pairs of sessions that used the same random number "seed" to generate random values. For example, in the first column for seed 1, the two sessions used the same random value sequences, but one of the sessions was done with a "grandfathered" treatment, and the other was done with an "auction" treatment. The numbers to the right show average prices by session. Note that the two treatments for seed 6 had identical average prices,

so these two will be removed from the analysis. The difference row shows treatment differences. Here notice that seeds 5 and 7 yield the same treatment difference of 0.44. These are the two smallest differences in absolute values, so they would get ranks 1 and 2, but since there is a tie, the ranks are recorded as 1.5 for each. Fill in the remaining entries in the table, absolute values of differences, ranks, and signed ranks, and compute the sum of signed ranks (adding together all of the numbers you end up with in the bottom row). Use table 13.6 from chapter 13 to show that the sum of signed ranks is less than the cutoff for $N = 8, p = 0.05$ (remember we removed one pair), and hence, that the null of no treatment effect cannot be rejected at the 5% level.

Seed No.	1	2	3	4	5	6	7	8	9
Auctions	23.00	23.33	22.67	23.22	23.67	23.56	24.44	22.44	23.89
Grandfathered	21.33	21.44	23.44	22.67	23.22	23.56	24.00	23.56	22.89
Difference	1.67	1.89	−0.78	0.56	0.44	0.00	0.44	−1.11	1.00
Abs. Value			0.78	0.56	0.44	0.00	0.44		
Rank			4	3	1.5	−	1.5		
Signed Rank			−4	3	1.5	−	1.5		

3. (non-mechanical) Speculate on what types of auction procedures might be considered by a seller to counter possible bidder collusion, or to take advantage of strong bidder risk aversion.

4. Consider the case of equally likely high or low demand lines, shown in figure 28.6. With banking, firms can save unused permits in low-demand periods with low use values. If the permit price is $10, what horizontal difference would represent permits sold (on the vertical auction supply) but not used? Conversely, if demand is high and banked permits are used in addition to those purchased that period, what horizontal difference would represent this drawdown of banked permits if the permit price is $10? Finally, explain how the permit banking could stabilize prices at $10 in all periods, with permits being purchased and saved in low-demand periods and being taken out of banks in high-demand periods. In other words, explain how banking can remove price variability when high- and low-demand conditions are known to be equally likely.

5. Consider the session auction revenues for the discriminatory and Shanghai auction treatments in table 28.2. What is the minimum of the binary win counts to be used for a Mann-Whitney test to evaluate the significance with which the null hypothesis of no effect can be rejected? Consult the Mann-Whitney table in chapter 13 to determine the p value.

6.   (non-mechanical) What does the argument you constructed in your answer
      to problem 4 imply about the effect of the ECR dashed-line step in figure 28.6
      (when bidders know that high and low demands tend to balance out)? Instead
      of having each demand be independently and randomly determined to be
      high or low with equal probabilities, what alternative demand or information
      structure might make for an interesting test of the ECR?

29

Combinatorial and Two-Sided Auctions

Markets are marvelously efficient in terms of aggregating preference intensities and arranging efficient patterns of production and trade. In some cases, however, it is natural to consider allocation methods that provide more options than are available in naturally occurring markets. This chapter considers mechanisms that let participants bid for combinations of licenses or commodities. For example, the US Federal Communications Commission (FCC) conducts auctions for communications bandwidth at diverse locations and frequency bands. These auctions have generally been run simultaneously for distinct licenses, with bid-driven price increases. An alternative to the standard auctions for separate licenses would be to allow bidders to submit bids for combinations or *packages* of licenses.

Packages can be desirable when bidders' values for combinations are higher than the sum of the separate values for the component licenses. If individuals bid on both local licenses and regional packages, a "clock" mechanism could raise the prices on licenses for which multiple bidders have submitted a bid, either singly or as part of a package bid. For example, suppose that one bidder bids on licenses A, B, and D singly and another bidder bids on an ABC package, with no individual bids received on C. Then the prices of A and B would be raised, since those licenses are targeted by two bids. But with only one bid on D and one license at location D, there is no excess demand and its price would not be raised. In this manner, bidders can indicate whether they are still willing to buy after prices are incremented in response to excess demand. Laboratory experiments have been used extensively to help design and evaluate alternative package bidding procedures, including the hierarchical package bidding (HPB) format used in a 2008 FCC auction. There is considerable current interest in auctions that permit the FCC to buy spectrum back from analog television stations and resell it to wireless providers, and two-sided auctions will also be discussed in the last part of the chapter.

Note to the Instructor: Some of the auction setups discussed in this chapter are available on the Veconlab Spectrum Auction program, on the Auctions menu. Please consult that menu for current default setup suggestions.

29.1 FCC Bandwidth Auctions and Package Bidding Alternatives

Government agencies in a number of countries have switched to using auctions to allocate portions of broadcast frequency bandwidth. These auctions typically involve a large number of licenses, where licenses are defined by geographic region and adjacent frequency intervals. The rationale for conducting auctions simultaneously is that there may be complementarities in valuation as bidders strive to obtain contiguous licenses. Adjacent bandwidth may allow the bidders to enjoy economies of scale and to provide more valuable services to consumers. For example, a person who signs up for cell phone service with a particular company would typically be willing to pay more if the company also provides reliable service, free of "roaming fees," in adjacent geographic areas. With a few exceptions, these simultaneous auctions are run as English auctions, with simultaneous ascending bids for each license. Bids are collected in a sequence of "rounds," with the high bidder on each license being designated as a provisional winner at that point.

After each round, bidders who are eligible to bid for a particular license but are not currently the high bidder, can maintain their active status by bidding above the current high bid by a specified increment. The activity rules can be complex, but the main idea is simple; they are intended to force firms to keep bidding in order to be eligible to bid in later rounds. For example, a bidder who obtained two activity units via up-front payments at the start of the auction could bid for at most two units in a given round, although it is up to the bidder to decide which two units to bid on. A bidder with an activity of two who only bids on a single unit (and is not the current high bidder on any other) would lose a unit of activity. This bidder would only be able to bid on a single unit in subsequent rounds, unless one of a limited number of activity "waivers" is used to regain the second activity unit. The purpose of this forced bid activity is to keep the high bids moving upward, so that valuation information is revealed during the course of the auction, and not in sudden bid jumps in the final rounds. As one auction expert put it: "you can't lie and wait like a snake in the grass to snipe at the end." The auctions stop when no bids for any license are changed in a given round, and the provisional bidders at that point become the auction winners.

This type of simultaneous, multi-round (SMR) auction has been used by the FCC to sell bandwidth for personal communications services, beginning in the 1990s. The amounts of money raised are surprisingly large, and similar auctions have been implemented in Europe and other places. The auctions have proved to be fast, efficient, and lucrative as compared with the administrative ("beauty-contest") allocations that were used previously; see chapter 12 on rent seeking for further discussion of the potential inefficiencies of these administrative procedures.

Table 29.1. An Example of an Exposure Problem in an Ascending Bid Auction

Values	Round 1	Round 2	Round 3
Bidder I $V_A = 10, V_B = 0, V_{AB} = 10$	**4 for A**, 0 for B	no change	**6 for A**
Bidder II $V_A = 0, V_B = 10, V_{AB} = 10$	0 for A, **4 for B**	no change	**6 for B**
Bidder III $V_A = 5, V_B = 5, V_{AB} = 30$	3 for A, 3 for B	**5 for A, 5 for B**	no change

There are, however, some reasons to question the efficiency of simultaneous auctions for individual licenses. A bidder's strategy in a single-unit, ascending-bid auction with a known private value is fairly simple; one must keep bidding actively until the high bid exceeds the bidder's private value. In this manner, the bidding will exclude the low-value buyers, and the prize will be awarded to the bidder with the highest value. The strategic environment is more interesting with common value elements, since information about the unknown common value might be inferred from observing when other bidders drop out.

Even with private values, the strategic bidding environment can be complicated by valuation complementarities, as can be seen by considering a simple example. Suppose that there are two contiguous licenses being sold, regions A and B, with three competing bidders, I, II, and III. Bidder I is a local provider in region A and has a private value of 10 for A and 0 for B. Conversely, bidder II is a local provider in the other region and has a value of 0 for A and 10 for B. The third bidder is a regional provider, with values of 5 for A alone, 5 for B alone, and 30 for the AB combination. These valuations are shown on the left in table 29.1, where the subscripts indicate the license(s) obtained: A, B, or the AB combination.

Consider the bidding sequence for the first three rounds is as shown in table 29.1. After the first round of bidding, bidder I is the high bidder for A, and bidder II is the high bidder for B. Suppose that the activity rule requires bidder III to raise these bids to 5 to stay active in each region, as shown in the outcome for round 2. In response, bidders I and II raise the bids for their preferred licenses to 6, to stay active, as shown in the round 3 column.

As round 4 begins, bidder III faces a dilemma if the private values of the first two bidders are not known. It makes no sense for bidder III to compete for a single license, since the bidding has already topped the bidder's private value of 5 for each single license. But to bid above this value on both licenses produces an *exposure problem*, since bidder III does not know whether the other two bidders will push the individual bids for the two licenses up to a level where the sum is greater than 30, which is the value of the AB combination to bidder III.

It is therefore dangerous for this bidder to compete for the AB combination and only end up winning one of them. If this exposure risk causes bidder III to drop out, then the outcome is inefficient in the sense that the total value (10 for bidder I, who gets license A, and 10 for bidder II, who gets license B) is less than the value of the AB combination (30) to bidder III. The resulting efficiency would be 20/30 or 66% in this case. One way to reduce the severity of the exposure problem is to allow bidders to withdraw bids during the auction. But opportunities for withdrawals are limited under FCC rules, and a bidder who defaults on a final winning bid is assessed a penalty, along with compensation to the FCC if the license is later sold below the price for which the bidder defaulted. These withdrawal limits are intended to force bidders to make serious bids that convey information about license valuations.

Since limited bid withdrawal options do not fully protect bidders who try to acquire combinations, the FCC has, on occasion, considered auctions with "package bidding." Such auctions, for example, would allow simultaneous bidding for A alone, for B alone, and for the AB package. When the bidding stops, the seller would then select the final allocation that maximizes the sales revenue, subject to the constraint that no unit is sold twice. With large numbers of licenses, this combinatorial bidding is complicated by the fact that it may be difficult to calculate the revenue-maximizing allocation of licenses, *at least in finite time*. Economists and operations researchers have worked on algorithms to deal with the revenue-maximization problem in a timely manner, and as a result, the FCC has considered and tested the use of combinatorial bidding for bandwidth licenses.

Package bidding, of course, may introduce problems that are not merely due to computation. Consider the example shown in table 29.2, where the valuations are as before, with the exception that the AB package for bidder III has been reduced from 30 to 16. In this example, the bidding in the first round is the same as before, except that bidder III bids 6 for the AB package instead of bidding 3 for each license. Since I and II each submit bids of 4 for their preferred licenses, the revenue-maximizing allocation after the first round would be to award the licenses separately, for a total revenue of 8. Bidder III tops this in the second round by bidding 10 for the AB package, and the other bidders respond with bids of 8 for A and 5 for B in round 3, for a total of 13. To stay competitive, bidder III responds with a bid of 15 for the AB package, which would be the revenue-maximizing allocation on the basis of bids after round 4.

In order for bidders I and II to obtain their preferred licenses, it would be necessary for them to raise the sum of their bids, but each would prefer that the other be the one to do so. For example, bidder I would like to maintain a bid of 8 and have bidder II come up to 8 to defeat the package bid of III. But bidder II would prefer to see some of the joint increase come from bidder 1. This coordination problem is magnified if there are greater numbers of bidders involved

Table 29.2. An Example of a "Threshold Problem" with Package Bidding

Values	Round 1	Round 2	Round 3	Round 4
Bidder I $V_A = 10, V_B = 0, V_{AB} = 10$	**4 for A**, 0 for B	no change	**8 for A**	no change
Bidder II $V_A = 0, V_B = 10, V_{AB} = 10$	0 for A, **4 for B**	no change	**5 for B**	no change
Bidder III $V_A = 5, V_B = 5, V_{AB} = 16$	6 for AB	**10 for AB**	No change	**15 for AB**
Suggested Prices	$P_A{=}4, P_B{=}4, P_{AB}{=}8$	$P_A{=}5, P_B{=}5,$ $P_{AB}{=}10$	$P_A{=}8, P_B{=}5,$ $P_{AB}{=}13$	$P_A{=}9, P_B{=}6,$ $P_{AB}{=}15$

and if they do not know one another's values. It is easy to imagine that coordination issues might result in a failure for regional bids to pass the total *threshold* of the winning national bid, as some bidders try to "free ride" and let others bear the cost of raising the bid total. A *coordination/threshold* failure in this example would result in allocating the AB package to bidder III, for a total value of 16, which is below the sum of private values (10 + 10) if the licenses are awarded separately to bidders I and II. In this case, the efficiency would be 16/20, or 80%.

One solution to the coordination/threshold problem is for the package bidding procedure to provide provisional prices after each round, which would help bidders see how high they must go to get into the action on a particular license or package. See the bottom row of table 29.2 for an example of how these prices might be computed. In each case, the prices are greater than or equal to the highest bid on each license. If the provisional winner is a package bid, then the sum of the price components is that package bid. If the provisional winners are individual bids, then the prices equal those bids. The prices provide important signals. For example, the prices after round 4, where the provisional winner is a package bid, would tell the individual license bidders how high they need to go (bids of 9 and 6) to stay competitive with the package bid, and in this sense, the prices help solve the coordination problem. In this case, the way the prices were determined was by dividing the "overhang" difference between the winning package bid of 15 and the sum of the two high license bids of 8 and 5. This difference, 15 − 8 − 5 = 2, so 1 was added to each of the two high license bids to get new prices of 8 + 1 = 9 and 5 + 1 = 6. Bidders who are not provisional winners typically have to bid the price plus a minimal bid increment.

One way that a large bidder could attempt to deter competition is to begin with a high "jump bid" on a package, e.g., going directly to a bid in the 11–14 range for the package AB for the setup in table 29.2, which could trigger coordination issues. In a more complicated design where bidders I and II have other licenses that they might pursue, the large jump bid on AB might cause these small bidders to focus their attention in a different direction, where the difference between their valuation and the current bid or suggested price is larger. Such behavior is sometimes called "straightforward bidding," and it provides a good description of behavior in laboratory experiments, although it is easy to imagine strategic reasons for bidding in a non-straightforward manner to avoid pushing up prices on licenses that a bidder is particularly interested in. For example, bidding experts often discuss "parking" by bidding "under" a strong dominant firm bidder to preserve activity and then switching to bid on one's preferred licenses late in the auction. This strategic parking would not be straightforward bidding. Parking is one reason that the FCC switched to "blind bidding" so that firms did not know the identities of other bidders, and could not identify dominant bidders like Verizon. In addition, the FCC has adopted limits to how much a bid can be increased in a single round, after observing and hearing economic "experts" brag about some obvious cases where dramatic jump bids caught small bidders off guard and forced them to let their activity drop.

McCabe, Rassenti, and Smith (1991) observed inefficiencies being created by occasional jump bids in laboratory experiments, and they devised a simple *clock auction* to alleviate this problem. The idea behind a clock auction is that it provides prices on each item being auctioned, and these prices are increased in a *mechanical* manner in response to excess demand for an item (instead of being determined by price bids submitted by bidders). After prices are announced for a round, each bidder indicates which items or packages are of interest, where the package price is just the sum of the clock prices for the items in that package. The clock provides incentives for bidders to truthfully reveal whether or not they are interested by not allowing them to come back in once they withdraw interest on an item or package.

The way a clock auction works is best explained by an example. For the valuation setup on the left side of table 29.2, suppose that the clock prices start at 6 for each item. At these prices, bidder I would indicate a willingness to buy A, bidder II to buy B, and bidder III would only be willing to buy the AB package (for a price of $2 \times 6 = 12$). By bidding on the package, bidder III avoids the exposure of bidding above 5 for either A or B singly, and the clock mechanism includes the package bid in the determination of excess demand for the items in this package. Thus, there would be excess demand for both items, with two bidders showing interest in each item. As a result, the clock would click up the price for each item to 7 and then 8, at which point bidder III would withdraw, since that bidder would not be willing to pay more than $2 \times 8 = 16$ for the AB package. The two items would sell for 8 to the small bidders, which is the efficient outcome in this case. Notice that the gradual price increases in the clock auction preclude jump bids, and these increases force the small bidders to raise their bids together to stay in the auction, thereby solving the threshold problem. To summarize:

> **Exposure and Threshold Coordination Problems:** *A bidder faces an "exposure problem" if its values for a combination of licenses exceed the sum of values on individual licenses in the package. When bidding surpasses individual license values but does not yet reach the package value, the exposure is the risk of winning only part of the desired package and paying more than the partial package is worth. This problem is solved when the bidder is able to bid on the package as a whole, which precludes the possibility of only obtaining a part of the package. But package bidding can create other problems, when the package bid is provisionally winning and each bidder on a part of the package would prefer that the other one raise their bid on "their" part in order to counter the package bid. The coordination problem can be alleviated by using an auction that provides prices that convey information about how high a bid on an individual license or package must be raised to be competitive in the next round of bidding.*

29.2 Experimental Tests of Package Bidding Alternatives

The discussion in the previous section does not begin to cover all of the alternative proposals that economists have put forth for running simultaneous multi-round auctions. Recently, some of the strongest interest has been in procedures that incorporate either computer-generated prices or clock-driven prices. In addition, there have also been proposals to follow the ascending price phase by a second phase, e.g., a sealed-bid "proxy" bidding phase with a second-price type of setup (Ausubel, Cramton, and Milgrom, 2004). Some of the proposed approaches have not been evaluated with experiments, but for those that have, the focus has been on performance in two dimensions: sales revenue and economic efficiency. In addition to revenue and efficiency, government policy makers are also concerned with other issues, e.g., that the selected auction format does not unfairly disadvantage small bidders. Moreover, high revenues are not considered to be per se beneficial if they result in losses incurred by winning bidders as a result of the exposure or other problems. The remainder of this section briefly summarizes the results of two evaluations of price-based mechanisms.

Porter et al. (2003) report results of 55 laboratory auctions used to compare the "combinatorial clock" with other methods of selling off multiple units simultaneously, with and without package bidding. The combinatorial auction was uniformly more efficient than the alternative considered in the test environment; in fact, the combinatorial clock yielded 100% efficiency in all but two of the auctions in which it was used.

Table 29.3. Bidder Values with Strong Complementarities: Non-existense of Competitive Equilibrium

Package	A	B	C	AB	BC	CA	ABC
Bidder I	3	3	0	30	3	3	30
Bidder II	0	3	3	3	30	3	30
Bidder III	3	0	3	3	3	30	30
Bidder IV	3	3	3	24	24	24	**36**

Since the clock raises prices until there is no excess demand, one can think of the combinatorial clock auction as a device for finding a competitive equilibrium set of prices. Nevertheless, value complementarities may preclude the existence of such prices (Bykowsky, Cull, and Ledyard, 2000), as the example in table 29.3 illustrates. Here there are three small bidders: bidder I who is interested in licenses A and B, bidder II who is interested in B and C, and bidder III who is interested in A and C. Think of these three bidders as being "regional" providers with partially overlapping interests. The large bidder IV is a "national" bidder who is interested in all three. This bidder's value of 36 for the ABC package produces the maximum efficiency in this example. At any price below 15 for each license, bidder I would bid on AB, bidder II would bid on BC, and bidder III would bid on CA. Thus there would be two bids on each license at prices below 15, and the clock would increment the prices for A, B, and C. But as soon as the prices pass 15, there would be no bidders at all, since the price of the paired packages would exceed 30 and the price of ABC would exceed 45, which exceeds bidder IV's value of 36 for this package (as shown by the boldfaced number in the lower right corner of the table). Thus, demand is greater than supply when the prices are less than or equal to 15, and supply is greater than demand when prices are above 15. This is the intuition for why there is no set of competitive prices (at which demand equals the 1 license supplied) for individual licenses in this example.

In contrast, non-clock-based proposals for package bidding allow bids on combinations with prices that are not the sum of the individual component prices. Note that by allowing prices for packages, we can ensure that demand equals supply, e.g., by putting a price of 35 on the ABC package and prices of 31 on the three paired packages, AB, BC, and AC. This example in table 29.3 suggests that there may be cases where allowing bidders to submit bids on packages may yield better outcomes than having the prices be determined by a clock mechanism.

The valuation structure in table 29.3 is the basis for an experiment in Brunner, Goeree, and Holt (2005) that compared clock-driven and bid-driven package bidding mechanisms. Groups of four bidders competed in a series of ten auctions, using randomly generated license valuations that were approximately centered around the valuations in table 29.3. When the random realizations caused the ABC value for bidder IV to be large enough relative to the two-license packages for the regional bidders, a competitive equilibrium would typically exist. The realized valuation profiles were reordered and selected so that competitive equilibrium prices did not exist in every other round. In the clock treatment, the price provided for each package at the start of each round was the sum of the clock prices for each license. In the non-clock package-bidding treatment, pseudo-competitive prices were computed and provided so that bidders would know approximately how high they needed to bid to "get into the action" in that round. These prices were essentially those computed from a mathematical programming problem, which are known as "RAD" prices (DeMartini et al., 1999). For the final five auctions, the efficiencies were higher with non-clock package bidding (98%) and the combinatorial clock (98%) than with a simultaneous multi-round auction that did not allow package bidding (92%). While these two methods of package bidding were comparable in terms of efficiency, the combinatorial clock yielded higher revenues than the non-clock package-bidding procedure. An experimental economist once remarked to the author that "the clock is a hungry animal" with reference to the revenue generation of a clock auction. The observation of relatively high efficiencies of both package bidding mechanisms in this simple environment with three licenses raises the issue of which auction procedures would be better in more complex settings with more licenses and more bidders.

29.3   Hierarchical Package Bidding and More Recent Auctions

One major obstacle with unrestricted package bidding is that the number of possible packages can be quite large, as can the number of possible allocations to bidders. Suppose that there are n licenses and b bidders. Then there are b possible ways to allocate the first license, and b ways to allocate the second license, due to the fact that the bidder who won the initial license could win the second as well. Thus, there are $b * b = b^2$ ways to allocate the first two licenses. Reasoning in this way, there are b^n possible allocations for all n licenses. Since natural logs and exponential functions are inverses and cancel each other out, it is possible to rewrite this number of allocations as: $b^n = e^{\ln(b^n)} = e^{n\ln(b)}$, which is an exponential function of the number of licenses. Such exponential functions become very large very quickly. With large numbers of possible allocations, there is no assurance that the FCC could take hundreds of bids from hundreds of bidders and find the allocation that maximizes sales revenue to determine provisional winners in a reasonable amount of time.

Goeree and Holt (2010) proposed a way to assign prices to a hierarchical package structure that limited the possible numbers of packages and was amenable to simple recursive (paper and pencil) calculations to determine provisional winners and associated prices for licenses and packages. The FCC termed this approach "Hierarchical Package Bidding," with the acronym HPB. The hierarchy with three tiers, for example, would have three layers of successively smaller packages: a national package consisting of all licenses, a layer of non-overlapping regional packages that span the set of all licenses, and a bottom layer of individual local licenses.

Most policy discussions of package bidding revolve around whether it favors large bidders, who generally prefer being able to bid on large combined licenses, and whether small businesses are disadvantaged as a result. This discussion has important efficiency ramifications, since it could be the case that local or regional carriers have cost advantages associated with local knowledge and experience, and those advantages could be overwhelmed by a strong bid on a national package. A major motivation for the HPB auction was to let *economic* considerations determine whether licenses would be sold individually, in regional packages, or as a national license that spans all geographic areas.

This endogenous packaging objective is implemented by first calculating the highest bid on each local license, e.g., the 12 individual local licenses labeled A through L in the third row of figure 29.1. The second row of the figure shows the three regional packages, ABCD, EFGH, and IJKL, each with four licenses. A regional license award could be justified if the highest bid on a regional package is greater than the sum of the high bids on each component license. The "regional revenue" is defined to be the maximum of (1) the highest bid on the regional package and (2) the sum of the highest bids on the local licenses in that region. Finally, the sum of the regional revenues is compared with the highest bid on the national package ABCDEFGHIJKL, and the winner in that comparison determines the allocation, either as a single national license or some combination of local and/or regional licenses. The distributions of the bidders' random value draws and the synergies associated with acquiring multiple licenses can be devised to ensure that the optimal allocation would sometimes be a national license and sometimes a mix of local and regional licenses.

The hierarchical calculations of the revenue-maximizing set of package and license awards are done recursively from the bottom up, so the calculations are trivial from a computational perspective and are intuitive from the perspective of individual bidders. Transparency and computational feasibility (with no need for numerical approximations) are major advantages of the HPB approach.

A very important aspect of the HPB structure is the calculation of prices for each license, with prices of packages being the sum of prices for component licenses. These prices tend to enhance competition and alleviate the threshold coordination problem. To see how prices are calculated, suppose that all provisionally winning bids are for individual licenses. In this case, the provisionally winning individual license bids constitute the license prices, and the prices for packages are calculated as sums of prices for the local license components. The package prices (sum of provisionally winning individual license bids) would reveal the threshold that a package bidder would need to cross in order to beat the individual license bidders. Price computations are more complicated when some packages are provisional winners. To see the intuition, consider the special case of a two-tier hierarchy, with a national package, composed of individual licenses, with no regional tier. In this case, the "overhang" between the provisionally winning national package bid and the sum of high bids on local licenses is divided among the localities in proportion to population. This allocation ensures that the high bids at the local level, augmented with a share of the overhang, will equal the provisionally winning national bid. Then those prices would convey information about how high the bids must go at the local level to be competitive with the provisionally winning national bid. For example, if the winning national bid on ABC is 14 and the highest individual license bids are 2 for A, 2 for B, and 4 for C, the overhang is $14 - 2 - 2 - 4 = 6$, which would be allocated equally if the three areas have equal populations. Thus the prices for individual licenses would each be raised by 2, to 4 for A, 4 for B, and 6 for C. An analogous allocation of overhangs at the regional level is used for a three-tier hierarchy.

The Goeree and Holt experimental design (which was cleared with the FCC in advance) involved 7 bidders. There were two types of bidders: small or "regional" bidders (labeled 1 through 6) and one "national" bidder (labeled 7), as shown in figure 29.1 with a national package ABCDEFGHIJKL and three non-overlapping regional packages with 4 individual licenses each.

The bidder ID numbers listed below each of the licenses show which bidders were given positive private base values for those licenses. In each session, bidder 7 was a national bidder with positive values and enough activity to bid on all 12 licenses listed in the figure. The interests of the 6 regional were staggered. For example, regional bidder 1 had positive private value draws for licenses A,

ABCDEFGHIJKL

ABCD				EFGH				IJKL			
A	B	C	D	E	F	G	H	I	J	K	L
1	1	1	1	3	3	3	3	5	5	5	5
6	6	2	2	2	2	4	4	4	4	6	6
7	7	7	7	7	7	7	7	7	7	7	7

Figure 29.1. Hierarchical Package Bidding with National, Regional, and Local Tiers: Bidder ID Numbers Indicate Licenses with Positive Private Base Values. *Notes:* Odd-numbered regional bidders, 1, 3, and 5, have interests that match the regional package definitions in the middle row. In half of the auctions, the regional package definitions were shifted to match the interests of even-numbered bidders. Bidder 7 has national interests in all 12 licenses.

B, C, and D, and bidder 2 had interests shifted to be C, D, E, and F. The interests of the odd-numbered regional bidders, 1, 3, and 5, match the regional package definitions, and the even-numbered regional bidders have interests that are split between two regional areas. In half of the auctions, the regional package definitions were shifted to match the interests of the even-numbered regional bidders. This treatment is referred to as "HPB even."

In addition to the 12 licenses in three tiers shown in the figure, there was a second group of 6 licenses for which the national bidder had no value; this second "band" served to equalize earnings to provide some positive earnings to regional bidders when a national bidder would win a national license. Importantly, base private values in either band were incremented sharply as more contiguous licenses were acquired, which is the way that "synergies" were induced.

The experiment compared three auction procedures. One was the simultaneous multi-round (SMR) auction discussed in the previous section, with separate auctions for individual licenses, connected only by activity calculations that implement "use or lose it" constraints on how many licenses a bidder can make at any given round. Bidders could not bid on packages. In contrast, bidders had full flexibility to bid on any combination of licenses in the modified package bidding treatment (MPB) that was a minor variation of the RAD package bidding setup, mentioned in the previous section. The most important aspect of the MPB format to keep in mind is that there was *full flexibility* in bidding on self-made packages, subject to activity constraints. A bidder could bid on packages with any number of elements, ranging from 2 to 12. Under HPB, the admissible packages have a strict hierarchical structure, with a single national package ABCDEFGHIJKL and three non-overlapping regional packages as shown in figure 29.1. Therefore, there are only four possible packages under HPB in this structure, as compared with the extremely large number of packages possible under MPB. The prices for each provisional allocation made under MPB were calculated using a mathematical programming procedure that was not explained to subjects.

In all three formats, bidders were able to create "custom" packages in order to see the values associated with winning combinations of licenses. Under MPB, bidders could bid on any of these self-created packages. In contrast, under SMR, these custom packages would be shown but could not be placed into the bidding basket. Finally, under HPB, bidders could put one or more of the four predefined packages into the bid basket, and attach price bids to those pre-defined packages but not for custom packages.

The experiment results are summarized in figure 29.2, with efficiency averages by treatment on the left and revenue averages on the right. In this case, revenues are expressed as a percentage of the revenue that results if all bids reveal values, a Walrasian revenue prediction described in the previous chapter. Experiment results reveal a clear advantage for the hierarchical package bidding

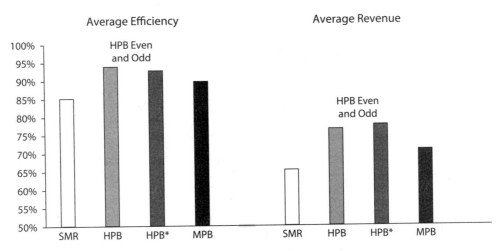

Figure 29.2. SMR with No Packages (white bars), HPB Hierarchical Package Bidding (gray bars), and MPB Flexible Package Bidding (dark bars). *Key:* HPB has packages aligned with regional interests of even-numbered bidders, and HPB* is for packages aligned with regional interests of odd-numbered bidders.

format, HPB, in an environment with value complementarities. Recall that the value structure was such that the pre-made packages did not match the preferred packages for half of the regional bidders in each treatment. HPB yielded significantly higher auction revenues and efficiencies, and lower numbers of unsold licenses. Unsold licenses were most common in the SMR auction, as bidders withdrew bids to avoid exposure.

Given the presence of value complementarities, the authors anticipated the lower efficiencies and revenues to be observed for the simultaneous multi-round auction format that has been extensively used by the FCC. What came as a surprise was the relative ranking of HPB and the more flexible MPB. A close examination of the bid data revealed a common pattern that contributed to this difference. Consider the setup in figure 29.3 in which the random draws were such that bidders 1, 2, and 3 (shown in bold) had the highest private value draws. Since their interests overlap and only one bidder can be awarded each license, these three bidders may not be able to challenge a national bidder 7 who is bidding on the package of all 12 licenses. An analysis of the bid data for flexible package bidding (MPB) indicated that "home-made" packages did tend to overlap, causing a "fitting problem" that made it difficult for strong regional bidders to win, even if an award to regional bidders was the efficient outcome. In a sense, the regional bidders in the flexible package MPB auction were "stepping on each other's toes." As a result, the number of licenses awarded to the national bidder was much higher than the optimal number under MPB, but not under SMR (no packages) or under hierarchical package bidding (with non-overlapping pre-made packages). More importantly, in rounds when the

A B C D E F G H I J K L

A	B	C	D	E	F	G	H	I	J	K	L
1	**1**	**1**	**1**	**3**	**3**	**3**	**3**	5	5	5	5
6	6	2	2	2	2	4	4	4	4	6	6
7	7	7	7	7	7	7	7	7	7	7	7

Figure 29.3. Coordination Issues with Flexible Packages: Overlapping Interests of Strong Regional Bidders Shown in Bold. *Notes:* If bidders with IDs 1, 2, and 3 (shown in bold) have sets of high random value draws and construct 4-license packages that match their interests in order to maximize synergies, the overlap may prevent them from challenging a strong national bidder (ID7).

national bidder won nothing, regional bidders were often unable coordinate their bids and avoid overlaps under MPB, while their coordination problems were virtually non-existent with hierarchically structured packages.

Pre-packaging has the obvious disadvantage that the chosen packages may not be optimal, but in a non-overlapping hierarchical structure they are chosen to "fit," which enables bidders to coordinate their bids and avoid threshold problems with positive effects for efficiencies and revenues. In addition, assignment and pricing is transparent and easily verifiable by bidders as the auction proceeds. Of course, one has to be careful in generalizing the relative performance of tiered and flexible package bidding to other environments. But if the number of licenses is increased, thereby increasing the potential complexity of the fitting problem, these results would seem to suggest that the simple HPB procedure would continue to offer advantages over a fully flexible form of package bidding such as MPB. To summarize:

Experiment Results for Package Bidding Comparisons: *Hierarchical Package Bidding (HPB) yielded significantly higher revenues and efficiencies than were observed in sessions with fully flexible package bidding (MPB) and in the sessions with no package bidding (SMR). The degraded performance of SMR, which was lower than both package bidding formats in both dimensions, was largely due to the exposure problem. The hierarchical package bidding format provided protection from the exposure, and the pre-made non-overlapping regional packages provided protection from overlapping bids by strong regional bidders using overlapping home-made packages in MPB.*

The FCC decided to use HPB in the "C block" of 700 MHz auction that was run in 2008, citing the simplicity of the pricing rule and the HPB performance in experiments. Overall, the auction raised a record (at that time) $19.6 billion. The high revenue was probably partly due to the low frequencies in the bands being auctioned, which has high penetration and propagation properties. There was room for improvement due to frictions caused by activity rules, penalties for dropping bids, and non-aligned geographic boundaries across frequency blocks (vastly different geographic sizes and subset misfits). Experiments could be useful in documenting these types of frictions if similar misalignments are likely to occur.

The use of HPB was the first major FCC auction that implemented package bidding, and when HPB was considered for a subsequent auction, one of the documents filed in favor was submitted by a very large bidder, AT&T, while documents filed in opposition cited the interests of small bidders, who generally have opposed package bidding. This opposition may well be in the interests of small bidders, but the other side of the coin is that HPD does not *require* licenses to be sold in large packages, since the sizes of the awards are determined endogenously by the bidding in the auction, i.e., by economic, not political, considerations. And the failure of large bidders (Verizon and AT&T) to actively participate in the recent FCC auction (to be discussed in the section that follows) might have been due to the absence of package bidding opportunities.

29.4 The FCC Incentive Auction

In 2012, the US Congress authorized the FCC to run a two-sided *incentive auction*, with the idea of spectrum being transferred from low-value uses (e.g., analog TV stations) to high-value wireless data provision. Two-sided auctions are common in laboratory experiments, i.e., Vernon Smith's double auction. In addition, the consignment section of the previous chapter describes two-sided auctions that have been successfully implemented for emissions permits. Two-sided auctions are also common in business, e.g., procurement. Nevertheless, a two-sided spectrum auction had never been tried. In the case of the incentive auction, there were large financial stakes, with the idea of repurposing about 100 Mhz of spectrum for more than $100 billion of spectrum being a focal point in initial discussions at a Stanford University planning conference a couple of years before the actual auction began in March 2016.

Given the novelty and the high stakes, the four-year planning period seems reasonable; you cannot just start a massive auction and change course, in the way that you can sell emissions permits one way in a given month and then add a fix in the next month. The two-sided incentive auction, however, lasted a full year, closing in March 2017, which is a very long time in light of changing economic conditions. During this time, some of the major potential bidders acquired spectrum from other sources. In any case, three major bidders (Verizon, AT&T, and Sprint) did not participate in the end. By the time the first phase of the auction finally started, Verizon had decided that they did not need the spectrum, or at least that was the reason given. AT&T did participate in the first stage, but then shed its activity as quickly as FCC rules would allow. This non-participation was a disaster for revenue generation, given that more than 90% of the industry cash flow was controlled by Verizon and AT&T at that time.

The FCC webpage summary presented a more positive view of the final results: "the auction used market forces to align the use of broadcast airwaves with 21st century consumer demands for video and broadband services. The auction preserves a robust broadcast TV industry while enabling stations to generate additional revenues that they can invest into programming and services to the communities they serve." Even though an additional 42 Mhz of spectrum could have been sold, the auction did repurpose about 84 Mhz of spectrum, with sales in excess of $19 billion, although $10 billion of that went to the sellers, the TV station owners. The net revenue of about $10 billion was a lot of money, but it was much lower than revenues from some other recent FCC auctions.

At a 2017 post-auction conference in Washington, DC, participants discussed ways in which the auction fell short of expectations. One person noted that falling demand was an issue as firms' needs changed over the five years of planning and running the auction. Although the auction cleared 84 Mhz of spectrum, many low-value TV stations failed to clear.

Instead of running a single auction with buyers and sellers, the FCC incentive auction went back and forth between a *reverse auction* (TV station sellers offering to sell) and a *forward auction* (wireless providers bidding to buy). Each phase could take a month or more. Many people at the DC conference complained that this switching process was unnecessarily delayed because each reverse auction phase had to start from the beginning and then run through 50 or so rounds of bidding to reach a stopping point. An increase in the minimum bid increments would speed things up, but one person noted that AT&T dropped out when the increment went from 5% to 10%. There was general agreement that it was essential to make the two-phase auction process go more quickly.

The incentive auction was quite simple in some respects, e.g., there was no package bidding, which could have depressed interest from large bidders. Even so, the two-stage auction process itself was quite complex. One staff person from the Commerce Committee exclaimed that it was the "most complicated auction in the history of humanity." He noted that when the post-auction repacking phase is included, the whole process will have taken more than a decade (from initial discussions to final spectrum being used), as compared with Obamacare, which only took 1–2 years. David Salant from the Toulouse School of Economics also commented on the complexity and uncertainty, and recounted that they might have to increase bids by large amounts in any given round. Complexity is a problem in that it reduces participation and adds uncertainty that depresses bids. An AT&T official at the conference remarked that "what people don't like is uncertainty."

Despite the "incentive auction" name, some significant incentive problems were noted. For example, there was a feeling that sellers might have exercised market power. One seller owned 12 TV stations in the Pittsburgh area and only ended up selling five of them. The concern was that this seller had submitted high bid prices on the ones not sold in order to raise the prices on the stations that were sold. This supply-reduction strategy was predicted in an earlier (non-experimental) analysis (Doraszelski et al., 2016). There were other price anomalies on the reverse side, with TV stations in Harrisburg, Pennsylvania selling for more than stations in the Philadelphia area, a difference that is difficult to justify. One of the auction design consultants conceded that they had devoted a lot of attention to incentives on the forward (buyer) side, but that more thought could have been given to seller incentives. Another participant stated that the auction put "too much power in the hands of large bidders, a single forward bidder had the opportunity to end a stage by hiding their activity in small markets." Each switch back and forth between forward and reverse phases would take months.

For all of its complexity, the incentive auction was incremental in the sense that it was built from procedures of the "tried and true" simultaneous multi-round auction (SMR) that has been used for forward auctions for decades, e.g., multiple rounds and activity rules. The two-sided nature was novel, and one person noted that there had not been time to look at experimental data, and therefore, the designers had to "rely on theory." But at the earlier Stanford conference, Ron Harstad (University of Missouri) had pointed out that the relevant theory was based on private values, whereas spectrum licenses have common value components. Some people at the DC conference wondered about totally different approaches, e.g., having a quick sealed-bid auction to collect bids to buy and sell to see what happens. David Salant noted that he always thought the forward and reverse auctions should "run in parallel." Tom Hazlett from Clemson University advanced the idea of "overlay licenses," which implement a Coaseian idea of giving rights to a single firm for a range of channels, so that the firm can negotiate directly with the current owners and clear them out. These are important areas for further research, and *this time* experiments might be useful. While attending an earlier Stanford planning conference, the author had asked about experiments, and was told that there were computer simulations instead. Later that day, the author related this situation to an FCC official and asked, "How do you sleep at night?"

29.5 Extensions

The first proposal for a hierarchical set of structured licenses can be found in Rothkopf, Pekeč, and Harstad (1998). See Klemperer (2002) for an authoritative discussion of the general considerations involved in designing multi-unit auctions. Noussair (2003) provides a clear description of the exposure and co-ordination/threshold problems that is targeted to a broad scientific audience. Kagel and Levin (2004) analyze the exposure problem and strategic demand reduction in a "clean" environment with small bidders simulated by computer. David Salant's more recent book, *A Primer on Auction Design, Management, and Strategy* (2014), has a chapter on combinatorial bidding that covers HPB and combinatorial clock auctions. A nice compilation of current research on spectrum auctions can be found in the collection of papers that is edited by Bichler and Goeree (2017) in the *Handbook of Spectrum Auction Design*. Overlay licenses are described in Thomas Hazlett's 2017 book and 2014 paper with the provocative title "Efficient Spectrum Allocation with Hold-Ups and without Nirvana."

Chapter 29 Problems

1. Consider the valuation setup with strong complementarities in table 29.3, with the only change being that the license C value for bidder III is raised from 3 to 10. What is the efficient allocation in this case? Is there a competitive equilibrium and associated prices for A, B, and C that yield this efficient allocation?

2. For the valuation setup in the previous problem, what would the observed efficiency be if bidder IV were to win all three licenses?

3. (non-mechanical) The two-phase auctions mentioned in section 29.3 have the property that there is an initial auction followed by a second auction, a kind of "final shootout." For simplicity, suppose that there is a single prize and the second phase is a sealed-bid first-price auction, with the prize going to the high bidder in the second stage. What can be done to keep the first phase from being ignored by bidders, i.e., what procedures could be adopted to induce bidders to compete seriously in the first phase?

4. Consider a three-stage hierarchy, with a national license ABCDE and two regional licenses, ABC and DE, and five local licenses. If the highest bids are 10 for A, 10 for B, and 15 for C, and the highest bid on ABC is 30, what is the "regional revenue" (defined in section 29.3) for this ABC region?

5. Consider a two-stage hierarchy with a package ABC and local licenses A, B, and C, each of which has the same population. If the high bids are 10 for A, 20 for B, 30 for C, and 69 for ABC, what would be the prices for the three individual licenses under hierarchical package bidding?

6.   (non-mechanical and open-ended) After reading section 29.4, please consider whether or how the positive experience with emissions *consignment* auctions discussed in the previous chapter might be applied to designing two-sided spectrum auctions. What problems might arise in terms of strategic manipulation?

30

Matching Mechanisms

Price-based allocations are considered to be undesirable or even unethical in situations for which priority should not be given to wealthy individuals, e.g., the assignment of dorm rooms, slots in popular classes, duty assignments for military school graduates, or urban high school admissions. In such cases, an allocation can be constructed from ranked preferences submitted by people on one or both sides of the market. For example, the assignment of medical students to residency programs is based on wish lists submitted by both hospitals and medical students. Such a system serves as a clearinghouse that takes the submitted rankings and arranges matches.

A good matching system will, if possible, provide incentives for participants to submit truthful rankings, and it will have a stability property, which is that no pair of people who are not already matched with each other would prefer to break off and be re-matched with each other (matches are stable).

This chapter describes the properties of the most commonly used matching mechanisms and evaluates their performance.

> **Note to the Instructor:** A class-appropriate version of the Veconlab Matching program, on the Auctions menu, is under development. Please consult that menu for current default setup suggestions. The instructor might also consider using the one-sided "serial dictator" procedure (section 30.3) to assign students to class tasks.

30.1 Matching Mechanisms: Priority Product and Deferred Acceptance

Previously, job offers to students in professional schools had been made earlier and earlier by competing employers. But early offers and acceptances may cause inefficiencies, since rushed decisions may be made in the face of uncertainties, and preferences can change. Matching mechanisms have been used to alleviate issues associated with early offers. Applications include residencies for medical students or high-status judicial clerkships for law students. In order for a matching mechanism to be successful in stopping the practice of extending early offers, it must be developed to prevent the "unraveling" process that can then result in inefficiently early matches.

In the market for federal appellate court clerks in the United States, for example, judges who compete with each other for the best law students began to make earlier and earlier offers. This process continued until offers were being extended and accepted two years prior to graduation, on the basis of first-year grades in law school (Roth, 1984).

An alternative, which is used in many medical labor markets, is to establish a matching mechanism. This matching mechanism uses computer programs to create matches based on the ranked preferences submitted by both prospective employers and employees. Such a system was adopted in the United States in the 1950s for medical residents, and a modified version of this system is still in use (Roth, 1984). The success of the US National Resident Matching Program was noticed in the UK, where unraveling and early offers were prevalent. The matching algorithms subsequently adopted in the UK varied from place to place. Some of these systems survived and prospered, whereas others failed as unraveling did not diminish. These performance differences cry out for further analysis.

This section describes the two prominent types of matching mechanisms, using a simplified version of the Kagel and Roth (2000) experiment to illustrate the properties of these mechanisms. Both of these algorithms begin by having each prospective employee ("worker") submit a ranked list of employers ("firms"), with number 1 being the person's top preference, etc. Similarly, each firm submits a ranked list of available workers.

These two systems primarily differ in how they use submitted rankings to determine which matches to implement. For one of the systems, the attractiveness of a specific worker-firm match is measured by the product of the rankings that each side gave to the other. This *priority product* is four, for example, if the worker is the firm's second choice and the firm is the worker's second choice. The process begins by enacting the match with the lowest priority product. The worker and firm for this match are then removed from the pool, with the next match being for the lowest priority product for the remaining workers and firms. With equal numbers of workers and positions, this will result in no unmatched workers or firms. The priority product mechanism has been used in various applications, e.g., matching students to schools. Variations of priority product mechanisms have also been used in laboratory experiments, e.g., to sort subjects to groups in voluntary contributions games (Page et al., 2005).

The other commonly used matching procedure is known as *deferred acceptance*. In this system, after firms have sent out offers to their most preferred worker, workers keep their best offer on the table and reject all others. Firms with rejected offers then send offers to their next most preferred worker. This process repeats until there are no rejections. A worker who keeps an offer on the table in one stage may reject it later if another offer arrives that is higher on the worker's preference list. As a result, each worker's satisfaction with the match on the table will not go down from one stage to the next.

It is useful to think of what deferred acceptance would mean in a marriage market. Suppose that each eligible man would make a list and propose to the woman at the top. Women who receive multiple proposals would provisionally accept the proposal that they prefer and reject the others. Men who are rejected will send out another proposal to the woman next on their list. Women with multiple proposals, including those who had previously accepted an offer, can re-choose as new proposals arrive, even if it means dumping a previous partner. A man who has been released in this manner would then continue making proposals by going further down on his list. This process continues until there are no more proposals. While this analogy conveys procedures vividly, the actual computerized matching mechanisms do not provide any between-round feedback, so there is no regret, disappointment, or emotional damage from rejection, as would surely develop in a marriage matching mechanism.

Even though both of these two systems are based on the same ranking inputs, the properties of the resulting matches can differ in important ways. Deferred acceptance takes actual preference rankings and produces an outcome that is *stable* in the sense that it would be impossible to find a worker and firm who would prefer to be matched with each other instead of keeping their assigned partners, as long as both sides are truthful. To see this, note that if a firm prefers a particular worker over the person with whom it is matched, then that other worker must be higher on its rank list. This means that the firm must have previously proposed a match to that preferred worker while coming down the list. Thus, the worker who rejected that offer previously would, if given the chance, reject it again. Therefore, even though firms may prefer to switch to a different worker after the matching is complete, that other worker will not want to leave their match.

In contrast, the priority-product system does not always produce a stable outcome, even if all firms and workers submit their true rankings, as will be shown next by looking at a simple example. Suppose that half of the workers are of high productivity, and a match with them is worth about $15 to each firm, regardless of the firm's own productivity. In contrast, a firm who was matched with a low-productivity worker earns $5. Similarly, half of the firms are high-productivity firms in the sense that a worker matched with one of them would earn $15. Matches with one of the low-productivity firms only provided the worker with $5 average earnings. The remainder of this section develops this example and uses it to compare the stability of matches determined by either deferred acceptance or priority products.

The experiment to be discussed in the next section involves one additional feature: actual match earnings are equal to the *partner's* productivity value, plus or minus a small, privately observed, random deviation that was less than $1

Table 30.1. Payoffs for Each Matching

F1 earnings	F2 earnings	F3 earnings	F4 earnings
W2 = 14.90	W2 = 15.50	W1 = 15.30	W1 = 14.90
W1 = 14.20	W1 = 14.60	W2 = 15.00	W2 = 14.50
W3 = 5.90	W3 = 5.30	W3 = 5.50	W4 = 5.50
W4 = 5.60	W4 = 4.60	W4 = 4.70	W3 = 5.10

W1 earnings	W2 earnings	W3 earnings	W4 earnings
F2 = 14.90	F2 = 15.50	F1 = 15.50	F1 = 15.50
F1 = 14.80	F1 = 14.60	F2 = 14.60	F2 = 15.10
F3 = 5.60	F3 = 5.00	F3 = 5.30	F3 = 4.90
F4 = 5.30	F4 = 4.60	F4 = 4.60	F4 = 4.50

in absolute value. These deviations are meant to model idiosyncratic match values, e.g., that a medical student may have a small preference for a hospital that is located near friends and relatives. A simplified version of the experiment setup, with four workers and four firms, is shown in table 30.1. For example, firm F1 would earn $14.90 if it is matched with worker W2, who is preferred to W1, W3, and W4 in that order. Similarly, the bottom part of the table indicates that worker W1's earnings are highest for a match with firm F2, who is preferred to F1, F3, and F4 in that order.

If the true rankings in the table are submitted under the *deferred acceptance procedure*, then firms 1 and 2 send initial offers to W2, who keeps F2 on the table and rejects F1. Firms 3 and 4 send offers to W1, who keeps F3 and rejects F4. The low-productivity workers get no offers in the first stage. In the second stage, the F1 offer that was rejected initially is re-sent to W1, who keeps it and rejects the offer it had with the low-productivity firm F3. At this stage, the high-productivity workers, W1 and W2, are provisionally matched with firms F1 and F2 respectively, and these pairings cannot be overturned by a low-productivity match. It can be shown that the process ends after matching W3 with F3 and matching W4 with F4 (problem 1). This setup is stable. Even though W4 would rather be matched with F3, the feeling is not mutual since F3 prefers W3 to W4. A similar observation can be made for the high-productivity firms and workers, so the outcome is stable.

Next consider the *priority-product system* under the assumption that participants submit their true preference rankings. Since W2 and F2 are each other's first choice, the priority product is 1 for this match, which can be shown to be the lowest product (problem 2). This match is then implemented, taking W2 and F2 out of the picture. Next consider whether F1 will be matched with W1. F1 is W1's second choice and vice-versa, hence the priority product is 4. This is higher than the product for F1 and W3, which is 3 since F1 is W3's first choice and W3 is F1's third choice. The product for F3 and W1 is also 3, and these are the lowest priority products remaining after those in the initial match (W2, F2) were removed. Thus the algorithm will match W1 and F3 in one pair, and W3 and F1 in another. The resulting match *is* unstable since both of the high-productivity agents (F1 and W1) would prefer to be matched with each other.

Importantly, deferred acceptance arrangements are protected against unraveling, whereas the priority-product system is not. To summarize:

> **Deferred Acceptance and Unraveling:** *If all firms and workers submit their true rankings, the deferred acceptance system will always generate a stable outcome in which pairs who are not currently matched with each other would not prefer to be re-matched with each other (no pairwise blocking). In contrast, the priority-product system (products of preference rankings) does not always produce a stable outcome even when workers and firms submit their true preferences.*

30.2 Experimental Analysis of Matching Mechanisms and Unraveling

An unstable system will create regret and dissatisfaction after matchings are announced, so participation may decline as workers and firms seek to match with each other bilaterally rather than through the matching system. Because these bilateral offers are not constrained to a single time period like the matching system was, the problem of early offers may arise.

Kagel and Roth (2000) designed an experiment, based on the example in the previous section, in which unraveling due to early matches would occur. The treatments were structured to determine whether the imposition of a centralized matching process (deferred acceptance or priority product) would replace the tendency for workers and firms to make and accept early offers.

The match values used in the experiment had a structure similar to that of table 30.1, but with six workers and six firms in each session. The way the process worked was that each firm could send an offer to a single worker in each period. There were three periods in the process, -2, -1, and 0 (graduation day). Such bilateral matches incur a cost of $2 if accomplished at time -2, and they incur a cost of $1 if accomplished at time -1. There is no cost for making a match in time 0. Anyone who arranged a match could not make or accept subsequent offers. In the treatment rounds with a matching system, it operated for unmatched people and firms in time 0. Anyone who remained unmatched after time 0 would earn $0.

Each session of the experiment began with ten rounds or repetitions of this three-period market process, with no matching mechanism. Match value profiles were shuffled randomly from round to round, but these profiles stayed the

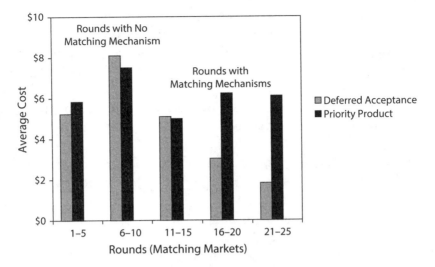

Figure 30.1. Average Costs of Early Matchings. *Source*: Kagel and Roth (2000).

same for the three periods (-2, -1, and 0) of a round. After ten rounds with no matching mechanism, one of the matching mechanisms (deferred acceptance or priority product) was implemented at period 0 of round 11 and in all subsequent rounds.

One way to evaluate the outcomes is to look at the average costs associated with early matches, which are graphed in figure 30.1. The first two sets of bars, on the left side, are for the ten rounds prior to the imposition of a matching mechanism in round 11. Notice that costs of early matching rise from the first five rounds to the second five rounds, which shows the unraveling as earlier matches generate higher costs. Then costs for both treatments fall in rounds 11–15, as each type of matching mechanism reduces costs when it is operative. But in the final rounds 16–25, costs continued to fall for the deferred acceptance sessions but not for the priority-product sessions (dark bars). Although not shown in the graph, the most costly matches in time period "-2" vanish completely for the final five rounds of the deferred acceptance sessions. To summarize:

Experimental Results and Matching Mechanisms: *Costly early matches were commonly observed in the absence of a centralized clearinghouse matching procedure. The imposition of a matching mechanism after a number of rounds without one tended to reduce early matching costs. With the priority-product mechanism, costly early matchings tended to reappear. Such unraveling was not observed with the deferred acceptance mechanism.*

This experiment is quite innovative from a methodological perspective. First, note that the sequential (within-subjects) design is used because the matching mechanisms were implemented in a natural experiment in settings where un-raveling had already occurred. Therefore, sequence effects were a focus of atten-tion, not a nuisance.

The second methodological point is that the experiments enabled the au-thors to observe the patterns of behavior *during the process of adjustment* at a level of detail that was not available from field data. In the experiments, the sudden availability of a matching mechanism option did not deter firms from making early offers just as they had before. What happened, however, was that workers who were not matched in periods -2 and -1 ended up trying the matching mechanism and obtaining good results, especially for the deferred acceptance mechanism. The priority-product mechanism did improve matching initially before unraveling re-emerged with an increase in costly early match agree-ments. In addition, the laboratory experiment was able to reproduce both the unraveling patterns observed in the field and the superior performance of the mechanism that generated stable matches.

Another reason that the experiment is important is that it is not necessarily the case that participants on both sides of a deferred acceptance mechanism do have an incentive to reveal their true preferences: "No stable matching mecha-nism exists for which stating the true preferences is a dominant strategy" (Roth and Sotomayor, 1990). In particular, it is possible to construct examples in which there is one stable match that players on one side unanimously prefer, and another that players on the other side prefer, which can create incentives for strategic misrepresentation. Since the stability property of the deferred ac-ceptance system is conditional on working with true preferences, it is impor-tant to evaluate the mechanism with actual subjects' choices that may waver from truthful revelation, either for strategic reasons or, more likely, for reasons associated with bounded rationality. The improved performance of deferred ac-ceptance in the Kagel and Roth experiment is reassuring, especially in terms of its stability and resistance to unraveling.

Without a laboratory experiment, one might wonder whether the previ-ously observed survival of the matching mechanism in two UK locations and its demise in two other locations had been due to actual differences in the mechanism details, or to chance differences in demographics, social norms, or some other unobserved aspect of the field context. The laboratory used subjects selected from the same pool for both treatments, so the performance differences could be attributed to the incentive properties of the two mech-anisms. Taken together, the laboratory experiment and the natural experi-ment provide a clearer picture of adjustment and behavior in these types of markets.

30.3 Serial Dictator (One-Sided) Matching Mechanisms

The matching mechanisms considered in the previous sections were two-sided in the sense that both workers and firms submitted priority lists. There are also matching mechanisms, however, that only take into account the preferences of one side. The simplest single-sided matching mechanism is a *serial dictator* process in which each person on one side of the market submits a preference list. Then those lists are put into a random order. The person at the top of the pile receives their top choice. The second person on the random order receives their top choice if it is available, otherwise they receive their second choice. At each stage, the person whose turn it is receives the best option that is not already taken by someone higher in the sequence. Notice that the truthful ranking incentives would not be altered by the use of an exogenous non-random order based on some other criterion such as seniority or prior performance in another dimension. To summarize:

> **Serial Dictatorship:** *When preference lists submitted by one side of the market are ordered and executed in sequence, it is in a person's interest to submit a truthful ranking, since the ranking list has no effect on when the person's turn comes up, and when their turn does come up, only a truthful ranking will ensure that the best available option at that point is obtained.*

The author uses this method to assign students to chapters of this book for purposes of making a class presentation after running an experiment for that chapter. Even so, students are often skeptical of arguments that it is in their own best interest to provide a truthful ranking. Such skepticism should be discussed and addressed by looking at examples.

Matching systems may start to break down, however, in the face of constraints and conflicting objectives, although such factors can be partially resolved, especially using single-sided matching mechanisms. For example, one issue with matching residents to hospitals these days is that some couples have joint location issues and wish to be matched with hospitals in the same geographic area. This degrades other aspects of the matching process, but it is important to the participants. Al Roth worked on a modification of the medical matching mechanism to accommodate joint locations.

In terms of the class chapter assignments with up to three students per chapter, a similar consideration arises when pairs of friends wish to work together on a presentation. That is accommodated with the serial dictator procedure by having pairs submit joint rankings, so that when a pair's turn comes up, the pair receives the best available option at that point. There can be frictions, however, if a pair's first choice already has two students assigned, so two slots are not available and the procedure has to go down the pair's joint preference list.

The use of a matching mechanism to assign class presentation topics illustrates another important point. Standard matching mechanisms are not market based, and thus are not sensitive to preference *intensities*. If two people rank the same option first, but if one person is almost indifferent between their first two options and the other has a strong preference, the rank-based system will not be responsive to this important information. Some modifications that let people express indifference have been proposed (Fragiadakis and Troyan, 2017), but further modifications that incorporate strong intensity effects would degrade some of the desired theoretical properties, like the incentive to submit truthful preference information. One way to introduce an intensity dimension is to let people come early, wait in line, or engage in some other costly activity that signals intensity. The trouble with this approach is revealed in the word "costly." These effort-based competitions are analogous to rent-seeking contests, which were evaluated in chapter 12 and shown to result in high amounts of wasteful, duplicated expenditures.

In contrast, market allocations are efficient precisely because bids and asks do reflect cost and value differences. The problem is that it is clearly inappropriate to use price-based allocations for some commodities, like slots in urban high schools for example. Holding an auction for such slots would result in income stratification and would be perceived as being grossly unfair. When a market mechanism is acceptable, however, the use of a rank-based mechanism instead can be detrimental for the reasons discussed above. For example, if half of the faculty commute by car and the other half would prefer taking a bus or walking to paying a fee that covers the costs of parking services, then a non-market (free) allocation of parking locations via serial dictatorship would misallocate a significant proportion of the parking passes! The point is that matching mechanisms are not used to promote efficient allocations, but rather, to promote fair access to limited resources or avoid conflicts with other objectives like racial balance or income stratification.

30.4 Extensions: Baseball, Dorm Rooms, School Choice, Deep Space, and Rush

The deferred acceptance procedure was developed by Gale and Shapley (1962), and the incentive properties of these mechanisms are discussed in more detail in Roth and Sotomayor (1990). Roth, Shapley, and coauthors have discussed applications to college admissions and marriage. Two-sided mechanisms are widely used in sorority rush, where "unraveling" might mean making a choice soon after arriving on campus, before having time to find a range of friends. The author's own experience was with a fraternity rush system that cleared during the first week of the first year, during a pre-class advising phase, which was preceded by summer rush parties in some selected cities for incoming students before they even arrived on campus!

There are a number of papers that present experiments that are motivated by specific matching systems used in the field. Nalbanthian and Schotter (1995) report experiments motivated by matching of free agents in professional baseball. In addition, there are many interesting experimental studies of one-sided matching problems, e.g., assigning students to dorm rooms when money-based allocations like auctions are precluded (Chen and Sonmez, 2004). Olsen and Porter (1994) study both auction-like and non-price rankings-based mechanisms that are motivated by the Jet Propulsion Lab's allocation of time slots on its Deep Space Network of antennas. Chen and Sonmez (2006) use experiments to evaluate a "Boston system" of assigning students to schools on the basis of submitted rankings. They conclude that alternatives like the Gale-Shapley mechanism would improve performance over priority-based systems like the Boston system.

The Boston system looks good on paper because students tend to get assigned to the school with their highest *stated* rank. For example, in a priority-based allocation, a student may not "waste" a top rank on an option that is popular with everyone else. Some press accounts openly warn students about stating their actual preferences if it means "shooting too high":

> Make a realistic, informed selection on the school you list as your first choice. It's the cleanest shot you will get at a school, *but if you aim too high you might miss.* Here's why: If the random computer selection rejects your first choice, your chances of getting your second choice school are greatly diminished. That's because you then fall in line behind everyone who wanted your second choice school as their first choice. (Tobin, *St. Petersburg Times*, September 14, 2003; italics added)

When students act on incentives to misstate their preference rankings under this system, this strategic misrepresentation makes the assignments based on *submitted* rankings look better than they really are, based on *actual* preference rankings.

Demographic quotas in terms of race and income are sometimes imposed to limit the scope of matches of students to public schools. Fragiadakis and Troyan (2016) have developed a dynamic quota mechanism that respects distributional quotas and improves the performance of rank-based matching mechanisms. Fragiadakis and Troyan (2017) also analyze serial dictator mechanisms that allow participants to put options into "indifference bins." This information can help the mechanism make better assignments. For example, if one person expresses indifference between options A and B, and another ranks A above B, then B could be assigned to the second person, even if the first person was higher in the random sequence of a serial dictator match. The indifference bin adjustment improves performance in the Fragiadakis and Troyan experiment. Featherstone and Roth (2017) report on a modified serial dictator mechanism that deals with diversity constraints and indifference in a setting that involves matching MBA students to foreign educational trips at the Harvard Business School (see problem 6).

Chapter 30 Problems

1.   Assuming that the preference orderings implied by table 30.1 are submitted under the deferred acceptance program, show that W3 is matched with F3 and W4 with F4.

2.   Please fill out the table of priority products for the example from table 30.1, assuming that the submitted preference orderings match those given in the table. The second column and row of the priority-product table have been completed for you, and this column and row can be crossed out since the top priority product (1) causes W2 and F2 to be matched first. Use the resulting table to determine all four matches.

	W1	W2	W3	W4
F1		2		
F2	2	1	6	8
F3		6		
F4		8		

3.   Your answer to the previous question should show F1 being matched with the low-productivity worker W3. If F1 had foreseen this unfortunate outcome, how could F1 have changed the submitted ranking of workers to achieve a better match with a high-productivity worker? Explain.

4.   Consider how you might modify one of the match value numbers in table 30.1 to show that the deferred acceptance algorithm does not necessarily generate an efficient outcome in the sense of maximizing total earnings of all firms and workers. To do this, consider a very large change in a match value that does not alter the outcome of the deferred acceptance procedure, but that does cause a firm to miss an extremely productive match with a worker who does not rank that firm very high.

5. (non-mechanical) Discuss whether or not a *pair of people* should report a truthful ranking of chapter presentation topics in an experimental economics class, when the matching is done with a serial dictator process.

6. (non-mechanical) Suppose that there are ten MBA students applying for an educational trip, and that some of the available trips are to Europe, and the rest are to Canada. Suppose further that past experience suggests that all participants prefer the European destinations to the Canadian destinations. If half of the students are white and half are non-white, how could a serial dictator mechanism be designed to ensure that not all of the European trips are allocated to one race or another?

APPENDIX 1

Hints on End of Chapter Problems

Chapter 1 Hints for Problems

1. Replace the P in the demand formula with the supply price: $1 + 0.2Q$ to obtain one equation in a single variable, Q, and solve for Q. Then determine P by substituting this value for Q into the right side of the supply price: $P = 1 + 0.2Q$.

2. At a price of 9, only the four high-value buyers would be willing to trade, but all sellers would be willing to sell.

3. The procedure is analogous to that for problem 2.

4. To find total surplus, note that the units traded at a price of 6 are those with costs of 2 and values of 10, so the surplus per unit traded is 8. Then figure out the total surplus for all units.

5. Think about how you might put buyers and sellers into separate groups to get more units traded. One group could have high values, high costs, and high prices, for example.

6. With high values of 10 and high costs of 8, the surplus for each unit traded is 2. A similar argument can be applied for the low values and low costs. In total, there are 8 units traded, so calculate the total surplus for these units and divide by the maximum obtained in your answer to problem 4.

Chapter 2 Hints for Problems

1. The values go from 10 down to 2, so the vertical intercept of the highest demand step is at 10. Start drawing a horizontal line at \$10 moving to the right from the vertical axis. Each person has a single card, so the step width is 1. Then add in the other steps going down the demand curve, in a manner that is analogous to the steps on the left side of figure 2.5. The supply curve starts with a step at 2, with other steps at 3, 4, . . . coming up the supply curve. Note that supply and demand are symmetric, and find the intersection or overlap.

For part b, there will be a step at 2, but no steps at 3, 4, and 5, so the steps go from 2 to 6 to 7, etc. See where this moves the intersection or overlap. Since this is a reduction in supply, the price should rise and quantity should fall.

2.   One way to move the average but not the intersection is to change an extreme value (e.g., raise the highest value or reduce the lowest value). For part b, try making a change that alters the prediction, and then make another adjustment that equalizes the averages.

3.   For supply, put more than half of the unit costs at the minimum, say $1, and then let supply rise for the remaining units. Fix demand to intersect supply well above the minimum cost, and do final adjustment by adding low-value units to the right side of demand to move the median value down enough to equal the median cost.

4.   As an approximation, let supply start at $0 and go up, let demand start at $24 and come down, and let them intersect at a quantity and price of $12. Then consider the standard welfare loss triangles associated with the quantity reduction to 6 units. You can draw a vertical line at a quantity of 6, and then look at the welfare loss triangle that results from its intersection with supply and demand.

Chapter 3 Hints for Problems

1.   Multiply 0.6 times the square root of 77, and continue from there.

2.   The line should curve up (increasing marginal utility), which would imply that the second dollar is worth more than the first, etc. (risk preference).

3.   Your answer should be the middle number in the list, given the symmetry of the possible payoffs around this middle value. You should do the calculations, with probabilities of 1/9, and check this intuition.

4.   One way to do the calculations is to enter the payoffs in column A of a spread sheet and then transform them into utilities with the following formula inserted in column B: $= (A1)^{\wedge}0.5$ for the case of square root utility. Remember to multiply utilities by probabilities and sum for each lottery, before comparing the resulting expected utilities (sums of products of probabilities and utilities).

5.   Consider what are the best and worst outcomes of the lottery.

6.   Think of the dollar numbers on the left side as prices.

7.   You can modify the table from problem 5 and list the commodity being valued on the right. Then explain what could be learned from the crossover point.

8.   The square root of 6 is about 2.5, as can be verified in Excel: $= 6\wedge(0.5)$.

9.   Maximizing the expected payoff expression provided in the statement of the problem is the same as maximizing the numerator, since the choice variable x does not appear in the denominator. It is convenient to multiply terms in the numerator to express it in an equivalent form: $Nx^{1-r} - x^{2-r}$. The calculus rule to be used is that the derivative of Cx^b is Cbx^{b-1} where C and b are constants (not functions of x). To take the derivative of a function with an exponent, the whole expression is multiplied by the exponent b, and the exponent is then reduced by 1. The derivative of x^2 is $2x$, for example. And the derivative of Nx^{1-r} is $N(1-r)x^{-r}$. Similarly, the derivative of $-x^{2-r}$ is $-(2-r)x^{1-r}$. Therefore, the derivative of the expected payoff is the sum of the derivatives of each part, which is: $N(1-r)x^{-r} - (2-r)x^{1-r}$. The term x^{-r} can be factored out of both of these terms, so that the derivative can be expressed: $[N(1-r) - (2-r)x] \, x^{-r}$. Setting this derivative equal to 0 and dividing both sides by x^{-r}, one obtains an equation that determines the best x given r. You can then solve for r and check your answer against equation (3.1) in the text.

Chapter 4 Hints for Problems

1.   You need to calculate the expected money payoff for each decision. This is done by summing products of probabilities and associated payoffs.

2.   Read the hint for problem 1, and remember that probabilities for the two possible payoffs must sum to 1.

3.   Which job posting do you think would receive more applications? For which job will the probability of being hired be lower? Then think about how those probabilities might be perceived by someone whose behavior is characterized by probability weighting with the standard shape that overweights low probabilities.

4.   The natural log and exponential functions are inverse functions, so that $\exp(\ln(x)) = x$. Inverse functions just "undo" each other, in the same way that that taking the square of the square root of x will yield x.

5.   No hint provided, this is open-ended.

Chapter 5 Hints for Problems

1. One way to answer this is to sketch a revised version of figure 5.1 and do some counting of *a* marbles, and then form the relevant ratio.

2. To get started, note there are two ways of getting a balanced sample, *ab* and *ba*. If the draws are from cup A, each of these possible orders has a probability of $(3/4)(1/4) = 3/16$, so $\Pr(s|A) = 2(3/4)(1/4) = 6/16$. You will also need to calculate the probability of getting a balanced sample of draws from cup B and then construct the ratio needed for Bayes' rule.

3. Think about what the *N* lottery might turn out to be.

4. No hint provided.

5. Think in terms of a ratio of numbers of burnt sides. For example, of the two pancakes with burnt sides, there are a total of three burnt sides. This should get you started.

6. On average, there should be about 10 women out of 1,000 with the cancer, of which 8 will show a positive test (80%). Of the 990 who do not have breast cancer, there is a 10% false positive rate, so how many false positives would be expected? Then use these numbers to determine the Bayes' rule probability. If your answer is above 10%, you are in good company with the German physicians, but you should recalculate.

7. First restrict attention to the bottom row (Moderate). Note that counting the H marks in the bottom row, you can conclude that $\Pr(H|Good) = 2/3$. Then calculate $\Pr(H|Bad)$ and then construct the relevant ratio to find the Bayes' rule probability $\Pr(Good|H)$. In the end, you will see that even though the outcome is H with the Moderate decision, it is best to switch to the Extreme decision after updating your beliefs about the state. Thus this is an example for which the "win/stay, lose/switch" strategy would not be advisable.

8. Does the observed sample look like the contents of one of the cups?

Chapter 6 Hints for Problems

1. For the row of table 6.1 with 20 chances of event A ($R = 0.2$), the bottom formula in equation (6.1) implies that $1 - (0.2)^2 = 1 - 0.04 = 0.96$, which matches the payoff for event B shown in the right column of the table in the 20 row. The other payoff number, 0.36, can be calculated analogously.

2. You should multiply the subjective probability (0.2 in this case) times the relevant number in the event A payoff column. If the probability of event A is 0.2, what is the probability of event B? This probability should be multiplied by the event B payoff in the relevant row.

3. Remember the argument that an extremely risk-averse person would tend to report equal probabilities for each event, regardless of their actual belief probability. Then think about someone who is extremely risk seeking.

4. Could a separate QSR procedure be used for the second event, with the third probability calculated as a residual? What problems might arise?

5. The derivative of $\ln(x)$ with respect to x is $1/x$ for $x > 0$.

6. You might consider what might happen if the subject reports a selling price $P > V$ and the "bid" is in between the two. Conversely, show that it would never be optimal to report a minimum selling price below V, using a similar argument. Remember that the random bid used in the BDM procedure does not depend on the minimum selling price that is set!

7. Think about changing the gamble on the right side to be an ambiguous gamble. Why would decisions in such a table potentially be affected by both ambiguity aversion and risk aversion? In that case, how could the original version of table 6.4 be used to "subtract off" risk aversion effects?

Chapter 7 Hints for Problems

1. Use your intuition, there is no trick.

2. The rational strategy (picking the more likely event every time after that event is apparent) will earn 20 cents with what probability? The matching strategy involves only choosing the more likely event 3/4 of the time instead of every time, so the chance of getting the 20-cent payoff is lower. What is this chance and what are the expected earnings?

3. This should be mechanical if you answered problem 2 correctly.

4. One way to do more simulations is to copy the block of code for the first simulation.

5. Imagine that for some random reason the first 14 realizations are all one event. Given the blocking, which event is more likely on the fifteenth round?

6.   Consider the top row of table 7.1, with AA observed and private information of *a*. In the cascade model, this is like seeing three *a* draws, so figure 5.3 applies, and the answer of 8/9 is found by using the counting heuristic. Other rows are done analogously.

7.   The intuition for your answer should be that the information implied by the first decision is just as good as the information implied by the third draw, which goes in the opposite direction, so the second draw can break a "tie" if it contains at least some information. To decide whether the second decision does contain information *when it matches the first*, consider two cases: (1) where the second person's information matches the first person's decision, so the second person's decision is clear; and (2) where the second person's information does not match the first decision, so the second person chooses randomly.

Chapter 8 Hints for Problems

1.   If both choose high effort the payoffs are $30 - 25 = 5$. Would either player have an incentive to defect to low effort?

2.   If they both push, do they each earn $6 or $8? Your answer fixes the payoffs for the Cooperate/Cooperate outcome. If one person pulls and the other pushes, the person who pulls gets $6 + $8 = $14.

3.   The off-diagonal payoffs for different suits are $0, and the payoffs are positive ($6 each or $8 each) for matching suits. After you fill in the payoffs, show that there are two Nash equilibria and then determine the type of game that this is.

4.   It is not a prisoner's dilemma if there are two Nash equilibria.

5.   Hints are already provided in the question.

6.   Hints are already provided in the question.

7.   If the average is 50, then 20 plus half of the average is 45.

8.   One possibility is to alter the payoffs in a manner that gets rid of losses, but that does not change the coordination game structure.

9.   Figure out which distances from the center line in the figure determine the deviation losses, and then try to relate that to the sign of the expected payoff difference when $p = 0.5$.

10. Try some parameter values for *V* to see if you end up with a coordination game, with two equilibria, one of which is better for both players. Be sure to write out the table of payoffs. Remember that a person who chooses a higher effort only receives the minimum of the two efforts.

Chapter 9 Hints for Problems

1. A first mover who choose *S* in the bottom game in figure 9.2 earned 80. By choosing *R*, the first mover earns either 20 or 90, but you have to calculate the probability of each. Note that 52% of the first mover choices were *S*, so that leaves 48% *R* choices. Of those, 12% received the *P* response. That is a rate of 12 out of 48, or 25% *P* responses given that the game gets to that point. This observation should be used in constructing an answer to the question.

2. To answer this, you have to show that a unilateral deviation by either player will not increase that player's payoff, assuming that the other does not deviate from the equilibrium being considered. Given that the first player is choosing *S*, which fixes the payoff for the second mover, it follows that a unilateral deviation for that player will have no effect on that player's payoff, so the second mover has no incentive to deviate. Next, check to see if the first mover has a unilateral incentive to deviate, given the second mover's decision of *P*.

3. The same hints for problem 2 apply.

4. The decision node for the first mover has three arrows, one for each decision, and each of these arrows leads to a node with two arrows for the two second mover decisions, accept or reject.

5. To ensure that this unequal split is a simple Nash equilibrium, the second mover has to have a strategy that involves rejection for other less favorable divisions. Such a strategy is not sequentially rational.

6. The number of quarters that can be taken at each node increases from 1 to 2, to 4, to 8, 16, and finally to 32 ($8). So the centipede will have six legs. The person who chooses to "take" will earn all of the quarters, so the payoffs at each node involve $0 for the person who does not take.

7. Use the analogue of the logit choice function in equation (9.4).

Chapter 10 Hints for Problems

1. If right chooses the black 8 with probability p, the relevant equation for the left side player is $8(1-p) = 2p$, which can be solved for p. The analysis for the other player is similar.

2. You can look at the expected payoff formulas in section 10.2, with the 2 numbers changed to 3.

3. Let p be the probability of high effort, so the relevant equation is $20p = 10p + 3(1-p)$, and solve for p. You can check to see if your answer is on the same side of 1/2 as the "diamond" on the vertical axis in figure 8.2 in chapter 8. That diamond symbol represents the mixed equilibrium, because the expected payoff difference for the two decisions is exactly 0 where the expected payoff difference line crosses the center vertical line.

4. There were 12 players, 5 decisions each, for a total of 60. Of those, 45 were East for Row or West for Column. This ratio, 45/60, can be compared to the probability of choosing one's preferred location, which we can denote by p. You calculate p by equating expected payoffs for one player to determine the other's p. By symmetry, you only need to do this for one of the players. Answer should be 4/5.

5. Can you think of any scenario in which the identities of the attackers and defenders are drawn from some pools in an unpredictable manner? In what kinds of settings are the identities pretty much the same every period?

6. The diamond on the center vertical line in figure 8.2 represents the mixed equilibrium, because the expected payoff difference for the two decisions is exactly 0 where the expected payoff difference line crosses the center vertical line. To explain stability, look at the directional arrows and the logic behind them based on signs of expected payoff differences.

7. When the call probability γ is low, the best response is to bluff ($\beta = 1$) at the top of the "box." Use analogous reasoning to find best-response call rates.

8. In the expected payoff calculations in the final section, replace each -1 term with $-L$, and replace -2 with $-2L$, where $L > 1$ indicates loss aversion. Loss aversion should raise bluff rates above the Nash prediction of 1/3 with $L = 1$, and lower call rates below the Nash prediction of 2/3.

Chapter 11 Hints for Problems

1. Suppose that each person knows the other's claim and that one of them is higher. Would the person with the high claim have an incentive to lower their claim, and how is that incentive related to the test for a Nash equilibrium?

2. Suppose that both claims are 0. What would the payoffs be, and would one player make more by choosing a claim of -1? What if both claims were -2?

3. Compare the risks of unilateral claim increases for the two-person and four-person games.

4. A claim of 200 would be the highest in all four rounds, so the minimum claim would be the other person's claim, shown in parentheses. Use that information and the penalty amount, 10, to calculate what the team's earnings would have been with this "go to the max" strategy.

5. Each player earns 0 if there is an exit. If there is no exit, what would sequential rationality imply about the decisions (cooperate or defect) in the final round?

6. More precision means less noise. One guess is that this would move the average claims down toward the Nash prediction, but this should be checked.

7. Use the definition of V to show that $V = 1 + \delta V$, and then solve for V to get the desired answer.

Chapter 12 Hints for Problems

1. When you check the profitability of a *unilateral* deviation to an effort of 2 or to an effort of 4, remember to assume that the other three people continue to choose 3. Of course, to be complete, all possible deviations would have to be considered, but you can get the idea by considering deviations to neighboring effort levels of 2 and 4.

2. The hint for problem 1 also applies here, but with different deviation efforts to be checked.

3. You can use equation (12.3).

4. You can use equation (12.3) again.

5.   There is no bias if the length of the number lists is proportional to the win probability.

6.   Remember to think in terms of reference points.

7.   Do you think the result would be an unusually large influx of pre-proposals? What else might go wrong?

Chapter 13 Hints for Problems

1.   Think about the effects of persistence across treatments and heterogeneity across groups.

2.   Since these are rankings, the order matters. This enumeration task can be done by trial and error, or with a system (find the three rankings that begin with a Q and the other three that begin with a D).

3.   The relevant formula has one factorial expression in the numerator and a product of two factorial expressions in the denominator.

4.   The hint is provided in the lengthy statement of the problem.

5.   Probabilities of sets of independent events are computed as products of probabilities of each outcome.

6.   The rank 7 in the bottom row of table 13.4 beats three of the rank observations on the right. Then use the $n_1 = n_2 = 4$ row of table 13.2 to read off the p value for a one-tailed test.

7.   The sum of signed ranks is: $-1 + 2 + 3 + 4 + 5 + 6 = 19$. Then use the $N = 6$ pairs row of table 13.6 to determine the p value for a Wilcoxon matched pairs test.

Chapter 14 Hints for Problems

1.   Consider the product of 0.5 and $10, and subtract the conflict cost, and then decide what is the lowest integer the responder would accept.

2.   The analysis based on "X" and "Y" in section 14.5 applies.

3.   Think about the phrase "minimal winning coalition" that is sometimes mentioned in the literature on legislative bargaining.

4.   What about a person who cares more about relative earnings (relative to the other person) than absolute earnings, i.e., a person for whom the envy component of inequity aversion is strong?

Chapter 15 Hints for Problems

1.   One approach is to find two "points" in the payoff space, one when the first mover passes nothing, and the other for when the first mover passes all $10 and nothing is returned. The budget line should connect these points, as could be checked by considering what happens when the proposer passes $5.

2.   As before, consider the two extremes, in which the responder passes nothing back and passes everything back.

3.   Vertical changes in the responder's payoff do not affect the proposer's utility. Think about whether this implies that the proposer indifference curves are vertical lines or horizontal lines.

4.   For concreteness, you could suppose that the responder is willing to give up a dollar if it provides the proposer with at least $3. What does this indicate about the slope of the responder's indifference curve?

5.   For the previous hint, think about how a reciprocating responder might be willing to give up more to reward a proposer.

6.   Remember to multiply the value of the product times the agent's share, before subtracting the effort cost. The fixed fee portion of the contract is like a fixed cost that does not affect the optimal amount of effort. So with a 0.8 share rate and an effort of 8, the numbers in the eighth row of table 15.3 imply that the agent earns $0.8*160 – 64 = 64$. In this same manner, you should be able to show that the worker only earns 63 for efforts of 7 and 9.

Chapter 16 Hints for Problems

1.   An answer requires an assumption about how the "big game" is divided. If you assume equal division, the hunter gives up 5,000 to hunt, and if each

household (including the hunter's) receives 2,500, then these numbers can be used to estimate the MPCR.

2. With group size 4, the return per person from a contribution of x after doubling is $2x/4$, and the MPCR is $2/4 = 0.5$. From this, you should be able to change the group size to 8 and show how the MPCR due to the doubling and dividing procedure is lower for the larger group. Be sure to compute the MPCR.

3. Recall that with group size 4 in Ensminger's design, the return per person from a contribution of x after begin doubled is $2x/4$, and the MPCR is $2/4 = 0.5$. Doubling the group size doubles the denominator, so the numerator in the MPCR calculation (the $2x$) has to be doubled. So instead of doubling the total contributions, they need to be increased by what factor to hold MPCR constant?

4. What about using a private room in which people can remove some of an endowment from an envelope and deposit it through a slot into a bucket? You need to figure out how to be sure that the perceived thickness of their endowment envelope after they emerge from the private room does not convey any information.

5. Person 1 gets $(2/3)$ of $(2 + 3 + 4) = 6$, etc.

6. For external, think about how much more each of the others receives if person 1 contributes 1 more token. For internal, think about how much more (if any) person 1 receives if person 1 contributes 1 more token.

7. This comparison is done in the text.

8. See if you can find 10 comparisons, just as you would for the "binary win to the right" comparisons that were done earlier in chapter 13 for the Jonckheere directional test.

9. Those would be the entries with equal internal and external returns.

10. Start by writing down your own definitions of warm glow and economic altruism, based on what you infer from the discussions in the text where formal definitions were not supplied. The word economic suggests the importance of price.

11. One of many possible approaches would be to think about alternative frames that involve "taking," "contributing," or "investing."

Chapter 17 Hints for Problems

1.   Use equation (17.6).

2.   Use equation (17.6) again.

3.   You need to figure out what the cost of volunteering is, using the information given, before using equation (17.6) again. It may be easiest to do the calculations in a spreadsheet, e.g., Excel.

4.   For $V = 2$ and $L = 0$ and $C = 0.25$, the ratio $C/(V - L)$ in equation (17.7) is 1/8. You can use this to calculate the probability of getting no volunteers at all for each group size.

5.   In considering the effects of risk aversion, you might consider the payoffs for each decision, volunteer or not, and then decide which action has payoffs that are safe (close together) and which has payoffs that are risky (farther apart).

6.   If they both volunteer, they each earn $V - C = 24$. If neither volunteers, they each earn 0. You can fill in the payoffs for the other two cells of the payoff matrix.

7.   First take equation (17.4) and express it in a manner so that $(1 - p)^{N-1}$ is on one side of the equation, and then raise both sides to the power $1/(N - 1)$, which should yield (17.5).

8.   The statement of the problem provides detailed help with the spreadsheet.

9.   It follows from the right side of (17.7) that the probability of getting no volunteer should converge to $\frac{C}{V-L}$ as N gets large. You can check this with the parameters used for the spreadsheet.

10.   No hint provided.

11.   Think of the initial belief as a flat line with height of 0.5. Find the expected payoff difference for this belief when $N = 2$ (from the straight dashed line). Then draw a vertical line with a horizontal coordinate that equals this expected payoff difference. Now go up the vertical line to see where it crosses the curved distribution function line. The crossing point determines the stochastic response probability to the original belief.

12.   The center vertical line is at a point where the expected payoff difference is 0, i.e., where each decision has the same expected payoff. Note that the absence

of a directional bias away from the center point would mean that each decision (volunteer, not volunteer) is equally likely if the expected payoffs of those decisions are equal. Would this be the case if the random errors associated with each decision do not have identical distributions with zero expected values?

Chapter 18 Hints for Problems

1. With four entrants, table 18.2 indicates that each earns $2.50, for a total of $10. But there are 12 people, and the eight non-entrants each earn $0.50, so the total social benefit is $10 + $4 = $14, which matches the number in table 18.3. The social benefits for other numbers of entrants can be calculated similarly.

2. Be sure to take the *average* of the payoffs for six and ten entrants (sum the payoffs and divide by 2). Then compare this with the sure payoff for eight entrants.

3. For intuition, think about why adding commuters to a congested road slows things down for the others, and what it means when there are lots of others.

4. With an entry fee of $3, the payoffs in the bottom row of table 18.2 would go down by $3. You need to calculate the number of entrants in order to calculate fee revenues.

5. The student might consult the instructions used for a card-based public goods experiment, which are provided in the instructions appendix for the contributions game for chapter 16. The idea is to deal two cards to each person, and then to collect cards played and redistribute them in the reverse order so that each person will receive their own card back for the beginning of the next round. Then you need to adapt that game to the Jamestown common field setting with starvation. This problem is open-ended and unstructured.

Chapter 19 Hints for Problems

1. Draw straight lines connecting the three ideal points. For any point in the interior of the resulting triangle, show that two voters can be made better off by moving toward one of the edges. Next, consider points on the three connecting lines and look for an alternative that would garner two votes.

2.   For middle-income voters, H > L > M, where the letters refer to high, low, and medium funding, and the > sign indicates preference. Write down the directional rankings for the other two voter income types. Start with a status quo of H, and then determine what option beats it in a vote, and then figure out what beats that, and so on until a cycle is discovered.

3.   Think of two groups that together constitute a majority and have strong preferences for local public goods, A for group I and B for group II. Suppose that group III has to pay one-third of the tax cost. If A and B are adopted, the sum of the benefits would have to be greater than 2/3 of the cost (the tax share for those two groups). How can you construct an example for which the total cost for all three groups is greater than the total benefit, but for which two of the groups would vote to provide the public good? Please provide numbers.

4.   First review the preference ranks you determined for problem 2, and then write down which voter types will vote sincerely for middle over high in the first stage. Then determine the outcome in the second stage. Then think about which group of people who voted sincerely for middle in the first stage might wish they had voted against middle in that stage.

5.   No hint provided.

6.   The median voter is at the middle of the three preferred points, *not* at the midpoint of the road (50). If the location is at 33, what is the total travel cost for the two people at the end? If the location is at 50, what is the total travel cost for all three people? For the final question, use your intuition to show that a movement away from the median at 33 toward the high travel cost person's preferred point will reduce total cost.

Chapter 20 Hints for Problems

1.   You need to check whether 6 is a best response if the other firm produces 3, and whether 3 is a best response when the other firm produces 6. So look in the 6 column to see if there is an asterisk for the payoff associated with an output of 3, etc.

2.   Don't be fooled by the triple asterisk. An output of 3 is not a best response if the other firm chooses 3 (look at the payoffs).

3.   Instead of outputs of 4 and 4, try 5 and 3.

4. The vertical dashed line would be at 6, and the *MR* line would intersect the horizontal axis halfway between 6 and 13 because it is twice as steep as residual demand.

5. To show that price is decreasing, all you need is to show that total quantity NQ^* is increasing. Remember that the equilibrium quantity per firm, Q^*, is a function of N (each firm produces less when there are more firms). So you first use (20.3) to express NQ^* in terms of N. The result will have an $N/(N+1)$ term. If you are not sure whether this is increasing in N, try some integers, say 2, 3, 4, to convince yourself.

6. If there are two firms, the second firm taking the other's output as given, divides remaining part of the horizontal line in half, so there are three equal parts, one for the other firm, one for the firm whose residual *MR* is shown in the figure, and one part to the right of the $MR = MC$ intersection. Derive the analogous condition when there are three firms.

Chapter 21 Hints for Problems

1. If the other seller keeps a price of $7 and you raise your price to $8 (recall that prices were constrained to be integers), how many units would you sell? What would your cost for those units be (remember that you sell your low-cost units first)? Given these answers, what would your earnings be for this unilateral price increase? Is the resulting earnings level below what you would earn by matching the $7 price and selling two units?

2. For the second part, be sure that your answer satisfies $F(0) = 0$, $F(20) = 1$ and $F(10) = 1/2$. Since prices are uniformly distributed, this function should be proportional to p.

3. At a price of $0, earnings will be $0, so all you have to do is show that a unilateral price increase yields positive earnings when the other seller stays with a price of $0. You have to use information about the other seller's capacity to show that a unilateral price increase will result in a positive sales quantity.

4. First redraw figure 21.4 with demand that is vertical at 6 units, and with total capacity for both sellers at 10 units. If one seller's price is higher, that high-price seller will sell X units, where $X < 3$ is determined by the residual demand left after the low price seller sells all 5 of its units. What is X? At what price would $6X = 3p$? At prices below that level, it is better to raise price to $6 and sell X units rather than split demand (3 each) at p. The rest is easy.

5. First redraw figure 21.4 with supply horizontal at $1 up to a quantity of 10. The residual demand X is the same as what you determined in problem 4. At what price, p, would you be indifferent to selling X units at $6 versus selling 3 units at p? Be sure to include the cost of $1 per unit in the equation that you use to solve for p, which is the lower limit of the range of randomization.

6. This hint will not make sense unless you draw and modify a supply-demand figure as you think about the problem. Give each seller some low-cost units, and then some units at a higher cost "step." You need a scenario in which a vertical demand comes down across the high-cost step, where the number of seller units between the crossing and the point to the right where supply becomes vertical is greater than any one seller's capacity *on that step*. For example, if each of 3 sellers have 2 units on this upper supply step, and if demand cuts the step in half, there are 3 units to the right of the crossing. In this case, no individual seller could pull their 2 units off the market and raise price. But if a merger between two of these sellers created a large seller with 4 units on the step, then that seller could raise price above the level of the step by pulling 4 units off. In order for this to be profitable, this seller would need to have one or more units at lower costs below the flat part of the high-cost step.

Chapter 22 Hints for Problems

1. Equation (22.1) is relevant.

2. Equations (22.2) and (22.3) are relevant.

3. Work with modified versions of equations (22.1)–(22.3).

4. If you answered this correctly, the quantity should be half of the monopoly quantity that you found in your answer to problem 3.

5. The optimal fee should force the downstream firm's profit down to a penny.

6. You can modify the formula in (22.7). To check, use the marginal intuition to equate marginal cost and the expected value of marginal revenue of ordering one more unit.

7. Even though the price of $30 is high, there will be losses if realized demand is low and 100 units have been purchased. How low?

8.   In particular, could loss aversion ever explain cases in which order quantities
       are *too high* relative to risk neutrality predictions?

9.   Losses can be prevented with fixed payments, but such a simple fix may not
       work if people still "code" bad outcomes as losses relative to a safe earnings
       level. Can you think of a better approach?

10.   Think of design changes that might preserve the two-treatment structure,
       but with predicted order quantities in some middle location for both treat-
       ments. Or perhaps alter demand in some manner so that predictions in both
       treatments are on the same side of the midpoint of demand. The brute force
       100×100 spreadsheet method might provide a framework for calculating pre-
       dicted orders in new designs.

Chapter 23 Hints for Problems

1.   You can look at the surrounding text discussion to find the exact values and
       costs.

2.   Profits are determined by the difference between the equilibrium price and
       the seller costs. Add up the unit profits for all units with positive profits (other
       units will not be produced).

3.   Answering this question requires a conjecture about the relative effects of
       buyer shopping and seller pricing decisions.

4.   Answering this involves thinking about adverse selection effects.

5.   Begin by sketching the modified Figure 23.2 and consider what might
       happen.

Chapter 24 Hints for Problems

1.   In the first period, prior to any dividends, the expected value will be the sum
       of expected dividends over five periods, plus the redemption value, which also
       equals the sum of expected dividends.

2.   To find the present value, divide by $(1 + r)^t$, using the r and t provided in the
       problem. The investment is worthwhile if the present value of the future pay-
       out is higher than the current cost.

3.   Given that excess cash can amplify bubbles, how can payments for accuracy be structured so that they do not affect cash balances during the market sequence?

4.   Think about why deception in general is undesirable in terms of having subjects believe that announced procedures are really being followed in the future.

5.   To obtain a flat fundamental value sequence, you need to set the redemption value to equal the present value of all "future" dividends if the market were to continue forever. This present value is the V in equation (24.1b). Then think about how the redemption value would need to be adjusted to generate an increasing fundamental value.

Chapter 25 Hints for Problems

1.   If both depositors withdraw, then each person's payoff is either $10 (with probability 1/2) or $0 (with probability 1/2), so the expected payoff is $5. This is higher than the $0 payoff that one receives from not withdrawing when the other person does withdraw.

2.   If both withdraw, then each person's payoff is either $10 (with probability 1/2) or $9 (with probability 1/2), so the expected payoff is $9.5. This is higher than the $9 insurance payoff that one receives from not withdrawing when the other person does withdraw.

3.   Assume that the other player is flipping a fair coin to decide whether to withdraw and compare one's own expected payoff from withdrawing or not.

4.   You will need to use the numbers of workers and firms to graph the market supply and demand functions for labor. Be sure to use units for integer increments on horizontal lines.

5.   The 42.5 peak price for the first session in the top row beats three of the session peaks in the bottom row (42, 36, 38.5). Proceed in this way to determine the total number of binary wins. The p value for a one-tailed test can be determined from the second-to-last row of table 13.2 for the Mann-Whitney test.

6.   Consider the effects of "excess cash" in the two treatments.

7.   Could there be anything analogous to or different from the "cooperation chasing" behavior that was observed for voluntary contributions games with free entry and exit that was discussed earlier in the chapter 16?

8.   In each case in the bottom row, the bank is borrowing from the DW window, so will the earnings go up or down as a result of the lowered cost of discount window borrowing?

Chapter 26 Hints for Problems

1.   You can complete the formula in steps by beginning with the difference (value minus bid) and copying this down the column to be sure that those differences diminish by 50 cents in each lower row, ending up with a difference of 0 when the bid is 8 (assuming that you entered a value of 8 in cell B1). Errors at this point are probably due to not using $B$1 reference with dollar signs that fix the reference as the difference formula is copied down. (The $ sign fixed reference notation, which works in Excel, might be different in other programs.) The notation in cell B3 should be: =($B$1 − A3) at this point. Once you are happy with the value minus bid differences, then go back to cell B3 and adjust to formula to include the probability of winning, which should be connected to the value minus bid difference with an asterisk * to indicate multiplication. Remember to use parentheses so that the difference is computed *before* multiplication occurs: =($B$1 − A3)*(A3/10) where the parentheses around the A3/10 term are not needed, although this second set of parentheses does make it easier to see logical structure of the formula. Then go to the cell with the new formula (with the value-bid difference multiplied by the probability of winning), and copy/drag it down.

2.   Review the second paragraph of section 26.4 to see how the probability of winning in a two-person auction needs to be entered.

3.   For the experiment that was already done, the first half used a no-regret setup and the second half was done with a regret treatment. The question is how to redesign the experiment so that regret is not confused with order and experience effects. To do this, you will need to have regret and no-regret done with the same level of experience. Think about how that might be done while still investigating risk aversion effects. What about adding more treatments?

4.   An auction payoff structure is the product of a probability of winning and a payoff conditional on winning. Try to relate this idea to the payoff for the bomb task. Then think about whether the similarity is a good thing or a bad thing.

Chapter 27 Hints for Problems

1. You must replace the multiplier of 1.5 with a symbol K, where $0 < K < 2$. If the bid b is accepted, then the seller value should, on average, be halfway between 0 and b, so the expected seller value for an accepted bid is $b/2$. Now find the function of b that represents the *buyer* expected value. If this buyer expected value is less than the bid b, then there will be a loss on average. To determine the optimal bid, use your intuition: what would *you* bid in this case?

2. If a bid of 50 will be rejected, what is the expected payoff for such a bid? If a bid of 100 will always be accepted, then the expected payoff for this bid is the difference between the expected buyer value and the bid of 100: (expected buyer value $- 100$). If the bid of 100 is accepted, then the seller value is anything between 50 and 100, so it is 75 on average. Multiply this expected seller value by 1.5 to get the expected buyer value and use that in the above formula in parentheses. If the result is a positive expected profit, then a bid of 100 is better than a bid of 50.

3. The procedure for finding the expected payoff is exactly analogous to the procedure used in the hint for problem 2 above, just use the 70 this time.

4. For the spreadsheet construction, hints on the relevant Excel commands are provided in the hint for problem 26.1 from the previous chapter (see above). The calculus method is described in the hint for problem 5 below.

5. You are given the answer (100) in the statement of problem 4, so now derive this with calculus. The seller value is at least 50, so from the bidder's point of view, the expected value to the owner for an accepted bid of b is going to be 50 plus half of the distance from 50 to b: $50 + \frac{(b-50)}{2}$. Therefore, the buyer expected value is 1.5 times this expected seller value, or $(50 + \frac{b-50}{2})\frac{3}{2}$. The payoff formula is the product of two terms: (expected buyer value $- b$)(probability of winning). The only remaining part of this formula to find is the probability of a bid b being accepted. This probability is $\frac{(b-50)}{50}$, which is 0 for a bid of 50 that will be rejected, and which is 1 for a bid of 100 that will be accepted for sure. Now you have all of the pieces for the payoff formula that is the above product of probability and payoff expressions. The final step is to multiply terms and take a derivative and set it equal to zero. Then you solve that equation for b, to show that $b = 100$. In other words, the optimal bid is at the upper end of the range of feasible bids in this case.

Chapter 28 Hints for Problems

1.   Be sure to include the $3 revenue from selling the permit when the second unit is not produced.

2.   The absolute values will all be positive. After these are ranked, the signed ranks will be given minus signs whenever the number in the difference row was negative. Then take the sum of signed ranks and check the Wilcoxon table in chapter 13.

3.   What is the main risk a bidder faces in a sealed-bid auction? How can this risk be reduced by altering the bid amount, and how does this risk reduction depend on the auction format used? Please use your intuition and imagine that you are a bid officer defending your proposed bid to company management officials.

4.   Hints are provided in the statement of the problem.

5.   The auction revenues for all three sessions with the discriminatory auction treatment are higher than the three auction revenues for the Shanghai auction treatment. So the binary win count in one direction is 0.

6.   This is an open-ended question. You might consider the effects of providing less information to subjects about possible demands, or maybe a structure with random switches between high and low demand that do not occur every period, e.g., some persistence. In any case, you need to explain why your proposed treatment would be an interesting experiment to run.

Chapter 29 Hints for Problems

1.   Since bidder III has the highest value for *C*, note who has the highest value for the combination *AB*. What would be the total value obtained by awarding the license *C* to bidder III and awarding the package *AB* to bidder 1? If *AB* goes to bidder I, you would need prices for *A* and *B* that sum to a number that is less than or equal to 30, which is bidder I's value for *AB*. The price of *C* would have to be below 10 for bidder III to be willing to buy it. But the prices would need to be high enough to prevent the other bidders from bidding on any combinations that would create excess demand. Can you find any such prices?

2.   Look at bidder IV's value for *ABC*.

3. Think about using an activity rule to keep bidders submitting bids until the bidding gets so high that the number of active bidders falls from 3 to 2. Why would those two bidders have wanted to stay active up to this point?

4. Think about whether it would be better for revenue to allocate the licenses as a package or as three separate licenses.

5. The "overhang" of $69 - 60$ must be allocated to the three licenses. This allocation should involve equal parts for each license since the three populations in those areas are equal.

6. Treat this problem as a challenge that was not undertaken by the people who designed the FCC incentive auction. Maybe construct some numerical examples of how a procedure might work. Is there a danger of sellers manipulating the process to get high prices?

Chapter 30 Hints for Problems

1. F4 is turned down by W1 and W2, F3 is first accepted by W1 and then dropped, and is turned down by W2. Both F3 and F4 next approach their third choices.

2. F1 ranks W1 second, and W1 ranks F1 second, so the priority product for the top left cell is 4. Fill out the other priority products and then enact them sequentially in order from a priority product of 1 (if any) to a priority product of 2 (if any) to 3 (if any) to 4, etc.

3. No hint provided.

4. Look for a match that was not made under deferred acceptance, and increase the value of another match by a large amount in a way that would not alter the submitted rankings.

5. If you think that submitting an untruthful list is better, be sure to document why this is the case in terms of a specific example.

6. Remember that the order in which submitted rankings are acted on does not necessarily have to be random (as mentioned in the first paragraph of section 30.3).

References

Ackert, Lucy and Richard Deaves (2010) *Behavioral Finance: Psychology, Decision-Making, and Markets*, Mason, OH: South-Western Cengage.

Akerlof, George A. (1970) "The Market for 'Lemons': Quality Uncertainty and the Market Mechanism," *Quarterly Journal of Economics*, 84, 488–500.

Allais, Maurice (1953) "Le Comportement De L'homme Rationnel Devant Le Risque, Critique Des Postulates Et Axiomes De L'ecole Americaine," *Econometrica*, 21, 503–546.

Alsopp, Louise and John D. Hey (2000) "Two Experiments to Test a Model of Herd Behavior," *Experimental Economics*, 3, 121–136.

Anderson, Lisa R. and Beth A. Freeborn (2010) "Varying the Intensity of Competition in a Multiple Prize Rent Seeking Experiment," *Public Choice*, 143(1–2), 237–254.

Anderson, Lisa R. and Charles A. Holt (1996a) "Classroom Games: Understanding Bayes' Rule," *Journal of Economic Perspectives*, 10, 179–187.

——— (1996b) "Classroom Games: Information Cascades," *Journal of Economic Perspectives,* 10, 187–193.

——— (1997) "Information Cascades in the Laboratory," *American Economic Review*, 87, 847–862.

Anderson, Lisa R., Charles A. Holt, and David Reiley (2008) "Congestion and Social Welfare," in *Experimental Methods, Environmental Economics*, T. L. Cherry, J. F. Shogren, and S. Kroll, eds., London: Routledge, 280–292.

Anderson, Lisa R. and Sarah L. Stafford (2003) "An Experimental Analysis of Rent Seeking Under Varying Competitive Conditions," *Public Choice*, 115, 199–216.

Anderson, Simon P., Jacob K. Goeree, and Charles A. Holt (1998) "Rent Seeking with Bounded Rationality: An Analysis of the All Pay Auction," *Journal of Political Economy*, 106, 828–853.

Andreoni, James (1995) "Cooperation in Public Goods Experiments: Kindness or Confusion," *American Economic Review*, 85(4), 891–904.

Andreoni, James, Marco Castillo, and Regan Petrie (2003) "What Do Bargainers' Preferences Look Like? Experiments with a Convex Bargaining Game," *American Economic Review*, 93(3), 972–985.

Andreoni, James and Alison Sanchez (2014) "An Experimental Analysis of the Cognitive Processes Underlying Beliefs and Perception Manipulation," Discussion Paper, UC San Diego.

Andreoni, James and Lise Vesterlund (2001) "Which Is the Fair Sex? Gender Differences in Altruism," *Quarterly Journal of Economics*, 116, 293–312.

Ansolabehere, Stephen, Shanto Iyengar, Adam Simon, and Nicholas Valentino (1994) "Does Attack Advertising Demobilize the Electorate?" *American Political Science Review*, 88, 829–838.

Aragones, Enriqueta and Thomas R. Palfrey (2004) "The Effect of Candidate Quality on Electoral Equilibrium: An Experimental Study," *American Political Science Review* 98 (February), 77–90.

Armantier, Olivier and Charles A. Holt (2017a) "Discount Window Stigma: An Experimental Investigation," Discussion Paper, Federal Reserve Bank of New York.

—— (2017b) "Endogenous Reference Price Auctions for a Diverse Set of Commodities: An Experimental Analysis," Discussion Paper, Federal Reserve Bank of New York.

Armantier, Olivier, Charles A. Holt, and Charles R. Plott (2013) "A Procurement Auction for Toxic Assets with Asymmetric Information," *American Economic Journal: Microeconomics*, 5(4), 2013, 142–162.

Armantier, Olivier and Nicolas Treich (2013) "Eliciting Beliefs: Proper Scoring Rules, Incentives, Stakes, and Hedging," *European Economic Review*, 62, 17–40.

Asparouhova, Elena, Peter Bossaerts, Nilanjan Roy, and William Zame (2016) "'Lucas in the Laboratory," *Journal of Finance*, 71(6), 2727–2780.

Assenza, Tiziana, Te Bao, Cars Hommes, and Domenico Massaro (2014) "Experiments on Expectations in Macroeconomics and Finance," in *Experiments in Macroeconomics*, volume 17 of *Research in Experimental Economics*, J. Duffy, ed., Emerald Insight, Series ISSN, Bingley, UK: Emerald Group Publishing 0193-2306.

Ausubel, Larry M., Peter Cramton, and Paul Milgrom (2004) "The Clock-Proxy Auction: A Practical Combinatorial Auction Design," Discussion Paper, University of Maryland.

Babcock, Linda, Maria Recalde, Lise Vesterlund, and Laurie Weingart (2017) "Gender Differences in Accepting and Receiving Requests for Tasks with Low Productivity," *American Economic Review*, 107(3), 714–747.

Badasyan, Narine, Jacob K. Goeree, Monica Hartmann, Charles A. Holt, John Morgan, Tanya Rosenblat, Marcos Servatka, and Dirk Yandell (2009) "Vertical Integration of Successive Monopolists: A Classroom Experiment," *Perspectives on Economic Education Research*, 5(1), 1–18.

Bagnoli, Mark and Michael McKee (1991) "Voluntary Contributions Games: Efficient Private Provision of Public Goods," *Economic Inquiry*, 29(2), 351–366.

Balkenborg, Dieter, Todd Kaplan, and Timothy Miller (2011) "Teaching Bank Runs with Classroom Experiments," *Journal of Economic Education*, 42(3), 224–242.

Ball, Sheryl B., Max H. Bazerman, and J. S. Caroll (1990) "An Evaluation of Learning in the Bilateral Winner's Curse," *Organizational Behavior and Human Decision Processes*, 48, 1–22.

Ball, Sheryl B. and Charles A. Holt (1998) "Classroom Games: Bubbles in an Asset Market," *Journal of Economic Perspectives*, 12, 207–218.

Bannerjee, Abhijit V. (1992) "A Simple Model of Herd Behavior," *Quarterly Journal of Economics*, 107, 797–817.

Barclay, Pat (2004) "Trustworthiness and Competitive Altruism Can Also Solve the 'Tragedy of the Commons,'" *Evolution and Human Behavior*, 25, 209–220.

Baron, David P. and John A. Ferejohn (1989) "Bargaining in Legislatures," *American Political Science Review*, 83, 1181–1206.

Barr, Abigal M. (2003) "Trust and Expected Trustworthiness: Experimental Evidence from Zimbabwean Villages," *Economic Journal*, 113, 614–630.

Basu, Kaushik (1994) "The Traveler's Dilemma: Paradoxes of Rationality in Game Theory," *American Economic Review*, 84(2), 391–395.

Battalio, Raymond C., John H. Kagel, and Komain Jiranyakul (1990) "Testing Between Alternative Models of Choice Under Uncertainty: Some Initial Results," *Journal of Risk and Uncertainty*, 3(1), 25–50.

Battalio, Raymond C., John H. Kagel, and Don N. MacDonald (1985) "Animals' Choices over Uncertain Outcomes: Some Initial Experimental Results," *American Economic Review*, 75, 597–613.

Baujard, Antoinette, Herrade Igersheim, Isabelle Lebon, Frédéric Gavrel, and Jean-François Laslier (2014) "Who's Favored by Evaluative Voting? An Experiment Conducted During the 2012 French Presidential Election," *Electoral Studies*, 34, 131–145.

Bazerman, Max H. and William F. Samuelson (1983) "I Won the Auction but Don't Want the Prize," *Journal of Conflict Resolution*, 27, 618–634.

Becker, Gordon M., Morris H. DeGroot, and Jacob Marschak (1964) "Measuring Utility by a Single-Response Method," *Behavioral Science*, 9, 226–232.

Becker, Tilman, Michael Carter, and Jorg Naeve (2005) "Experts Playing the Traveler's Dilemma," Discussion Paper, Honhenheim University.

Bediou, Benoit, Irene Comeig, Ainhoa Jaramillo-Gutierrez, and David Sander (2013) "The Role of 'Perceived Loss' Aversion on Credit Screening: An Experiment," *Spanish Journal of Finance and Accounting*, 157, 83–98.

Berg, Joyce E., John W. Dickhaut, and Kevin A. McCabe (1995) "Trust, Reciprocity, and Social History," *Games and Economic Behavior*, 10, 122–142.

Bergstrom, Ted, Rod Garratt, and Greg Leo (2015) "Let Me or Let George? Motivations of Competing Altruists," Discussion Paper, UC Santa Barbara.

Bernanke, Ben S. (2009) "The Federal Reserve's Balance Sheet: An Update," Speech at the Federal Reserve Board Conference on Key Developments in Monetary Policy, Washington, DC, available at http://www.federalreserve.gov/newsevents/speech/bernanke20091008a.htm.

Bernoulli, Daniel (1738) "Specimen Theoriae Novae De Mensura Sortis (Exposition on a New Theory on the Measurement of Risk)," *Comentarii Academiae Scientiarum Imperialis Petropolitanae*, 5, 175–192, translated by L. Sommer in *Econometrica*, 1954, 22, 23–36.

Bichler, Martin and Jacob K. Goeree (2017) *Handbook of Spectrum Auction Design*, Cambridge, UK: Cambridge University Press.

Bikhchandani, Sushil, David Hirschleifer, and Ivo Welch (1992) "A Theory of Fads, Fashion, Custom, and Cultural Change as Informational Cascades," *Journal of Political Economy*, 100, 992–1026.

Binmore, Ken, Lisa Stewart, and Alex Voorhoeve (2012) "How Much Ambiguity Aversion? Finding Indifferences between Ellsberg's Risky and Ambiguous Bets," *Journal of Risk and Uncertainty*, 45, 215–238.

Binswanger, Hans P. (1980) "Attitudes toward Risk: Experimental Measurement in Rural India," *American Journal of Agricultural Economics*, 62, 395–407.

Bohr, Clement E., Charles A. Holt, and Alexandra V. Schubert (2017) "Saving, Spending, and Speculation," Discussion Paper, University of Virginia.

Bolton, Gary E. and Axel Ockenfels (2000) "ERC: A Theory of Equity, Reciprocity, and Competition," *American Economic Review*, 90(1), 166–193.

Bolton, Gary E., Axel Ockenfels, and Ulrich W. Thonemann (2012) "Managers and Students as Newsvendors," *Management Science*, 58(12), 2225–2233.

Bornstein, Gary, Tamar Kugler, and Anthony Ziegelmeier (2004) "Individual and Group Decisions in the Centipede Game: Are Groups More 'Rational' Players?" *Journal of Experimental Social Psychology*, 40, 599–605.

Bornstein, Gary and Ilan Yaniv (1998) "Individual and Group Behavior in the Ultimatum Game: Are Groups More 'Rational' than Individuals," *Experimental Economics*, 1, 101–108.

Bostian, AJ A., Jacob K. Goeree, and Charles A. Holt (2005) "Price Bubbles in an Asset Market Experiment with a Flat Fundamental Value," Discussion Paper, University of Virginia.

Bostian, AJ A. and Charles A. Holt (2013) "Classroom Clicker Games: Wisdom of the Crowds and the Winner's Curse," *Journal of Economic Education*, 44, 217–229.

Bostian, AJ A., Charles A. Holt, and Angela M. Smith (2008) "The Newsvendor Pull-to-Center Effect: Adaptive Learning in a Laboratory Experiment," *Manufacturing and Service Operations Management*, 10(4), 590–608.

Boulou-Resheff, Beatrice and Charles Holt (2017) "Inventory Management with Carryover in a Laboratory Setting: Going Beyond the Newsvendor Paradigm," Working Paper, University Paris 1, Sorbonne.

Brams, Steven J. and Peter C. Fishburn (1978) "Approval Voting," *American Political Science Review*, 72, 831–847.

—— (1988) "Does Approval Voting Elect the Lowest Common Denominator?" *Political Science and Politics*, 21, 277–284.

Brañas-Garza, Pablo, Praveen Kujal, and Balint Lenkei (2016) "Cognitive Reflection Test: Whom, How, When..." Research Report, Behavioral Economics Group, Middlesex University, London.

Brañas-Garza, Pablo and John Smith (2016) "Cognitive Abilities and Economic Behavior," *Journal of Economic Behavior and Organization*, 64, 1–4.

Brandts, Jordi, and David Cooper (2006) "A Change Would Do You Good, . . . An Experimental Study On How to Overcome Coordination Failure in Organizations," *American Economic Review*, 96(2), 669–693.

—— (2007) "It's What You Say, Not What You Pay: An Experimental Study of Manager-Employee Relationships in Overcoming Coordination Failure," *European Economic Review*, 5(6), 1223–1268.

Brandts, Jordi, and Charles A. Holt (1992) "An Experimental Test of Equilibrium Dominance in Signaling Games," *American Economic Review*, 82, 1350–1365.

Breaban, Adriana and Charles F. Noussair (2017) "Emotional State and Market Behavior," *Review of Finance*, 1–31.

Breit, William and Kenneth Elzinga (1978) *Murder at the Margin* (under the pseudonym Marshall Jevons), Princeton, NJ: Princeton University Press.

Brennan, Geoffrey and James M. Buchanan (1984) "Voter Choice: Evaluating Political Alternatives," *American Behavioral Scientist*, 28(2), 185–201.

Brown, Alexander L., Colin F. Camerer, and Zhikang Eric Chua (2009) "Learning and Visceral Temptation in Dynamic Saving Experiments," *Quarterly Journal of Economics*, 124(1), 197–231.

Brown, Alexander L. and Paul J. Healy (2018) "Separated Choices," *European Economic Review*, 101, 20–34.

Brunner, Christoph, Jacob Goeree, and Charles A. Holt (2005) "Bid Driven Versus Clock Driven Auctions with Package Bidding," Discussion Paper, presented at the Southern Economic Association Meetings in Washington, DC, November 2005.

Bryant, John (1983) "A Simple Rational Expectations Keynes-Type Model," *Quarterly Journal of Economics*, 98, 525–528.

Buchan, Nancy R., Rachel T. A. Croson, and Sara Solnick (2008) "Trust and Gender: An Examination of Behavior and Beliefs in the Investment Game," *Journal of Economic Behavior and Organization*, 68(3–4), 466–476.

Burtraw, Dallas, Jacob Goeree, Charles Holt, Erica Myers, Karen Palmer, and William Shobe (2009) " Collusion in Auctions for Emissions Permits: An Experimental Analysis," *Journal of Policy Analysis and Management*, 28(4), 672–691.

Burtraw, Dallas, Charles A. Holt, Erica Myers, Jacob Goeree, Karen Palmer, and William Shobe (2010) "Price Discovery in Emission Permit Auctions," in R. M. Isaac and D. A. Norton, eds., *Experiments on Energy, the Environment, and Sustainability, Research in Experimental Economics, Vol. 14*, Bingley, UK: Emerald Group Publishing, 11–36.

Bykowsky, Mark, Robert Cull, and John Ledyard (2000) "Mutually Destructive Bidding: The FCC Auction Design Problem," *Journal of Regulatory Economics*, 17(3), 205–228.

Caginalp, Gunduz, David Porter, and Vernon L. Smith (2001) "Financial Bubbles: Excess Cash, Momentum, and Incomplete Information," *Journal of Psychology and Financial Markets*, 2, 80–99.

Camerer, Colin F. (1989) "An Experimental Test of Several Generalized Utility Theories," *Journal of Risk and Uncertainty*, 2, 61–104.

———— (1995) "Individual Decision Making," in *The Handbook of Experimental Economics*, J. H. Kagel, and A. E. Roth, eds., Princeton, NJ: Princeton University Press, 587–703.

———— (2003) *Behavioral Game Theory*, Princeton, NJ: Princeton University Press.

———— (2016) "The Promise and Success of Lab-Field Generalizability in Experimental Economics: A Critical Reply to Levitt and List," in *Handbook of Experimental Economic Methodology*, G. R. Frechette and A. Schotter, eds., Oxford, UK: Oxford University Press, 249–295.

Camerer, Colin F. and Teck-Hua Ho (1999) "Experience Weighted Attraction Learning in Normal-Form Games," *Econometrica*, 67, 827–874.

Camerer, Colin F., Teck-Hua Ho and Juin-Kuan Chong (2004) "A Cognitive Heirarchy Model of Games," *Quarterly Journal of Economics*, 119(3), 861–898.

Capen, E. C., R. V. Clapp, and W. M. Campbell (1971) "Competitive Bidding in High Risk Situations," *Journal of Petroleum Technology*, 23, 641–653.

Capra, C. Monica, Irene Comeig, and Matilde O. Fernandez Banco (2014) "Entrepreneurship and Credit Rationing: How to Screen Successful Projects," in *Entrepreneurship, Innovation and Economic Crisis, Lessons for Research, Policy and Practice*, K. Rudiger et al., eds., Cham, Switzerland: Springer International, 139–148.

Capra, C. Monica, Jacob K. Goeree, Rosario Gomez, and Charles A. Holt (1999) "Anomalous Behavior in a Traveler's Dilemma?" *American Economic Review*, 89, 678–690.

———— (2002) "Learning and Noisy Equilibrium Behavior in an Experimental Study of Imperfect Price Competition," *International Economic Review*, 43(3), 613–636.

Capra, C. Monica and Charles A. Holt (2000) "Classroom Experiments: A Prisoner's Dilemma," *Journal of Economic Education*, 21(3), 229–236.

Capra, C. Monica, Tonomi Tanaka, Colin F. Camerer, Lauren Feiler, Veronica Sovero, and Charles Noussair (2009) "The Impact of Simple Institutions in Experimental Economies with Poverty Traps," *Economic Journal*, 119(539), 977–1009.

Cardenas, Juan Camilo (2003) "Real Wealth and Experimental Cooperation: Evidence from Field Experiments," *Journal of Development Economics*, 70(2), 263–289.

Cardenas, Juan Camilo, John K. Stranlund, and Cleve Willis (2000) "Local Environmental Control and Institutional Crowding Out," *World Development*, 28(10), 1719–1733.

—— (2002) "Economic Inequality and Burden-sharing in the Provision of Local Environmental Quality," *Ecological Economics*, 40, 379–395.

Carpenter, Jeffrey, Steven Burks, and Eric Verhoogen (2005) "Comparing Students to Workers: The Effects of Stakes, Social Context, and Demographics on Bargaining Outcomes," in *Field Experiments in Economics*, J. Carpenter, G. Harrison, and J. List, eds., Greenwich, CT: JAI Press, 261–290.

Carpenter, Jeffrey, Eric Verhoogen, and Steven Burks (2005) "The Effect of Stakes in Distribution Experiments," *Economics Letters*, 86, 393–398.

Cason, Timothy N. (1995) "Cheap Talk and Price Signaling in Laboratory Markets," *Information Economics and Policy*, 7, 183–204.

—— (2000) "The Opportunity for Conspiracy in Asset Markets Organized with Dealer Intermediaries," *Review of Financial Studies*, 13(2), 385–416.

Cason, Timothy N. and Douglas D. Davis (1995) "Price Communications in Laboratory Markets: An Experimental Investigation," *Review of Industrial Organization*, 10, 769–787.

Castillo, Marco, Regan Petrie, and Anya Samek (2017) "Time to Give: A Field Experiment on Intertemporal Charitable Giving," Discussion Paper, Texas A&M.

Celen, Bogachan and Shachar Kariv (2004) "Distinguishing Informational Cascades from Herd Behavior in the Laboratory," *American Economic Review*, 94(3), 484–497.

Chamberlin, Edward H. (1948) "An Experimental Imperfect Market," *Journal of Political Economy*, 56, 95–108.

Charness, Gary and Uri Gneezy (2012) "Strong Evidence for Gender Differences in Risk Taking," *Journal of Economic Behavior and Organization*, 83(1), 50–58.

Charness, Gary, Edi Karni, and Dan Levin (2013) "Ambiguity Attitudes and Social Interactions: An Experimental Investigation," *Journal of Risk and Uncertainty*, 46, 1–25.

Charness, Gary and Dan Levin (2005) "When Optimal Choices Feel Wrong: A Laboratory Study of Bayesian Updating, Complexity, and Affect," *American Economic Review*, 95, 1300–1309.

Chen, Yan, Peter Katuscak, and Emre Ozdenoren (2013) "Why Can't a Woman Bid Like a Man," *Games and Economic Behavior*, 77(1), 181–213.

Chen, Yan and Charles R. Plott (1996) "The Groves-Ledyard Mechanism: An Experimental Study of Institutional Design," *Journal of Public Economics*, 59, 335–364.

Chen, Yan and Tayfun Sonmez (2004) "An Experimental Study of House Allocation Mechanisms," *Economics Letters*, 83(1), 137–140.

—— (2006) "School Choice: An Experimental Study," *Journal of Economic Theory*, 127, 202–231.

Cherry, Todd L., Peter Frykblom, and Jason F. Shogren (2002) "Hardnose the Dictator," *American Economic Review*, 92(4), 1218–1221.

Christiansen, Nels, Sotiris Georganas, and John H. Kagel (2014) "Coalition Formation in a Legislative Voting Game," *American Economic Journal: Microeconomics*, 6(1), 182–204.

Christie, William G. and Roger D. Huang (1995) "Following the Pied Piper: Do Individual Returns Herd Around the Market?" *Financial Analysts Journal*, 51, 31–37.

Cinyabuguma, Matthias, Talbot Page, and Louis Putterman (2005) "Cooperation under the Threat of Expulsion in a Public Goods Experiment," *Journal of Public Economics*, 89, 1421–1435.

Cipriani, Marco, Ana Fostel, and Dan Houser (2012) "Leverage and Asset Prices: An Experiment," George Mason Working Paper in Economics, 12–05.

Cipriani, Marco and Antonio Guarino (2005) "Herd Behavior in a Laboratory Financial Market," *American Economic Review*, 95(5), 1227–1443.

Clement, Douglas, ed. (2007) "Interview with Eugene Fama," *The Region*, Federal Reserve Bank of Minneapolis, December.

Cochard, Francois, Phu Nguyen-Van, and Marc Willinger (2004) "Trust and Reciprocity in a Repeated Investment Game," *Journal of Economic Behavior and Organization*, 55(1), 31–44.

Cohen, Mark and David Scheffman (1989) "The Antitrust Sentencing Guideline: Is the Punishment Worth the Cost," *Journal of Criminal Law*, 27, 330–336.

Cohn, Alain, Jan Engelmann, Ernst Fehr, and Michel André Maréchal (2015) "Evidence for Countercyclical Risk Aversion: An Experiment with Financial Professionals," *American Economic Review*, 105(2), 860–885.

Colander, David, Sieuwerd Gaastra, and Casey Rothschild (2010) "The Welfare Costs of Market Restrictions," *Southern Economic Journal*, 77(1), 213–223.

Comeig, Irnene, Esther Del Brio, and Matilde O. Fernandez Blanco (2014) "Financing Successful Small Business Projects," *Management Decision*, 52(2), 365–377.

Comeig, Irene, Charles A. Holt, and Ainhoa Jaramillo (2016) "Dealing with Risk: Gender, Stakes, and Probability Effects," Discussion Paper, University of Virginia.

Composti, Jeanna (2003) "Asymmetric Payoffs in a Soccer Field Experiment," Distinguished Majors Thesis, University of Virginia.

Cooper, Russell, Douglas V. DeJong, Robert Forsythe, and Thomas W. Ross (1996) "Cooperation without Reputation: Experimental Evidence from Prisoners' Dilemma Games," *Games and Economic Behavior*, 12(2), 187–218.

Cooper, Russell and Andrew John (1988) "Coordinating Coordination Failures in Keynesian Models," *Quarterly Journal of Economics*, 103, 441–464.

Coppinger, Viki M., Vernon L. Smith, and Jon A. Titus (1980) "Incentives and Behavior in English, Dutch and Sealed-Bid Auctions," *Economic Inquiry*, 18, 1–22.

Coppock, Lee and Charles A. Holt (2014) "Teaching the Crisis: A Leverage Experiment," Discussion Paper, University of Virginia.

Coppock, Lee, Charles A. Holt, Anna Rorem, and Sijia Yang (2015) "Economics in 10,000 Words: A Picture Is Worth a Thousand Words," Discussion Paper, University of Virginia.

Corgnet, Brice, Hernán-González, Praveen Kujal, and David Porter (2014) "The Effect of Earned Versus House Money on Price Bubble Formation in Experimental Asset Markets," *Review of Finance*, 1–34.

Coricelli, Giogio, Dietmar Fehr, and Gerlinde Fellner (2004) "Partner Selection in Public Goods Experiments," *Journal of Conflict Resolution*, 48, 356–378.

Coursey, David L., John L. Hovis, and William D. Schulze (1987) "The Disparity between Willingness to Accept and Willingness to Pay Measures of Value," *Quarterly Journal of Economics*, 102, 679–690.

Cox, James C. (2004) "How to Identify Trust and Reciprocity," *Games and Economic Behavior*, 46, 260–281.

Cox, James C., Bruce Roberson, and Vernon L. Smith (1982) "Theory and Behavior of Single Object Auctions," in *Research in Experimental Economics, Vol. 2*, V. L. Smith, ed., Greenwich, CT: JAI Press, 1–43.

Cox, James C., Vernon L. Smith, and James M. Walker (1985) "Expected Revenue in Discriminative and Uniform Price Sealed-Bid Auctions," in *Research in Experimental Economics, Vol. 3*, V. L. Smith, ed., Greenwich, CT: JAI Press, 183–232.

—— (1988) "Theory and Individual Behavior of First-Price Auctions," *Journal of Risk and Uncertainty*, 1, 61–99.

Cox, James C. and Sadiraj Vjollca (2001) "Risk Aversion and Expected Utility Theory: Coherence for Small and Large Scale Gambles," Discussion Paper, University of Arizona.

Crockett, Sean and John Duffy (2015) "An Experimental Test of the Lucas Asset Pricing Model," Discussion Paper, University of California, Irvine.

Crosetto, Paolo and Antonio Filippin (2013) "The 'Bomb' Elicitation Task," *Journal of Risk and Uncertainty*, 47(1), 31–65.

Croson, Rachel T. A. (1996) "Partners and Strangers Revisited," *Economics Letters*, 53, 25–32.

Croson, Rachel T. A. and Melanie Marks (2000) "Step Returns in Threshold Public Goods: A Meta- and Experimental Analysis," *Experimental Economics*, 2(3), 239–259.

Croson, Rachel and Karen Donohue (2002) "Experimental Economics and Supply-Chain Management," *Interfaces*, 32(5), 74–82.

—— (2005) "Upstream Versus Downstream Information and Its Impact on the Bullwhip Effect," *System Dynamics Review*, 21(3), 249–260.

—— (2006) "Behavioral Causes of the Bullwhip Effect and the Observed Value of Inventory Information," *Management Science*, 52(3), 323–336.

Cummings, Ronald, Charles. A. Holt, and Susan K. Laury (2004) "Using Laboratory Experiments for Policy Making: An Example from the Georgia Irrigation Reduction Auction," *Journal of Policy Analysis and Management*, 3(2), 241–263.

Dal Bó, Pedro and Guillaume R. Fréchette (2011) "The Evolution of Cooperation in Infinitely Repeated Games: Experimental Evidence," *American Economic Review*, 101(1), 411–429.

Darley, John M. and Bibb Latane (1968) "Bystander Intervention in Emergencies: Diffusion of Responsibility," *Journal of Personality and Social Psychology*, 8, 377–383.

Davis, Douglas D. and Charles A. Holt (1993) *Experimental Economics*, Princeton,NJ: Princeton University Press.

—— (1994a) "Market Power and Mergers in Markets with Posted Prices," *RAND Journal of Economics*, 25, 467–487.

—— (1994b) "The 1994 Virginia Senate Market," unpublished draft, Virginia Commonwealth University.

—— (1994c) "Equilibrium Cooperation in Three-Person, Choice-of-Partner Games," *Games and Economic Behavior*, 7, 39–53.

—— (1996) "Price Rigidities and Institutional Variations in Markets with Posted Prices," *Economic Theory*, 9(1) 63–80.

—— (1998) "Conspiracies and Secret Price Discounts," *Economic Journal*, 108, 736–756.

Davis, Douglas D. and Robert Reilly (1998) "Do Too Many Cooks Always Spoil the Stew? An Experimental Analysis of Rent Seeking and the Role of a Strategic Buyer," *Public Choice*, 95, 89–115.

Davis, Douglas D. and Arlington W. Williams (1991) "The Hayek Hypothesis in Experimental Auctions: Institutional Effects and Market Power," *Economic Inquiry*, 29, 261–274.

Davis, Douglas D. and Bart Wilson (2002) "Collusion in Procurement Auctions: An Experimental Examination," *Economic Inquiry*, 40(2), 213–230.

Dechenaux, Emmanuel, Dan Kovenock, and Roman M. Sheremeta (2015) "A Survey of Experimental Research on Contests, All-Pay Auctions, and Tournaments," *Experimental Economics*, 18(4), 609–669.

Deck, Cary A., Jungmin Lee, Javier A. Reyes, and Christopher C. Rosen (2013) "A Failed Attempt to Explain Within Subject Variation in Risk Taking Behavior Using Domain Specific Risk Attitudes," *Journal of Economic Behavior and Organization*, 87, 1–24.

DeJong, David V., Robert Forsythe, and Russell Lundholm (1985) "Ripoffs, Lemons, and Reputation Formation in Agency Relationships: A Laboratory Market Study," *Journal of Finance*, 40, 809–820.

DeMartini, Christine, Anthony M. Kwasnica, John O. Ledyard, and David Porter (1999) "A New and Improved Design for Multi-Object Iterative Auctions," Caltech Social Science Working Paper 1054, revised March 1999.

de Palma, Andre, Moshe Ben-Akiva, David Brownstone, Charles Holt, Thierry Magnac, Daniel McFadden, Peter Moffatt, Nathalie Picard, Kenneth Train, Peter Wakker, and Joan Walker (2008) "Risk, Uncertainty, and Discrete Choice Models," *Marketing Letters*, 8, nos. 3–4 (July), 269–285.

Devenow, Andrea and Ivo Welch (1996) Rational Herding in Financial Economics," *European Economic Review*, 40, 603–615.

Diamond, Douglas W. and Philip H. Dybvig (1983) "Bank Runs, Deposit Insurance, and Liquidity," *Journal of Political Economy*, 91(3), 401–419.

Diekmann, Andreas (1985) "Volunteer's Dilemma," *Journal of Conflict Resolution*, 29, 605–610.

—— (1986) "Volunteer's Dilemma: A Social Trap without a Dominant Strategy and Some Empirical Results," in *Paradoxical Effects of Social Behavior: Essays in Honor of Anatol Rapoport*, A. Diekmann, and P. Mitter, eds., Heidelberg: Physica-Verlag, 187–197.

Dimmock, Stehen G., Roy Kouwenberg, and Peter P. Wakker (2016) "Ambiguity Attitudes in a Large Representative Sample," *Management Science*, 62(5), 1363–1380.

Dohmen, Thomas, Armin Falk, David Huffman, Uwe Sunde, Jurgen Schupp, and Gert Wagner (2011) "Individual Risk Attitudes: Measurements, Determinants, and Behavioral Consequences," *Journal of the European Economic Association*, 9(3), 522–550.

Doraszelski, Ulrich, Katja Seim, Michael Sinkinson, and Peichun Wang (2016) "Strategic Supply Reduction as Rent-Seeking Behavior," Discussion Paper, UPenn, Wharton.

Dorsey, Robert and Laura Razzolini (2003) "Explaining Overbidding in First-Price Auctions Using Controlled Lotteries," *Experimental Economics*, 6, 123–140.

Downs, Anthony (1957) "An Economic Theory of Political Action in a Democracy," *Journal of Political Economy*, 65, 135–150.

Drehmann, Mathias, Jörg Oechssler, and Andreas Roider (2005) "Herding and Contrarian Behavior in Financial Markets: An Internet Experiment," *American Economic Review*, 95(5), 1403–1426.

Drichoutis, Andreas and Jason L. Lusk (2016) "What Can Multiple Price Lists Really Tell Us about Risk Preferences?" *Journal of Risk and Uncertainty*, 53, 89–106.

Duffy, John (2016) "Macroeconomics: A Survey of Laboratory Research," in *Handbook of Experimental Economics, Vol. 2*, J. Kagel and A. Roth, eds., Princeton, NJ: Princeton University Press, 1–90.

Duffy, John and Daniela Puzzello (2017) "Monetary Policies in the Laboratory," presentation made at the 2017 Economic Science Association Meetings, Richmond, Virginia.

Duffy, John and Margit Tavits (2008) "Beliefs and Voting Decisions: A Test of the Pivotal Voter Model," *American Journal of Political Science*, 52(3), 603–618.

Dufwenberg, Martin, Tobias Lindqvist, and Even Moore (2005) "Bubbles and Experience: An Experiment," *American Economic Review*, 95(5), 1731–1737.

Durham, Yvonne (2000) "An Experimental Examination of Double Marginalization and Vertical Relationships," *Journal of Economic Behavior and Organization*, 42(2), 207–229.

Eavey, Cheryl L. and Gary J. Miller (1984) "Fairness and Majority Rule Games with a Core," *American Journal of Political Science*, 28(3), 570–586.

Eckel, Catherine C. (2016) "Review of *Risky Curves: On the Empirical Failure of Expected Utility Theory*," *Economics and Philosophy*, 32(3), 540–548.

Eckel, Catherine C. and Sascha C. Füllbrunn (2015) "Thar SHE Blows? Gender Competition, and Bubbles in Experimental Asset Markets," *American Economic Review*, 105(2), 905–920.

——— (2017) "Hidden vs. Known Gender Effects in Experimental Asset Markets," *Economics Letters*, 156, 7–9.

Eckel, Catherine C. and Philip Grossman (1998) "Are Women Less Selfish Than Men? Evidence from Dictator Games," *Economic Journal*, 108, 726–735.

——— (2002) "Sex Differences and Statistical Stereotyping in Attitudes toward Financial Risk," *Evolution & Human Behavior*, 23(4), 281–295.

——— (2008a) "Subsidizing Charitable Contributions: A Natural Field Experiment Comparing Matching and Rebate Subsidies," *Experimental Economics*, 11(3), 234–252.

——— (2008b) "Differences in the Economic Decisions of Men and Women: Experimental Evidence," in *Handbook of Experimental Economics Results*, Vol. 1, C. R. Plott and V. L. Smith, eds., New York: Elsevier, 509–519.

——— (2008c) "Forecasting Risk Attitudes: An Experimental Study Using Actual and Forecast Gamble Choices," *Journal of Economic Behavior and Organization*, 68(1), 1–17.

Eckel, Catherine C. and Charles A. Holt (1989) "Strategic Voting Behavior in Agenda-Controlled Committee Experiments," *American Economic Review*, 79, 763–773.

Eckel, Catherine C., Cathleen Johnson, and Claude Montmarquette (2005) "Savings Decisions of the Working Poor: Short and Long-Term Decisions," in *Research in Experimental Economics Vol. 10: Field Experiments in Economics*, G. Harrison, J. Carpenter, and J. List, eds., Bingley, UK: Emerald Group Publishing, 219–260.

Eckel, Catherine C., Cathleen Johnson, Claude Montmarquette, and Christian Rojas (2007) "Debt Aversion and the Demand for Loans for Postsecondary Education," *Public Finance Review*, 35(2), 233–262.

Eckel, Catherine C. and Rick K. Wilson (2004) "Is Trust a Risky Decision?" *Journal of Economic Behavior and Organization*, 55(4), 447–465.

Edwards, Ward (1962) "Subjective Probabilities Inferred from Decisions," *Psychological Review*, 69(2), 109–135.

Ehrhart, Karl-Martin and Claudia Keser (1999) "Mobility and Cooperation: On the Run," CIRANO Working Paper 99s-24.

Ellickson, Robert C. (1993) "Property in Land," *Yale Law Journal*, 102, 1315–1400.

Ellsberg, Daniel (1961) "Risk, Ambiguity, and the Savage Axioms," *Quarterly Journal of Economics*, 75, 643–669.

Engelbrecht-Wiggans, Richard and Elena Katok (2008) "Regret and Feedback Information in First Price Sealed-Bid Auctions," *Management Science*, 54(4), 808–819.

Engelbrecht-Wiggans, Richard, John A. List, and David H. Reiley (2006) "Demand Reduction in Multi-unit Auctions with Varying Numbers of Bidders: Theory and Field Experiments," *International Economic Review*, 47(1), 203–231.

Ensminger, Jean (2004) "Market Integration and Fairness: Evidence from Ultimatum, Dictator, and Public Goods Experiments in East Africa," in *Foundations of Human Sociality: Economic Experiments and Ethnographic Evidence from Fifteen Small-Scale Societies*, Henrich, Boyd, Bowles, Camerer, Fehr, and Gintis, eds., Oxford, UK: Oxford University Press, 356–381.

Falk, Armin (2007) "Gift Exchange in the Field," *Econometrica*, 75(5), 1501–1511.

Falk, Armin and Ernst Fehr (2003) "Why Labour Market Experiments?" *Labour Economics*, 10, 399–406.

Falk, Armin, Ernst Fehr, and Christian Zehnder (2006) "Fairness Perceptions and Reservation Wages: Behavioral Effects of Minimum Wages," *Quarterly Journal of Economics*, 121(4), 1347–1381.

Falk, Armin and Michael Kosfeld (2006) "The Hidden Costs of Control," *American Economic Review*, 96(5), 1611–1630.

Fantino, Edmund (1998) "Behavior Analysis and Decision Making," *Journal of the Experimental Analysis of Behavior*, 69, 355–364.

Featherstone, Clayton R. and Alvin E. Roth (2017) "Strategy-proof Mechanisms that Deal with Indifferences and Complicated Diversity Constraints: Matching MBAs to Countries at Harvard Business School," mimeo, Department of Economics, Stanford University.

Fehl, Katrin, Daniel J. van der Post, and Dirk Semmann (2011) Co-evolution of Behaviour and Social Network Structure Promotes Human Cooperation," *Ecology Letters*, 14, 546–551.

Fehr, Ernst and Simon Gächter (2000) "Cooperation and Punishment in Public Goods Experiments," *American Economic Review*, 90(4), 980–994.

Fehr, Ernst, Georg Kirchsteiger, and Arno Riedl (1993) "Does Fairness Prevent Market Clearing? An Experimental Investigation," *Quarterly Journal of Economics*, 108, 437–459.

Fehr, Ernst, Alexander Klein, and Klaus Schmidt (2001) "Fairness, Incentives, and Contractual Incompleteness," Working Paper 72, Institute of Empirical Research in Economics, University of Zurich.

Fehr, Ernst and John List (2004) "The Hidden Costs and Returns of Incentives— Trust and Trustworthiness Among CEOs," *Journal of the European Economic Association*, 2(5), 743–771.

Fehr, Ernst and Klaus M. Schmidt (1999) "A Theory of Fairness, Competition, and Cooperation," *Quarterly Journal of Economics*, 114, 769–816.

—— (2003) "Theories of Fairness and Reciprocity—Evidence and Economic Applications," in *Advances in Economics and Econometrics*, M. Dewatripont, L. Hansen, and S. J. Turnovsky, eds., Cambridge: Cambridge University Press, 208–257.

Fehr-Duda, Helga, Thomas Epper, Adrian Bruhin, and Renate Schubert (2011), "Risk and Rationality: The Effects of Mood and Decision Rules on Probability Weighting," *Journal of Economic Behavior and Organization*, 78(1), 14–24.

Feltovich, Nicholas J. (2006) "Slow Learning in the Market for Lemons: A Note on Reinforcement Learning and the Winner's Curse," *Computational Economics: A Perspective from Computational Intelligence*, DOI: 10.4018/9781591406495.ch007.

Fenig, Guidon, Mariya Mileva, and Luba Petersen (2017) "Deflating Asset Price Bubbles with Leverage Constraints and Monetary Policy," Working Paper, Simon Fraser University.

Fershtman, Chaim and Uri Gneezy (2001) "Discrimination in a Segmented Society: An Experimental Approach," *Quarterly Journal of Economics*, 116(1), 351–377.

Fey, Mark, Richard D. McKelvey, and Thomas R. Palfrey (1996) "An Experimental Study of Constant-Sum Centipede Games," *International Journal of Game Theory*, 25, 269–287.

Filippin, Antonio and Paolo Crosetto (2016) "A Reconsideration of Gender Differences in Risk Attitudes," *Management Science*, 62(11), 3138–3160.

Filiz-Ozbay, Emel and Erkut Y. Ozbay (2007) "Auctions with Anticipated Regret: Theory and Experiment," *American Economic Review*, 97(4), 1407–1418.

Finley, Grace, Charles A. Holt, and Emily Snow (2018) "The Welfare Costs of Price Controls and Rent Seeking in a Class Experiment," forthcoming in *Experimental Economics*.

Forrester, J. (1961) *Industrial Dynamics*, New York: MIT Press.

Forsythe, Robert, Joel L. Horowitz, N. E. Savin, and Martin Sefton (1988) "Fairness in Simple Bargaining Games," *Games and Economic Behavior*, 6, 347–369.

Forsythe, Robert, Roger Myerson, Thomas Rietz, and Robert Weber (1993) "An Experiment on Coordination in Multi-Candidate Elections: The Importance of Polls and Election Histories," *Social Choice and Welfare*, 10, 223–247.

—— (1996) "An Experimental Study of Voting Rules and Polls in Three-way Elections," *International Journal of Game Theory*, 25, 355–383.

Forsythe, Robert, Forrest Nelson, George R. Neumann, and Jack Wright (1992) "Anatomy of an Experimental Political Stock Market," *American Economic Review*, 82, 1142–1161.

Fragiadakis, Daniel E. and Peter Troyan (2016) "Improving Matching Under Hard Distributional Constraints," *Theoretical Economics*, 12(2), 863–908.

—— (2017) "Designing Mechanisms to Focalize Welfare-Improving Strategies," Discussion Paper, Texas A&M.

Franzen, Axel (1995) "Group Size and One Shot Collective Action," *Rationality and Society*, 7, 183–200.

Frechette, Guillaume R., John H. Kagel, and Steven F. Lehrer (2003) "Bargaining in Legislatures: An Experimental Investigation of Open versus Closed Amendment Rules," *American Political Science Review*, 97, 221–232.

Friedman, Daniel, R. Mark Isaac, Duncan James, and Shyam Sundar (2014) *Risky Curves: On the Empirical Failure of Expected Utility Theory*, London: Routledge.

Friedman, Milton and Rose D. Friedman (1989) *Free to Choose*, New York: Harcourt Brace.

Friedman, Milton and Leonard J. Savage (1948) "The Utility Analysis of Choices Involving Risk," *Journal of Political Economy*, 56(4), 279–304.

Füllbrunn, Sascha C. and Charles A. Holt (2017) "Gender Sorting and Bidding," Draft, Radbound University.

Füllbrunn, Sascha C., Dirk-Jan Janssen, and Utz Weitzel (2016) "Does Risk Aversion Cause Overbidding? New Evidence from First Price Sealed Bid Auctions," Discussion Paper 16-3, Radboud University.

Gächter, Simon and Manfred Königstein (2009) "Designing a Contract: A Simple Principal-Agent Problem as a Classroom Experiment," *Journal of Economic Education*, 40(2), 173–187.

Gale, David and Lloyd Shapley (1962) "College Admissions and the Stability of Marriage," *American Mathematical Monthly*, 69, 9–15.

Garber, Peter (1989) "Tulipmania," *Journal of Political Economy*, 97(3), 535–560.

Gardner, Roy, Elinor Ostrom, and James M. Walker (1990) "The Nature of Common-Pool Resource Problems," *Rationality and Society*, 2, 335–358.

Gerber, Alan S. and Donald P. Green (2000) "The Effects of Canvassing, Telephone Calls, and Direct Mail on Voter Turnout: A Field Experiment," *American Political Science Review*, 94, 653–663.

—— (2004) "Reclaiming the Experimental Tradition in Political Science," in *State of the Discipline*, Vol. 3, I. Katznelson and H. Milnor, eds., New York: Norton, 805–832.

Gibbons, Jean D. and S. Chakraborti (2014) *Nonparametric Statistical Inference*, Fourth edition, revised and expanded, New York: Taylor & Francis.

Gigerenzer Gerd and Ulrich Hoffrage (1995) "How to Improve Bayesian Reasoning Without Instruction: Frequency Formats," *Psychological Review*, 102, 684–704.

—— (1998) "Using Natural Frequencies to Improve Diagnostic Inferences," *Academic Medicine*, 73, 538–540.

Giusti, Giovanni, Janet Hua Jiang, and Yiping Xu (2016) "Interest on Cash, Fundamental Value Process and Bubble Formation," *Journal of Behavioral and Experimental Finance*, 11, 44–50.

Gneezy, Uri and Jan Potters (1997) "An Experiment on Risk Taking and Evaluation Periods," *Quarterly Journal of Economics*, 112(2), 631–645.

Goeree, Jacob K. and Charles A. Holt (1999a) "Employment and Prices in a Simple Macro-Economy," *Southern Economic Journal*, 65(3), 637–647.

—— (1999b) "Rent Seeking and the Inefficiency of Non-Market Allocations," *Journal of Economic Perspectives*, 13, 217–226.

—— (2000) "Asymmetric Inequality Aversion and Noisy Behavior in Alternating-Offer Bargaining Games," *European Economic Review*, 44, 1079–1089.

—— (2001) "Ten Little Treasures of Game Theory, and Ten Intuitive Contradictions," *American Economic Review*, 90(5), 1402–1422.

—— (2003) "Learning in Economics Experiments," in *Encyclopedia of Cognitive Science*, Vol. 2, L. Nadel, ed., London: Nature Publishing Group, McMillan, 1060–1069.

—— (2004) "A Model of Noisy Introspection," *Games and Economic Behavior*, 46(2), 281–294.

—— (2005a) "An Experimental Study of Costly Coordination," *Games and Economic Behavior*, 51(2), 349–364.

—— (2005b) "An Explanation of Anomalous Behavior in Models of Political Participation," *American Political Science Review*, 99(2), 201–213.

—— (2010) "Hierarchical Package Bidding: A Paper & Pencil Combinatorial Auction," *Games and Economic Behavior*, 70, 146–169.

Goeree, Jacob K., Charles A. Holt, and Susan K. Laury (2002) "Private Costs and Public Benefits: Unraveling the Effects of Altruism and Noisy Behavior," *Journal of Public Economics*, 82, 257–278.

—— (2003) "Altruism and Error in Public Goods Experiments: Implications for the Environment," in *Recent Advances in Environmental Economics*, J. List and A. de Zeeuw, eds., Northampton, MA: Edward Elgar, 309–339.

Goeree, Jacob K., Charles A. Holt, and Thomas R. Palfrey (2002) "Quantal Response Equilibrium and Overbidding in Private-Value Auctions," *Journal of Economic Theory*, 104(1), 247–272.

—— (2003) "Risk Averse Behavior in Asymmetric Matching Pennies Games," *Games and Economic Behavior*, 45, 97–113.

—— (2016) *Quantal Response Equilibrium: A Stochastic Theory of Games*, Princeton, NJ: Princeton University Press.

—— (2017) "Stochastic Game Theory for Social Science: A Primer on Quantal Response Equilibrium," forthcoming in *Handbook of Experimental Game Theory*, M. Capra, R. Croson, M. Rigdon, and T. Rosenblat, eds., Northampton, MA: Edward Elgar.

Goeree, Jacob, Charles Holt, William Shobe, Karen Palmer, and Dallas Burtraw (2010) "An Experimental Study of Auctions versus Grandfathering to Assign Pollution Permits," *Journal of the European Economic Association*, 8(2–3), 514–525.

Goeree, Jacob K., Charles A. Holt, and Angela M. Smith (2017) "An Experimental Examination of the Volunteer's Dilemma," *Games and Economic Behavior*, 102(C), 305–315.

Goeree, Jacob K., Theo Offerman, and Randolph Sloof (2013) "Demand Reduction and Preemptive Bidding in Multi-Unit License Auctions," *Experimental Economics*, 16(1), 52–87.

Goeree, Jacob K., Thomas R. Palfrey, Brian W. Rogers, and Richard D. McKelvey (2007) "Self-correcting Information Cascades," *Review of Economic Studies*, 74(3), 733–762.

Grether, David M. (1980) "Bayes' Rule as a Descriptive Model: The Representativeness Heuristic," *Quarterly Journal of Economics*, 95, 537–557.

—— (1992) "Testing Bayes' Rule and the Representativeness Heuristic: Some Experimental Evidence," *Journal of Economic Behavior and Organization*, 17, 31–57.

Grether, David M. and Charles R. Plott (1984) "The Effects of Market Practices in Oligopolistic Markets: An Experimental Examination of the *Ethyl* Case," *Economic Inquiry*, 24, 479–507.

Groneck, Max, Alexander Ludwig, and Alexander Zimper (2017) "The Impact of Biases in Survival Beliefs on Savings Behavior," SAFE Working Papers, No. 169.

Großer, Jens and Thomas R. Palfrey (2014) "Candidate Entry and Political Polarization: An Antimedian Voter Theorem," *American Journal of Political Science*, 58(1), 127–143.

—— (2017) "Candidate Entry and Political Polarization: An Experimental Study," forthcoming in *American Political Science Review*.

Großer, Jens and Arthur Schram (2006) "Neighborhood Information, Exchange and Voter Participation: An Experimental Study," *American Political Science Review*, 100(2), 235–248.

Gunnthorsdottir, Anna, Daniel Houser, and Kevin McCabe (2007) "Disposition, History and Contributions in Public Goods Experiments," *Journal of Economic Behavior and Organization*, 62(2), 304–315.

Güth, Werner, Rolf Schmittberger, and Bernd Schwarze (1982) "An Experimental Analysis of Ultimatum Bargaining," *Journal of Economic Behavior and Organization*, 3, 367–388.

Guzik, Victor S. (2004) "Contextual Framing Effects in a Common Pool Resource Experiment," Economics Honors Thesis, Middlebury College.

Hardin, Garrett (1968) "The Tragedy of the Commons," *Science*, 162, 1243–1248.

Harper, D.G.C. (1982) "Competitive Foraging in Mallards: 'Ideal Free' Ducks," *Animal Behavior*, 30, 575–584.

Harrison, Glenn (1989) "Theory and Misbehavior in First-Price Auctions," *American Economic Review*, 79(4), 749–762.

Harrison, Glenn W., Morten P. Lau, and Elisabet E. Rutstrom (2007) "Estimating Risk Attitudes in Denmark: A Field Experiment," *Scandinavian Journal of Economics*, 109(2), 341–368.

Harrison, Glenn W. and John List (2004) "Field Experiments," *Journal of Economic Literature*, 42, 1009–1055.

Harrison, Glenn W., Eric Johnson, Melayne M. McInnes, and Elisabet Rutstrom (2005) "Risk Aversion and Incentive Effects: Comment," *American Economic Review*, 95(3), 897–901.

Harrison, Glenn W., Jimmy Martinez-Correa, and J. Todd Swarthout (2014) "Eliciting Subjective Probabilities with Binary Lotteries," *Journal of Economic Behavior and Organization*, 101, 128–140.

Harsanyi, John C. and Reinhard Selten (1988) *A General Theory of Equilibrium Selection in Games*, Cambridge, MA: MIT Press.

Haruvy, Ernan, Yaron Lahav, and Charles N. Noussair (2007) "Traders' Expectations in Asset Markets: Experimental Evidence," *American Economic Review*, 97(5), 1901–1920.

Haruvy, Ernan and Charles N. Noussair (2006) "The Effect of Short Selling on Bubbles and Crashes in Experimental Spot Asset Markets," *Journal of Finance*, 61(3), 1119–1157.

Hawkes, Kristen (1993) "Why Hunter-Gatherers Work," *Current Anthropology*, 34(4), 341–361 (with commentaries and author's reply).

Hay, George A. and Daniel Kelley (1974) "An Empirical Survey of Price Fixing Conspiracies," *Journal of Law and Economics*, 17, 13–38.

Hazlett, Thomas W. (2014) "Efficient Spectrum Allocation with Hold-Ups and without Nirvana," George Mason Law and Economcis Research Paper 14–16.

—— (2017) *The Political Spectrum: The Tumultuous Liberation of Wireless Technology, from Herbert Hoover to the Smartphone*, New Haven, CT: Yale University Press.

Hazlett, Thomas W. and Robert J. Michaels (1993) "The Cost of Rent-Seeking: Evidence from Cellular Telephone License Lotteries," *Southern Economic Journal*, 59(3), 425–435.

Healy, Andrew and Jennifer Pate (2009) "Asymmetry and Incomplete Information in an Experimental Volunteer's Dilemma," in *18th World IMACS Congress and MODSIM09 International Congress on Modelling and Simulation*, R. S. Anderssen, R. D. Braddock, and L.T.H. Newham, eds., 1459–1462. ISBN: 978-0-9758400-7-8.

Henrich, Joseph, Robert Boyd, Samuel Bowles, Colin Camerer, Ernst Fehr, Herbert Gintis, and Richard McElreath (2001) "In Search of Homo Economicus: Behavioral Experiments in 15 Small-Scale Societies," *American Economic Review*, 91(2), 73–84.

Hertwig, Ralph and Andreas Ortmann (2001) "Experimental Practices in Economics: A Methodological Challenge to Psychologists?" *Behavioral and Brain Sciences*, 24(3), 383–403.

Hewett, Roger, Charles A. Holt, Georgia Kosmopoulou, Christine Kymn, Cheryl X. Long, Shabnam Mousavi, and Sudipta Sarangi (2005) "A Classroom Exercise: Voting by Ballots and Feet," *Southern Economic Journal*, 72(1), 252–263.

Hey, John D. (1995) "Experimental Investigations of Errors in Decision Making under Risk," *European Economic Review*, 39, 633–640.

Hey, John D. and Valentino Dardanoni (1988) "Optimal Consumption Under Uncertainty: An Experimental Investigation," *Economic Journal*, 98(390), 105–116.

Hillman, Arye L. and Dov Samet (1987) "Dissipation of Contestable Rents by Small Numbers of Contenders," *Public Choice*, 54(1), 63–82.

Ho, Teck-Hua and Juanjuan Zhang (2008) "Designing Contracts for Boundedly Rational Customers: Does the Framing of the Fixed Fee Matter?" *Management Science*, 54(4), 686–700.

Hodgson, Ashley (2014) "Adverse Selection in Health Insurance: A Classroom Experiment," *Journal of Economic Education*, 45(2), 90–100.

Hoffman, Elizabeth, Kevin McCabe, Keith Shachat, and Vernon L. Smith (1994) "Preferences, Property Rights, and Anonymity in Bargaining Games," *Games and Economic Behavior*, 7, 346–380.

Hoffman, Elizabeth, Kevin McCabe, and Vernon L. Smith (1996a) "On Expectations and Monetary Stakes in Ultimatum Games," *International Journal of Game Theory*, 25(3), 289–301.

—— (1996b) "Social Distance and Other-Regarding Behavior in Dictator Games," *American Economic Review*, 86(3), 653–660.

Hoffman, Elizabeth and Mark Spitzer (1982) "The Coase Theorem: Some Experimental Tests," *Journal of Law and Economics*, 25, 73–98.

—— (1985) "Entitlements, Rights and Fairness: An Experimental Examination of Subjects' Concepts of Distributive Justice," *Journal of Legal Studies*, 14, 259–297.

Holt, Charles A. (1985) "An Experimental Test of the Consistent-Conjectures Hypothesis," *American Economic Review*, 75, 314–325.

—— (1986) "Preference Reversals and the Independence Axiom," *American Economic Review*, 76, 508–515.

—— (1989) "The Exercise of Market Power in Laboratory Experiments," *Journal of Law and Economics*, 32, S107–S131.

—— (1992) "ISO Probability Matching," Discussion Paper, University of Virginia.

—— (1996) "Classroom Games: Trading in a Pit Market," *Journal of Economic Perspectives*, 10, 193–203.

Holt, Charles A. and Lisa R. Anderson (1999) "Agendas and Strategic Voting," *Southern Economic Journal*, 65, 622–629.

Holt, Charles A. and Douglas D. Davis (1990) "The Effects of Non-Binding Price Announcements on Posted Offer Markets," *Economics Letters*, 34, 307–310.

Holt, Charles A., Cathleen Johnson, Courtney Mallow, and Sean Sullivan (2012) "Tragedy of the Common Canal," *Southern Economic Journal*, 78(4), 1142–1162.

Holt, Charles, Cathleen Johnson, and David Schmidtz (2015) "Prisoner's Dilemma Experiments," in *The Prisoner's Dilemma*, M. Peterson, ed., Cambridge UK: Cambridge University Press, 243–264.

—— (2017) "Endogenous Group Formation in Social Dilemma Experiments," Discussion Paper, University of Virginia.

Holt, Charles A., Andrew Kydd, Laura Razzolini, and Roman Sheremeta (2016) "The Paradox of Misaligned Profiling: Theory and Experimental Evidence," *Journal of Conflict Resolution*, 60(3), 482–500.

Holt, Charles A., Loren Langan, and Anne Villamil (1986) "Market Power in Oral Double Auctions," *Economic Inquiry*, 24, 107–123.

Holt, Charles A. and Susan K. Laury (1997) "Classroom Games: Voluntary Provision of a Public Good," *Journal of Economic Perspectives*, 11, 209–215.

—— (2002) "Risk Aversion and Incentive Effects," *American Economic Review*, 92(5), 1644–1655.

—— (2005) "Risk Aversion and Incentive Effects: New Data without Order Effects," *American Economic Review*, 95(3), 902–912.

—— (2008) "Theoretical Explanations of Treatment Effects in Voluntary Contributions Experiments," in *Handbook of Experimental Economics Results*, Vol. 1, C. Plott and V. Smith, eds., New York: Elsevier, 846–855.

—— (2014) "Assessment and Estimation of Risk Preferences," *Handbook of the Economics of Risk and Uncertainty*, Vol. 1, M. Machina and K. Viscusi, eds., Oxford: North Holland, ch. 4, 135–201.

Holt, Charles and Evan Zuofu Liao (2016) "The Pursuit of Revenue Reduction: An Experimental Analysis of the Shanghai License Plate Auction," Discussion Paper, University of Virginia.

Holt, Charles A., Megan Porzio, and Michelle Yingze Song (2017) "Price Bubbles, Gender, and Expectations in Experimental Asset Markets," *European Economic Review*, 100, 72–94.

Holt, Charles A. and Alvin E. Roth (2004) "The Nash Equilibrium: A Perspective," *Proceedings of the National Academy of Sciences, U.S.A.*, 101(12), 3999–4002.

Holt, Charles A. and David Scheffman (1987) "Facilitating Practices: The Effects of Advance Notice and Best-Price Policies," *RAND Journal of Economics*, 18, 187–197.

Holt, Charles A. and Roger Sherman (1990) "Advertising and Product Quality in Posted Offer Experiments," *Economic Inquiry*, 28(3), 39–56.

—— (1994) "The Loser's Curse," *American Economic Review*, 84, 642–652.

—— (1999) "Classroom Games: A Market for Lemons," *Journal of Economic Perspectives*, 13(1), 205–214.

—— (2014) "Risk Aversion and the Winner's Curse," *Southern Journal of Economics*, 81, 7–22.

Holt, Charles A. and William Shobe (2017) "Consignment Auctions," draft presented at California Institute of Technology.

Holt, Charles, William Shobe, Dallas Burtraw, Karen Palmer, and Jacob Goeree (2007) *Auction Design for Selling CO2 Emissions Allowances Under the Regional Greenhouse Gas Initiative*, report on the Regional Greenhouse Gas Initiative website: http://www.rggi .org/docs/rggi_auction_final.pdf

Holt, Charles A. and Angela Smith (2009) "An Update on Bayesian Updating," *Journal of Economic Behavior and Organization*, 69, no. 2 (February), 125–134.

—— (2016) "Belief Elicitation with a Simple Lottery Choice Menu: Invariant to Risk Preferences," *American Economic Journal: Microeconomics*, 8(1), 110–139.

Holt, Charles A. and Angela M. Smith (2017) "Rent Dissipation: Effects and Explanations," Discussion Paper presented at the 2018 Public Choice Meetings, University of Virginia.

Holt, Charles A. and Fernando Solis-Soberon (1992) "The Calculation of Equilibrium Mixed Strategies in Posted-Offer Auctions," in *Research in Experimental Economics*, Vol. 5, R. M. Isaac, ed., Greenwich, CT: JAI Press, 189–229.

Holt, Charles A. and Sean Sullivan (2017) "Permutation Tests for Economics Experiments: A User's Guide," Discussion Paper presented at the October 2017 North American ESA Meeting, Richmond, Virginia.

Hommes, Cars H., Joep Sonnemans, Jan Tuinstra, and Henk van de Velden (2005) "Coordination of Expectations in Asset Pricing Experiments," *Review of Financial Studies*, 18(3), 955–980.

Hossain, Tanjim and Ryo Okui (2013) "The Binarized Scoring Rule," *Review of Economic Studies*, 80, 984–1001.

Hotelling, Harold (1929) "Stability in Competition," *Economic Journal*, 39, 41–57.

Houser, Daniel, Daniel Schunk, and Joachim Winter (2010) "Distinguishing Trust from Risk: An Anatomy of the Investment Game," *Journal of Economic Behavior and Organization*, 74 (1–2), 72–81.

Hück, Steffen, Hans-Theo Normann, and Jörg Oechssler (1999) "Learning in Cournot Oligopoly: An Experiment," *Economic Journal*, 109, 80–95.

Hück, Steffen and Jörg Oechssler (2000) "Information Cascades in the Laboratory: Do They Occur for the Right Reasons?" *Journal of Economic Psychology*, 21, 661–671.

Hung, Angela A. and Charles R. Plott (2001) "Information Cascades: Replication and an Extension to Majority Rule and Conformity Rewarding Institutions," *American Economic Review*, 91, 1508–1520.

Hussam, Reshmaan, David Porter, and Vernon L. Smith (2008) "Thar She Blows: Can Bubbles Be Rekindled with Experienced Subjects?" *American Economic Review*, 98(3), 924–937.

Isaac, R. Mark and Duncan James (2000) "Just Who Are You Calling Risk Averse?" *Journal of Risk and Uncertainty*, 20(2), 177–187.

Isaac, R. Mark and Charles R. Plott (1981) "The Opportunity for Conspiracy in Restraint of Trade," *Journal of Economic Behavior and Organization*, 2, 1–30.

Isaac, R. Mark and Stanley Reynolds (1988) "Appropriability and Market Structure of a Stochastic Invention Model," *Quarterly Journal of Economics*, 4, 647–672.

Isaac, R. Mark and James M. Walker (1985) "Information and Conspiracy in Sealed Bid Auctions," *Journal of Economic Behavior and Organization*, 6, 139–159.

—— (1988a) "Communication and Free-Riding Behavior: The Voluntary Contributions Mechanism," *Economic Inquiry*, 26, 585–608.

—— (1988b) "Group Size Hypotheses of Public Goods Provision: The Voluntary Contributions Mechanism," *Quarterly Journal of Economics*, 103, 179–199.

Kachelmeier, Steven J. and Mohamed Shehata (1992) "Examining Risk Preferences under High Monetary Incentives: Experimental Evidence from the People's Republic of China," *American Economic Review*, 82, 1120–1141.

Kagel, John H. (1995) "Auctions: A Survey of Experimental Research," in *The Handbook of Experimental Economics*, J. H. Kagel and A. E. Roth, eds., Princeton, NJ: Princeton University Press, 501–585.

Kagel, John H. and David Levin (1986) "The Winner's Curse and Public Information in Common Value Auctions," *American Economic Review*, 76, 894–920.

—— (2002) *Common Value Auctions and the Winner's Curse*, Princeton, NJ: Princeton University Press.

—— (2004) "Multi-Unit Demand Auctions with Synergies: Behavior in Sealed-Bid versus Ascending-Bid Uniform-Price Auctions," *Games and Economic Behavior*, 53(2), 170–207.

Kagel, John H. and Alvin E. Roth (1995) *The Handbook of Experimental Economics*, Princeton, NJ: Princeton University Press.

—— (2000) "The Dynamics of Reorganization in Matching Markets: A Laboratory Experiment Motivated by a Natural Experiment," *Quarterly Journal of Economics*, 115, 201–237.

—— (2016) *The Handbook of Experimental Economics*, Vol. 2, Princeton, NJ: Princeton University Press.

Kahneman, Daniel, Jack L. Knetsch, and Richard H. Thaler (1991) "The Endowment Effect, Loss Aversion, and Status Quo Bias: Anomalies," *Journal of Economic Perspectives*, 5, 193–206.

Kahneman, Daniel and Amos Tversky (1973) "On the Psychology of Prediction," *Psychological Review*, 80, 237–251.

—— (1979) "Prospect Theory: An Analysis of Decision under Risk," *Econometrica*, 47, 263–291.

Karlan, Dean S. (2005) "Using Experimental Economics to Measure Social Capital and Predict Financial Decisions," *American Economic Review*, 95(5), 1688–1699.

Karlan, Dean and John A. List (2007) "Does Price Matter in Charitable Giving? Evidence from a Large-Scale Natural Feld Experiment," *American Economic Review*, 97(5), 1774–1793.

Keck, Steffen, Enrico Diecidue, and David Budescu (2014) "Group Decisions Under Ambiguity: Convergence to Neutrality," *Journal of Economic Behavior and Organization*, 103, 60–71.

Ketcham, Jon, Vernon L. Smith, and Arlington W. Williams (1984) "A Comparison of Posted-Offer and Double-Auction Pricing Institutions," *Review of Economic Studies*, 51, 595–614.

Keynes, John Maynard (1965, originally 1936) *The General Theory of Employment, Interest, and Money*, New York: Harcourt, Brace & World.

King, Ronald, Vernon Smith, Arlington Williams, and Mark Van Boening (1993) "The Robustness of Bubbles and Crashes in Experimental Stock Markets," in *Nonlinear Dynamics and Evolutionary Economics*, I. Prigogine, R. Day, and P. Chen, eds., Oxford, UK: Oxford University Press, 183–200.

Kirchler, Michael, Jürgen Huber, and Thomas Stöckl (2012) "Thar She Bursts: Reducing Confusion Reduces Bubbles," *American Economic Review*, 102(2), 865–883.

Kirstein, A. and Roland Kirstein (2009) "Iterative Reasoning in an Experimental 'Lemons' Market," *Homo Oeconomicus* 26(2), 179–213.

Klemperer, Paul (2002) "What Really Matters in Auction Design," *Journal of Economic Perspectives*, 16, no. 1 (Winter), 169–189.

—— (2010) "The Product-Mix Auction: A New Auction for Differentiated Goods," *Journal of the European Economic Association*, 8(2–3), 526–536.

Kloosterman, Andrew and Jack Fanning (2017) "A Simple Test of the Coase Conjecture: Fairness in Dynamic Bargaining," draft, presentation at the 2017 Economic Science Association meeting in Richmond, Virginia.

Kosfeld, Michael, Markus Heinrichs, Paul J. Zak, Urs Fischbacher, and Ernst Fehr (2005) "Oxytocin Increases Trust in Humans," *Nature Letters*, 435(2), 673–677.

Kousser, J. M. (1984) "Origins of the Run-off Primary," *Black Scholar*, 23–26.

Kragt, M., L.T.H. Newham, and A. J. Jakeman (2009) "A Bayesian Network Approach to Integrating Economic and Biophysical Modelling," in *18th World IMACS Congress and MODSIM09 International Congress on Modelling and Simulation*. Modelling and Simulation Society of Australia and New Zealand and International Association for Mathematics and Computers in Simulation, R. S. Anderssen, R. D. Braddock, and L.T.H. Newham, eds., pp. 2377–2383. ISBN 978-0-9758400-7-8. http://www.mssanz.org.au/modsim09/F12/kragt.pdf.

Krueger, Anne O. (1974) "The Political Economy of the Rent-seeking Society," *American Economic Review*, 64, 291–303.

Kruse, Jamie B., Stevan Rassenti, Stanley Reynolds, and Vernon Smith (1994) "Bertrand-Edgeworth Competition in Experimental Markets," *Econometrica*, 62, 343–372.

Kübler, Dorothea and Georg Weizsäcker (2004) "Limited Depth of Reasoning and Failure of Cascade Formation in the Laboratory," *Review of Economic Studies*, 71(2), 425–441.

Lahav, Yaron (2011) "Price Patterns in Experimental Asset Markets with Long Horizon," *Journal of Behavioral Finance*, 12, 20–28.

Laslier, Jean-François and Karine Van der Straeten (2008) "A Live Experiment on Approval Voting," *Experimental Economics*, 11, 97–105.

Laury, Susan K. and Charles A. Holt (2008) "Voluntary Provision of Public Goods: Experimental Results with Interior Nash Equilibria," in *Handbook of Experimental Economics Results*, Vol. 1, C. R. Plott and V. L. Smith, eds., New York: Elsevier, 792–801.

Ledyard, John O. (1995) "Public Goods: A Survey of Experimental Research," in *A Handbook of Experimental Economics*, A. Roth and J. Kagel, eds., Princeton, NJ: Princeton University Press, 111–194.

Ledyard, John O. and Kristin Szakaly-Moore (1994) "Designing Mechanisms for Trading Pollution Rights," *Journal of Economic Behavior and Organization*, 25, 167–196.

Lee, Hau L., V. Padmanabhan, and Seungjin Whang (1997a) "Information Distortion in a Supply Chain: The Bullwhip Effect," *Management Science*, 43(4), 546–558.

—— (1997b) "The Bullwhip Effect in Supply Chains," *Sloan Management Review*, 38(3), 93–102.

Leo, Greg (2017a) "Taking Turns," *Games and Economic Behavior*, 102, 525–547.

—— (2017b) "Complainer's Dilemma," draft presented at the 2017 Social Dilemmas Workshop, Amherst, Massachusetts.

Levine, David K. and Thomas R. Palfrey (2007) "The Paradox of Political Participation? A Laboratory Study," *American Political Science Review*, 101, 143–158.

Levine, Michael E. and Charles R. Plott (1977) "Agenda Influence and Its Implications," *Virginia Law Review*, 63, 561–604.

Levitt, Stephen D. and John A. List (2007) "What Do Lab Experiments Measuring Social Preferences Reveal About the Real World?" *Journal of Economic Perspectives*, 21(2), 153–174.

Levy, Matthew R. and Joshua Tasoff (2016) "Exponential-Growth Bias and Lifecycle Consumption," *Journal of the European Economic Association*, 14, 545–583.

Li, Yi (2017) "A New Incentive Compatible Payoff Mechanism for Eliciting both Utility and Probability Weighting," Discussion Paper, Georgia State University.

List, John A. and Todd L. Cherry (2000) "Learning to Accept in Ultimatum Games: Evidence from an Experimental Design That Generates Low Offers," *Experimental Economics*, 3, 11–29.

List, John A. and David Lucking-Reiley (2000) "Demand Reduction in Multi-Unit Auctions: Evidence from a Sportscard Field Experiment," *American Economic Review*, 90(4), 961–972.

—— (2002) "The Effects of Seed Money and Refunds on Charitable Giving: Experimental Evidence from a University Capital Campaign," *Journal of Political Economy*, 110(1), 215–233.

List, John A. and Michael K. Price (2005) "Conspiracies and Secret Price Discounts in the Marketplace: Evidence from a Field Experiment," *Rand Journal of Economics*, 36(3), 700–719.

List, John A., Sally Sadoff, and Mathis Wagner (2011) "So You Want to Run an Experiment, Now What? Some Simple Rules of Thumb for Optimal Experimental Design," *Experimental Economics*, 14, 439–457.

List, John A., Azeem Shaikh, and Yang Xu (2016) "Multiple Hypothesis Testing in Experimental Economics," Discussion Paper, University of Chicago.

Luce, R. Duncan (1959) *Individual Choice Behavior*, New York: John Wiley & Sons.

Lucking-Reiley, David (1999) "Using Field Experiments to Test Equivalence Between Auction Formats: Magic on the Internet," *American Economic Review*, 89(5), 1063–1080.

—— (2000) "Vickrey Auctions in Practice: From Nineteenth-Century Philately to Twenty-First Century E-Commerce," *Journal of Economic Perspectives*, 14, no. 3 (Summer), 183–192.

Lynch, Michael, Ross M. Miller, Charles R. Plott, and Russell Porter (1986) "Product Quality, Consumer Information and 'Lemons' in Experimental Markets," in *Empirical Approaches to Consumer Protection Economics*, P. M. Ippolito and D. T. Scheffman, eds., Washington, DC: Federal Trade Commission, Bureau of Economics, 251–306.

Mackay, Charles (1995) *Extraordinary Popular Delusions and the Madness of Crowds*, Hertfordshire, UK: Wordsworth Editions Ltd. (originally 1841).

Mago, Shakun, Anya Samak, and Roman Sheremeta (2016) "Facing Your Opponents: Social Identification and Information Feedback in Contests," *Journal of Conflict Resolution*, 60(3), 459–481.

Markowitz, Harry (1952) "The Utility of Wealth," *Journal of Political Economy*, 60, 150–158.

Marwell, Gerald and Ruth E. Ames (1981) "Economists Free Ride, Does Anyone Else? Experiments on the Provision of Public Goods, IV," *Journal of Public Economics*, 15, 295–310.

McCabe, Kevin A., Steven J. Rassenti, and Vernon L. Smith (1991) "Testing Vickrey's and Other Simultaneous Multiple Unit Versions of the English Auction," in *Research in Experimental Economics*, Vol. 4, R. M. Isaac, ed., Stamford, CT: JAI Press, 45–79.

McDowell, Robin (2003) "Going Once, Going Twice . . . ," *GMDA News*, 2(2), 1.

McKelvey, Richard D. and Peter C. Ordeshook (1979) "An Experimental Test of Several Theories of Committee Decision-Making under Majority Rule," in *Applied Game Theory*, S. J. Brams, A. Schotter, and G. Schwodiauer, eds., Wurzburg: Physica Verlag, 152–167.

McKelvey, Richard D. and Thomas R. Palfrey (1992) "An Experimental Study of the Centipede Game," *Econometrica*, 60, 803–836.

—— (1995) "Quantal Response Equilibria for Normal Form Games," *Games and Economic Behavior*, 10, 6–38.

—— (1998) "Quantal Response Equilibria for Extensive Form Games," *Experimental Economics*, 1, 9–41.

McLaughlin, Kevin and Daniel Friedman (2016) "Online Ad Auctions: An Experiment," Discussion Paper, University of California, Santa Cruz.

Meissner, Thomas (2016) "Intertemporal Consumption and Debt Aversion: An Experimental Study," *Experimental Economics*, 19(2), 281–298.

Merill Edge (2016) "Merrill Edge Report: Fall 2016," https://olui2.fs.ml.com/Publish/ Content/application/pdf/GWMOL/Merrill_Edge_Report_Fall_2016.pdf.

Miller, Ross M. and Charles R. Plott (1985) "Product Quality Signaling in Experimental Markets," *Econometrica*, 53, 837–872.

Millner, Edward L. and Michael D. Pratt (1989) "An Experimental Investigation of Efficient Rent Seeking," *Public Choice*, 62, 139–151.

—— (1991) "Risk Aversion and Rent Seeking: An Extension and Some Experimental Evidence," *Public Choice*, 69, 81–92.

Morton, Rebecca B. and Thomas A. Rietz (2008) "Majority Requirements and Minority Representation," *New York University Annual Survey of American Law*, 63(4), 691–726.

Nadler, Amos, Peiran Jiao, Veronika Alexander, Paul J. Zak, and Cameron J. Johnson (2016) "The Bull of Wall Street: Experimental Analysis of Testosterone in Asset Trading," Discussion Paper 806, University of Oxford, Department of Economics.

Nagel, Jack (1984) "A Debut for Approval Voting," *Political Science and Politics*, 17, 62–65.

Nagel, Rosemarie (1995) "Unraveling in Guessing Games: An Experimental Study," *American Economic Review*, 85, 1313–1326.

—— (1999) "A Survey of Experimental Beauty-Contest Games," in *Games and Human Behavior: Essays in Honor of Amnon Rapoport*, I. E. D. Budescu, I. Erev, and R. Zwick, eds., Hillside, NJ: Erlbaum, 105–142.

Nagel, Rosemarie and Fang Fang Tang (1998) "Experimental Results on the Centipede Game in Normal Form: An Investigation of Learning," *Journal of Mathematical Psychology*, 42, 356–384.

Nalbanthian, Haig and Andrew Schotter (1995) "Matching and Efficiency in the Baseball Free Agent System: An Experimental Examination," *Journal of Labor Economics*, 13, 1–31.

Nash, Betty Joyce (2007) "The Supreme Court Rules on Resale Price Pacts," *Region Focus*, Federal Reserve Bank of Richmond, Fall.

Nash, John F. (1950) "Equilibrium Points in N-Person Games," *Proceedings of the National Academy of Sciences, U.S.A.*, 36, 48–49.

Nelson, Julie (2016) "Not So Strong Evidence for Gender Differences in Risk Taking," *Feminist Economics*, 22(2), 114–122.

Neugebauer, Tibor and Reinhard Selten (2006) "Individual Behavior of First-Price Auctions: The Importance of Information Feedback in Computerized Experimental Auctions," *Games and Economic Behavior*, 54(1), 183–204.

Niederle, Muriel and Lise Vesterlund (2007) "Do Women Shy Away from Competition? Do Men Compete Too Much?," *Quarterly Journal of Economics*, 122, 1067–1101.

Niemi, Richard G. and Larry M. Bartels (1984) "The Responsiveness of Approval Voting to Political Circumstances," *Political Science and Politics*, 17, 571–577.

Noussair, Charles (2003) "Innovations in the Design of Bundled-Item Auctions," *Proceedings of the National Academy of Sciences*, 100(19), 10590–10591.

Noussair, Charles and Kenneth Matheny (2000) "An Experimental Study of Decisions in Dynamic Optimization Problems," *Economic Theory*, 15(2), 389–419.

Noussair, Charles N. and Steven Tucker (2006) "Futures Markets and Bubble Formation in Experimental Asset Markets," *Pacific Economic Review*, 11(2), 167–184.

Ochs, Jack (1994) "Games with Unique, Mixed Strategy Equilibria: An Experimental Study," *Games and Economic Behavior*, 10, 202–217.

—— (1995) "Coordination Problems," in *The Handbook of Experimental Economics*, J. H. Kagel and A. E. Roth, eds., Princeton, NJ: Princeton University Press, 195–249.

Oechssler, Jörg (2010) "Searching beyond the Lamppost: Let's Focus on Economically Relevant Questions," *Journal of Economic Behavior and Organization*, 73(1), 65–67.

Olsen, Mark and David Porter (1994) "An Experimental Examination into Design of Decentralized Methods to Solve the Assignment Problem With and Without Money," *Economic Theory*, 4, 11–40.

Orbell, John M. and Robyn M. Dawes (1993) "Social Welfare, Cooperators' Advantage, and the Option of Not Playing the Game," *American Sociological Review*, 58(6), 787–800.

Orbell, John M., Peregrine Schwartz-Shea, and Randy T. Simmons (1984) "Do Cooperators Exit More Readily than Defectors?" *American Political Science Review*, 78(1), 147–162.

Ostrom, Elinor and Roy Gardner (1993) "Coping with Asymmetries in the Commons: Self-Governing Irrigation Systems Can Work," *Journal of Economic Perspectives*, 7(4), 93–112.

Ostrom, Elinor, Roy Gardner, and James K. Walker (1994) *Rules, Games, and Common-Pool Resources*, Ann Arbor: University of Michigan Press.

Ostrom, Elinor and James K. Walker (1991) "Communication in a Commons: Cooperation without External Enforcement," in *Laboratory Research in Political Economy*, T. Palfrey, ed., Ann Arbor: University of Michigan Press, 289–322.

Ostrom, Elinor, James Walker, and Roy Gardner (1992) "Covenants With and Without a Sword: Self-Governance Is Possible," *American Political Science Review*, 86(2), 404–417.

Otsubo, Hironori and Amnon Rapoport (2008) "Dynamic Volunteer's Dilemmas over a Finite Horizon: An Experimental Study," *Journal of Conflict Resolution*, 52(6), 961–984.

Ouwersloot, Hans, Peter Nijkam, and Piet Rietveld (1998) "Errors in Probability Updating Behaviour: Measurement and Impact Analysis," *Journal of Economic Psychology*, 19, 535–563.

Page, Talbot, Louis Putterman, and Bulent Unel (2005) "Voluntary Association in Public Goods Experiments: Reciprocity, Mimicry, and Efficiency," *Economic Journal*, 115(506), 1032–1053.

Palan, Stefan (2013) "A Review of Bubbles and Crashes in Experimental Asset Markets," *Journal of Economic Surveys*, 27(3), 570–588.

Palfrey, Thomas R. (2009) "Laboratory Experiments in Political Economy," *Annual Review of Political Science*, 12, 379–388.

Palfrey, Thomas R. and Howard Rosenthal (1983) "A Strategic Calculus of Voting," *Public Choice*, 41, 7–53.

—— (1985) "Voter Participation and Strategic Uncertainty," *American Political Science Review*, 79, 62–78.

Palfrey, Thomas R. and Stephanie W. Wang (2009) "On Eliciting Beliefs in Strategic Games," *Journal of Economic Behavior and Organization*, 71, 98–109.

Pallais, Amanda (2005) "The Effect of Group Size on Ultimatum Bargaining," Unpublished student thesis, University of Virginia.

Parco, James E., Amnon Rapoport, and William E. Stein (2002) "Effects of Financial Incentives on the Breakdown of Mutual Trust," *Psychological Science*, 13, 292–297.

Peterson, Steven P. (1993) "Forecasting Dynamics and Convergence to Market Fundamentals: Evidence from Experimental Asset Markets," *Journal of Economic Behavior and Organization*, 22(3), 269–284.

Petrie, Regan, Susan Laury, and S. Hill (2004) "Crops, Water Usage, and Auction Experience in the 2002 Irrigation Reduction Auction," Water Policy Working Paper No. 2004-014.

Peysakhovich, Alexander and David G. Rand (2016) "Habits of Virtue: Creating Norms of Cooperation and Defection in the Laboratory," *Management Science*, 62(3), 631–647.

Plott, Charles R. (1983) "Externalities and Corrective Policies in Experimental Markets," *Economic Journal*, 93, 106–127.

—— (1986) "The Posted-Offer Trading Institution," *Science*, 232, 732–738.

—— (1989) "An Updated Review of Industrial Organization: Applications of Experimental Methods," in *Handbook of Industrial Organization*, Vol. 2, R. Schmalensee and R. D. Willig, eds., Amsterdam: Elsevier Science, 1111–1176.

Plott, Charles R. and Michael E. Levine (1978) "A Model of Agenda Influence on Committee Decisions," *American Economic Review*, 68, 146–160.

Plott, Charles R. and Jin Li (2009) "Tacit Collusion in Auctions and Conditions for Facilitation and Prevention: Equilibrium Selection in Laboratory Experimental Markets," *Economic Inquiry*, 47(3), 425–448.

Plott, Charles R. and Vernon L. Smith (2008) *Handbook of Economic Results*, Vol. 1, Amsterdam: North Holland.

Ponti, Giovanni (2002) "Cycles of Learning in the Centipede Game," *Games and Economic Behavior*, 30, 115–141.

Popova, Uliana (2006) "Equilibrium Analysis of Signaling with Asymmetric Information in a Poker," Distinguished Majors Thesis, Economics Department, University of Virginia.

Porter, David, Stephen Rassenti, Anil Roopnarine, and Vernon Smith (2003) "Combinatorial Auction Design," *Proceedings of the National Academy of Sciences*, 100(19), 11153–11157.

Porter, David, Stephen Rassenti, William Shobe, Vernon Smith, and Abel Winn (2009) "The Design, Testing, and Implementation of Virginia's NOx Allowance Auction," *Journal of Economic Behavior and Organization*, 69(2), 190–200.

Porter, David P. and Vernon L. Smith (1995) "Futures Contracting and Dividend Uncertainty in Experimental Asset Markets," *Journal of Business*, 68(4), 509–541.

Porter, Robert H. and J. Douglas Zona (1993) "Detection of Bid Rigging in Procurement Auctions," *Journal of Political Economy*, 101, 518–538.

Post, Emily (1927) *Etiquette in Society, in Business, in Politics, and at Home*, New York: Funk and Wagnalls.

Potters, Jan, Casper G. de Vries, and Frans van Winden (1998) "An Experimental Examination of Rational Rent-Seeking," *European Journal of Political Economy*, 14, 783–800.

Prelec, Drazen (1998) "The Probability Weighting Function," *Econometrica*, 66(3), 497–527.

Price, Curtis R. and Roman M. Sheremeta (2015) "Endowment Origin, Demographic Effects, and Individual Preferences in Contests," *Journal of Economics and Management Strategy*, 24(3), 597–619.

Rabin, Matthew (2000) "Risk Aversion and Expected Utility Theory: A Calibration Theorem," *Econometrica*, 68, 1281–1292.

Rabin, Matthew and Richard Thaler (2001) "Risk Aversion," *Journal of Economic Perspectives*, 15(1), 219–232.

Rassenti, Stephen J., Vernon L. Smith, and Bart J. Wilson (2001) "Turning Off the Lights: Consumer Allowed Service Interruptions Could Control Market Power and Decrease Prices," *Regulation*, 70–76.

Reiley, David H. (2005) "Experimental Evidence on the Endogenous Entry of Bidders in Internet Auctions," *Experimental Business Research, Vol. 2: Economic and Managerial Perspectives*, A. Rapoport and R. Zwick, eds., Norwell, MA, and Dordrect, Netherlands: Kluwer Academic Publishers, 103–121.

—— (2006) "Field Experiments on the Effects of Reserve Prices in Auctions: More Magic on the Internet," *RAND Journal of Economics*, 37(1), 195–211.

Reiley, David H., Michael B. Urbancic, and Mark Walker (2008) "Stripped-down Poker: A Classroom Game to Illustrate Equilibrium Bluffing," *Journal of Economic Education*, 39(4), 323–341.

Reinhart, Carmen and Kenneth Rogoff (2011) *This Time Is Different: Eight Centuries of Financial Folly*, Princeton, NJ: Princeton University Press.

Reynolds, Stanley S. and Bart J. Wilson (2005) "Market Power and Price Movements over the Business Cycle," *Journal of Industrial Economics*, 53(2), 145–174.

Riahia, Dorra, Louis Levy-Garboua, and Claude Montmarquette (2013) "Competitive Insurance Markets and Adverse Selection," *Geneva Risk and Insurance Review*, 38, 87–113.

Robbett, Andrea and Peter Hans Matthews (2017) "Partisan Bias and Expressive Voting," Working Paper, Middlebury College.

Romer, David (1996) *Advanced Macroeconomics*, New York: McGraw-Hill.

Rosenthal, Robert W. (1982) "Games of Perfect Information, Predatory Pricing, and the Chain Store Paradox," *Journal of Economic Theory*, 25, 92–100.

Roth, Alvin E. (1984) "The Evolution of the Labor Market for Medical Interns and Residents: A Case Study in Game Theory," *Journal of Political Economy*, 92, 991–1016.

Roth, Alvin E., Vesna Prasnikar, Masahiro Okuno-Fujiwara, and Shmuel Zamir (1991) "Bargaining and Market Behavior in Jerusalem, Ljubljana, Pittsburgh, and Tokyo: An Experimental Study," *American Economic Review*, 81, 1068–1095.

Roth, Alvin E. and Marilda A. Oliveira Sotomayor (1990) *Two-Sided Matching: A Game-Theoretic Model and Analysis*, Econometric Society Monographs. Cambridge, UK: Cambridge University Press.

Rothkopf, Michael H., Aleksandar Pekeč, and Ronald M. Harstad (1998) "Computationally Manageable Combinational Auctions," *Management Science*, 44, 1131–1147.

Rothschild, Michael and Joseph Stiglitz (1976) "Equilibrium in Competitive Insurance Markets: An Essay on the Economics of Imperfect Information," *Quarterly Journal of Economics*, 90(4), 629–649.

Salant, David (2014) *A Primer on Auction Design, Management, and Strategy*, Cambridge, MA: MIT Press.

Samuelson, William and Richard Zeckhauser (1988) "Status Quo Bias in Decision Making," *Journal of Risk and Uncertainty*, 1, 7–59.

Sanfey, Alan G., James K. Rilling, Jessica A. Aronson, Leigh E. Nystrom, and Jonathan D. Cohen (2003) "The Neural Basis of Economic Decision Making in the Ultimatum Game," *Science*, 300(13), 1755–1758.

Savage, Leonard J. (1971) "Elicitation of Personal Probabilities and Expectations," *Journal of the American Statistical Association*, 66, 783–801.

Schechter, Laura (2007) "Traditional Trust Measurement and the Risk Confound: An Experiment in Rural Paraguay," *Journal of Economic Behavior and Organization*, 62(2), 272–292.

Schotter, Andrew and Isabel Trevino (2014) "Belief Elicitation in the Lab," *Annual Review of Economics*, 6, 103–128.

Schreck, Michael Joseph (2013) "Individual Decisions in Group Settings: Experiments in the Laboratory and Field," Doctoral Dissertation, University of Virginia.

Schweitzer, Maurice and Gerard Cachon (2000) "Decision Bias in the Newsvendor Problem with a Known Demand Distribution: Experimental Evidence," *Management Science*, 46(3), 404–420.

Sefton, Martin (1992) "Incentives in Simple Bargaining Games," *Journal of Economic Psychology*, 13, 263–276.

Selten, Reinhard (1965) "Spieltheoretische Behandlung eines Oligopolmodells mit Nachfragetragheit," parts I–II, *Zeitschrift für die Gesamte Staatswissenschaft*, 121, 301–324, 667–689.

Selten, Reinhard and Joacim Buchta (1999) "Experimental Sealed Bid First Price Auctions with Directly Observed Bid Functions," in *Games and Human Behavior: Essays in Honor of Amnon Rapoport*, I.E.D. Budescu, I. Erev, and R. Zwick, eds., Hillside, NJ: Erlbaum, 101–116.

Selten, Reinhard, Abdolkarim Sadrieh, and Klaus Abbink (1999) "Money Does Not Induce Risk Neutral Behavior, But Binary Lotteries Do Even Worse," *Theory and Decision*, 46, 211–249.

Sheremeta, Roman M. (2013) "Overbidding and Heterogeneous Behavior in Contest Experiments," *Journal of Economic Surveys*, 27(3), 491–514.

——— (2015) "Behavioral Dimensions of Contests," in *Companion to the Political Economy of Rent Seeking*, R. Congleton and A. Hillman, eds., Northampton, MA: Edward Elgar, 150–164.

Sheremeta, Roman M. and Jingjing Zhang (2010) "Can Groups Solve the Problem of Over-bidding in Contests?" *Social Choice and Welfare*, 35(2), 175–197.

Sherstyuk, Katerina (1999) "Collusion without Conspiracy: An Experimental Study of One-Sided Auctions," *Experimental Economics*, 2, 59–75.

Shobe, William, Charles Holt, and Thaddeus Huetteman (2014) "Elements of Emission Market Design: An Experimental Analysis of California's Market for Greenhouse Gas Allowances," *Journal of Economic Behavior and Organization*, 107, 402–420.

Shobe, William, Karen Palmer, Erica Myers, Charles Holt, Jacob Goeree, and Dallas Burtraw (2010) "An Experimental Analysis of Auctioning Emission Allowances Under a Loose Cap," *Agricultural and Resource Economics Review*, 39(2) 162–175.

Shogren, Jason F., Seung Y. Shin, Dermot J. Hayes, and James B. Kliebenstein (1994) "Resolving Differences in Willingness to Pay and Willingness to Accept," *American Economic Review*, 84, 255–270.

Sieberg, Katri, David Clark, Charles Holt, Tim Nordstrom, and William Reed (2010) "Asymmetric Power in Single-Stage Conflict Bargaining Games," draft presented at the 2010 Economic Science Association Meetings, Copenhagen.

——— (2013) "An Experimental Analysis of Asymmetric Power in Conflict Bargaining," *Games and Economic Behavior*, 4(3), 375–397.

Siegel, Sidney (1956) *Nonparametric Statistics for the Behavioral Sciences*, New York: McGraw-Hill.

Siegel, Sidney and John Castellan, Jr. (1988) *Nonparametric Statistics for the Behavioral Sciences*, New York: McGraw-Hill.

Siegel, Sidney and Donald A. Goldstein (1959) "Decision-Making Behavior in a Two-Choice Uncertain Outcome Situation," *Journal of Experimental Psychology*, 57, 37–42.

Siegel, Sidney, Alberta Siegel, and Julia Andrews (1964) *Choice, Strategy, and Utility*, New York: McGraw-Hill.

Slonim, Robert and Alvin E. Roth (1998) "Learning in High Stakes Ultimatum Games: An Experiment in the Slovak Republic," *Econometrica*, 66, 569–596.

Smith, Adam (1976, originally 1776) *The Wealth of Nations*, E. Cannan, ed., Chicago: University of Chicago Press.

Smith, Vernon L. (1962) "An Experimental Study of Competitive Market Behavior," *Journal of Political Economy*, 70, 111–137.

——— (1964) "The Effect of Market Organization on Competitive Equilibrium," *Quarterly Journal of Economics*, 78, 181–201.

——— (1981) "An Empirical Study of Decentralized Institutions of Monopoly Restraint," in *Essays in Contemporary Fields of Economics in Honor of E.T. Weiler, 1914–1979*, J. Quirk and G. Horwich, eds., West Lafayette, IN: Purdue University Press, 83–106.

Smith, Vernon L., Gerry L. Suchanek, and Arlington W. Williams (1988) "Bubbles, Crashes, and Endogenous Expectations in Experimental Spot Asset Markets," *Econometrica*, 56, 1119–1151.

Smith, Vernon L. and James M. Walker (1993) "Monetary Rewards and Decision Cost in Experimental Economics," *Economic Inquiry*, 31, 245–261.

Stahl, Dale O. and Paul W. Wilson (1995) "On Players' Models of Other Players: Theory and Experimental Evidence," *Games and Economic Behavior*, 10, 208–254.

Starmer, Chris and Robert Sugden (1989) "Violations of the Independence Axiom in Common Ratio Problems: An Experimental Test of Some Competing Hypotheses," *Annals of Operations Research*, 19, 79–102.

—— (1991) "Does the Random-Lottery Incentive System Elicit True Preferences? An Experimental Investigation," *American Economic Review*, 81, 971–978.

Steiglitz, Ken and Daniel Shapiro (1998) "Simulating the Madness of Crowds: Price Bubbles in an Auction-Mediated Robot Market," *Computational Economics*, 12, 35–59.

Sterman, John D. (1989) "Modeling Managerial Behavior: Misperceptions of Feedback in a Dynamic Decision Making Experiment," *Management Science*, 35(3), 321–339.

Stöckl, Thomas, Michael Kirchler, and Huber Jürgen (2015) "Multi-Period Experimental Asset Markets with Distinct Fundamental Value Regimes," *Experimental Economics*, 18(2), 314–334.

Straub, Paul G. (1995) "Risk Dominance and Coordination Failures in Static Games," *Quarterly Review of Economics and Finance*, 35(4), 339–363.

Surowiecki, James (2004) *The Wisdom of Crowds: Why the Many Are Smarter Than the Few and How Collective Wisdom Shapes Business, Economies, Societies and Nations*, New York: Penguin Random House.

Svorenčík, Andrej and Harro Mass, eds. (2016) *The Making of Experimental Economics, Witness Seminar on the Emergence of a Field*, New York: Springer.

Sylwester, Karolina and Gilbert Roberts (2010) "Contributors Benefit Through Reputation-Based Partner Choice in Economic Games," *Biology Letters*, 6, 659–662.

Thaler, Richard H. (1988) "Anomalies: The Winner's Curse," *Journal of Economic Perspectives*, 2, 191–202.

—— (1989) "Anomalies: The Ultimatum Game," *Journal of Economic Perspectives*, 2, 195–206.

—— (1992) *The Winner's Curse*, New York: Free Press.

Tiebout, Charles M. (1956) "A Pure Theory of Local Expenditures," *Journal of Political Economy*, 64, 416–424.

Tobin, T. (2003) "Yep, It's Complicated," *St. Petersburg Times*, September 14.

Trautmann, Stefan T. and Gijs van de Kuilen (2015a) "Belief Elicitation: A Horse Race among Truth Serums," *Economic Journal*, 125, 2116–2135.

—— (2015b) "Ambiguity Attitudes," in *The Wiley-Blackwell Handbook of Judgement and Decision Making*, G. Karen and G. Wu, eds., Oxford, UK: Wiley-Blackwell, 89–116.

Tullock, Gordon (1967) "The Welfare Costs of Tariffs, Monopolies, and Thefts," *Western Economic Journal*, 5(3), 224–232.

—— (1980) "Efficient Rent Seeking," in *Towards a Theory of the Rent-Seeking Society*, J. M. Buchanan, R. D. Tollison, and G. Tullock, eds., College Station: Texas University Press, 97–112.

Tulman, Sarah Ann (2013) "Altruism(?) in the Presence of Costly Voting: A Theoretical and Experimental Analysis," Doctoral Dissertation, University of Virginia.

Tversky, Amos and Daniel Kahneman (1992) "Advances in Prospect Theory: Cumulative Representation of Uncertainty," *Journal of Risk and Uncertainty*, 5, 297–323.

Tversky, Amos and Richard H. Thaler (1990) "Anomalies: Preference Reversals," *Journal of Economic Perspectives*, 4, 201–211.

Van Boening, Mark, Arlington W. Williams, and Sean LaMaster (1993) "Price Bubbles and Crashes in Experimental Call Markets," *Economics Letters*, 41, 179–185.

van Dijk, Frans, Joep Sonnemans, and Frans van Winden (2002) "Social Ties in a Public Good Experiment," *Journal of Public Economics*, 85(2), 275–299.

Van Huyck, John B., Raymond C. Battalio, and Richard O. Beil (1990) "Tacit Coordination Games, Strategic Uncertainty, and Coordination Failure," *American Economic Review*, 80, 234–248.

—— (1991) "Strategic Uncertainty, Equilibrium Selection, and Coordination Failure in Average Opinion Games," *Quarterly Journal of Economics*, 91, 885–910.

Van Huyck, John B., Joseph P. Cook, and Raymond C. Battalio (1997) "Adaptive Behavior and Coordination Failure," *Journal of Economic Behavior and Organization*, 32, 483–503.

Vickrey, William (1961) "Counterspeculation and Competitive Sealed Tenders," *Journal of Finance*, 16(1), 8–37.

von Neumann, John and Oscar Morgenstern (1944) *Theory of Games and Economic Behavior*, Princeton, NJ: Princeton University Press.

Vulkan, Nir (2000) "An Economist's Perspective on Probability Matching," *Journal of Economic Surveys*, 14(1), 101–118.

Walker, James M., Roy Gardner, and Elinor Ostrom (1990) "Rent Dissipation in a Limited-Access Common-Pool Resource: Experimental Evidence," *Journal of Environmental Economics and Management*, 19, 203–211.

Wang, Jianxin, Dan E. Houser, and Hui Xu (2017) "Do Females Always Generate Small Bubbles? Experimental Evidence from U.S. and China," ISES, George Mason University.

Weizsäcker, Georg (2003) "Ignoring the Rationality of Others: Evidence from Experimental Normal-Form Games," *Games and Economic Behavior*, 44, 145–171.

Welch, Ivo (1992) "Sequential Sales, Learning, and Cascades," *Journal of Finance*, 47, 695–732.

Werden, Gregory J. (1989) "Price-Fixing and Civil Damages: Setting the Record Straight," *Antitrust Bulletin*, 24, 307–335.

Williams, Arlington W. (1987) "The Formation of Price Forecasts in Experimental Markets," *Journal of Money, Credit, and Banking*, 19(1), 1–18.

Wilson, Rick K. (1988) "Forward and Backward Agenda Procedures: Committee Experiments on Structurally Induced Equilibrium," *Journal of Politics*, 48, 390–409.

—— (2005) "Classroom Experiments: Candidate Convergence," *Southern Economic Journal*, 71, 913–922.

Wilson, Robert B. (1969) "Competitive Bidding with Disparate Options," *Management Science*, 15, 446–448.

Wolfers, Justin and Eric Zitzewitz (2004) "Prediction Markets," *Journal of Economic Perspectives*, 18(2), 107–126.

Xiao, Erte and Daniel Houser (2005) "Emotion Expression and Human Punishment Behavior," *Proceedings of the National Academy of Sciences*, 102(20), 7398–7401.

Yoder, R. D. (1986) "The Performance of Farmer-Managed Irrigation Systems in the Hills of Nepal," PhD Dissertation, Cornell University.

Zauner, Klaus G. (1999) "A Payoff Uncertainty Explanation of Results in Experimental Centipede Games," *Games and Economic Behavior*, 26, 157–185.

Zelland, David (2013) "All-In Auctions for Water," *Journal of Environmental Management*, 115, 78–86.

Zizzo, Daniel J., Stephanie Stolarz-Fantino, Julie Wen, and Edmund Fantino (2000) "A Violation of the Monotonicity Axiom: Experimental Evidence on the Conjunction Fallacy," *Journal of Economic Behavior and Organization*, 41(3), 263–276.

APPENDIX 2

Instructions for Class Experiments

Instructions for the Instructor

This final part of the book contains instructions for running some of the experiments in class, without the help of computers. Since web-based programs are so convenient, the only instructions included here are those for which doing things "by hand" is particularly effective for teaching, e.g., the pit market for chapter 1, or the various voting experiments. The instructions are adapted for classroom use, sometimes with more context and social interaction than would be appropriate for research. Many of the experiments can be done with groups or with one-shot interactions, which conserves class time.

For the experiments that are run by hand, there are several hints for making this process more useful:

- The class can often be divided into teams, or groups of 3–5 players. Such grouping makes it easier to collect and process decisions in a larger class, and it may facilitate learning from group discussions. Teams are especially useful when decisions are made in sequence, which would slow down the process of collecting data from many different players.
- Some games, like the pit market that comes first, are hard to do with groups, but the decentralized nature of trading makes this market possible to run "by hand" with as many as 60 students. With very large classes, use teams and have each team send a representative to the trading floor to negotiate and report back between rounds.
- Most of the record sheets on the following pages have enough space for recording decisions for several rounds of repeated play, but often the

main point can be made quickly with just a round or two of decision making, which helps prevent that *Groundhog Day* feeling.

- For competitive markets and auctions, it is typically not necessary to provide cash or other motivations for participants. When incentives are desired, as in games involving bargaining and fairness issues, one option is for the instructor to select one person afterward at random and pay a small percentage of their earnings in cash, e.g., with the announcement that "each dollar equals one penny." Another option is to let earnings be converted into points that can be used as lottery tickets for a prize that is awarded every week or two in class.

- For some of the games, it is sufficient for the instructor to pick pairs of people and let them reveal their decisions, with the class watching. This may provide an important element of participatory learning, without necessarily calling on each student or doing lots of repetition. For example, the card-based games for chapters 8 and 9 can be done this way, as can the "Stripped Down Poker" game described for chapter 10 and in Reiley, Urbancic, and Walker (2008).

- It is generally best to have someone read the instructions out loud, so that there is common knowledge about procedures and so that everyone finishes instructions at the same time.

- Some of the experiments require playing cards, which can be purchased for a low cost at any convenience store, and/or ten-sided dice, which are available from game stores. It makes sense to pick up one deck of cards for each ten students in the class, along with some extra ten-sided dice to allow for an occasional loss.

Many of the instructions that follow are loosely adapted from those used in research and teaching experiments done with various coauthors. These include: Holt (1996) for the pit market experiment in chapter 1, Laury and Holt (2002) for the lottery-choice experiment in chapter 3, Holt and Laury (1997) for the voluntary contributions game in chapter 16, and Goeree and Holt (1999b) for the lobbying game in chapter 12. The voting instructions for voting games in chapter 19 were adapted from Holt and Anderson (1999), Hewett et al. (2005), and Wilson (2005).

Pit Market Instructions (Chapter 1)

We are going to set up a market in which the people on my right will be buyers, and the people on my left will be sellers. There will be equal numbers of buyers and sellers. Several assistants have been selected to help record prices. I will now give each buyer and seller a numbered playing card. Some cards have been removed from the deck(s), and all remaining cards have a number. Please hold your card so that others do not see the number. The sellers' cards are red (hearts or diamonds), and the buyers' cards are black (clubs or spades). Each card represents a "unit" of an unspecified commodity that can be bought by buyers or sold by sellers.

Trading: Buyers and sellers will meet in the center of the room (or other designated area) and negotiate during a 5-minute trading period. When a buyer and a seller agree on a price, they will come together to the front of the room to report the price, which will be announced to all. Then the buyer and the seller will turn in their cards, return to their original seats, and wait for the trading period to end. There will be several market periods.

Sellers: You can each sell a single unit of the commodity during a trading period. The number on your card is the dollar cost that you incur if you make a sale, and you will not be allowed to sell below this cost. Your earnings on the sale are calculated as the difference between the price that you negotiate and the cost number on the card. If you do not make a sale, you do not earn anything or incur any cost in that period. Think of it this way: it's as if you knew someone who would sell you the commodity for a price that equals your cost number, so you can keep the difference if you are able to resell the commodity for a price that is above the acquisition cost. Suppose that your card is a 2 of hearts and you negotiate a sale price of \$3. Then you would earn: $3 - 2 = \$1$. You would not be allowed to sell at a price below \$2 with this card (2 of hearts). If you mistakenly agree to a price that is below your cost, then the trade will be invalidated when you come to the front desk; your card will be returned and you can resume negotiations.

Buyers: You can each buy a single unit of the commodity during a trading period. The number on your card is the dollar value that you receive if you make a purchase, and you will not be allowed to buy at a price above this value. Your earnings on the purchase are calculated as the difference between the value number on the card and the price that you negotiate. If you do not make a purchase, you do not earn anything in the period. Think of it this way: it's as if you knew someone who would later buy the unit from you at a price that equals your value number, so you can keep the difference if you are able to buy the unit at a price that is below the resale value. Suppose that your card is a 9 of spades and you negotiate a purchase price of $4. Then you would earn: $9 - 4 = \$5$. You would not be allowed to buy at a price above $9 with this card (9 of spades). If you mistakenly agree to a price that is above your value, then the trade will be invalidated when you come to the front desk; your card will be returned and you can resume negotiations.

Recording Earnings: Some buyers and sellers may not be able to negotiate a trade, but do not be discouraged since new cards will be passed out at the beginning of the next period. Remember that earnings are zero for any unit not bought or sold (sellers incur no cost and buyers receive no value). When the period ends, I will collect cards for the units not traded, and you can calculate your earnings while I shuffle and redistribute the cards. Your total earnings equal the sum of earnings for units traded in all periods, and you can use the Earnings Record Form on the next page to keep track of your earnings. Sellers use the left side of the Earnings Record Form, and buyers use the right side. At this time, please draw a diagonal line through the side that you will *not* use. All earnings are hypothetical. Please do not talk with each other until the trading period begins. Are there any questions?

Final Observations: When a buyer and a seller agree on a price, both should *immediately* come to the front table to turn in their cards together, so that we can verify that the price is neither lower than the seller's cost nor higher than the buyer's value. If there is a line, please wait together. After the price is verified, the assistant at the board will write the price and announce it loudly. Then those two traders can return to their seats to calculate their earnings. The assistants should come to their positions in the front of the room. Buyers and sellers, please come to the central trading area NOW, and begin calling out prices at which you are willing to buy or sell. The market is open, and there are 5 minutes remaining.

Your Name: _____

| Seller Earnings | Buyer Earnings |
| (Sellers use this side) | (Buyers use this side) |

First Period

_____ — _____ = _____   _____ — _____ = _____
(price)   (cost)   (earnings)   (value)   (price)   (earnings)

Second Period

_____ — _____ = _____   _____ — _____ = _____
(price)   (cost)   (earnings)   (value)   (price)   (earnings)

Third Period

_____ — _____ = _____   _____ — _____ = _____
(price)   (cost)   (earnings)   (value)   (price)   (earnings)

Fourth Period

_____ — _____ = _____   _____ — _____ = _____
(price)   (cost)   (earnings)   (value)   (price)   (earnings)

Fifth Period

_____ — _____ = _____   _____ — _____ = _____
(price)   (cost)   (earnings)   (value)   (price)   (earnings)

Sixth Period

_____ — _____ = _____   _____ — _____ = _____
(price)   (cost)   (earnings)   (value)   (price)   (earnings)

Total earnings, for all periods: _____   Total earnings, for all periods: _____

Lottery-Choice Instructions (Chapter 3)

Your decision sheet on the next page shows ten decisions listed in the left column. Each decision is a paired choice between "Option A" and "Option B." You will make ten choices and record these in the far right column, but only one of them will be used in the end to determine your earnings. Before you start making your ten decisions, please let me explain how these choices will affect your earnings, which will be hypothetical unless otherwise indicated.

Here is a ten-sided die that will be used to determine payoffs; the faces are numbered from 1 to 10 (the "0" face of the die will serve as 10). After you have made all of your choices, we will throw this die twice, once to select one of the ten decisions to be used, and a second time to determine what your payoff is for the option you chose, A or B, for the particular decision selected. Even though you will make ten decisions, only one of these will end up affecting your earnings, but you will not know in advance which decision will be used. Obviously, each decision has an equal chance of being used in the end.

Now, please look at Decision 1 at the top. Option A pays $2.00 if the throw of the ten-sided die is 1, and it pays $1.60 if the throw is 2–10. Option B yields $3.85 if the throw of the die is 1, and it pays $0.10 if the throw is 2–10. The other Decisions are similar, except that the chances of the higher payoff for each option increase as you move down the table. For Decision 10 in the bottom row, the die will not be needed since each option pays the highest payoff for sure, so your choice here is between $2.00 and $3.85.

To summarize, you will make ten choices. You may choose A for some decision rows and B for other rows, and you may make your decisions in any order. When you are finished, we will come to your desk and throw the ten-sided die to select which of the ten Decisions will be used, i.e., which row in the table will be relevant. Then we will throw the die again to determine your money earnings for the Option you chose for that Decision. Earnings for this choice will be added to your previous earnings (if any).

So now please look at the empty boxes on the right side of the record sheet. You will have to write a decision, A or B in each of these boxes, and then the die throw will determine which one is going to count. We will look at the decision that you made for the choice that counts, and circle it, before throwing the die again to determine your earnings for this part. Then you will write your earnings at the bottom of the page. Are there any questions?

Name/ID: _____

	Option A	Option B	Your Choice A or B
Decision 1	$2.00 if throw of die is 1 $1.60 if throw of die is 2–10	$3.85 if throw of die is 1 $0.10 if throw of die is 2–10	
Decision 2	$2.00 if throw of die is 1–2 $1.60 if throw of die is 3–10	$3.85 if throw of die is 1–2 $0.10 if throw of die is 3–10	
Decision 3	$2.00 if throw of die is 1–3 $1.60 if throw of die is 4–10	$3.85 if throw of die is 1–3 $0.10 if throw of die is 4–10	
Decision 4	$2.00 if throw of die is 1–4 $1.60 if throw of die is 5–10	$3.85 if throw of die is 1–4 $0.10 if throw of die is 5–10	
Decision 5	$2.00 if throw of die is 1–5 $1.60 if throw of die is 6–10	$3.85 if throw of die is 1–5 $0.10 if throw of die is 6–10	
Decision 6	$2.00 if throw of die is 1–6 $1.60 if throw of die is 7–10	$3.85 if throw of die is 1–6 $0.10 if throw of die is 7–10	
Decision 7	$2.00 if throw of die is 1–7 $1.60 if throw of die is 8–10	$3.85 if throw of die is 1–7 $0.10 if throw of die is 8–10	
Decision 8	$2.00 if throw of die is 1–8 $1.60 if throw of die is 9–10	$3.85 if throw of die is 1–8 $0.10 if throw of die is 9–10	
Decision 9	$2.00 if throw of die is 1–9 $1.60 if throw of die is 10	$3.85 if throw of die is 1–9 $0.10 if throw of die is 10	
Decision 10	$2.00 if throw of die is 1–10	$3.85 if throw of die is 1–10	

Ink Bomb Instructions (Chapter 3)

This task provides you with a single choice, with a resulting money payoff that depends on your choice and on a random event. The money that you earn from this task will be paid to you in cash. You will see a line with 12 numbered boxes:

☐ 1   ☐ 2   ☐ 3   ☐ 4   ☐ 5   ☐ 6   ☐ 7   ☐ 8   ☐ 9   ☐ 10   ☐ 11   ☐ 12

You will decide which ones to mark. After you have made your choices, a 12-sided die, with sides marked 1 through 12, will be thrown for you at your desk. Each integer: 1, 2, . . . ,12, is equally likely. If the throw matches one of the boxes that you have checked, then your earnings will be $0 for this task. If the throw corresponds to one of the boxes that you did NOT check, then you earn an integer number of dollars that equals the number of checked boxes.

Think of it this way: each box contains a dollar bill, but one of the boxes also contains an ink bomb that destroys all of the dollars in all boxes if the box with the bomb is opened. So if you decide to check more boxes, your earnings *if you win* will be greater, but there will be a greater chance of triggering the ink bomb and earning *nothing at all*.

Since you will only do the task once, we want to be sure that you understand the procedures, so please indicate the correct answer to the following questions:

1. If you do not check any boxes:

    __(A) your earnings will be $0 regardless of the die throw
    __(B) your earnings will depend on the throw of the die

2. If you check all 12 boxes:

    __(A) your earnings will be $0 regardless of the die throw
    __(B) your earnings will depend on the throw of the die

3. If you check N boxes (where N is some number between 1 and 12):

    __ (A) there are N chances in 12 that your earnings will be $0
    __ (B) there are N chances in 12 that your earnings will be N.

    We will come to your desk to check your answers and answer any questions.

Now please indicate with a **dark x** the boxes you wish to mark:

☐ 1 ☐ 2 ☐ 3 ☐ 4 ☐ 5 ☐ 6 ☐ 7 ☐ 8 ☐ 9 ☐ 10 ☐ 11 ☐ 12

We will throw a 12-sided die and if the outcome matches a box you checked, you earn \$0, but if the outcome does not match a box you chose, you earn the number of dollars that equals the number of boxes marked.

Name: _____ Die throw: ____ Earnings: ____

Paired Lottery Choices (Chapter 4)

Your decision sheet shows six decisions listed on the left, which are labeled 1, 2, . . . , 6. Each decision is a paired choice between a randomly determined payoff described on the left and another one described on the right. You will make six choices and record these in the final column, but only one of them will be used in the end to determine your earnings. Before you start making your six choices, please let me explain how these choices will affect your earnings. Unless otherwise indicated, all earnings are hypothetical.

Here are six-sided and ten-sided dice that will be used to determine payoffs; the faces are numbered from 1 to 6 and 0 to 9. After you have made all choices, we will throw the six-sided die to select one of them. Even though you will make six decisions, only one of these will end up affecting your earnings, but you will not know in advance which decision will be used. Obviously, each decision has an equal chance of being used in the end. After we select the payoff-relevant decision, we throw the ten-sided die two times to determine what your payoff is for the option you chose (on the left side or on the right side) for the particular decision selected.

Now look at Decision 1 at the top. If this were the decision that we ended up using, we would throw the ten-sided die two times. The two throws will determine a number from 0 to 99, with the first throw determining the "tens" digit and the second one determining the "ones" digit. The option on the left side pays $6.00 if the throw of the ten-sided die is 0–99, and it pays $0.00 otherwise. Since all throws are between 0 and 99, this option provides a sure $6.00. The option on the right pays $8.00 if the throw of the die is 0–79, and it pays $0.00 otherwise. Thus, the option on the right side of Decision 1 provides 80 chances out of 100 (a four-fifths probability) of getting $8.00. The options for the other decision rows are similar, but with differing payoffs and chances of getting each payoff.

To summarize, you will make six choices: one in each row. Make your choice by putting a check by the option (left or right) that you prefer. There should be one check mark in each row. You may make your decisions in any order. When finished, mark your choices, L for left or R for right, and we will come to your desk and throw the six-sided die to select which of the decisions will be used. We will circle that decision before throwing the ten-sided die again to determine your money earnings for the Option you chose for that decision.

Before you begin, let me mention that different people may make different choices for the same decision, in the same manner that one person may purchase commodities that differ from those purchased by someone else. We are interested in *your* preferences, i.e., which option you prefer to have, so please think carefully about each decision, and please do not talk with others in the room. Are there any questions?

Name: _____

	Left Side	Right Side	Your Choice L or R
Decision 1	$6.00 if throw of die is 0–99	$8.00 if throw of die is 0–79 $0.00 if throw of die is 80–99	
Decision 2	$6.64 if throw of die is 0–89 $0.25 if throw of die is 90–99	$5.47 if throw of die is 0–89 $2.75 if throw of die is 90–99	
Decision 3	$6.00 if throw of die is 0–24 $0.00 if throw of die is 25–99	$8.00 if throw of die is 0–19 $0.00 if throw of die is 20–99	
Decision 4	$3.89 if throw of die is 0–89 $25.00 if throw of die is 90–99	$5.11 if throw of die is 0–89 $6.00 if throw of die is 90–99	
Decision 5	$8.38 if throw of die is 0–66 $1.25 if throw of die is 67–99	$6.18 if throw of die is 0–66 $3.25 if throw of die is 67–99	
Decision 6	$4.25 if throw of die is 0–66 $9.50 if throw of die is 67–99	$4.85 if throw of die is 0–66 $5.90 if throw of die is 67–99	

Stripped Down Poker Instructions (Chapter 10)

This is a simplified game of poker, and we need two willing volunteers to play. We also need a volunteer to keep records. Please raise your hand if you are willing to help out or play. (You do not need to know anything about the rules of an actual poker game.)

Each person starts with $10 in one-dollar bills, which I will provide to them at this time. Each round begins with each person putting a stake of $1 on the table. Please put your stake in now for the first round.

Here I have a short deck with 4 aces and 4 kings, all other cards removed. Each time we draw from the deck, we will return the card before drawing again, so draws will be with replacement.

The person on my right will be the first mover, and the other person will be the second mover. The first mover will be given a card draw privately, the second mover will not receive a card and will not see the first mover's card until the end.

If the first mover receives an ace, the card will win. If the first mover receives a king, the card will lose, unless someone folds first.

I will begin by drawing a card to give to the first mover, who decides whether to fold and lose the $1 stake, or to raise by putting a second dollar on the table.

If the first mover folds, the second mover collects both dollars on the table.

If the first mover raises by putting a second dollar on the table, then the second mover can either fold and lose their $1 stake, or call the first mover's raise and put a second dollar on the table.

If the second mover folds at this point, the first mover collects the money on the table (the stakes and their own dollar used to raise). If the second mover calls, the fist mover's card is revealed. An ace for the first mover wins the $4 on the table, and a king loses the $4 to the second mover. Note that the second mover does not receive a card, so the outcome depends on the first mover's card.

Lobbying Game Instructions (Chapter 12)

This is a simple card game. Each of you has been assigned to a team of investors bidding for a local government communications license that is worth $16,000. The government will allocate the license by choosing randomly from the applications received. The paperwork and legal fees associated with each application will cost your team $3,000, regardless of whether you obtain the license or not. (Think of this $3,000 as the opportunity cost of the time and materials used in completing the required paperwork.) Each team is permitted to submit any number of applications, up to a limit of 13 per team. Each team begins with a working capital of $100,000.

There will be four teams competing for each license, each of which is provided with 13 playing cards of the same suit. Your team will play *any* number of these cards by placing them in an envelope provided. Each card you play is like a lottery ticket in a drawing for a prize of $16,000. All cards that are played by your team and the other three teams will be placed on a stack and shuffled. Then one card will be drawn from the deck. If that card is of your suit, then your team will win $16,000. Otherwise you receive nothing from the lottery. Whether or not you win, your earnings will decrease by $3,000 for each card that you play. To summarize, your earnings are calculated:

> Earnings = $16,000 if you win the lottery
>   − $3,000 times the number of cards you play (win or lose).

Earnings are negative for the teams that do not win the lottery, and negative earnings are indicated with a minus sign in the record table below. The cumulative earnings column on the right begins with $100,000, reflecting your initial financial capital. Earnings are hypothetical (in case you were wondering) and should be added to or subtracted from this amount. Are there any questions?

Round	Number of Cards Played	Cost per Card Played	Total Cost	License Value	Your Earnings	Cumulative Earnings $100,000
1		$3,000		$16,000		

The next lottery is for a second license. Your team begins again with 13 cards, but the cost of each card played is reduced to $1,000, due to a government

efficiency move that requires less paperwork for each application. This license
is worth $16,000 as before, whether or not your team already acquired a license.

Round	Number of Cards Played	Cost per Card Played	Total Cost	License Value	Your Earnings	Cumulative Earnings
2		$1,000		$16,000		

In the next lottery, the value of the license may differ from team to team. Your
team begins again with 13 cards, and the cost of each card played remains at
$1,000. Your instructor will inform you of the license value, which you should
write in the appropriate place in the table below.

Round	Number of Cards Played	Cost per Card Played	Total Cost	License Value	Your Earnings	Cumulative Earnings
3		$1,000				

In the final round, the license will be worth the same to you as it was in round 3,
but there is no lottery and no application fee. Instead, I will conduct an auction
by starting with a low price of $8,000 and calling out successively higher prices
until there is only one team actively bidding. The winning team will have to pay
the amount of its final bid. The losing teams do not have to pay anything for
the license that they did not purchase; the winning team earns an amount that
equals its license value minus the price paid. The revenue from the auction will
be divided equally among the four teams, e.g. for a winning bid of $8,000, you
would receive a $2,000 share and only have to pay a net amount of $6,000, etc.

Round	Your Earnings (Your license value minus your bid if you win, $0 otherwise)	Cumulative Earnings
4		

Play-or-Keep Contributions Game (Chapter 16)

Each of you will now be given four playing cards, two of which are red (hearts or diamonds) and two of which are black (clubs or spades). All of your cards will be the same number.

The exercise will consist of a number of rounds. At the start of a round, I will come to each of you in order, and you will play *two* of your four cards by placing these two cards face down on top of the stack in my hand.

Your earnings in dollars are determined by what you do with your red cards. For each red card that you keep in a round you will earn four dollars for the round, and for each black card that you keep you will earn nothing. Red cards that are placed on the stack affect everyone's earnings in the following manner. I will count the total number of red cards in the stack, and everyone will earn this number of dollars. Black cards placed on the stack have no effect on the count. When the cards are counted, I will not reveal who made which decisions. To summarize, your earnings for the round will be calculated:

Your Earnings = $4 times the # of red cards you kept
+ $1 times the total # of red cards I collect.

At the end of the round, I will return your own cards to you by coming to each of you in reverse order and giving you the top two cards, face down, from the stack in my hand. Thus, you begin the next round with two cards of each color, regardless of which cards you just played.

After round 5 (or perhaps sooner), I will announce a change in the earnings for each red card you keep. Even though the value of red cards kept will change, red cards placed on the stack will always earn one dollar for each person.

Use the space below to record your decisions, your earnings, and your cumulative earnings. (Optional: At the end of the game, one person will be selected at random and will be paid a percentage of earnings to be announced by the instructor.) All earnings are hypothetical for everyone else. Are there any questions?

Play-or-Keep Contributions Game

Name/ID: _____

Record Sheet: Earnings from Each Red Card Collected = $1.00

Round	Number of Red Cards Kept	Value of Each Red Card Kept	Earnings from Red Cards Kept	Earnings from Red Cards Collected	Total Earnings	Cumulative Earnings
1		$4				
2		$4				
3		$4				
4		$4				
5		$4				
6		$2				
7		$2				
8		$2				
9		$2				
10		$2				

Voting: Three Class Experiments (Chapter 19)

Voting Instructions for Choosing a Location

(Requires 100 numbered index cards and a method for randomly selecting candidates)

We will conduct a series of elections involving two candidates, with the winning candidate determined by majority rule (a coin flip is used to break ties). There will be two candidates in each election, an incumbent and a challenger, and everyone else will be a voter.

Here is a stack of index cards, with numbers between 1 and 100, which will be shuffled before distributing one to each person. The address number on your card (between 1 and 100) is your preferred location. As a voter, your earnings are calculated as 100 minus the absolute distance between your preferred location and the winning location. YOU SHOULD ONLY VOTE IF YOU RECEIVED A CARD.

Prior to voting, each candidate proposes a location, and a show of hands determines the winner. Suppose that you are a voter with a preferred location of 20 and that the winning candidate's position was 75, which produces an absolute difference of 55. As a voter, you would enter the letter "V" in the Role column for the row and record your earnings, 100 minus the absolute distance (100 − 55 = 45), as:

Period	Role (C or V)	Voter-Preferred Location	Winning Location	Absolute Distance	Earnings V: 100 − Distance C: 100 if wins
1	V	20	75	55	45
2	C proposes 45	NA	45	NA	100
3	C				

At the beginning, two people will be randomly selected to be candidates, who will write "C" in the role column, including the candidate's proposed location beside the "C." Then these proposed "platforms" will be announced. There is to be no other discussion between voters or candidates.

The winning candidate earns 100 and the loser earns 0, irrespective of the preferred locations marked on their cards. Then the winning candidate continues again in period 2 as an "incumbent," and the loser becomes a voter with a desired location as marked on their card. In the table above, the person is a candidate (role C) in period 2, and proposes 45, which is the winning location, so the person earns 100 for being the winning candidate. Notice that their preferred location and the absolute distance do not affect the candidate's earnings, as indicated by the NA in those columns. The person's preferred location will matter if that person returns to being a voter after losing an election.

As stated above, in each period, the previous winning candidate continues as an incumbent, but an incumbent can announce a new platform location. Than a challenger, selected at random from among the voters, announces a proposed alternative position. The incumbent chooses first, and the challenger has to choose a different proposed location in the range from 1 to 100. This process will continue for several periods, until an announcement about a change in procedures is made.

Announcement (made by instructor after several periods): The preferred locations for some voters will change, due to a redistricting that excludes some voters and adds some new voters. I will pass out new cards that reflect the location preferences of the voters in the new district, and I will retrieve the old index card for those new voters.

Name: _____

Decision Sheet for Voting on a Location

Period	Role (C or V)	Voter-Preferred Location	Winning Location	Absolute Distance	Earnings Voter: 100 – Distance Candidate: 100 if wins
1					
2					
3					
4					
5					
6					
7					
8					
9					
10					
11					
12					
13					
14					
15					

Instructions for Voting with Your Feet

Instructor: This requires several decks of cards, numbered cards only, sufficient to give each person 2–3 cards each. You will also need 5 manila envelopes for recording community names, some open classroom corner space, or weather permitting, a nice outdoor location.

Cards and Preferences: This is a simulation in which you will choose where to reside. There are five communities with locations and names, as marked on the manila envelopes. I will now assign each of you to one of these communities and give each person several numbered playing cards. The cards have a number and a suit (hearts, clubs, diamonds, and spades). The suit corresponds to a particular type of public good. The number reflects the intensity of your preference for that type of public good. For example, if your cards are 3 of hearts, 6 of hearts, and 4 of diamonds, then your intensity is $3 + 6 = 9$ for hearts, but only 4 for diamonds. The intensity will affect your value for each of the four possible public goods. The person in the above example would have no value for spade or club public goods. Your earnings will be the value to you of the public good level that your community chooses, minus the associated tax cost.

Voting: Your community must choose to provide *one and only one* of these four public goods. In addition, the community must decide on the level of that good to provide. For example, a community may decide to provide a level of 6 diamonds (which means the levels of the other three goods are zero). A high level corresponds to a more elaborate provision of that public good, e.g., the park is open later or the golf course has 18 holes instead of 9. These decisions will be made by a series of votes that follow an initial discussion of preference and negotiations. All votes will be decided on the basis of majority rule. These votes and discussions will be coordinated by a mayor, someone who I will now appoint for each community. The mayor will chair meetings, announce the community's choice of public good, and the individual tax rate (individual cost). If the mayor moves to another community, he/she should appoint another mayor before leaving. In the event of a tie, the mayor, who votes individually as well, can cast a second vote to break the tie.

Earnings: In general, you prefer for the community to provide a high level of the public good that you value most, as denoted by the suit for which you have a high card number (or sum of numbers). Your benefit will be $1 for each additional unit of the public good provided of a given suit, up to the intensity (your sum of card numbers) for that suit. You will be happiest if the community chooses a level that corresponds exactly to the total number that you have for that suit. The cost of providing the public good is $2 times the level provided. All members of the community must share this cost equally, so a high provision results in higher taxes. Therefore, you would not want your community to choose a level of the public good that is higher than the number you have for that suit. For example, a person with 9 hearts would receive $5 if the community provides 5 units of hearts, $9 if the community provides 9 units of hearts, but only $9 if the community provides more than 9 units of hearts. The tax cost for providing 9 hearts would be 9 × 2 = $18, which would be divided equally among members of the community, so larger communities are good if they provide the public good that you value.

Example 1. Suppose your cards are 8 of hearts and 2 of spades, and suppose your community has 4 people. If the community decides on a level of 6 hearts, then the cost of this decision is 2 × 6 = 12, which is to be divided by the number of people (4), yielding an individual cost of $3. Your payoff is $6 (although you have an 8 of hearts card, the community only voted to provide 6 hearts) minus the individual cost, $3, which equals $3. An increase in the level from 6 to 10 hearts would raise your benefit to $8 (limited by your 8 hearts card) minus your share of the cost, which is (10 × 2)/(4 people) = $5, so your new earnings would be $8 − $5 = $3.

Example 2. Now it is your turn. A person is in a community of 5 people which chooses Clubs at a level of 5 for a total cost of ____ and a per-person tax of ___. A person with the following cards: 9 Hearts and 10 Clubs would receive a benefit of ____, pay a tax of ____ , and hence would have a net benefit of ____.

Choosing Communities: The residents of each community will make their decision (suit and level) by voting, in a meeting conducted by the mayor. After all communities have made decisions, the mayors will announce their public good suit and level, and the associated tax. We will write these results on the board for each of the communities. Then people will be free to switch to a community with a public good decision more to their liking, with the understanding that newly configured communities will vote again at the start of the round on the type and level of the good to be provided. You can only switch once between each round. You may move communities to improve your well-being, or you may choose to remain in a community. Communities may dissolve and re-emerge during this process.

Records: Using the table provided, please record your cards (suits and numbers), community/location, the community decision, and your payoffs after each round.

Your name: _____

Your Card Distribution

	First Card	Second Card	Third Card	Fourth Card
SUIT (H, D, S, C)				
NUMBER				

Round	Community Name	Community Decision	Your Benefits	Your Costs ($2 × Level)/N	Earnings (Benefits – Tax Cost)
1					
2					
3					
4					
5					
TOTAL					

Agenda Voting Instructions for Program Expenditures

> **Instructor:** *This experiment requires a deck or two of playing cards, depending on the class size. As indicated in chapter 19, the setup is for multiples of 7 voters, e.g., 7, 14, 21, etc. Extra students can work in pairs.*

This is a simple exercise to illustrate the effects of different political institutions. At this time, each of you will be given two playing cards that determine whether or not you benefit from a variety of proposals. The group will vote to select among the proposals, using majority rule, with ties decided by the coin flip. IF YOU ARE WORKING WITH OTHERS IN A GROUP, YOU SHOULD ONLY VOTE IF YOU RECEIVED AN ENVELOPE WITH CARD(S). ONLY ONE VOTE PER GROUP.

There are two potential projects, a "Highway" and a "School." Each project, if adopted, will cost each person $200 in taxes, regardless of how they voted. The benefits to you depend on which cards you have. If one of your cards is a spade, you are a School person, and will receive a benefit of $300 if a school is built, so the benefit net of your tax share is $300 − $200 = $100. If one of your cards is a heart, then you are a Highway person and you will receive a benefit of $300 if the highway is built, again with the benefit net of taxes equal to $100. If you have both a heart and a spade, then your net benefit with both projects is: $300 − $200 + $300 − $200 = $200. If you do not have a spade and the group votes only to build a school, then your benefit is −$200, the tax cost. Your net benefit is also −$200 if you do not have a heart and the group votes only to build a highway. Finally, a club card has no direct effect on your earnings, so if you have a club and a spade, you receive a net benefit of $300–$200 if only school is adopted, and you receive $0–$200 if only highway is adopted. Similarly, if you have a club and a heart, you receive a net benefit of $300–$200 if only highway is adopted, and you receive $0–$200 if only school is adopted. At this time, please look at your cards and write down your net earnings for each of the four possibilities:

Highway only:	$_____ − $200 =	_____
School only:	$_____ − $200 =	_____
Both Highway and School:	$_____ − $400 =	_____
Neither:	$_____ − $0 =	_____

Negative earnings may be possible for some voters; losses will be subtracted and gains will be added to determine total earnings. These earnings are hypothetical and are used for purposes of discussion only (in the absence of other announcements).

Agenda 1

The first two votes determine which projects will be options on the final vote. The final vote will determine which projects are funded, and therefore, earnings are determined by the final vote. First, raise your hand if you want to fund the highway.

Your vote:

Yes _____ (fund highway)

No _____ (not fund highway)

Next, raise your hand if you want to fund the school, whether or not the highway was funded. Remember, only one vote per group if you are working with others.

Your vote:

Yes _____ (fund school)

No _____ (not fund school)

At this point, we have agreed to fund the following project(s): _____. Finally, we will decide whether to fund this project (or these projects as a package) or to go back to the initial situation of funding neither. First raise your hand if you prefer to fund neither project. Next raise your hand if you want to fund the project(s) approved thus far.

Your vote:

Fund neither _____

Fund package _____.

Now record your earnings.

Project(s) that are funded with Agenda 1 are: _____

Your earnings for Agenda 1: $ _____

Agenda 2

We will start over with a new agenda, and your earnings will be calculated in the same way as before, but separately from those of Agenda 1. (Imagine that you have moved to a new town just in time for the voting.) First, you will choose between funding neither project or just funding the highway. Raise your hand if you want to fund only the highway; now raise your hand if you want to fund neither.

Your vote:

Fund highway only _____

Fund neither _____

Next, you will choose between _____ (the winner of the previous vote) and to fund the school only. Raise your hand if you want to fund _____ (the winner of the previous vote); now raise your hand if you want to fund the school only.

Your vote:

Fund previous winner ____

Fund school only ____

Finally, you will choose between _____ (the winner of the previous vote) and to fund both projects. Raise your hand if you want to fund _____ (the winner of the previous vote); now raise your hand if you want to fund both projects.

Your vote:

Fund previous winner ____

Fund both projects ____

Now record your earnings.

The project(s) funded with Agenda 2 are: _____.

Your earnings for Agenda 2: $ ____

Takeover Game Instructions (Chapter 27)

In a minute I will divide the class into two-person teams. Half of the teams will be buyers and half will be sellers in a market. I will choose teams and indicate your role: buyer or seller. Each team should have a single copy of these instructions, with the role assignment, buyer or seller, written next to the place for your names below.

Each seller team is the owner of a business. The money value of the business to the seller is only known by the seller. Each of the buyers will be matched with one of the sellers. The buyer will make a single bid to buy the business. The buyer does not know the value of the business to the seller, but the buyer is a better manager and knows that he or she can increase the profits to 1.5 times the current level. After receiving the buyer's bid, the seller must decide whether or not to accept it. The seller will earn the value of the business if it is not sold, and the seller will earn the amount of the accepted bid if the business is sold. The buyer will earn nothing if the bid is rejected. If the buyer's bid is accepted, the buyer will earn 1.5 times the seller's value, minus the bid amount.

A ten-sided die will be thrown twice to determine the value to the seller, in thousands of dollars. This will be done for each seller individually. The die is numbered from 0 to 9; the first throw determines the tens digit and the second determines the ones digit, so the seller value is equally likely to be any integer number of thousands of dollars, from 0 to 99 thousand dollars. If the firm is purchased by the buyer, it will be worth 1.5 times the seller value, so the value to the buyer will be between 0 and 148.5 thousand dollars.

The record sheet provided can be used to record the outcomes. For simplicity, records will be in thousands of dollars, i.e., a 50 means 50 thousand. I will begin by coming to each seller's desk to throw the ten-sided die twice, and you can record the resulting seller value in column (1). Then each buyer, not knowing the seller's value, will decide on a bid and record it in column (2). I will then match the buyers and sellers randomly, and each buyer will communicate the bid to the corresponding seller, who will say yes (accept) or no (reject). The seller's decision is recorded in column (3). If the bid is accepted, then the seller will communicate the seller's value to the buyer, who will multiply it by 1.5 and enter the sum in column (4). Finally, buyers and sellers calculate their earnings in the appropriate column (5) or (6). The buyer earns either 0 (if no purchase) or the difference between the buyer value (4) and the accepted bid. The seller either earns the seller value (if no sale) or the amount of the accepted bid. Each of you will begin with an initial cash balance of 500 thousand dollars; gains are added to this amount and losses are subtracted. You can keep track of your cumulative cash balance in column (7), and all earnings are hypothetical. Are there any questions?

Now we will come around to the desks of sellers and throw the dice to determine the seller values. While we are doing this, those of you who are buyers should now decide on a bid and enter it in column (2). For those of you who are buyers, the seller value column (1) is blank. You only find out the seller value after you announce your bid. Those of you who are sellers have no decision to make at this time. When you hear the buyer's bid, you must choose between keeping the value of the firm (determined by the dice throws) or giving it up in exchange for the buyer bid amount. Please only write in the first row of the table at this time. (After everyone has been matched and has calculated their earnings for the first set of decisions, you will switch roles, with buyers becoming sellers, and vice versa. The second set of decisions will be recorded in the bottom row.)

Your Role: _____   Your Name: _____

Record Sheet for Takeover Game

	(1)	(2)	(3)	(4)	(5)	(6)	(7)
Round	Seller's Value	Buyer's Bid	Seller's Decision (accept or reject)	Value to Buyer 1.5*(1)	Seller's Earnings (1) or (2)	Buyer's Earnings (4) − (2) or 0	Cash Balance 500 (thousand)
1							
2							

Index